Neurocytology

Ennio Pannese

Neurocytology

Fine Structure of Neurons, Nerve Processes, and Neuroglial Cells

2nd fully revised and updated edition

Springer

Ennio Pannese
Former Professor of Human Anatomy and Neurocytology
and Head of the Institute of Histology, Embryology
and Neurocytology
University of Milan
Milan
Italy

Pannese, Neurocytology 1st edition: Georg Thieme Verlag, Stuttgart, 1994
Digitization of electron microscopy figures by: Studio Macor, Milano

ISBN 978-3-319-06855-8 ISBN 978-3-319-06856-5 (eBook)
DOI 10.1007/978-3-319-06856-5
Springer Cham Heidelberg New York Dordrecht London

Library of Congress Control Number: 2014956380

Printed on acid-free paper

Springer is part of Springer Science+Business Media (www.springer.com)

To the memory of Angelo Cesare Bruni (1884–1955), Angelo Bairati (1911–1994), and Rodolfo Amprino (1912–2007) with deep gratitude for what they taught me

Preface

Progress in experimental science does not usually take place at a constant pace, but is characterized by periods of intense growth, often related to the introduction of new techniques, alternating with periods of critical reflection. Such has been the case for cytological research on the nervous system. The decades around the end of the nineteenth and the beginning of the twentieth century were golden ones for these studies; work carried out then not only contributed significantly to our knowledge of the cells of the nervous system, but to the progress of cytology in general. Cellular studies on the nervous system led in that period to the discovery of the ergastoplasm and the Golgi apparatus, which were first detected in neurons.

A period of stasis followed, which lasted till the middle of the twentieth century. At that time a burst of investigative activity produced many significant new findings. The renewed progress depended initially on use of the transmission electron microscope and subsequently on the availability of other new techniques (scanning electron microscopy, freeze-fracturing, cell organelle isolation by differential centrifugation, autoradiography, tracing techniques, immunocytochemistry, etc.). It thus became possible to analyze the fine structure of nerve and neuroglial cells and to begin to define in a detailed and precise way the organization of the nervous system at a cellular level. At more or less the same time, although largely independently, the responses of individual nerve cells to various stimuli were being recorded by sophisticated physiological techniques. This parallel progression of morphological and physiological research led to an appreciation of how the characteristics of individual nerve cells and their precise organization are important in the functioning of the nervous system.

The first edition of this book, published 20 years ago, was written during the above mentioned burst of cytological research on the nervous system. Since the first edition was published, the introduction of new microscopies and especially the growth of molecular biology have produced a wealth of new knowledge, in particular on the intercellular communication in the nervous system and on the roles of neuroglial cells. These achievements made it necessary to update the entire text.

All chapters have been thoroughly revised and updated. While some sections have not changed appreciably, others have been almost entirely rewritten. Furthermore, in consideration of the growing interest in the aging process and the considerable progress that has been made in this field, subsections dealing with age-related changes have been added to all main sections. I wish to express my gratitude to all who, by personal communication or published reviews, pointed out inaccuracies in the first edition. In consideration of their observations, I have modified the text at appropriate points. As a consequence of the changes and additions, 127 references (present in the first edition) have been removed, and more than 650 new ones have been added. Four hundred and thirty of the latter are related to papers appeared over the last 20 years (i.e., after the publication of the first edition). The total number of references has increased from about 1,500 to about 2,000. References to the foundations of the discipline have been retained. Inevitably, despite painstaking and laborious scrutiny of the very numerous studies published in the field, valuable publications in one or more areas will have been omitted. Sincere apologies are due to authors who may have been inadvertently overlooked.

Notwithstanding the extensive revision and updating, several policies of the first edition have been retained in the new edition. Thus, the sections and subsections into which the text is divided are linked by numerous cross-references, which serve to supply the reader with a full overview of the subject under study, and at the same time avoid too much repetition. Some analytical information is presented in table form so as to lighten the text. Few abbreviations have been used to reduce to the minimum the need to ping-pong between the page being read and the list of abbreviations. Because English is not my first language, this new edition, like the first, was written as simply as possible. It is my hope that this simplicity has rendered the text clear and unambiguous.

The aim of the new edition remains the same: to provide a systematic survey of the organization of the nervous system at the cellular level, in a historical perspective. The major new findings are correlated with the classical notions of light microscopy and accounts of recent results are preceded by notes on the more important past contributions. This apposition of recent knowledge with long established notions is not simply intended to provide a more complete exposition, but also to emphasize that modern developments are rooted in the past. The inclusion of important early contributions also has the aim to correcting an attitude which today is too common. Young investigators often seem unaware of the steps by which we have reached our present state of knowledge, believing that all that is important has been discovered in the last 20 or so years. With inadequate instruments and means, but inspired by a passion for knowledge, our predecessors managed to establish an impressive body of fundamental insights not only into the cytology of the nervous system, but into all the disciplines nowadays known as the neurosciences. To acknowledge those who pioneered new areas, I have, whenever possible, cited the first paper (or papers) which appeared. Today few seem to think this is important, and only the most recent papers are generally cited.

I hope that the present book constitutes a useful starting point for research in the neurosciences. It may be of use especially to investigators engaged in studies of cellular neuropathology, neurochemistry, neurophysiology, and molecular neurobiology, providing them with essential information on the structure of nerve and neuroglial cells, and their relationships. It should also stimulate the integration, much to be desired, of the various branches of the neurosciences. The text should also be useful to workers in morphological fields other than the nervous system as a reference and teaching aid.

I am aware that active involvement by young investigators is essential for the continuity of research. Having arrived at the end of my research career, I have made every effort to complete this new edition in the hope that it encourages young researchers to further advance our knowledge on the cytology of the nervous system.

It is a pleasure to express my sincere thanks to the following colleagues who generously provided micrographs to illustrate the book: J. E. Bruni; D. Cantino and M. Testa; B. Ceccarelli, R. Fesce, F. Grohovaz, N. Iezzi and F. Torri-Tarelli; G. Gabella; S. Iurato, S. Colucci and A. Zambonin; S. Matsuda; S. Matsuda and M. Ledda; Y. Matsuda; E. Mugnaini and P. T. Massa; A. Peters; E. Reale and L. Luciano; L. Roncali, D. Cantino and B. Nico. Some of these are unfortunately no longer with us. In particular, I wish to mention Professor Enrico Reale with whom I carried out a series of studies on neurocytology. I also acknowledge Professor A. Calligaro, who kindly allowed me to reproduce three drawings by C. Golgi in the care of the Museum for the History of the University of Pavia.

In preparing the first edition I had efficient technical and secretarial help, and a generous contribution from an Italian bank. By contrast, for this new edition, which was prepared after my retirement, I had no such assistance. This contributed to the delay in publication. I did receive help, however, from Professor Liliana Luciano, who was able to get numerous hard-to-find articles and helped to prepare the new figures; from my daughter, who taught me elements of computer use, and arranged the tables; and from Mr. D. Ward, who helped me with the English. Their valuable assistance was much appreciated. I am indebted to my wife, whose forbearance was essential in allowing me to complete the book. Finally, I wish to thank Mrs. Antonella Cerri and Mr. Andrea Ridolfi of Springer Italy for their effective help in resolving the problems that arose during the preparation of the book.

Milan, Italy Ennio Pannese

Contents

Abbreviations

cm	Centimeter
CNS	Central nervous system
D	Dalton
DNA	Deoxyribonucleic acid
E face (of the plasma membrane)	Inner face of the outer (=External) leaflet of the split plasma membrane
GABA	γ-aminobutyric acid
h	Hour
kD	Kilodalton
m	Meter
MAP, MAPs	Microtubule-associated protein(s)
mm	Millimeter
mM	Millimolar
μm	Micrometer
ms	Millisecond
μs	Microsecond
mV	Millivolt
nm	Nanometer
P face (of the plasma membrane)	Outer face of the inner (=Protoplasmic) leaflet of the split plasma membrane
PNS	Peripheral nervous system
RNA	Ribonucleic acid
s	Second
S	Svedberg unit of sedimentation coefficient

I. Neurons and Interneuronal Connections: A Historical Overview

Among the components of the nervous tissue not visible to the naked eye, those first described were nerve fibers. A. Van Leeuwenhoek [1632–1723] observed these fibers in the peripheral nerves (1718) and interpreted them as "very minute vessels," i.e., as hollow tubes with a fluid content. A more correct description was given by F. Fontana [1730–1805], who interpreted them as thin solid cylinders (1781). Important contributions were subsequently given by R. Remak [1815–1865] and J.E. Purkinje[1] [1787–1869]. Remak (1836; 1837) described the unmyelinated nerve fibers which today bear his name and drew attention to their lack of a white outer layer, which is present in other nerve fibers. In each nerve fiber of the PNS, Remak recognized a flat ribbon, which he called the *Primitivband* and which corresponds to the transparent central axis described by Purkinje (1838) in myelinated nerve fibers. This axial component of the nerve fiber was later (1839) given the Latin term *cylinder axis* by J. F. Rosenthal [1817–1887], a pupil of Purkinje. Near the end of the century (1896) R. A. von Koelliker[2] [1817–1905] coined the term axon (*Neuraxon*) for this component.

The objects which were later designated nerve cell bodies were detected in invertebrates by R. J. H. Dutrochet [1776–1847], who described them as *cellules globuleuses* (1824) and subsequently in both invertebrates and vertebrates (1833; 1836) by C. G. Ehrenberg [1795–1876], who termed them *Kugeln* (club-shaped bodies); however, their precise significance was not recognized by these observers, but rather by Purkinje and his pupil G. G. Valentin [1810–1883]. Valentin (1836) demonstrated that some *Kugeln* bore a tail-like process (*schwanzförmige Verlängerung*), which most likely

[1] The real name of this great Czech investigator was Purkyně. Up to 1850 he used the version Purkinje, which corresponded to the pronunciation of his surname in German, at that time the only language of scientific discourse used in central Europe. Purkinje returned to the Czech version of his name when he initiated a campaign for the development of science in his country and began to encourage the use of the Czech language in order to make it easier for his fellow citizens to gain access to scientific knowledge. In spite of his decision to use the Czech version of his name, this author is always cited in the literature as Purkinje.

[2] In his publications this author used both forms Koelliker and Kölliker. Since in his letters he always signed himself Koelliker, I have used this form here.

E. Pannese, *Neurocytology: Fine Structure of Neurons, Nerve Processes, and Neuroglial Cells*, 2nd Edition,
DOI 10.1007/978-3-319-06856-5_1,

corresponded to the proximal segment of a large dendrite. Remak (1837) and Purkinje (1838) also observed that these processes sprouted from the nerve cell body, but the term dendrites was only introduced in 1889 by W. His [1831–1904].

Whereas Valentin (1836) maintained that nerve cell bodies and nerve fibers were contiguous but separate entities, cell-fiber continuity was detected by A. Hannover [1814–1894] in vertebrates (1840) and by H. L. F. Helmholtz [1821–1894] in invertebrates (1842) and later received general acceptance mainly due to Koelliker.

The clear distinction between axon and dendrites is mainly due to R. Wagner [1805–1864], Remak, and O. F. K. Deiters [1834–1863]. Wagner (1846) identified the axon and dendrites in the large nerve cells of the electric lobes of torpedoes: a variable number of processes, often branched and consisting of the same granular material as the nerve cell body, arose from the latter, but a single process in each cell looked different from the others, being longer, paler, less granular, unbranched, and of uniform thickness. By examining large mammalian neurons, Remak (1855) arrived at the same conclusion as Wagner (1846). Finally, the work of Deiters (1865) elevated the distinction between axon and dendrites to the status of a general law. Dying at 29 years of age, Deiters left a manuscript – edited in 1865 by his mentor M. Schultze [1825–1874] – in which he described the results of his studies, unfortunately incomplete, on the nervous system of man and other mammals. According to Deiters, the nerve cell possesses a body, a single axon, and several dendrites. The body consists of a mass of cytoplasm of granulo-fibrillar appearance and of a nucleus containing a prominent nucleolus. The dendrites (Fig. I.1), which Deiters called *Protoplasmafortsätze*, arise by gradual transition from the cell body and show the same fine granulo-fibrillar structure as the cell body; they repeatedly divide and become progressively thinner toward their ends, eventually disappearing into the ground substance of the nervous system. The axon (Fig. I.1), which may take origin directly from the nerve cell body or from one of the dendrites, is more homogeneous and more refractive than the latter; it is always unbranched, has a fairly uniform thickness, is smooth surfaced, and presents an unmyelinated proximal segment, beyond which it becomes enveloped by the myelin sheath. Deiters also described, and illustrated in his plates, fine axonal processes which, he claimed, sprouted from the dendrites and interconnected the nerve cells (Fig. I.1). Almost certainly, these fine axonal processes were actually preterminal segments of afferent axons synapsing on the dendrites (Van der Loos 1967).

The results described so far were mainly obtained using two techniques. The first involved the fixation, embedding, and cutting of nervous tissue, followed by staining of the resulting sections with hematoxylin or carmine. This procedure was reasonably adequate for studying the structure of other tissues but was unsuitable for investigating nervous tissue, since it only revealed incomplete images of nerve cells. In small nerve cells only the nucleus, surrounded by a narrow rim of cytoplasm, was visible; while in large nerve cells in addition to the nucleus, only the perikaryon and the initial segments of dendrites could be seen (Fig. I.2). Hematoxylin or carmine staining revealed so little of the small densely packed nerve cells found in certain regions of the brain that their true nature could not be discerned. For this reason these cells were referred to generically as "granules." The other procedure consisted in the immersion of nervous tissue blocks in reagents such as chromic acid or potassium dichromate solutions, which served both to fix and to harden the material. This was followed by mechanical isolation of individual nerve cells using needles under the microscope. While this technique had at least the merit of revealing most of the nerve cell (Fig. I.1), it also had considerable limitations. It could be applied only to the largest nerve cells and never allowed complete isolation even of these for the following

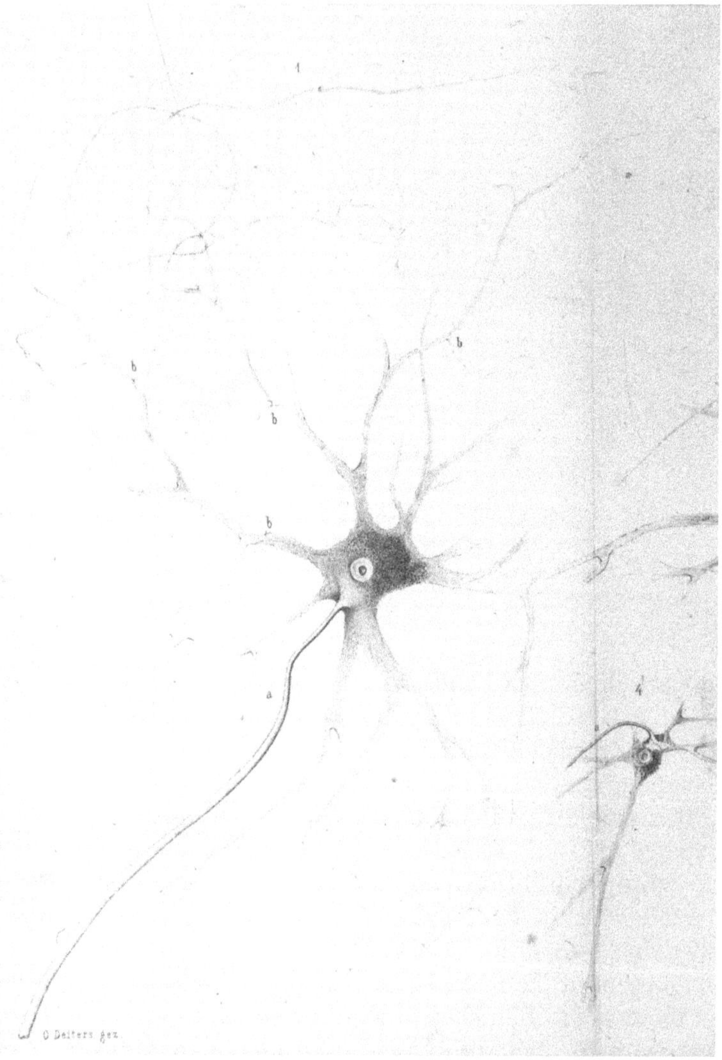

Fig. I.1 Motoneuron isolated by Deiters from the ventral horn of the spinal cord, probably of ox. In the cell body are evident an accumulation of pigment and the nucleus containing a prominent nucleolus. Dendrites divide repeatedly and become progressively thinner toward their ends, whereas the axon (**a**) appears unbranched and shows a rather uniform thickness. Fine axonal processes, which according to Deiters originate from the dendrites, are shown (**b**) (Drawing by Deiters published in 1865, 2 years after his death)

reasons: the fine terminal segments of the cell processes being teased out were inevitably broken and detached from their thicker and stronger proximal segments; fine terminal axonal segments belonging to other nerve cells were liable to remain attached to dendrites of the cell being dissected out. The erroneous conclusion of Deiters noted above, that the nerve cell had a second system of fine axonal processes in addition to the main axon, was due to this type of procedure which he used to study nerve cells.

In 1873 C. Golgi [1843–1926] invented the *reazione nera* (black reaction), which was of vital importance for the development of our understanding of the structure and organization of nervous tissue (Pannese 1996, 1999, 2007). The procedure was as follows. Blocks of freshly removed nervous tissue were hardened and fixed in an aqueous

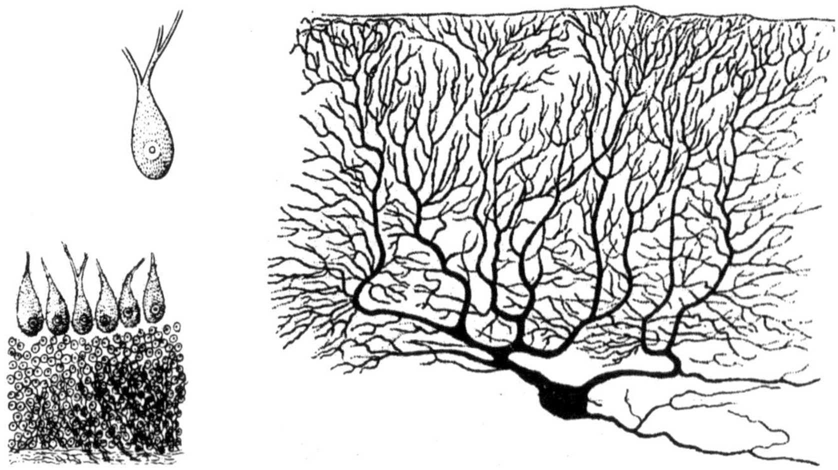

Fig. I.2 Images illustrating the impact of the black reaction on our knowledge of nerve cells. *Left*: Purkinje cells revealed using the procedures available prior to the invention of the black reaction. Drawing made by Purkinje for the meeting of German naturalists and physicians in Prague in 1837 (From Purkinje 1838). *Right*: A Purkinje cell impregnated using the black reaction (From Koelliker 1896)

solution of potassium dichromate and successively immersed in a solution of silver nitrate. The blocks were then dehydrated and cut without embedding, thus obtaining very thick sections. Microscopic examination of material thus prepared allowed Golgi to see the entire nerve cells intensely stained in black standing out against a light yellow background. Only a small proportion (1–5 %) of the nerve cells present in the tissue were impregnated, but these were often shown in their entirety, i.e., with all their processes (Figs. I.2 and I.3). The technique is therefore a partial one in that it does not reveal all the cells that make up the nervous tissue. It was this selectivity – at first sight a defect – that was one of the great advantages of the black reaction. In fact, to follow the entire course of a long axonal process, it was necessary to prepare very thick sections. If the technique had impregnated all the nerve cells present in such a section, the observer would have been unable to find his way in the inextricable tangle of nerve cell processes. By using his technique, Golgi (1882) obtained a number of results of major importance. He (a) established that the axon gives off lateral branches, today known as axon collaterals, while previously it was thought that the axon was always unbranched (see Fig. I.1); (b) showed the previously unsuspected variety of nerve cell types (see Chap. IV); and (c) demonstrated that dendrites are not in continuity with the dendrites of other nerve cells but end freely. The systematic use of the black reaction, initiated by Golgi and continued by other researchers, revealed that the CNS consists mainly of cells, while previously it was thought that the nerve cells were immersed in an amorphous ground substance, considered to occupy more than 50 % of the volume of the gray matter. Later the black reaction made it possible to discover the internal reticular apparatus (see Sect. V.A.3).

For a long time nerve cells were thought not to be independent units; their bodies were believed to be at the nodes of a syncytial network formed by anastomosis between dendritic branches [J. von Gerlach, 1820–1896] or interconnected through fine fibrils (neurofibrils) running without interruption from one nerve cell to another [I. von Apáthy, 1863–1922]. In the last 15 years of the nineteenth century, however, a number of authors, working independently of each other, explicitly questioned the syncytial conception of nervous tissue. On the basis of the results he obtained investigating the development of

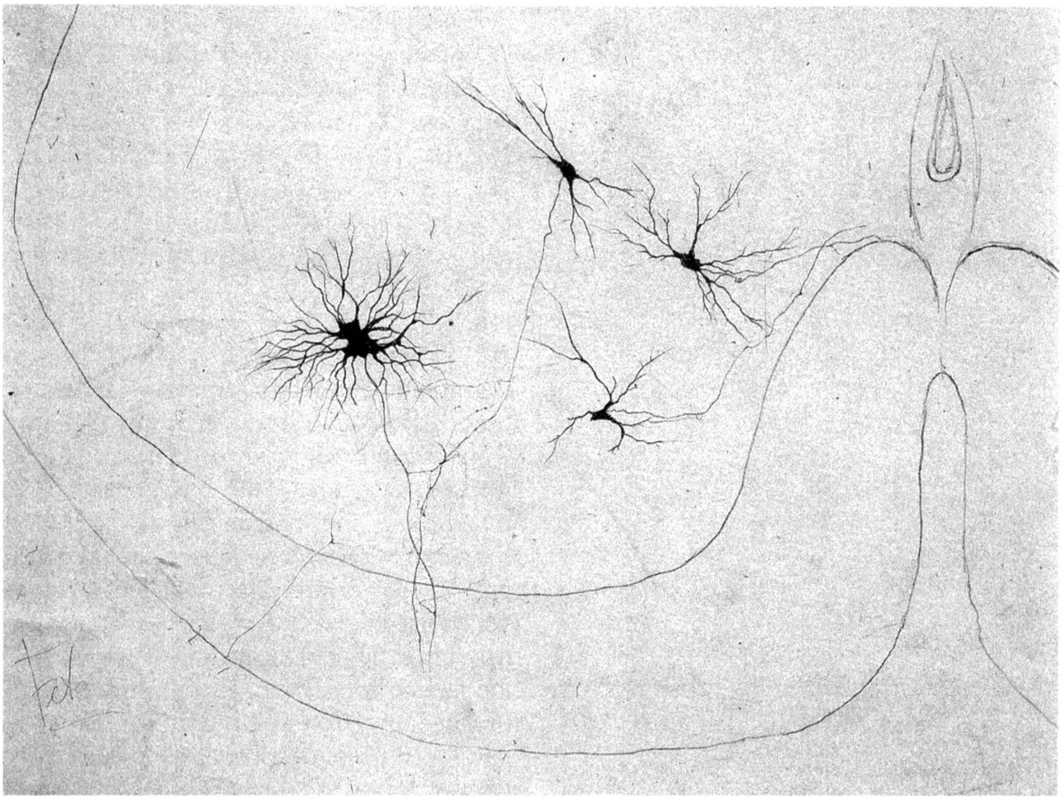

Fig. I.3 Drawing by Golgi showing multipolar neurons impregnated using his black reaction. Ventral horn of the spinal cord of a mammalian fetus

nerve fibers His (1886) concluded that every nerve fiber is a direct outgrowth of a single nerve cell; the latter is the genetic, nutritive, and functional center of the fiber; all other connections to the fiber are indirect or are formed secondarily. His drew attention to the fact that in the PNS nerve fibers end freely in motor end plates or in sense organs. Not finding evidence for Gerlach's syncytial network in the CNS, His proposed that also here nerve fibers end freely. A. H. Forel [1848–1931] confirmed the branching of axons, but never observed unequivocal images of axons fused into a net-like structure. Moreover, the experiments by B. A. von Gudden [1824–1886], in whose laboratory Forel had worked for 5 years, showed that, after lesions to a group of nerve cells, degeneration was confined to these cells and their fibers, while neighboring, uninjured nerve cells remained unaffected. On the basis of his and Gudden's findings, Forel (1887) advanced the idea that each fiber belongs to a single nerve cell and that nerve cells are connected by contact and not by cytoplasmic continuity; he further proposed that this contact is sufficient to transmit excitation from a nerve fiber to the next.

The work of these authors led to the formulation of a new conception of the structural organization of nervous tissue. According to this theory, nervous tissue does not consist of a syncytial network but of distinct entities, which closely contact each other, but are not in cytoplasmic continuity (Fig. I.4). These basic units were called neurons (1891) by H. W. G. von Waldeyer [1836–1921]. The most forceful advocate of this conception, which is generally known as the neuron theory, was certainly S. Ramón y Cajal [1852–1934].

In substance, the neuron theory extended the cell theory to nervous tissue. When it is

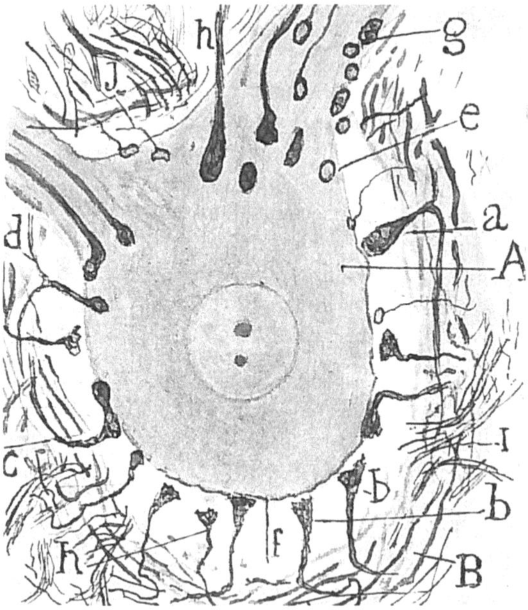

Fig. I.4 Large motoneuron of the ventral horn of the cat spinal cord with presynaptic endings of afferent axons applied to its surface (Drawing published by Ramón y Cajal (1934) in the paper which summarized the evidence that the nervous tissue consists of distinct units which are in contiguity, but not continuity, with one another)

recalled that the theory that all plant and animal organisms are composed of cells (cell theory) was enunciated in 1838–1839 by M. J. Schleiden [1804–1881] and T. Schwann [1810–1882], it is clear that there was considerable delay before the theory was extended to nervous tissue. There were several reasons for this. First of all, while many tissues are made up of cells having a regular shape and microscopic size, the elements of the nervous tissue often have extremely irregular shapes and are usually much larger than typical cells of other tissues. The individual cells of many tissues are generally totally contained within one microscopic field, whereas the nerve cell is rarely visualized in its entirety in a single histological section, mainly because of its axon length. Furthermore, many authors were of the opinion that the nerve impulse could spread more easily through a continuous syncytial reticulum than within a tissue consisting of a great number of discrete entities.

The controversy between "neuronists" (Ramón y Cajal, His, Forel, Koelliker, G. M. Retzius [1842–1919], A. Van Gehuchten [1861–1914], M. von Lenhossék [1863–1937], E. Tanzi [1856–1934], E. Lugaro [1870–1940], and others) and "reticularists" (Apáthy, A. Bethe [1872–1954], Golgi, A. S. Dogiel [1852–1922], H. Held [1866–1942], J. Boeke [1874–1956], P. Stöhr Jr. [1891–1979], and others) continued for many years and was at times acrimonious. It is noteworthy that it was Golgi himself who provided one of the main research techniques, the black reaction, for establishing the neuron theory, which he fought against so tenaciously throughout his professional career. The wealth of evidence which accumulated over time settled the controversy in favor of the neuron theory. Some of the main evidence in this context was (a) the findings of His (1889) that nervous tissue developed from individual cells (neuroblasts); (b) the physiological observations on the basis of which C. S. Sherrington [1857–1952] in the seventh edition of Foster's *A Text Book of Physiology* (Foster and Sherrington 1897) introduced the term synapse to refer to the region of contact between one nerve cell and the next, specialized for the transmission of signals; (c) the demonstration that a nerve impulse caused the release of acetylcholine at the neuromuscular junction (Dale et al. 1936); (d) the studies by Waller (1850, 1852a) [A. V. Waller, 1816–1870] and Forel (1887) on the consequences of sectioning and injuring the axon; (e) the demonstration that neurofibrils do not run without interruption from one nerve cell to another and do not even enter the presynaptic bouton; and (f) the electron microscope observation that there is a discontinuity between the pre- and postsynaptic neurons (Fig. VI.1), each of which is bounded by its own plasma membrane (see Sect. VI.B.3). The latter observation constituted the definitive direct evidence in favor of the neuron theory.

After this theory had received almost general assent, investigators of the nervous system remained sharply divided on the mechanism of synaptic transmission. A lively debate on this topic took place during the 1930s and 1940s. Certain authors claimed that transmission was due to a direct current flow from the pre- to the postsynaptic neuron; others maintained that transmission was mediated by a chemical substance released from the presynaptic neuron which initiated the current flow in the postsynaptic neuron. The experiments of Kuffler (1942a, b) [S. W. Kuffler, 1913–1980] and Fatt and Katz (1951, 1952) settled this controversy in favor of chemical transmission. Some years later, however, using intracellular recording techniques, Furshpan and Potter (1957, 1959) discovered that at the giant motor junction of the crayfish transmission was electrical (see Sect. VI.I for further details). Later, other examples of electrical transmission were described, and accordingly, it was established that in the nervous system, there are both chemical synapses and electrical junctions. In vertebrates, chemical synapses are much more abundant than electrical junctions.

Other brief historical notes on individual aspects of nerve and neuroglial cells may be found at the beginning of the relevant sections.

II. Some Evolutionary Aspects and General Features of Neurons

The ability to react to environmental stimuli is a general property of all organisms, both unicellular and multicellular. In the latter, specialization of cellular function is the rule, and the groups of cells able to react to stimuli may be relatively distant from the point of stimulation. Under such conditions, the ability to react to stimuli has been considerably enhanced by the development and refinement of devices for signal propagation.

The earliest signal propagation phenomena probably arose in epithelial tissues (Horridge 1968), where the cells are in close contact with each other, and would have been facilitated by the development of specialized intercellular junctions. Coordinated ciliary movement is one of the better known consequences of signal transmission through a layer of epithelial cells.

Neurons, i.e., cells specialized for the reception, conduction[1], and transmission of signals, would have evolved from epithelial cells. While an individual epithelial cell can only conduct signals over a very short distance, a single nerve cell with its elongated processes can conduct signals rapidly between distant points. The appearance of neurons, therefore, brought a real advance to the process of signal propagation. While small organisms may not need high-speed signal propagation, this is an absolute necessity in larger organisms, for example, for activities such as prey capture or escape from predators. Hence, nerve cell differentiation was probably an essential precondition for size increase in organisms.

Signal transmission between the cells involved in the reception of and reaction to environmental stimuli probably first occurred by an electrotonic mechanism. In the nervous system of coelenterates, which is the simplest in the animal kingdom, many neurons are electrically coupled. Although an electrotonic mechanism of signal transmission similar to that found in epithelial tissues still operates in many neurons (see Sect. VI.1), even in mammals, comparative studies on homologous nervous structures of different species indicate an evolutionary trend toward a decrease in the number and proportion of electrically coupled neurons (Shapovalov 1980). It seems that very early in the course of evolution nerve cells developed the capacity to influence the activity of other cells by a chemical mechanism, i.e., by the release from axon terminals of physiologically active substances synthesized by the nerve cells themselves.

Certain neurons which transmit signals via a chemical mechanism synthesize relatively large amounts of the messenger substances, which are released from

[1] In accordance with Lugaro's (1917) proposal, the term "conduction" is here employed to indicate the intracellular propagation of signals and the term "transmission" to indicate the intercellular transference of signals.

E. Pannese, *Neurocytology: Fine Structure of Neurons, Nerve Processes, and Neuroglial Cells*, 2nd Edition,
DOI 10.1007/978-3-319-06856-5_2,

axon terminals into the extracellular space of tissues or into the general circulation. Via the blood stream, these substances may reach very distant targets. The secretory activity predominates in these nerve cells, whereas impulse conduction probably serves only to trigger the release of the messenger substances (Cross 1974). Such cells are in certain respects similar to endocrine gland cells and are known as neurosecretory neurons. In other neurons, the biosynthesis of messenger substances is greatly diminished in correlation with the establishment of precise anatomical connections between neurons and their targets, i.e., with the formation of synaptic junctions. Target cells are thus directly influenced by messenger substances, only tiny amounts of which are required to exert their effect. Bioelectric activity has become the dominant feature of such neurons. Even primitive invertebrates such as coelenterates are provided with elementary synaptic contacts in addition to electrically coupled nerve cells and neurosecretory neurons (Scharrer 1976). A large percentage of these synapses are peptidergic (Grimmelikhuijzen et al. 1992).

While in more primitive invertebrates neurosecretory cells constitute a very large proportion of neurons (e.g., more than 50 % of all neurons are neurosecretory in the ganglia of annelids), in the course of evolution the proportion of neurosecretory cells decreased considerably. The responses evoked by neurosecretory signals are not particularly rapid; moreover, they are diffuse responses since they usually involve many cells. On the other hand, the establishment of precise anatomical connections between neurons (or between a neuron and a nonneuronal element), i.e., the formation of synaptic junctions, allows the rapid and precisely localized transmission of signals. In the course of evolution, as the number of neurons increased and ever more complex integrative centers developed, this second mode of signal transmission, which is certainly more efficient than the neurosecretory mechanism, became more common. However, neurosecretory neurons have not disappeared completely; they still operate even in higher vertebrates, though in much reduced numbers. The development of an endocrine system proper also contributed to the reduction in neurosecretory cell numbers. This system is absent in the primitive invertebrates, makes its first appearance in arthropods, and is highly developed in vertebrates. Certain functions primitively carried out by neurosecretory neurons were later taken over by the endocrine system. The appearance of synaptic transmission was a fundamental evolutionary advance not only on neurosecretory communication but also on electrotonic transmission. Compared with the latter, synaptic transmission is more effective and selective and allows the transmission of inhibitory influences in a simpler and more efficient manner.

For an account of the mechanisms by which high-speed signal propagation has been attained in the course of evolution, see Sects. VII.B.2 and VII.B.4h.

The appearance of synaptic junctions led to the early establishment of the two-neuron reflex arc with a consequent division of labor between neurons, one of which became specialized in the reception of stimuli, the other in signal transmission to a nonneuronal element. Association neurons with connecting functions, also called interneurons, were later interposed between the sensory and effector nerve cells of the two-neuron arcs. With their appearance, one of the main characteristics of nerve cells, namely, their tendency to form chains, became evident. The interposition of an interneuron results in greater time elapsing between a stimulus and the ensuing response but allows the modification of impulses before their transmission to the effector neuron; in other words, while two-neuron chains allow generally stereotyped responses, the presence of one or more interneurons enhances the versatility and flexibility of the system. Sensory, effector, and association neurons are the three fundamental classes always present in metazoans beginning with the platyhelminths. In the course of evolution, the proportion of interneurons to sensory and effector neurons gradually increased, so that this ratio may be used as an indicator of the evolutionary stage of the nervous system of a given species.

The fundamental physiological properties of nerve cells are very similar in the simplest and the most complex organisms. All neurons in fact (a) react to various physical and chemical stimuli giving rise to signals (excitability), (b) convey these signals at high speed (conductivity), and (c) transmit them to other neurons or to nonneuronal cells (e.g., muscle or gland cells) thereby influencing their activity.

Although they vary considerably in shape and size (see Chap. III), the great majority of vertebrate neurons possess a body, which accommodates the nucleus, and two kinds of process (dendrites and the axon). Various parts of the neuron show degrees of functional specialization. For example, the cell body is the metabolic center of the cell, in which many of the extramitochondrial proteins required by the neuron are synthesized; dendrites receive most of the impulses that arrive to the neuron; the initial segment of the axon generates the action potential; the axon conveys this to its terminals; and the latter deliver the impulses to other neurons or effector cells.

III. Shape and Size of Neurons

Neuronal form and function are closely related since the shape of a neuron determines its connections with other neurons and influences the way in which it processes synaptic information. These interrelationships account for the importance of studying neuronal morphology and the factors which control it.

While in a nonneuronal tissue most cells have a roughly similar shape, nerve cells exhibit a great variety of forms. To simplify, we may say that most neurons possess a body and two types of process, called dendrites and axons (Fig. I.1). The cell body, also called the soma, consists of a nucleus and its surrounding cytoplasm, named the perikaryon.

In vertebrates most neurons are provided with several dendrites, but only one axon. Dendrites, so named because their ramifications recall the branches of a tree, were in the past often referred to as protoplasmic processes (see Chap. I). They are essentially direct extensions of the perikaryon, have irregular contours, and become narrower as they extend further from the cell body, giving off their branches at acute angles (Fig. I.1). The field of dendrite ramification is limited to the neighborhood of the cell body. The dendritic tree displays a characteristic pattern in each type of neuron.

The axon arises either directly from the perikaryon or from the proximal portion of a dendrite by way of a small conical elevation called the axon hillock. The axon has smooth contours, is relatively uniform in diameter throughout its length (Fig. I.1), and usually gives off very few collateral branches along its course but is richly branched near its terminal field. As a rule, axonal branches arise from their parent stem at right or obtuse angles. Dendrites and axons differ not only in their morphology but also in their structure and molecular composition (see Sect. V.D.2, and also Kamiguchi and Lemmon 1998). The main differences between dendrites and axons are listed in Table III.1.

While a number of characteristics discernible with the available techniques allow us in most cases to distinguish clearly between dendrites and axons in the vertebrate nervous system, this may be difficult in invertebrates. Dendritic and axonal structure and function are often intermingled in the processes of invertebrate neurons. For this reason, the term neurite has come into use to refer to all neuronal processes. However, the term neurite has been used for over a century with the same meaning as axon, and thus, the current trend to use neurite as a general term for all neuronal processes may cause confusion. The term neurite will not be employed here firstly to avoid confusion and also because this book deals mainly with the structure of the vertebrate nervous system, where, as noted above, it is generally possible to distinguish dendrites from axons.

E. Pannese, *Neurocytology: Fine Structure of Neurons, Nerve Processes, and Neuroglial Cells*, 2nd Edition,
DOI 10.1007/978-3-319-06856-5_3,

Table III.1 Major differences between dendrites and axons

	Dendrites	Axons
Diameter	Decreases with distance from the cell body	Remains fairly constant throughout the length of the axon
Branching	Numerous branches originate near the cell body; acute-angled branching	Branching is usually restricted to the terminal field; right-angled branching
Spines	Present in many, but not all, neurons	Generally absent; may rarely be present on the initial segment only
Myelin sheath	Extremely rare	Often present
Golgi apparatus	Present in primary dendrites	Absent
Nissl bodies and free ribosomes	Usually abundant	Rare
Microtubules	Oriented in either direction	Uniformly oriented with plus end distal to the cell body
Microtubule-associated protein 2 (MAP-2)	Detectable immunologically, sometimes intense immunoreactivity	Usually not detectable immunologically, sometimes very weak immunoreactivity
Tau-1 protein	Absent or rare	Present, sometimes abundant
Neurofilaments	Unphosphorylated or poorly phosphorylated	Extensively phosphorylated, particularly in the distal axon
Glutamate receptor subunit (GluR1)	Detectable immunologically	Not detectable immunologically
Growth-associated protein 43 (GAP43)	Not detectable immunologically	Detectable immunologically
Membrane proteins (ion channels and cell adhesion proteins)	Distribution differs in dendrites and axons	
Spectrin isoforms	That present in dendrites differs from that found in axons	

Several studies have sought to identify the roles that various factors play in controlling neuronal shape. The shape of each neuron is controlled both by its genetic program and by the surrounding environment. The relative influence exerted by these intrinsic and extrinsic factors seems to vary with the neuronal class, and a wide spectrum of possibilities exists. For example, in some neurons intrinsic factors determine the basic features of the cell morphology, as shown by the fact that when these neurons are removed at an early developmental stage and grown in dissociated cell cultures, they exhibit the main features of corresponding neurons grown in situ (rat cortical neurons, Dichter 1978; rat hippocampal neurons, Banker and Cowan 1979 and Dotti et al. 1988; sympathetic preganglionic neurons of the chick spinal cord, Honig and Hume 1986). In such cases, extrinsic factors only mold the details of dendritic and axonal morphology (see Rakic 1974 and Jacobson 1978 for reviews). In other neurons, while intrinsic factors still exert a major influence, it would seem that extrinsic factors play a much greater role than in the neurons mentioned above. Thus, certain nerve cells removed at an early developmental stage and grown in dissociated cell cultures exhibit one or more basic features that are clearly different from those of the same neurons grown in situ (sympathetic neurons of rat superior cervical ganglion, Bruckenstein and Higgins 1988 and Tropea et al. 1988; neurons of rat nodose ganglion, De Koninck et al. 1993). In these neurons extrinsic factors do not simply mold the details of dendritic and axonal morphology, but influence basic features such as the extension of dendrites.

Glial cells, afferent innervation, synaptic inputs, trophic factors, and hormones are among the extrinsic factors known to influence neuronal shape. In vitro studies have shown that glial cells are able to influence the features of the dendritic tree. For example, they can stimulate the extension of dendrites (mesencephalic and striatal neurons of rat, Denis-Donini et al. 1984 and Chamak et al. 1987; Purkinje cells of mouse, Buard et al. 2010; neurons from various regions of the rat CNS, Fallon 1985; sympathetic neurons of rat and chick, Tropea et al. 1988, Johnson et al. 1989, and Clendening and Hume 1990) or, conversely, they can prevent dendrite extension (sensory neurons of rat nodose ganglion, De Koninck et al. 1993). Glial cells probably influence the expression of several morphological traits in neurons not only in culture but also in vivo. The influence of the afferent innervation on neuronal shape is illustrated by the following findings. If cerebellar granule cells fail to develop because of a genetic abnormality or are destroyed during development (e.g., by X-ray irradiation), the Purkinje cell dendrites are deprived of a large percentage of their normal afferent connections. Under such conditions, these dendrites do not develop fully and display an immature branching pattern in the adult while retaining their unmistakable shape (Altman and Anderson 1972; Rakic and Sidman 1973b). The dendritic tree of hippocampal pyramidal cells that have been deprived of their normal afferent connections during development is similar to that of control neurons; however, the number and size of the terminal dendritic segments are reduced. Mauthner cells deprived of vestibular afferents exhibit a reduced dendritic arborization in the region which normally receives vestibular terminals; conversely, superinnervation of these cells by vestibular axons is followed by a localized enhancement of dendritic branching in the region receiving the extra terminals (Goodman and Model 1988). The following experiments show that synaptic inputs can contribute to molding the dendritic tree. Unilateral acoustic deprivation in the developing chick affects dendritic development in the nucleus laminaris neurons that normally receive inputs from the plugged ear (Gray et al. 1982). Deprivation of synaptic activity induces strong impairment of the dendritic growth in rat CA1 pyramidal cells (Groc et al. 2002). In vivo and in vitro observations have revealed that trophic factors promote the extension, elongation, and branching of dendrites in developing neurons of the PNS (Snider 1988; De Koninck et al. 1993; Lein et al. 1995; Mertz et al. 2000; Niblock et al. 2000). Finally, it has been shown that hormones play a role in the development (Rami et al. 1986; Gould and Butcher 1989) and remodeling (Williams and Truman 2005) of the dendritic tree of certain neurons. Other experiments indicating that extrinsic factors influence the pattern and extent of the dendritic tree are reported in Sect. V.C.3. Several classes of molecules are involved in the regulation of dendritic length and branching. These include neurotrophins, cell surface receptors, regulators of microtubule and actin, signaling molecules, calcium signaling proteins, and transcription factors (for a review, see Jan and Jan 2010). With regard to the axon, extrinsic factors can influence both the direction (Van der Loos 1965; Globus and Scheibel 1967c) and extent (Baird et al. 1992; Qian et al. 1992) of axonal growth; such factors are target-derived signals, growth factors, adhesion molecules, and extracellular matrix components (e.g., see Bixby and Harris 1991; Reichardt and Tomaselli 1991; Cohen-Cory and Fraser 1995).

Unlike cells of other tissues, whose volume varies little with respect to the mean, nerve cells differ greatly in size. Even within a single organism neurons may display a very large size range. For example, in a single human individual, the cell bodies of the smallest neurons are 5–8 μm in diameter [e.g., cerebellar granule cells (Fig. IV.6), granule cells of the olfactory bulb (Fig. IV.10), dwarf (or arachniform) cells of the cerebral cortex, bipolar cells of the retina, stellate cells of the substantia gelatinosa of the spinal cord], whereas the cell bodies of the largest neurons exceed 100 μm in diameter [e.g., spinal cord motoneurons (Fig. IV.2), spinal ganglion cells (Fig. IV.9), Betz cells in the cerebral cortex]. Even neurons of the same type may vary greatly in size within a single organism; for example, the cell body of spinal ganglion pseudounipolar neurons may range from 10 to 80 μm in diameter in a rat and from 15 to 120 μm in a human subject. These size differences become more striking when the volume of the cell body is considered rather than the diameter; it has been calculated that in humans the soma volume of a cerebellar granule cell may be 300 μm^3, whereas that of a Betz cell in the cerebral cortex may reach 200,000 μm^3 (Haug 1982).

Among the vertebrates in general, the following nerve cells reach a considerable size: Mauthner cells of the medulla of teleosts, lungfish, and amphibians; Müller cells of cyclostomes; and some neurosecretory cells of the spinal cord of several fish. The nerve cells of some invertebrates, such as those of the visceral ganglion of sea hares (Aplysia), have a body which may attain 1 mm in diameter.

Observations on vertebrate neurons innervating the periphery have revealed a correlation between the volume of the nerve cell body and the extent of its peripheral field of innervation (Levi 1906, 1908; Hahn 1912; Terni 1914; Donaldson and Nagasaka 1918; Franca-Netto 1951). Enriques (1908) reached similar conclusions concerning the neurons of the nervous ganglia of invertebrates. The following study by Terni (1920) confirmed this correlation. The amputation of the lizard tail is followed by the regeneration of skin, muscles, and supporting tissues, whereas the spinal cord and spinal ganglia do not regenerate. The last three pairs of spinal ganglia left in situ cranial to the amputation plane provide sensory innervation to the regenerated part of the tail. Within these ganglia, the bodies of the neurons, whose peripheral field is thus greatly enlarged, significantly increase in size. Further evidence that the size of the nerve cell body correlates with the extent of its peripheral target have been obtained in other neurons (e.g., neurons of the rat superior cervical ganglion that innervate the submandibular gland, Voyvodic 1989; neurons of the rat pelvic ganglion that innervate bladder smooth muscle, Gabella et al. 1992). Some data suggest that neurotrophins produced by cells of the peripheral target are involved in the size increase of the nerve cell body (Steers et al. 1991).

The body size of homologous neurons innervating the periphery is usually greater in large animals than in small ones (Hardesty 1902; see also Purves 1988). This size difference is probably due to the fact that the extent of the peripheral field innervated by a given type of neuron is usually greater in large animals. It is also true that certain types of nerve cell whose axon is confined to the CNS and which do not therefore have direct connections with the periphery [e.g., pyramidal cells of the cerebral cortex (Fig. IV.4) and Purkinje cells of the cerebellar cortex (Fig. IV.5)] have larger bodies in large animals than in small ones. Here, too, the size difference is probably related to varying extent of the field innervated by the axon.

The correlation between the size of a nerve cell body and the extent of the field innervated by its axon may explain why, in some species, neurons enormously larger than most of the others are present. Such neurons in fact innervate particularly large fields (e.g., Mauthner cells, Rohon-Beard cells, neurons innervating the electric organs of certain fish). The size and complexity of the dendritic tree also increase with the animal size (Barasa 1960; Purves and Lichtman 1985; Snider 1987).

However, the extent of the field of innervation is probably not the only factor influencing neuronal size. Among neurons involved in the control of specific aspects of sexual behavior, for instance, some are larger in males than in females. Belonging to this category are certain neurons of the rat preoptic area (Gorski et al. 1980), the motoneurons of the spinal nucleus of the bulbocavernosus of the rat (Breedlove and Arnold 1981), certain of those of the nucleus robustus archistriatalis of the canary (DeVoogd and Nottebohm 1981), and certain neurons of the toad pretrigeminal nucleus (Schmidt 1982a). Androgen treatment of adult females leads to the enlargement of the neurons in the latter nucleus; these may then reach the size of the corresponding neurons in the male (Schmidt 1982b). In the canary, certain neurons of the telencephalic higher vocal center enlarge their somata in response to testosterone treatment (Johnson and Bottjer 1993). Thus, the size of given neurons may be under the influence of sex hormones (see also Ulibarri et al. 1995; Keast and Saunders 1998). In certain adult mammals the cyclical variation of the levels of circulating sex hormones brings about coincident changes in the soma size and dendritic extent of particular neurons (Forger and Breedlove 1987). These observations reveal a previously unsuspected plasticity in the CNS of adult mammals under normal conditions. Thyroid hormone also seems to influence the size of specific nerve cells, at least during critical periods of development (e.g., neurons of the mesencephalic nucleus of the trigeminal nerve of some amphibians, Kollros 1977; cholinergic basal forebrain neurons of the rat, Gould and Butcher 1989). Finally, there are studies which suggest that neuronal activity levels can affect cell size (Pasic and Rubel 1989).

It is to be noted that the volume of the nerve cell body does not always give an adequate idea of the true size of the nerve cell since, especially in large neurons, the volume of the axon and its branches may considerably exceed that of the cell body. The volume of the peripheral axonal branch (excluding the central branch and all collateral and terminal ramifications) of a large neuron from a human ganglion amounts to 125 times the volume of the corresponding cell body (Heidenhain 1911). In a neuron of the human L4 spinal ganglion with a 50-μm diameter cell body and an axon of 5 μm in diameter and 1.5 m long, the ratio between the volume of the axon and the volume of the cell body is 500:1 (Devor 1999). In a giant Betz pyramidal cell, the axon volume may be approximately 350 times the volume of the corresponding cell body (Haug 1982). In the smallest neurons, on the other hand, the volumes of axon and cell body are more closely comparable. For example, the ratio of axon volume to cell body volume is about 2:1 in a cerebellar granule cell.

As we have seen, the axon length can be considerable. The axon of a neuron of the blue whale (the largest mammal) may be over 10 m long. With regard to the axon diameter and the factors that control it, note first that this has great functional importance, since diameter is a principal determinant of the conduction velocity, i.e., of the velocity at which nerve impulses are propagated along the axon (see Sect. VII.B.4h). Usually, the axon diameter correlates directly with the nerve cell body volume (Marinesco 1909; Ramón y Cajal 1909). Based on this fact, it was thought for many years that the axon diameter was controlled by intrinsic (=genetic) factors. Other observations, however, have made it clear that also factors extrinsic to the neuron play an important role in the control of the axon diameter. The whole question of the regulation of the axon diameter now appears much more complicated than previously thought. Optic nerve axons from visually deprived rats are significantly narrower than control axons (Fernández et al. 1993). This suggests that, at least in certain neurons, functional activity can influence the axon diameter. When a rat neuron is disconnected from its target cells, the diameter of its axon diminishes, suggesting that interactions between a neuron and its target cells may be involved in the control of the axon diameter (Hoffman et al. 1988). The following studies suggest that interactions between an axon and its supporting cells also play a role in this regulation. (a) In a given motoneuron axon of the rat, the peripheral portion (myelinated by Schwann cells) in the ventral root is thicker than the central portion (myelinated by oligodendrocytes) in the spinal cord (Fraher 1978). (b) When a segment of a nerve from a Trembler mouse is grafted onto a normal nerve, normal axons passing through the grafted segment lack myelin or are hypomyelinated by the Trembler Schwann cells and axon diameter is also reduced, but is restored to normal in the distal nerve stump where the axon is normally myelinated (Aguayo et al. 1979). (c) In cultures of rat spinal ganglion neurons, myelinated axons have a larger diameter than bare axons (Windebank et al. 1985). (d) In normal lizard axons that have both myelinated and unmyelinated portions, the axon diameter is significantly greater in the myelinated than in the unmyelinated portions (Pannese et al. 1988d). (e) The axons of the rat retinal ganglion cells are narrow in the unmyelinated portion of the optic nerve, but increase substantially in size in the myelinated domain of the nerve (Hernández et al. 1989). (f) Following X-ray irradiation of the optic nerve and tract on one side of the rat brain, axons crossing the chiasm have an unmyelinated segment and a myelinated segment; in the same axon, the former is smaller in diameter than the latter (Colello et al. 1994). (g) In human demyelinating neuropathies, a marked reduction in axon diameter occurs locally in regions of segmental demyelination (Prineas and McLeod 1976). The mechanisms by which intrinsic and extrinsic factors affect axon diameter are poorly understood. Several studies suggest that these factors influence axon diameter by modulating the packing density of neurofilaments (Yamasaki et al. 1991; Cole et al. 1994; Sánchez et al. 1996), the phosphorylation of neurofilaments (de Waegh et al. 1992; Cole et al. 1994; Nixon et al. 1994), and the stoichiometric proportions of the neurofilament subunit proteins (Wong et al. 1995; Marszalek et al. 1996; Xu et al. 1996). The myelin-associated glycoprotein (MAG) is probably involved in this process (Dashiell et al. 2002; see also Sects. VII.B.4d and VIII.F.2).

The relationship between neuronal size and neuronal function has been well studied in somatic motoneurons, i.e., neurons which innervate striated muscle fibers. Usually a single motoneuron of the spinal cord and brainstem innervates more than one striated muscle fiber. The number of muscle fibers innervated by a single motoneuron varies from muscle to muscle and species to species and ranges from less than ten to many hundreds (see Table III.2). The functional set consisting of an individual motoneuron of the spinal cord or brainstem together with the muscle fibers innervated by it is called a motor unit. This term was introduced by Sherrington (1925) to define the axon of a motoneuron together with the muscle fibers it innervates. As commonly used today, this term does not correspond exactly to Sherrington's original definition. As a rule, the size of a motoneuron is directly related to the number of muscle fibers it supplies, while the excitability of a motoneuron varies inversely with its size (Henneman 1957; Henneman et al. 1965). Therefore, a small motoneuron innervates few muscle fibers, thus generating a modest amount of force in the muscle, but has a low excitation threshold (=high excitability). Conversely, a large motoneuron innervates a large number of muscle fibers generating considerable force in the muscle, but requires an intense stimulus for activation. Thus, weak levels of stimulation produce a very precise motor activity, while intense stimulations result in a large increase in generated force.

Because of their highly irregular shape, neurons usually have a very large surface area, and this is one of their distinguishing features. A large surface area facilitates exchange between the neuron and its environment and makes possible the formation of numerous interneuronal connections. The

Table III.2 Motor units

Muscle	Family	Species	Common name	Mean number of muscle fibers per motor unit	Author(s)
Sterno-cutaneous	Ranidae	*Rana temporaria*	Common frog	25	Reichert (1851)
Abductor digiti minimi of the foot	Ranidae	[a]	Frog	40	Tergast (1873)
Cutaneous dorsi	Ranidae	[a]	Frog	20	Lucas (1909)
Extrinsic eye muscles	Bovidae	*Ovis ammon aries*	Sheep	6	Tergast (1873)
Diaphragm	Muridae	*Rattus norvegicus*	Rat	25	Krnjević and Miledi (1958)
Soleus	Muridae	*Rattus norvegicus*	Rat	110	Brown et al. (1976)
	Leporidae	*Oryctolagus cuniculus*	Rabbit	170	Bixby et al. (1980)
Sartorius	Leporidae	*Oryctolagus cuniculus*	Rabbit	112	Van Harreveld (1947)
Extensor digitorum longus	Leporidae	*Oryctolagus cuniculus*	Rabbit	370	Bixby et al. (1980)
Pronator	Leporidae	*Oryctolagus cuniculus*	Rabbit	255	Bixby et al. (1980)
Sartorius	Canidae	*Canis familiaris*	Dog	50	Tergast (1873)
Brachial biceps	Canidae	*Canis familiaris*	Dog	104	Tergast (1873)
Tenuissimus	Felidae	*Felis catus*	Cat	20	Porter and Hart (1923)
Soleus	Felidae	*Felis catus*	Cat	120	Clark (1931)
	Felidae	*Felis catus*	Cat	140–190[b]	Burke et al. (1974)
Extensor digitorum longus	Felidae	*Felis catus*	Cat	165	Clark (1931)
Anterior tibial	Felidae	*Felis catus*	Cat	110	Hagbarth and Wohlfart (1952)
Gastrocnemius (medial head)	Felidae	*Felis catus*	Cat	400–800[b]	Burke and Tsairis (1973)
	Cercopithecidae	*Papio papio, Papio hamadryas, Papio anubis*	Baboon	790	Wray (1969)
Abductor pollicis brevis	Cercopithecidae	*Papio papio, Papio hamadryas, Papio anubis*	Baboon	220	Wray (1969)
Lateral rectus of the eye	Hominidae	*Homo sapiens*	Man	9	Björkman and Wohlfart (1936); Feinstein et al. (1955)
Platysma	Hominidae	*Homo sapiens*	Man	25	Feinstein et al. (1955)

(continued)

Table III.2 (continued)

Muscle	Family	Species	Common name	Mean number of muscle fibers per motor unit	Author(s)
Deltoid	Hominidae	*Homo sapiens*	Man	70	Gath and Stålberg (1981)
Brachial biceps	Hominidae	*Homo sapiens*	Man	72	Gath and Stålberg (1981)
Brachioradialis	Hominidae	*Homo sapiens*	Man	>410	Feinstein et al. (1955)
First dorsal interosseus of the hand	Hominidae	*Homo sapiens*	Man	340	Feinstein et al. (1955)
First lumbrical of the hand	Hominidae	*Homo sapiens*	Man	107	Feinstein et al. (1955)
Anterior tibial	Hominidae	*Homo sapiens*	Man	609	Feinstein et al. (1955)
	Hominidae	*Homo sapiens*	Man	124	Gath and Stålberg (1981)
Gastrocnemius (medial head)	Hominidae	*Homo sapiens*	Man	1,784	Feinstein et al. (1955)

[a]Species not indicated in the original paper
[b]The original papers did not report means but only the ranges indicated

axon and dendrites are the major contributors to the extended neuronal surface area. For the neuron examined quantitatively (see earlier, this chapter) by Heidenhain (1911), his calculations indicated that the surface area of the axon was more than 800 times that of the cell body. However, at that time the slender projections arising from the perikaryon of the spinal ganglion neurons had not yet been recognized (see later, this chapter). Because of these projections, the surface area of the cell body is greater than that calculated by Heidenhain. Therefore, it is likely that in this neuron the ratio between the surface area of the axon and that of the cell body is about 500:1. Concerning dendrites only, note that in the cat the ratio of the surface area of the dendritic tree to that of the cell body is 5:1 in the giant cells of the reticular formation (Mannen 1966), 25:1 in the pyramidal neurons of the somatosensory cortex (Mungai 1967), and 32:1 in spinal motoneurons (Ulfhake and Kellerth 1981); the same ratio is 8–10:1 in the neurons of the guinea pig inferior mesenteric ganglion (Ermilov et al. 2000).

These data show that the perikaryon generally contributes little to the total surface area of the neuron. However, devices which increase the perikaryal surface area have been described, particularly for neurons that do not possess dendrites. In vertebrates, the perikaryon of spinal ganglion neurons is endowed with many slender projections (Figs. III.1, V.5, and VII.3) whose length ranges between 0.3 and 3 μm and whose transverse diameter averages 0.2 μm (Pannese et al. 1990a, b; Pannese 2002). These projections increase the surface area of the perikaryon by about 40 % in the cat and rabbit and by about 30 % in the gecko and lizard (Pannese et al. 1983, 1985). Projections akin to these have been described in the neurons of the trigeminal mesencephalic nucleus; in the mouse and hamster, they increase the surface area of the perikaryon by 12–14 % (Hinrichsen and Larramendi 1970; Alley 1974).

In preparations of sensory ganglia stained by silver impregnation and examined under the light microscope, in addition to typical pseudounipolar neurons, nerve cells have been observed with short and thick projections or with long and fine expansions ending in club-shaped enlargements or, finally, with fenestrations situated in the superficial region of the perikaryon (Fig. III.2). More detailed descriptions of these structures are found in Disse (1893), Levi (1908), Held (1909a), and Ramón y Cajal (1909). They were more frequently detected in various pathological states or under particular experimental conditions (e.g., in grafted or injured sensory ganglia) but were also observed under normal conditions. Nageotte (1907) interpreted these formations as expressions of the regenerative activity of the nerve cell. Since these processes do not usually contact other neurons and, therefore, do not play a role in the conduction of nervous signals, Nageotte (1907) named them paraphytes thereby distinguishing them from orthophytes (axons and dendrites), i.e., the processes which conduct nervous signals. Levi (1908), however, attributed a trophic role to these atypical processes, suggesting that by amplifying the surface area of the neuron, they facilitated the metabolic exchange between the neuron and its environment. References to paraphytes are rarely encountered in the more recent literature, probably because of the widespread belief that these processes largely arise from the shrinkage of the nerve cell body or other technical artifacts.

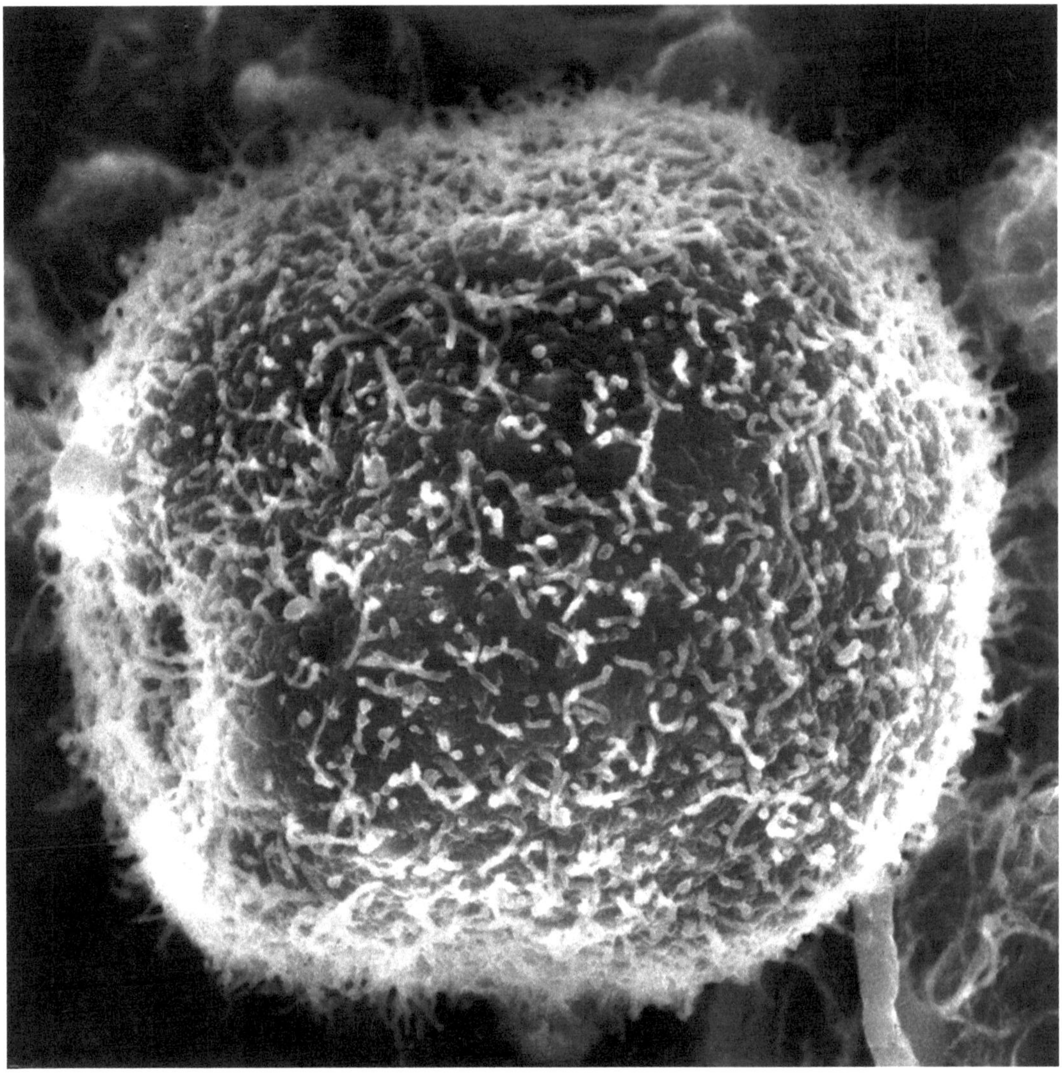

Fig. III.1 Perikaryal projections. Scanning electron micrograph of the perikaryal surface of a sensory neuron exposed after removal of all enveloping sheaths. The surface is rendered irregular by the presence of numerous projections. Rabbit spinal ganglion, ×4,300

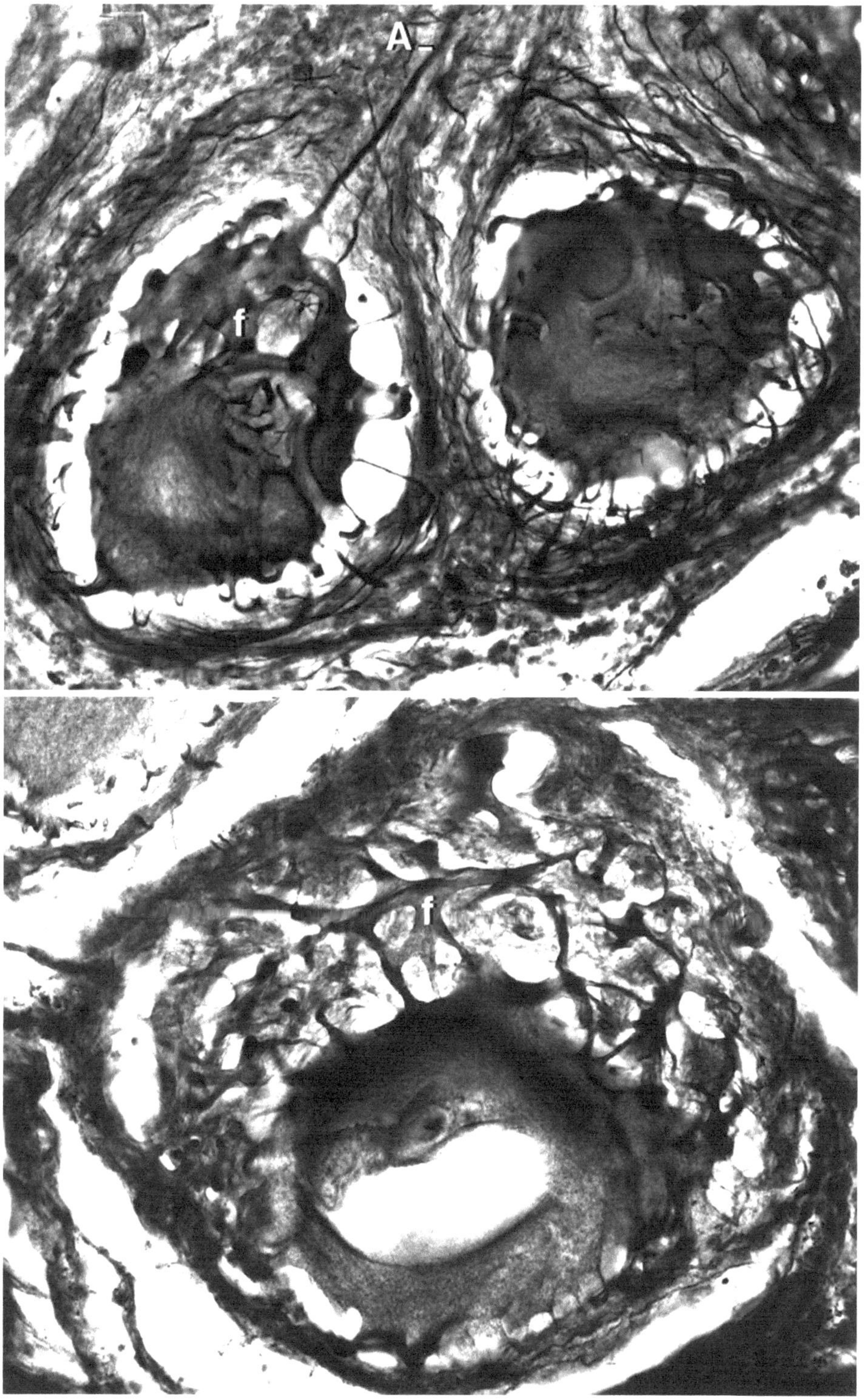

Fig. III.2 Paraphytes. The large sensory neurons shown in these light micrographs display numerous perikaryal projections of varying length and thickness as well as fenestrations (**f**) situated in the superficial regions of their perikarya. **A** axon. Cajal method, horse spinal ganglion, ×400

IV. Different Types of Neuron

Despite numerous attempts, the great variety in number, length, and arborization pattern of processes and in the shape, size, and position of the cell body has interfered with the efforts to produce a satisfactory classification of nerve cells. Here, therefore, a list of the most common types of neuron, distinguished by the characteristics of their processes, is given.

Most nerve cells in vertebrates have several dendrites but only one axon. These multipolar neurons correspond well to the typical nerve cell so often referred to when describing the nerve cell in a general way.

Golgi (1882) distinguished two types of multipolar neuron on the basis of the configuration of the axon. The axon of type I cells sends out few collateral branches, retaining its distinct identity, and becomes myelinated. In contrast, the axon of type II cells, which usually remains unmyelinated, divides repeatedly shortly after leaving the soma to give rise to a complex ramification in the neighborhood of the cell body. It may be recalled here that Golgi believed that the highly ramified axons of type II cells formed, in the gray matter, a continuous neural network, which he called *rete nervosa diffusa* (diffuse nervous network, Fig. IV.1). He also believed that the few collateral branches of the axons of type I cells contributed to the formation of the same network. This simple classification is still valid today, particularly since the differing morphologies of the two types of nerve cell reflect their different functional roles. The correlation between shape and function becomes clearer by complementing Golgi's original description with the following remarks. The axon of a type I cell may vary in length, but it always reaches a territory different from that in which the dendrites of the cell arborize (Figs. IV.2, IV.4, IV.5, and IV.6). By contrast, the axon of a type II cell divides repeatedly in the neighborhood of its own neuronal body, i.e., in the same circumscribed field in which the dendrites of that cell ramify (Fig. IV.3).

Examples of type I cells are the motoneurons of the spinal cord (Fig. IV.2) and brainstem, the large pyramidal cells (Fig. IV.4), the large cells of Clarke's column, and the Purkinje cells (Fig. IV.5) of the cerebellar cortex. All these cells have axons which terminate in fields distinct from those containing the corresponding bodies and dendrites. The cell body and dendrites of a motoneuron lie in the spinal cord or brainstem, while the axon leaves the CNS and ends in the periphery. The cell body and dendrites of a large pyramidal cell lie in the cerebral cortex, while the axon leaves the cortex and terminates in other regions of the CNS (e.g., the brainstem or the spinal cord). The cell body and dendrites of a large cell of Clarke's column lie in the spinal gray matter, while its axon terminates in the cerebellar cortex. The Purkinje cell has its soma and dendrites in the cerebellar cortex, while its axon ends

E. Pannese, *Neurocytology: Fine Structure of Neurons, Nerve Processes, and Neuroglial Cells*, 2nd Edition, DOI 10.1007/978-3-319-06856-5_4,

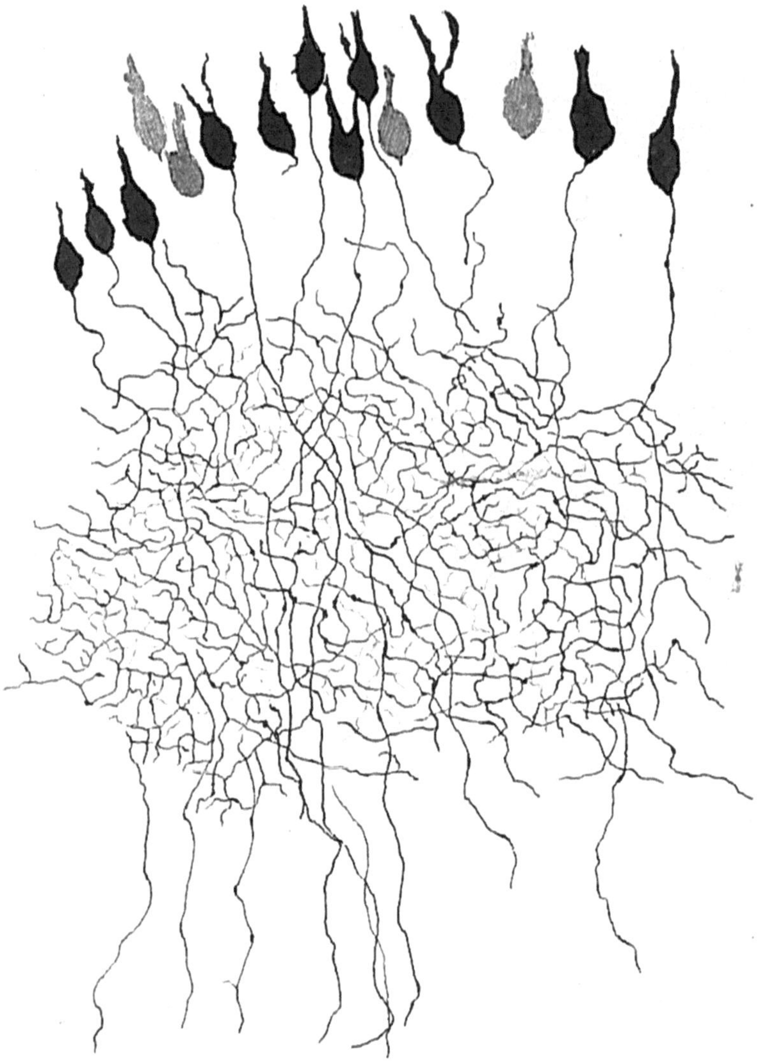

Fig. IV.1 Drawing by Golgi showing the *rete nervosa diffusa* (diffuse nervous network) impregnated using his black reaction. Fascia dentata of a mammal

in the central nuclei of the cerebellum. Examples of type II cells are the stellate cells in the cerebral cortex, those in the reticular formation of the brainstem, and the Golgi cells in the cerebellar cortex. The axonal arborization of each of these neurons remains in the same field as its dendrites (Fig. IV.3).

It is evident from these morphological differences that the two types of neuron have distinct functional roles. The type I cell performs mainly integrative and projection functions; it collects information from a large number of afferents and, by means of its axon, conveys its integrated responses to more or less distant points. The type II cell, whose axon and dendrites arborize in the same circumscribed field, receives the same inputs as do the neurons on which its axon terminates, so that it has a mainly local, modulatory role (Palay and Chan-Palay 1977).

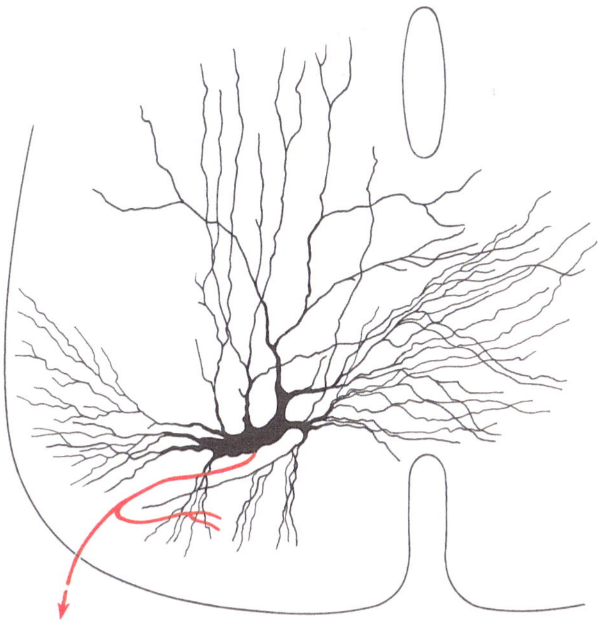

Fig. IV.2 Multipolar neuron (Golgi type I cell). Drawing of a motoneuron of the spinal cord as seen in a Golgi preparation. The cell body and dendrites of this neuron lie in the ventral horn of the spinal cord, while the axon (*red*) leaves the CNS and terminates in the periphery

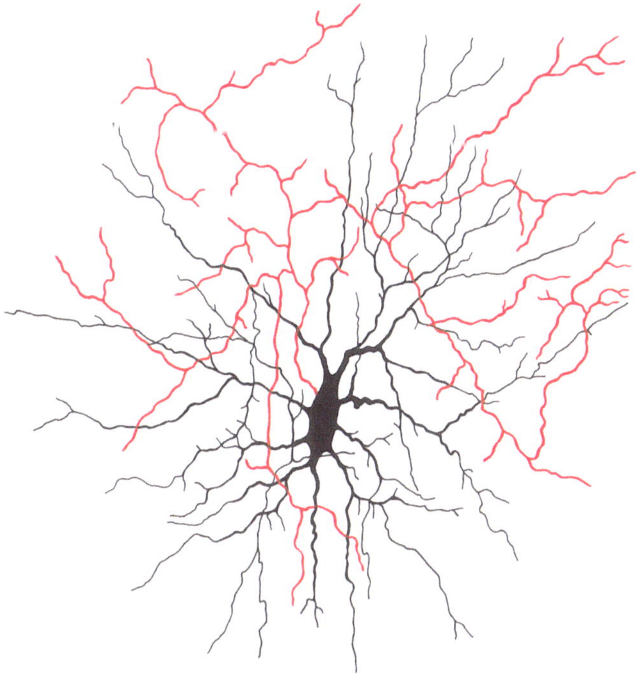

Fig. IV.3 Multipolar neuron (Golgi type II cell). Drawing of a neuron of the lentiform nucleus as seen in a Golgi preparation. The axon (*red*) divides repeatedly in the neighborhood of its own neuronal body, i.e., in the same circumscribed field in which the dendrites of this neuron ramify

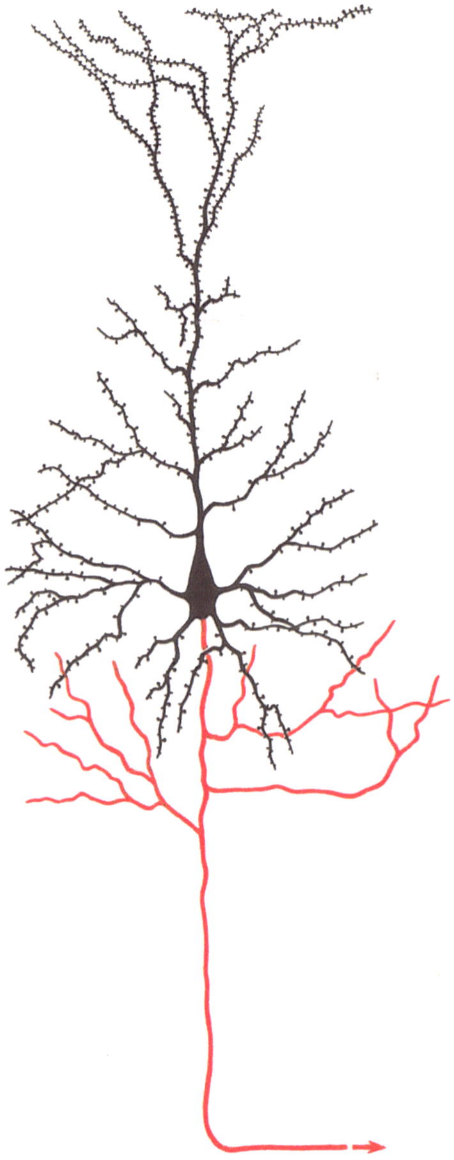

Fig. IV.4 Multipolar neuron (Golgi type I cell). Drawing of a large pyramidal cell of the cerebral cortex as seen in a Golgi preparation. The cell body and dendrites of this neuron lie in the cerebral cortex, while the axon (*red*) leaves the cortex and terminates in the brainstem or spinal cord

Not all neurons, however, are multipolar cells. There are neurons that lack dendrites and bear only axons. Examples are the olfactory receptor neurons of vertebrates (Fig. IV.7); the receptor cells of the retina (rods and cones); the primary sensory cells of many invertebrates (Fig. IV.7), whose bodies lie within the integument or directly beneath it; the neurons of the mesencephalic nucleus of the trigeminal nerve of mammals; and those of sensory ganglia. The latter may have a globular, oval, or spindle-shaped body with two axons (bipolar sensory neurons, Fig. IV.8), or they may have a single axon which divides into two branches (pseudounipolar sensory neurons, Fig. IV.9).

The peripheral process of bipolar sensory neurons and the peripheral branch of the single process of pseudounipolar sensory neurons, both of which conduct signals toward the cell body, are often referred to as dendrites. This usage follows the terminology employed by Ramón y Cajal. Cajal interpreted these processes as dendrites in order to reconcile the existence of bipolar and pseudounipolar sensory neurons with the law of dynamic polarization of the neuron. According to this law [originally formulated by Van Gehuchten (1891)], signaling within a neuron flows in a unidirectional manner from the dendrites and cell body, which receive inputs from other neurons, to the axon and from there to the presynaptic terminals. However, work in the second half of the twentieth century showed that (a) impulses can travel along axons and dendrites in both directions and (b) the directionality of signal conduction is not an intrinsic property of the neuron but depends on the asymmetry of synapses. Furthermore, there are clear structural and molecular differences between axons and dendrites (see Table III.1, Sect. V.D.2, and also Kamiguchi and Lemmon 1998), and both the peripheral process of bipolar sensory neurons and the peripheral branch of the single process of pseudounipolar sensory neurons have the structural and molecular characteristics of axons.

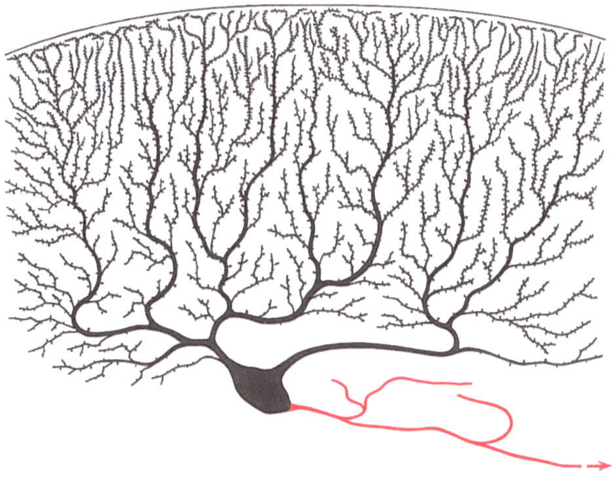

Fig. IV.5 Multipolar neuron (Golgi type I cell). Drawing of a Purkinje cell of the cerebellar cortex as seen in a Golgi preparation. The cell body and dendrites of this neuron lie in the cerebellar cortex, while the axon (*red*) leaves the cortex and terminates in the central nuclei of the cerebellum

Bipolar sensory neurons are numerous in the sensory ganglia of cyclostomes, selachians, and teleosts as well as the cochlear and vestibular ganglia of higher vertebrates, while pseudounipolar sensory neurons are found in all the other sensory ganglia of higher vertebrates. Signal conduction is faster and more reliable through pseudounipolar than through bipolar sensory neurons (Lüscher et al. 1994). Thus, the phylogenetic evolution from a bipolar to a pseudounipolar morphology results in improved function. The shape of a sensory ganglion cell faithfully reflects its functional role. For example, a spinal ganglion neuron connects the surface of the body or the muscles or the joints with the spinal cord (Fig. IV.9). This cell transduces stimuli of various types into trains of nerve signals and conveys them to the CNS with retention of the topographic order (Palay and Chan-Palay 1977). The central process of this neuron bifurcates into ascending and descending branches which send collaterals into the dorsal gray column (Fig. IV.9), thus distributing sensory information in an orderly way to other neurons.

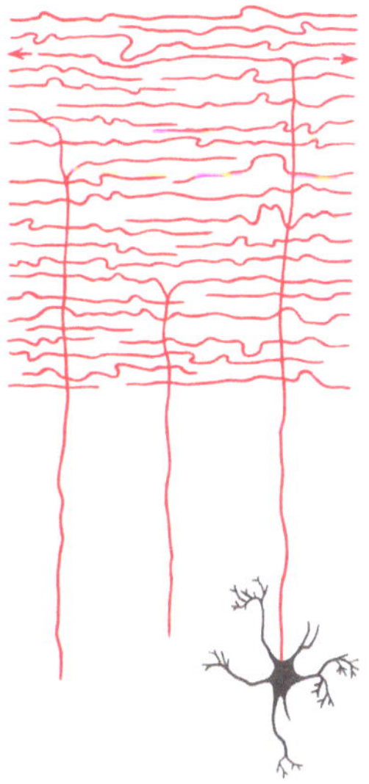

Fig. IV.6 Multipolar neuron (Golgi type I cell). Drawing of a granule cell of the cerebellar cortex as seen in a Golgi preparation. The cell body and dendrites of this neuron lie in the inner cortical layer, while the axon (*red*) extends into the outer cortical layer

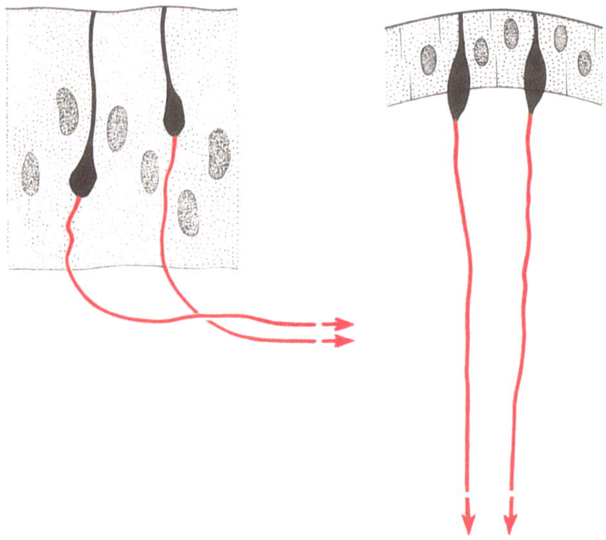

Fig. IV.7 Neurons bearing only axonal processes. *Left*: Drawing of two olfactory receptor neurons as seen in a Golgi preparation. Axons are in *red*. The nuclei of some supporting cells of the olfactory epithelium are also shown. *Right*: Drawing of two primary sensory neurons of an invertebrate (gastropod) as seen in a Golgi preparation. Axons are in *red*. The nuclei of some integumental cells are also shown

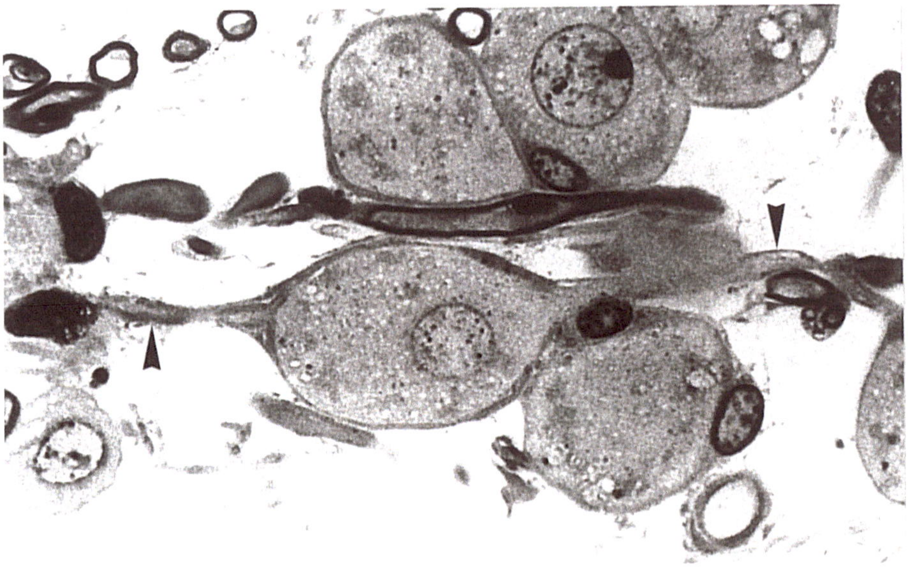

Fig. IV.8 Neurons bearing only axonal processes. Bipolar sensory neuron, whose two axons are marked with *arrowheads*. Human cochlear ganglion, ×1,750 (Courtesy of S. Iurato, S. Colucci, and A. Zambonin)

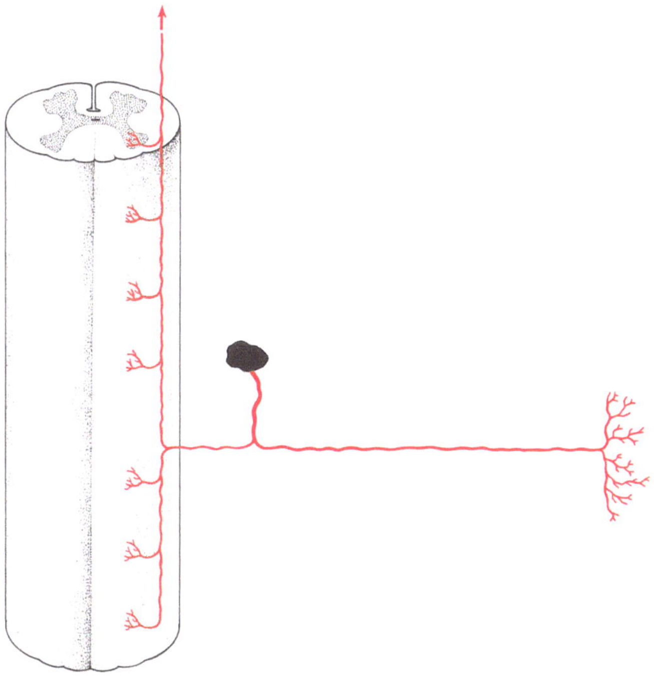

Fig. IV.9 Neurons bearing only axonal processes. Drawing of a pseudounipolar sensory neuron of a spinal ganglion as seen in a Golgi preparation. This neuron has a single axon (*red*) which divides into a peripheral and a central branch. The former terminates in the periphery, while the central branch enters the spinal cord where it bifurcates into an ascending and a descending secondary branch. Each of these sends collaterals into the dorsal gray column

There are also neurons lacking axons; these include the granule cells of the olfactory bulb (Fig. IV.10), and certain retinal cells (e.g., a type of horizontal cell and many types of amacrine cells). Axonless neurons are also present in invertebrates.

The motoneurons of the ventral ganglia of certain invertebrates (e.g., polychaetes, leeches, and crustaceans) present singular characteristics. They have a globular or club-shaped body which contains the nucleus and sends out only one process; the latter runs for a variable distance (up to several hundred micrometers) and then expands in diameter and gives rise to several dendritic branches and to a single axon (Fig. IV.11). The dendrites ramify in the central core of the ganglion, participating in a dense meshwork of processes called the neuropil, while the axon leaves the ganglion and ends inside a muscle.

Bipolar and multipolar neurons are also found in the peripheral plexuses of invertebrates. It is doubtful, however, whether these two cell types, in which a clear distinction between axon and dendrites cannot be made, correspond to the bipolar and multipolar neurons of vertebrates.

As some of the examples reported above illustrate, knowledge of the morphology of a neuron can be very useful in interpreting its function, particularly when its connections are known in detail. Another interesting aspect of the relation between shape and

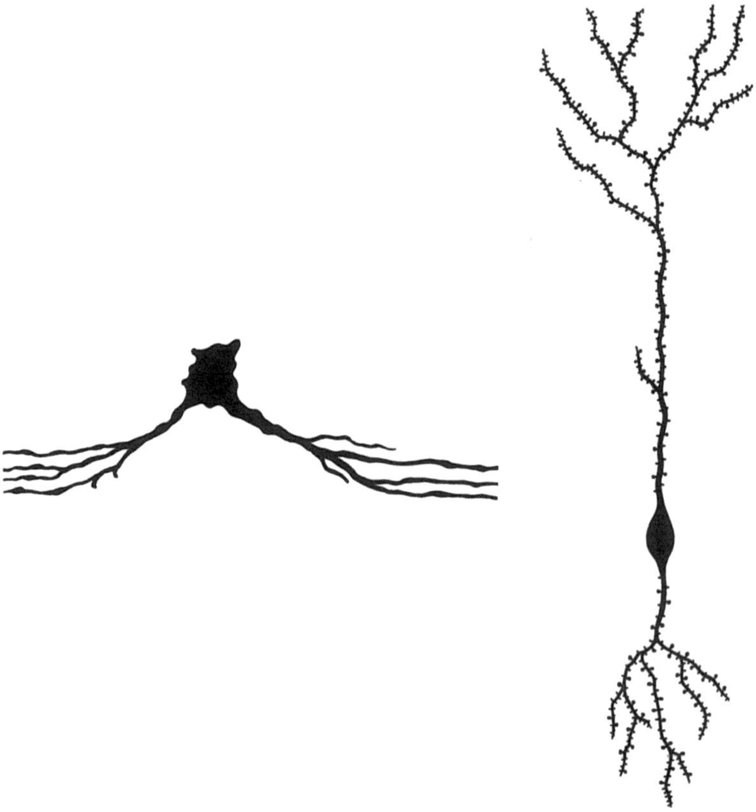

Fig. IV.10 Neurons lacking axonal processes. *Left*: Drawing of an amacrine cell of the retina as seen in a Golgi preparation. *Right*: Drawing of a granule cell of the olfactory bulb as seen in a Golgi preparation. Note that many spines (gemmules) project from the dendrites of this neuron

function in neurons regards the possible correlation between morphology and electrophysiological properties. The clearest examples of this correlation are to be found in the process of metamorphosis of certain insects. For example, nerve cells that survive metamorphosis converting from slow to fast motoneurons undergo specific changes in electrophysiological properties which are temporally correlated with an extensive remodeling of the dendritic morphology (Duch and Levine 2000). The development of the technique of intracellular filling, which allows to evaluate the physiological properties of neurons before they are filled to study their morphology (for a review, see Thomson and Armstrong 2011), gave a considerable impulse to studies on the correlation between the morphology and electrophysiological properties of vertebrate neurons. A number of sites at which such a correlation has been demonstrated are listed in Table IV.1.

An attempt to classify neurons on the basis of dendrite configuration was made by Ramón-Moliner (1968), who recognized isodendritic, allodendritic, and idiodendritic neurons. However, this classification does not appear satisfactory, particularly because it groups highly diverse neurons into the same category (Peters et al. 1991).

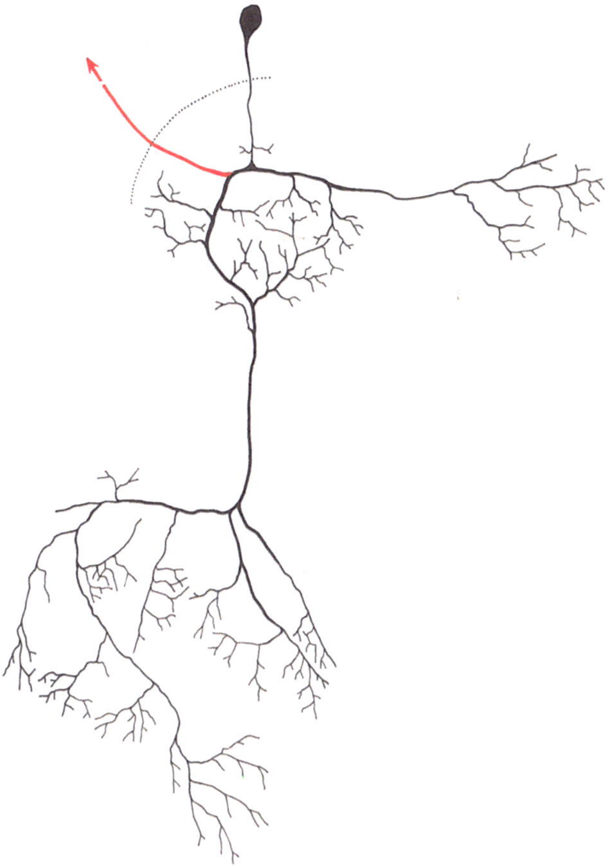

Fig. IV.11 Motoneuron from a ventral ganglion of an invertebrate (crustacean). Only one process originates from the body of this neuron; the process in turn gives rise to several dendritic branches and to a single axon (*red*). The dendrites ramify in the central core of the ganglion, while the axon leaves the ganglion and terminates inside a muscle

Table IV.1 A partial list of the sites at which a correlation between morphology and electrophysiological properties of vertebrate neurons has been demonstrated

Site	Family	Species	Common name	Author(s)
Sympathetic ganglia	Caviidae	*Cavia aperea porcellus*	Guinea pig	Boyd et al. (1996)
Myenteric plexus	Muridae	*Rattus norvegicus*	Rat	Browning and Lees (1996)
	Muridae	*Mus musculus*	Mouse	Nurgali et al. (2004)
Enteric submucosal plexus	Caviidae	*Cavia aperea porcellus*	Guinea pig	Lomax et al. (2001)
Spinal cord	Muridae	*Rattus norvegicus*	Rat	Lu et al. (2001)
Medulla oblongata	Muridae	*Rattus norvegicus*	Rat	Sedlacek et al. (2007)
Amygdala	Muridae	*Rattus norvegicus*	Rat	Washburn and Moises (1992)
	Muridae	*Rattus norvegicus*	Rat	Rainnie et al. (1993)
Neostriatum	Muridae	*Rattus norvegicus*	Rat	Kawaguchi (1993)
Hypothalamus	Muridae	*Rattus norvegicus*	Rat	Chevaleyre et al. (2001)
	Canidae	*Canis familiaris*	Dog	Sha et al. (2004)
Globus pallidus	Caviidae	*Cavia aperea porcellus*	Guinea pig	Nambu and Llinás (1997)
Basal forebrain	*Muridae*	*Rattus norvegicus*	Rat	Duque et al. (2007)
Cerebral cortex	*Felidae*	*Felis catus*	Cat	Deschênes et al. (1979)
	[a]	[a]	Monkey	Hamada et al. (1981)
	Caviidae	*Cavia aperea porcellus*	Guinea pig	McCormick et al. (1985)
	Muridae	*Rattus norvegicus*	Rat	Mason and Larkman (1990)
	Muridae	*Rattus norvegicus*	Rat	Protopapas and Bower (2000)
	Muridae	*Rattus norvegicus*	Rat	Liu et al. (2000)

[a]Family and species not indicated in the original paper

V. The Structure of Neurons

A. The Perikaryon

1. The Nissl Substance

This material, earlier observed by Arndt (1875), Key and Retzius (1876), and Flemming (1882), was carefully described by Nissl (1889, 1894)[1], who established the importance of alcohol fixation for its visualization. Under the light microscope, this cytoplasmic component occurs in the form of bodies which are stained intensely by means of basic dyes (e.g., methylene blue, toluidine blue, thionin, and cresyl violet). This material, also known as chromophilic substance, chromidial substance, tigroid bodies, etc., exhibits distinctive and consistent characteristics in different types of neuron: for example, it appears as large, rhomboid masses separated by light channels in the motoneurons of the spinal cord (Fig. V.1) and brainstem and as small granules in some neurons of the sensory ganglia (Fig. V.1) and in the Mauthner cells of teleosts. In light microscope preparations, it seems to be absent from certain small neurons, such as cerebellar cortex granule cells, retina bipolar cells, and substantia gelatinosa small neurons (see later, this section).

Since fixation and postmortem phenomena may influence the appearance of the Nissl substance, some authors regarded the Nissl bodies seen in stained sections as artifacts due to the precipitation of some substance normally dispersed within the cytoplasm (see, among others, Held 1895). However, Nissl bodies were also detected in living neurons (Palay and Wissig 1953; Hild 1954; Deitch and Murray 1956), whereupon their preexistence in vivo could no longer be doubted. Using cytochemical techniques, it has been shown that ribonucleoproteins are among the main components of Nissl bodies (Brachet 1940; Gersh and Bodian 1943); this accounts for their strong basophilia.

In electron micrographs (Figs. V.2 and V.3), each Nissl body appears to consist of a number of cisternae whose membranes are studded with polysomes (i.e., cisternae of the granular endoplasmic reticulum), and free polysomes occurring in the cytoplasmic matrix between the cisternae (Palay and Palade 1955). Most of the ribosomes in each Nissl body are set into the free polysomes occurring in the cytoplasmic matrix between the cisternae. Such polysomes usually consist of a few (5 or 6) ribosomes, whereas membrane-attached polysomes are composed of a greater number (up to 30). In the motoneurons of the spinal cord and brainstem, each Nissl body consists of numerous,

[1] In reality, Nissl's first observations on the objects now known as Nissl bodies were reported in an essay on the pathological changes of nerve cells in the cerebral cortex. For this essay, F. Nissl [1860–1919] was awarded a prize in a scientific competition set up in 1884 by the Medical Faculty of Munich University (Kreutzberg 1984). This text remains unpublished.

E. Pannese, *Neurocytology: Fine Structure of Neurons, Nerve Processes, and Neuroglial Cells*, 2nd Edition,
DOI 10.1007/978-3-319-06856-5_5, © Springer International Publishing Switzerland 2015

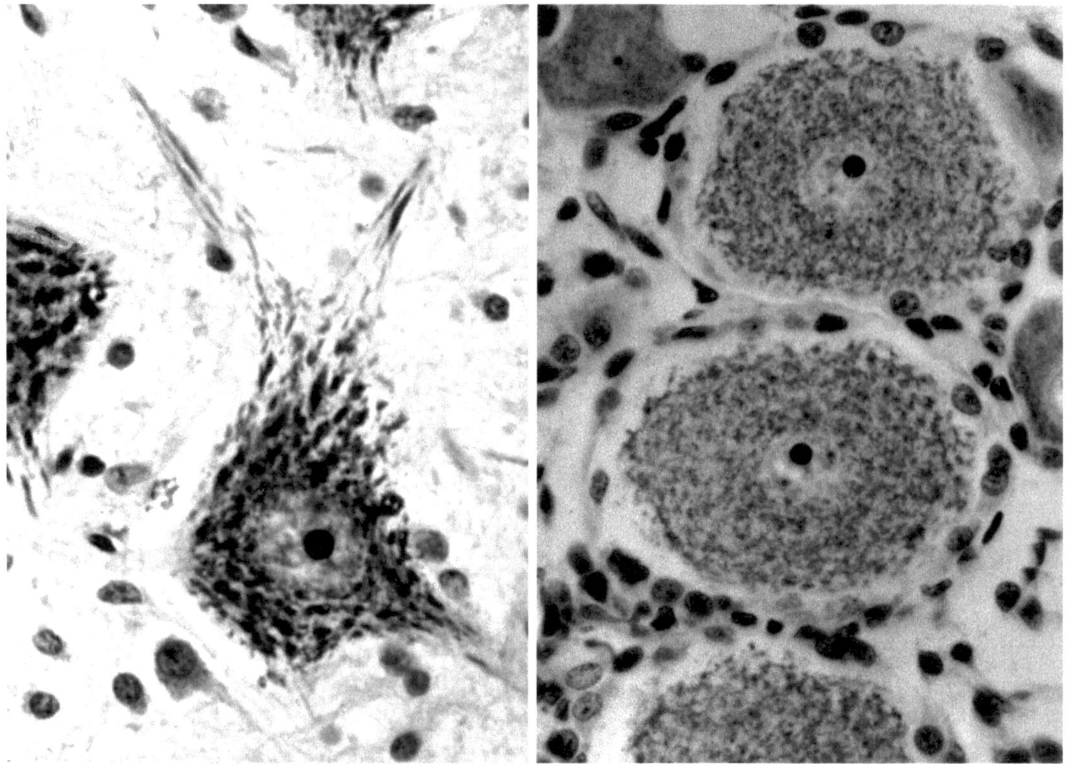

Fig. V.1 Nissl substance. In motoneurons (*left*), Nissl bodies appear as large, rhomboid masses within the perikaryon and as elongated bodies in the initial portions of dendrites, while in sensory neurons (*right*), most Nissl bodies appear as small granules. A large, deeply stained nucleolus can be seen in the nucleus of both types of neuron. Light micrographs of sections stained with cresyl violet, spinal cord motoneurons and spinal ganglion neurons of man, ×675

broad, and sometimes fenestrated cisternae of the granular endoplasmic reticulum which are oriented parallel to each other (Fig. V.2) thereby forming a highly regular stack; anastomoses between cisternae are frequent. In addition to these well-ordered Nissl bodies, isolated cisternae and small groups of randomly distributed cisternae can also be found. In the spinal ganglion neurons, the cisternae of the Nissl bodies often lack a preferential orientation (Fig. V.3), while in the small- and medium-sized pyramidal neurons of the cerebral cortex, the cisternae of the granular endoplasmic reticulum usually occur isolated or in small groups. In the granule cells of the cerebellar cortex, in which the Nissl substance cannot be appreciated by light microscopy, only discrete tubules of the granular endoplasmic reticulum and free polysomes can be found (Fig. V.4). In summary, electron microscope studies have established that (a) the granular endoplasmic reticulum and free polysomes, i.e., the components of the Nissl substance, are present in all neurons; (b) in all nerve cells most polysomes occur free in the cytoplasmic matrix and only a small proportion are attached to the cisternae of the endoplasmic reticulum; and (c) only the quantities of polysomes and cisternae and the organizational pattern of the latter can differ from one type of neuron to another. Thus, the failure to detect the Nissl substance in light microscope preparations of certain neurons is due to the presence of too few polysomes and to the consequent insufficient take up of basic dye.

There are a number of important differences between the structure of neuronal Nissl

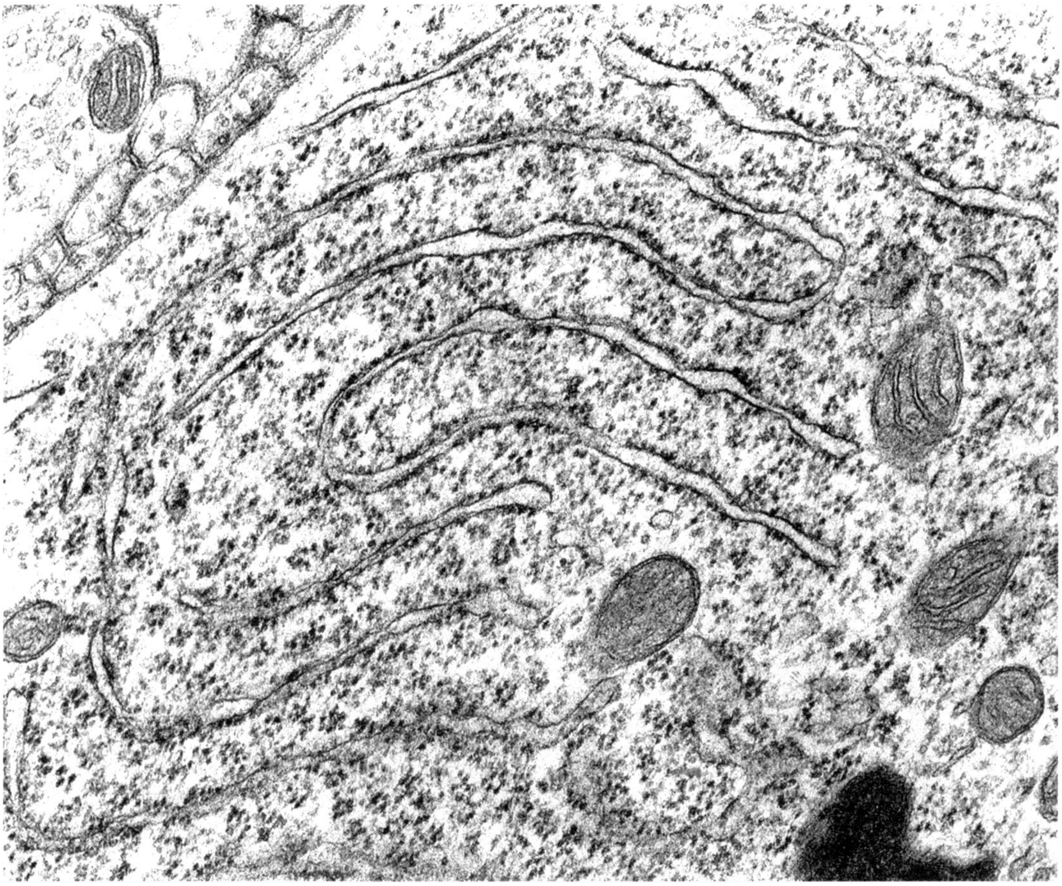

Fig. V.2 Nissl substance. In the Nissl body of a motoneuron shown here, the cisternae of the granular endoplasmic reticulum are arranged in a fairly ordered array. In addition to the polysomes attached to the outer surface of the membranes limiting the cisternae, there are many polysomes which occur free in the cytoplasmic matrix between the cisternae. Ventral horn of the rat spinal cord, ×50,000

bodies and that of the ergastoplasm of the protein-synthesizing gland cells. For example, in the exocrine pancreatic cells the membranes limiting the ergastoplasmic cisternae are uniformly studded with attached polysomes, and only a few polysomes occur free in the cytoplasmic matrix, whereas in the Nissl bodies large areas of the cisternal membranes are devoid of ribosomes, and the small polysomes occurring free in the cytoplasmic matrix between the cisternae are very numerous.

Results of studies carried out in the course of neuronal development and during axon regeneration suggest that the two classes of polysomes in the nerve cell may synthesize proteins playing different roles. The immature neuron (the neuroblast of classical terminology) as yet unintegrated into a neuronal circuit and the nerve cell engaged in axon regeneration both contain free polysomes almost exclusively (Pannese 1963a, 1968, 1974). Since in both of these situations the nerve cell is mainly engaged in the synthesis of new protoplasm and its synaptic activity is missing or deficient, free polysomes were assumed to synthesize mostly structural proteins, i.e., proteins for the maintenance and renewal of the protoplasm in both the perikaryon and cell processes (Pannese 1963a). This hypothesis has received support from in vitro studies carried out in cell-free systems containing either free polysomes or membrane-attached polysomes. By this technique, it was shown that neurofilament

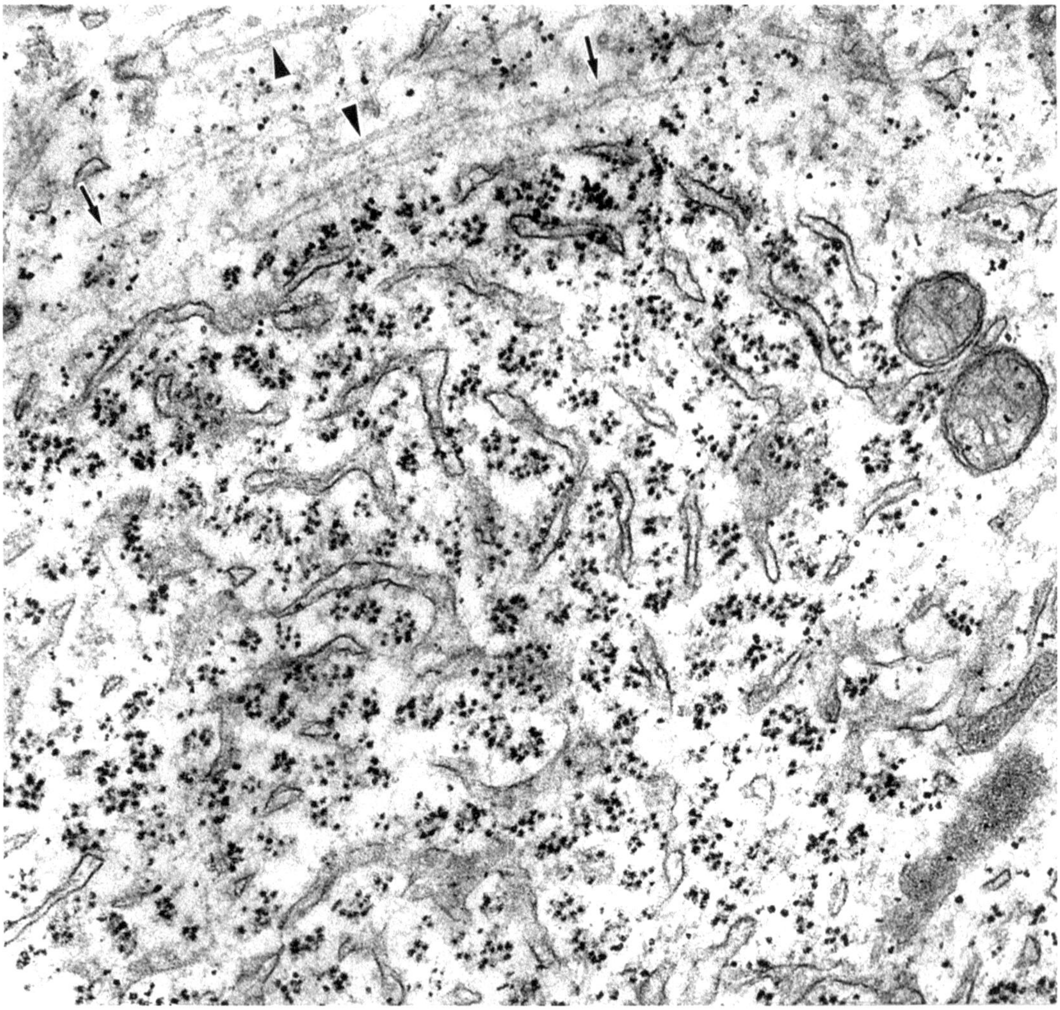

Fig. V.3 Nissl substance. In the Nissl body of a sensory neuron shown here, the cisternae of the granular endoplasmic reticulum lack a preferential orientation. As in the Nissl body shown in Fig. V.2, many polysomes occur free in the cytoplasmic matrix between the cisternae. Microtubules (some indicated by *arrowheads*) and neurofilaments (some *arrowed*) can be seen in the cytoplasm surrounding the Nissl body. Rabbit spinal ganglion, ×53,000

proteins are synthesized on the free polysome population and that there is no detectable synthesis of such proteins on the membrane-attached polysomes (Strocchi et al. 1982). These latter might be engaged as usual in the synthesis of enzymes, since some enzymes (e.g., acetylcholinesterase and acid phosphatase) have been localized by means of cytochemical techniques within the cisternae of the granular endoplasmic reticulum. It is relevant to note here that the abundant polysomes present in the perikaryon of large neurons are

Fig. V.4 Granule cells of the cerebellum. The nucleus, which occupies most of the cell body, displays dense clumps of chromatin, some of which lie adjacent to the nuclear envelope. Note that the Nissl substance, which in these cells cannot be appreciated by light microscopy, consists of a small number of discrete tubules of the granular endoplasmic reticulum (some *arrowed*) and free polysomes. Golgi complexes (**G**) and mitochondria can also be seen in the cytoplasmic rim. Rat cerebellar cortex, ×20,000

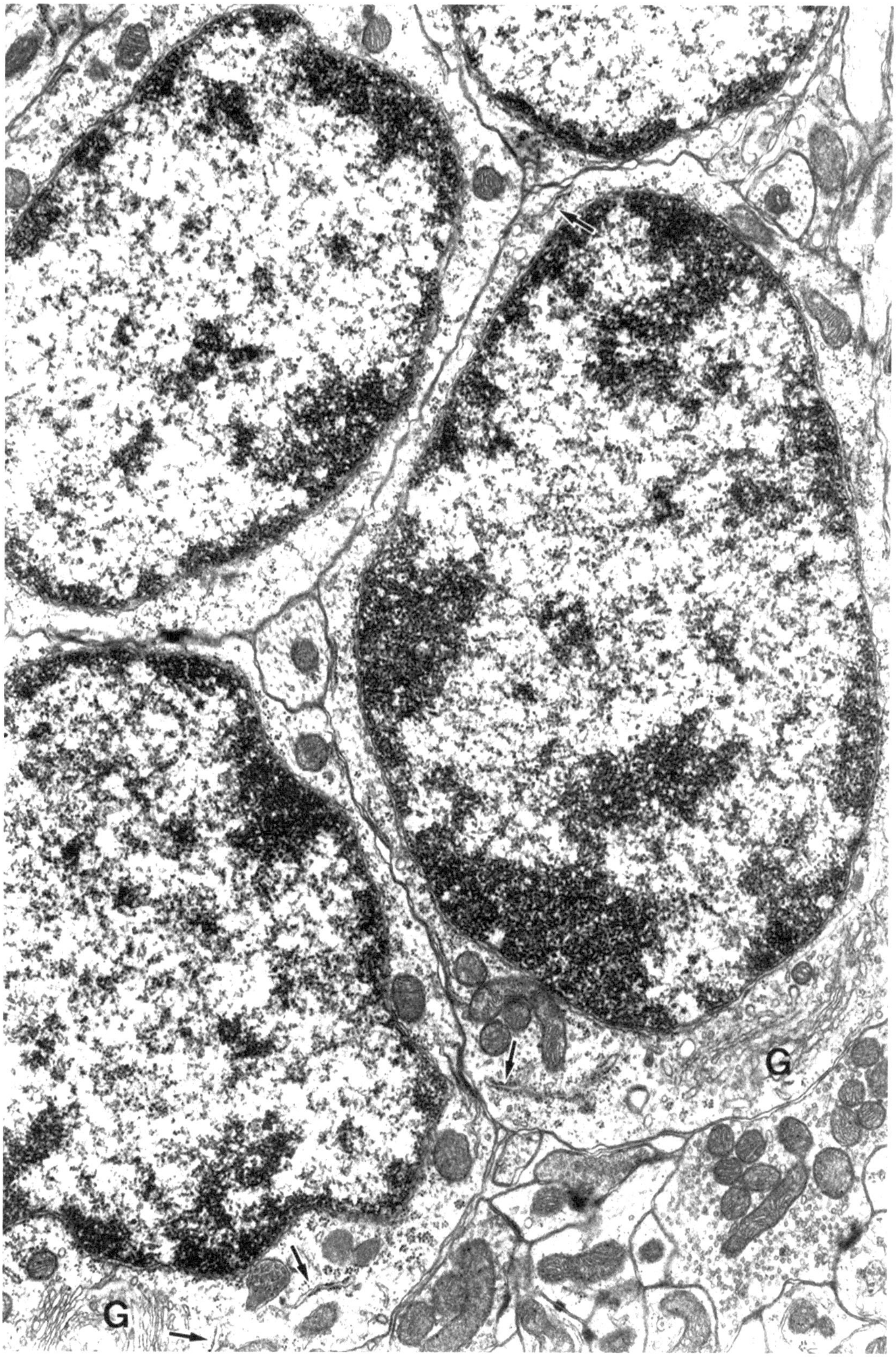
G
G

in relation to the very high rate of protein synthesis that occurs there. The perikaryon synthesizes not only its own proteins but also many of those present in neuronal processes, and as noted in Chap. III, the volume of the processes is much greater than that of the perikaryon in large neurons. Furthermore, the specific proteins present in the plasma membrane of the neuron (e.g., receptors and ion channels) – which are synthesized in the perikaryon – have a limited half-life and undergo continual turnover. For example, membrane Na^+ channels are thought to be replaced about every 3 days (Schmidt and Catterall 1986).

2. The Agranular Reticulum

Within the perikaryon, this organelle consists of irregularly branching and anastomosing tubules and cisternae devoid of polysomes (Fig. V.5), which are continuous at many sites with elements of the granular endoplasmic reticulum. In the reticulum of the neuronal perikaryon, therefore, some regions have attached polysomes, whereas others lack polysomes. The agranular tubules and cisternae are apparently dispersed at random. Flattened cisternae lying close to and parallel with the plasma membrane (Fig. VII.2) are present in many neurons (Rosenbluth 1962b; Kaiserman-Abramof and Palay 1969). No organelle is ever found interposed between these cisternae and the plasma membrane (Fig. VII.2), they being only between 10 and 60 nm apart. These cisternae, which have been called subsurface cisternae by Rosenbluth (1962b) and hypolemmal cisternae by Kaiserman-Abramof and Palay (1969), often appear in continuity with the elements of the granular or agranular reticulum lying more deeply in the neuronal perikaryon. In most neurons, subsurface cisternae are neither very numerous nor very large, but in Purkinje cells they are so highly developed, not only in the perikaryon, but also in the dendrites (Fig. V.15) and axon, that they can be used to identify isolated profiles of these neurons in thin sections (Palay and Chan-Palay 1974).

Calcium deposits have been located within elements of the agranular endoplasmic reticulum (Duce and Keen 1978; Burton and Laveri 1985; Takei et al. 1992; for a review, see Kostyuk and Verkhratsky 1994). Thus, it has been suggested that in the neuron this reticulum is able to store and release Ca^{2+} and hence play a role in Ca^{2+} buffering (Pozzo-Miller et al. 1997), as indeed it is thought to do in other cell types. More recent evidence suggests that the agranular reticulum consists of several functionally distinct compartments, although it forms a network of tubules and cisternae continuous throughout the neuron. The various compartments probably have different complements of receptors and ion channels. As a result of this functional heterogeneity, the agranular reticulum seems to be able to generate, within a single neuron, many different Ca^{2+} signals so as to regulate several Ca^{2+}-dependent processes simultaneously (for a review, see Blaustein and Golovina 2001). It has been suggested that subsurface cisternae are functionally coupled to the plasma membrane microdomains lying close to them, and there is evidence that both contain clusters of specific ion transporters. These units could be sites where Ca^{2+} move between the agranular reticulum and the extracellular fluid or sites of signal initiation (for a review, see Blaustein and Golovina 2001). Some compartments of the agranular reticulum contact mitochondria. It has been hypothesized that juxtamitochondrial cisternae are involved in (a) the synthesis and transport of phospholipids to mitochondria, and (b) the transfer of Ca^{2+} between endoplasmic reticulum and mitochondria. Cisternae of the agranular endoplasmic reticulum appear frequently arranged in stacks in Purkinje cells. Some stacks likely exist in living Purkinje cells, but massive stacks formation is probably a response to hypoxic conditions (Takei et al. 1994).

3. The Golgi Apparatus

This organelle was originally described by Golgi (1898a, b) in nerve cells (Fig. V.6),

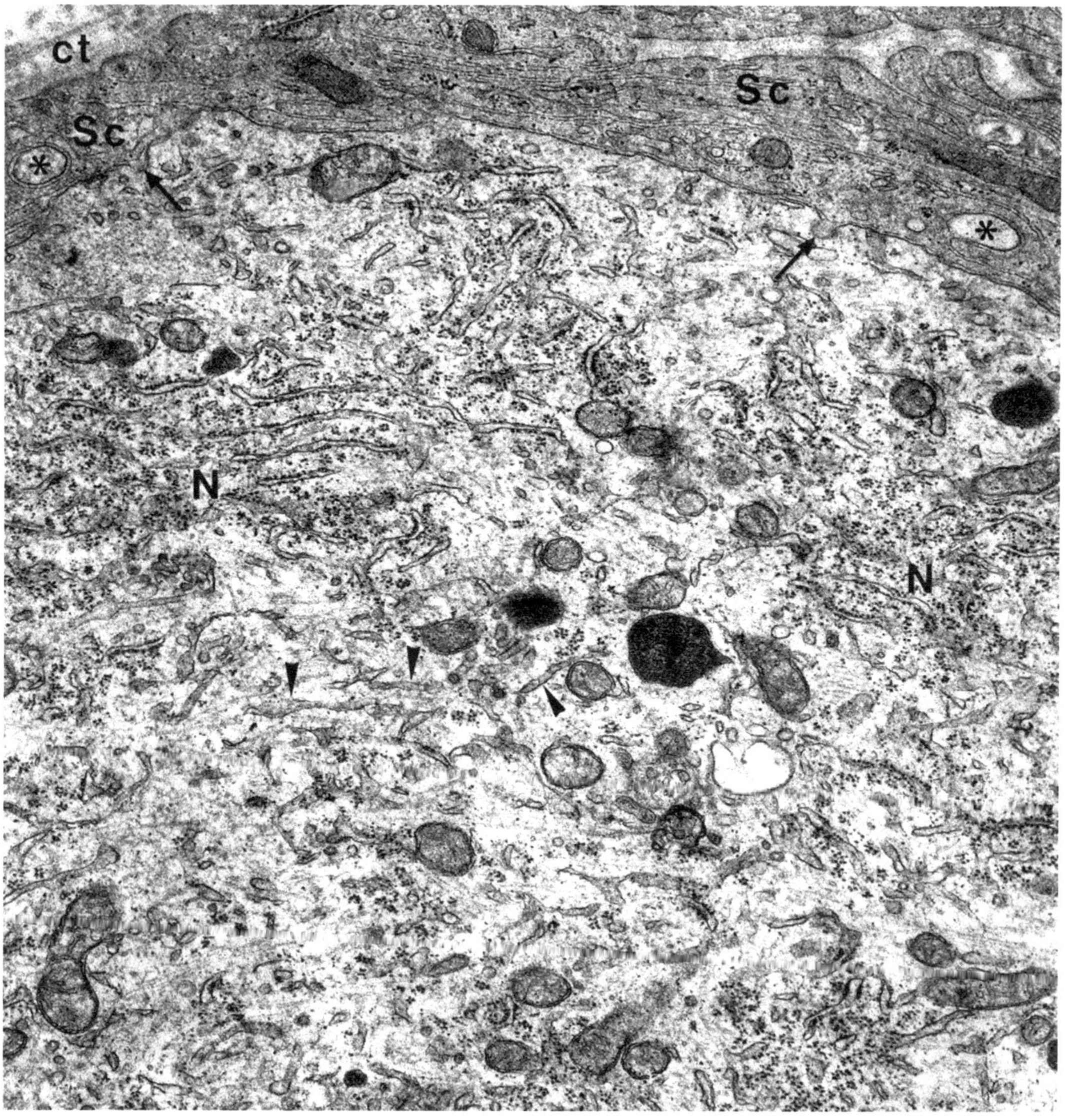

Fig. V.5 Perikaryal portion of a spinal ganglion neuron. Nissl bodies (**N**), profiles of the agranular endoplasmic reticulum (some of which are indicated with an *arrowhead*), mitochondria, and dense bodies can be seen in the neuronal perikaryon. Microtubules and neurofilaments are interposed between the other organelles. * indicates cross sections of projections which arise from the neuronal perikaryon at other levels; they thus appear as discrete entities completely embedded in the satellite cell sheath (**Sc**). *Arrows* point to pinocytotic vesicles along the neuronal plasma membrane. **ct** connective tissue. Rabbit spinal ganglion, ×28,000

and named the internal reticular apparatus by him. When examined under the light microscope after silver impregnation, in thick sections it appears as a continuous, loose-meshed network of irregular, tortuous ribbons distributed through most of the perikaryon, except near the cell surface. Numerous authors questioned the reality of the apparatus described by Golgi and the ensuing debates gave rise to long-lasting and at times acrimonious controversies regarding its real existence. It is now universally recognized that the Golgi apparatus is not the result of technical

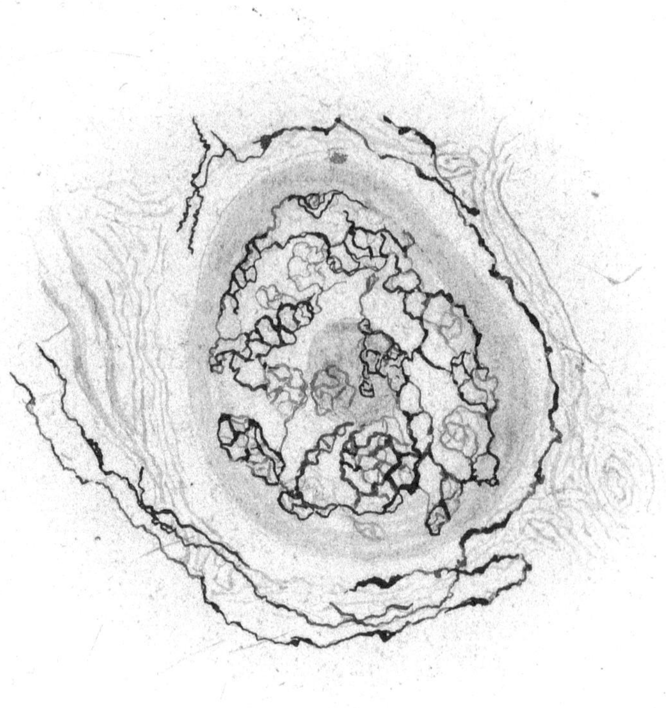

Fig. V.6 Drawing by Golgi showing the internal reticular apparatus impregnated using his rapid method. Spinal ganglion of a mammal

artifacts, but a genuine organelle present in all eukaryotic cells.

When a nerve cell is examined in semithin sections by light microscopy (Fig. V.7) or in thin sections under the electron microscope, only a few segments of the continuous network detectable in thick sections by light microscopy can be seen. These segments appear as discrete entities. However, when the same type of neuron is examined in thick sections under the high-voltage electron microscope, such segments appear to be connected to each other (Rambourg et al. 1973). Likewise, the structure resulting from the reconstruction of a series of thin sections is a continuous network. In electron micrographs of thin sections, each segment of the Golgi apparatus appears to consist of a stack of five to seven flattened, smooth-surfaced cisternae (Fig. V.7). This whole structure is usually called a Golgi complex. A distinguishing feature of the Golgi complex is its lack of either free or attached ribosomes. The cisternae are thin in their central portions and slightly expanded at their periphery and are closely spaced. Although cisternae belonging to the same stack are interconnected, very few anastomoses can be seen between them in single thin sections. The Golgi complex is usually slightly curved (Fig. V.7), with its concave surface facing the nucleus. Hence, two surfaces or faces (a convex face, also called outer or forming or cis, and a concave face, also called inner or maturing or trans) can be distinguished.

This organelle exhibits a morphological and cytochemical polarity which is particularly evident in gland cells and less evident in neurons. The cisternae at the two surfaces of the same stack differ somewhat in their structure; the outermost ones are highly fenestrated, whereas the innermost ones show very few perforations. In thin sections grazing the external surface of the Golgi complex, regularly spaced, circular perforations can be seen

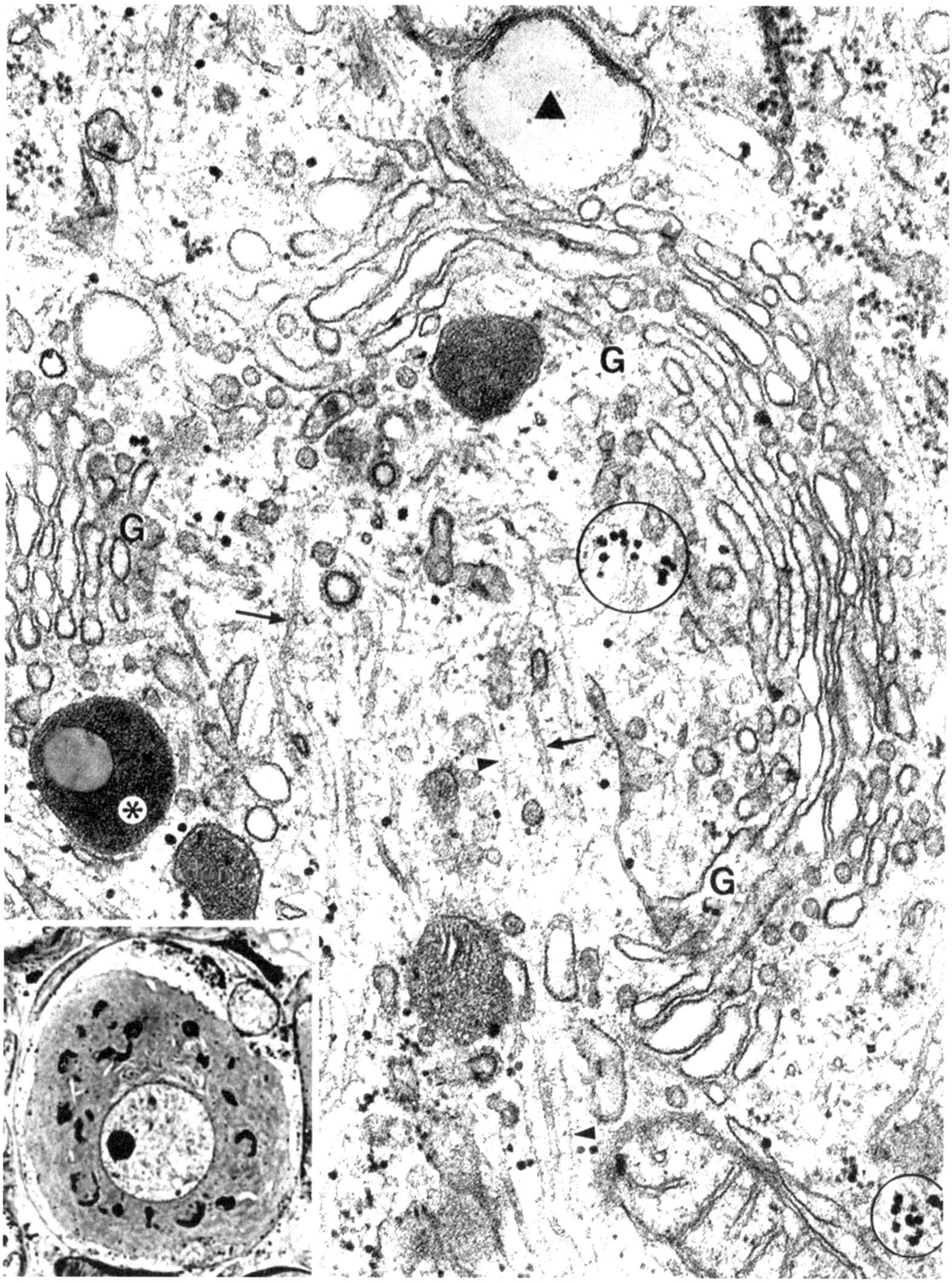

Fig. V.7 Golgi apparatus. The *inset* at the *lower left* shows a light micrograph of a sensory neuron stained with the Golgi method. Whereas in thick sections the Golgi apparatus appears as a continuous network, in a semithin section like that shown here only segments of this network can be seen appearing as discrete entities. Spinal ganglion of an adult rat, ×1,200 (Courtesy of S. Matsuda). The remainder of the figure shows an electron micrograph of a single segment of the Golgi apparatus (**G**) from the same type of neuron. The segment appears to consist of a stack of curved cisternae with many associated vesicles. The cisternae are flattened and smooth surfaced. In addition to the Golgi complex, the figure shows a lipofuscin granule (*), a lipid droplet (▲), glycogen particles (some *encircled*), polysomes, microtubules (some *arrowed*), and neurofilaments (some indicated with *arrowheads*). Spinal ganglion of an adult fowl, ×57,500

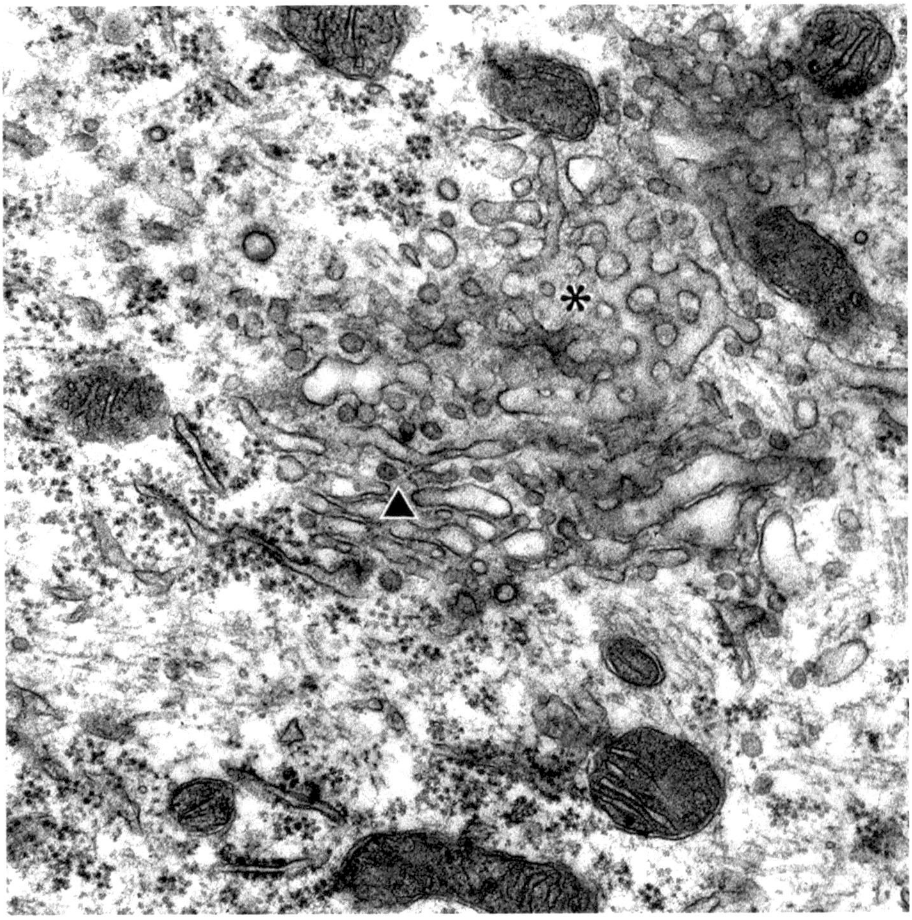

Fig. V.8 Golgi apparatus. In the *upper half* of the field, the outermost cisterna (*) of a Golgi complex has been grazed so that its circular perforations are apparent. In the *middle* of the field, the cisternae have been sectioned almost perpendicular to their surfaces, so that they appear stacked (▲). Rabbit spinal ganglion, ×50 000

in the outermost cisterna (Fig. V.8). In thin sections almost perpendicular to the plane of the outermost cisterna, the latter appears as a chain of circular or oval profiles, i.e., it is apparently discontinuous. Reconstructions from series of thin sections perpendicular to the surfaces of the Golgi complex demonstrate the continuity of cisternae around these perforations. These features can also be observed in freeze-fractured preparations and in thick sections examined under the high-voltage electron microscope. Golgi cisternae not only differ somewhat in structure, they also display different cytochemical properties according to their location within the stack. When the tissue is subjected to prolonged osmium impregnation, only the outermost cisternae become impregnated both in nonneuronal cells (Friend and Murray 1965; Friend 1969) and in neurons (Rambourg et al. 1973). In tissue stained by acid phosphatase, thiamine pyrophosphatase, or adenosine monophosphatase histochemical techniques, the reaction products are confined to the innermost cisternae (neurons, Osinchak 1964, Novikoff 1967, Lane 1968; nonneuronal cells, Friend and Farquhar 1967, Friend 1969). Cytochemical studies on nonneuronal cells have shown that maturing face cisternae also contain glycoproteins (Neutra and Leblond 1966a,b). As a rule, numerous vesicles are associated with the Golgi complex (Fig. V.7). Some are 20–60 nm in diameter and have a smooth or coated limiting membrane; others are larger and smooth surfaced. A network of agranular tubules occurs near the maturing face of the Golgi complex. This structure had been interpreted by Novikoff (1967) as a specialized region of the endoplasmic reticulum in close topographical relation to the Golgi apparatus, but distinct from it. This structure had been referred to by Novikoff (1967) as the GERL system (Golgi-associated endoplasmic reticulum from which lysosomes form) to call attention to its role in the formation of lysosomes. At present, it is thought that this network of tubules is part of the Golgi complex and is the common site of exit of all materials processed through the Golgi apparatus. The acronym GERL has been replaced by the term trans-Golgi network (Griffiths and Simons 1986).

Several functions have been ascribed to the Golgi apparatus mainly as a result of cytochemical studies carried out on gland cells: (a) the condensation and segregation of secretory products destined for release outside the cell, (b) the terminal glycosylation of glycoproteins that will be incorporated into the plasma membrane, and (c) the segregation of hydrolytic enzymes into vesicles. Although the neuronal Golgi apparatus has not been the object of such an extensive series of cytochemical studies as that carried out on its counterpart in gland cells, it is assumed to have similar functions. First of all, there are neurons, called neurosecretory cells (see Chap. II), which produce polypeptide hormones and release them into the extracellular space; in these neurons the hormone-containing vesicles take origin from the Golgi apparatus (Scharrer and Brown 1961). Furthermore, the segregation of proteins destined for fast axonal transport (see Sect. V.D.4) takes place within the Golgi apparatus in all neurons (Hammerschlag et al. 1982). It is thought that several kinds of transmitters can be synthesized and recycled within the synaptic terminals (see Sect. VI.E); however, the enzymes necessary for the synthesis of these transmitters are manufactured in the neuronal perikaryon, being segregated into vesicles by the Golgi apparatus and later transported along the axon to the synaptic terminals (see Sect. V.D.4). Cytochemical studies on a fraction of the Golgi apparatus isolated from nonneuronal cells and consisting mainly of membranous elements have shown that this fraction contains glycosyltransferases. These enzymes are responsible for the addition of terminal hexoses to glycoproteins (Fleischer et al. 1969; Bergeron et al. 1973). Glycoproteins assembled within the neuronal Golgi apparatus, as well as those assembled within the Golgi apparatus of other cell types, are eventually delivered to the plasma membrane (see Sect. V.E). Finally, coated vesicles (about 50 nm in diameter) containing hydrolytic enzymes arise from the trans-Golgi network and transport these enzymes to late endosomes or autophagic vacuoles (see Sect. V.A.5).

4. Mitochondria

In living nerve cells, staining with Janus green B readily reveals the presence of these organelles, which appear round, more or less elongated, or branched. In single sections, branched mitochondria can be observed very rarely, but in serial sections their number is shown to be considerable (about one-third of the total in the neuronal perikarya of the lateral geniculate nucleus in the rat, according to Karlsson 1966b). Perikaryal mitochondria range from 0.1 to 0.5 μm in width and from a fraction of one to several micrometers in length. In electron micrographs, perikaryal mitochondria appear to lie mainly in the spaces between Nissl bodies (Fig. V.5), and only occasionally are found inside them. In living neurons cultured in vitro, mitochondria have been observed to change their shape, size, and position (Pomerat et al. 1967). They have been seen to move from one region of the perikaryon to another along the narrow channels which separate the Nissl bodies and to migrate from the perikaryon into one process as well as in the opposite direction. Mitochondrial movements are probably not random. Mitochondria provide most of the energy necessary to meet cellular requirements. There is evidence that within the cell mitochondria are localized at sites of high activity and energetic demand (Park et al. 2001; Hollenbeck and Saxton 2005). Mitochondrial movements could therefore be related to such needs, which would vary with time. Mitochondria move along cytoskeletal elements, and motor proteins (e.g., kinesins and dynein) generate the force required for that movement. Neuronal mitochondria differ neither in structure nor, probably, in metabolic role from their counterparts in other cell types. However, they display the following peculiarities, whose significance is unclear: dense granules are not common in their matrix, and long mitochondria often have longitudinally oriented cristae.

5. Lysosomes and Peroxisomes

Using the differential centrifugation technique, a fraction rich in hydrolytic enzymes was isolated from cell homogenates, and the organelles comprising this fraction were called lysosomes to draw attention to their richness in such enzymes (de Duve et al. 1955). When the isolated lysosome fraction was examined under the electron microscope, the organelles comprising it were recognized as a new particulate cell component of highly heterogeneous shape, size, and internal structure. Thus, while other cytoplasmic organelles are usually identified on the basis of their shape, size, and internal structure, the identification of lysosomes rests essentially on biochemical criteria, i.e., on the occurrence of lytic enzymes. Lysosomes were identified in neurons initially by means of the light microscope (Novikoff 1960; Becker et al. 1960; Barron and Sklar 1961); later, their structural features were studied under the electron microscope.

Our knowledge of lysosome function derives principally from studies on non-neuronal cells. Lysosomes are concerned with the intracellular degradation of many exogenous substances and endogenous cell components. Like all cells, neurons take up exogenous substances in their cytoplasm. Experiments using tracer proteins have shown that these are taken up into neurons by endocytosis (Rosenbluth and Wissig 1964; Brightman 1965b; Holtzman and Peterson 1969). Multiple vesicles (100–200 nm in diameter), arising from the plasma membrane by endocytosis, fuse to form a vacuole called early endosome. However, some authors maintain that endocytic vesicles fuse with a preexisting vacuole to form this organelle. The early endosome, whose internal pH is 6.0–6.5, is present in the peripheral region of the perikaryon. Materials within the early endosome are subsequently transferred to a late endosome (internal pH 5.0–6.0), which lies more deeply within the perikaryon. It is not clear whether early endosomes move deeper in the perikaryon and become late endosomes, or whether they remain in the peripheral region of the perikaryon while their contents are transported to late endosomes by carrier vesicles. It has been suggested that multivesicular bodies perform this supposed transport function. Multivesicular bodies (Figs. V.12 and VII.19), first observed in neurons by Palay and Palade (1955), are membrane-bounded spherical vacuoles, about 500 nm in diameter, containing a varying number of small vesicles embedded in a more or less dense matrix. Sometimes, a patch of the membrane surrounding a multivesicular body appears flattened and lined by a fuzzy coat (Fig. V.12).

When endogenous cell components (e.g., mitochondria, endoplasmic reticulum cisternae, free ribosomes, and glycogen particles) are no longer fully functioning, they become surrounded by a membrane. In this way, an autophagic vacuole, also called autophagosome, is formed. First observed in liver cells, autophagic vacuoles (Fig. V.10) were subsequently detected in neurons by Novikoff and coworkers. Hydrolase-containing vesicles, about 50 nm in diameter, are associated with the Golgi complex (see Sect. V.A.3; see also Novikoff et al. 1971; Pannese et al. 1976). These vesicles, previously called primary lysosomes, coalesce with a late endosome or with an autophagic vacuole discharging their enzymes into the latter. In this way, a lysosome (0.25–1 μm in diameter, internal of pH 4.0–5.0) is formed. Lysosomes, previously called secondary lysosomes, are the site of digestion of sequestered materials (of exogenous or endogenous origin). Lysosomes have a content of variable density and structure; their structural heterogeneity is due to the variety of materials that they can contain and their stage of digestion at the time of fixation. The small molecules produced by lysosomic digestion are transferred to the cytoplasmic matrix and may be metabolized or used to build new macromolecules. Indigestible residues may persist within a lysosome, which therefore becomes a residual body. Lipofuscin granules are the best known example of residual bodies. These pigment granules, which range from 1 to 3 μm in diameter, appear as golden brown particles in unstained preparations examined with the light microscope and exhibit autofluorescence in ultraviolet light; they stain with Sudan black and are positive after staining with the periodic acid-Schiff reagent. When observed under the electron microscope (Figs. V.7 and V.10), lipofuscin granules appear as irregularly shaped, membrane-bounded bodies which exhibit a complex structure (Hess 1955); they consist of a dense, granular material, membranous arrays, and one or more homogeneous, spherical vacuoles of low density, which are often peripherally placed (Samorajski et al. 1965; Pineda et al. 1967; Glees and Gopinath 1973). Some reports suggest that the internal structure of lipofuscin granules increases in complexity with age (Samorajski et al. 1964, 1968; Brizzee and Johnson 1970; van den Bosch de Aguilar and Vanneste 1981; Hervonen et al. 1986; Ledda et al. 1999). The accumulation of lipofuscin with age is discussed in Sect. V.A.10.

Membrane-bounded, spherical or ovoid bodies (0.2–1.0 μm diameter) with a low or medium density content are found in neurons, as in all eukaryotic cells. These organelles were initially called microbodies, but after it was shown that they are rich in oxidative enzymes (e.g., urate oxidase, D-amino acid oxidase, and catalase), they were renamed peroxisomes. Within these bodies, a paracrystalline inclusion, called the nucleoid, may occasionally be present. The nucleoid consists of urate oxidase. Peroxisomes, that originate from the endoplasmic reticulum, are generally more numerous in nonneuronal cells of the nervous system (e.g., astrocytes, oligodendrocytes, satellite cells of sensory ganglia, and Schwann cells) than in neurons. Since peroxisomes display a rather wide range of size and may or may not contain a nucleoid, they can only be identified with certainty by means of the histochemical demonstration of their constituent enzymes. For more details on the presence of peroxisomes in the nervous tissue, see Holtzman (1982).

6. Neuromelanin Pigment

Neurons within the substantia nigra and locus coeruleus of several mammals, including humans, usually contain quantities of cytoplasmic granules of neuromelanin pigment. This pigment may also be found in other sites within the CNS and PNS (e.g., brainstem nuclei, and sensory and autonomic ganglia). These granules, which range from 1 to 2.5 μm in diameter, appear as dark brown particles in unstained preparations examined under the light microscope. They do not stain with the periodic acid-Schiff reagent, do not autofluoresce in ultraviolet light, and are both argentaffin and argyrophilic. When observed under the electron microscope, neuromelanin granules show some structural similarities to lipofuscin granules. They appear as irregularly shaped bodies consisting of three components: (a) a finely granular matrix of medium density; (b) a very dense, coarsely granular material; and (c) one or more homogeneous spherical globules of low density (Moses et al. 1966).

7. Microtubules and Neurofilaments

While microtubules can be found in all eukaryotic cells, neurofilaments are unique to neurons. Neuronal microtubules have been referred to as neurotubules by some authors. This term is not used in the present volume for the following reasons: (a) the term was originally used (De Robertis and Schmitt 1948) to refer to morphological entities found in preparations of fragmented nerves; these entities, interpreted as axonal constituents, were most likely collagen fibrils, as was later shown by studies carried out on thin sections (Schmitt and Geren 1950); (b) the use of the term neurotubules suggests that neuronal microtubules are substantially different from those found in most other cells; this, however, has been shown not to be the case. As in all other cells, under the electron microscope neuronal microtubules appear as long, unbranched entities, 20–26 nm in diameter, which in cross section show a dense wall about 6 nm thick and a lighter core. Neurofilaments are unbranched elements, about 10 nm in diameter and of indefinite length; when viewed in transverse section at high resolution, they appear to consist of an outer dense layer about 3 nm thick and a lighter core. In electron micrographs, perikaryal microtubules and neurofilaments (Figs. V.3 and V.7) appear to lie in the spaces between much more prominent and much more electron dense organelles (e.g., Nissl bodies, mitochondria, Golgi complexes). Microtubules may appear as individual units or arranged in groups of two to three elements, while neurofilaments occur individually scattered or loosely aggregated into small bundles. Both microtubules and neurofilaments can be seen funneling from the perikaryon into the bases of the axon and dendrites.

In fresh material and in preparations stained with methylene blue, thionin, or impregnated with silver or gold, light microscopists, beginning with Remak (1843, 1844), were able to detect fibrils within the neuronal cytoplasm. These were called neurofibrils by Bethe (1900). Neurofibrils lie in the spaces between Nissl bodies and form a network throughout the perikaryon, from which they may be traced into the axon and dendrites. Many light microscopists viewed neurofibrils as neuron-specific organelles which form the structural basis for the conduction of signals (see, among others, Schultze 1871; Apáthy 1897). This idea had a large following and was definitely dismissed only after the demonstration that invertebrate giant axons from which the axoplasm had been extruded and replaced by saline solution were able to conduct signals (Baker et al. 1962). It can therefore be understood why for about 50 years, up to the last decade of the nineteenth century, there flourished a vast literature on neurofibrils. A large part of this literature now has only historical interest, such as that concerning (a) the classification of neurons according to the arrangement of these structural components, (b) the controversy between authors claiming that neurofibrils are independent structures and those who maintained the existence of an anastomosing neurofibrillar network throughout the nervous system, (c) the actual presence of neurofibrils in living neurons.

Among structures revealed within neurons with the electron microscope, microtubules and neurofilaments seem to be the ideal candidates to correspond with neurofibrils, both because they are thin and extremely elongated and because their general distribution within the neuron matches that of the neurofibrils seen by light microscopy. In 1950, Schmitt advanced the hypothesis that neurofibrils consist of bundles of the 10–20-nm-thick filaments seen by him under the electron microscope in the axoplasm isolated from invertebrate giant nerve fibers. Later attempts were made to establish whether the procedures employed to demonstrate neurofibrils selectively stained only microtubules, neurofilaments alone, or both. The results obtained by Boycott et al. (1961), Gray and Guillery (1961), Potter (1971), and Gambetti et al. (1981) suggest that the impregnation procedures commonly used to reveal neurofibrils stain microtubules weakly or not at all, but do stain structures rich in neurofilaments. Delving deeper into the problem, Autilio-Gambetti et al. (1986) showed that the silver-binding sites reside in a highly charged unique domain near the carboxyl-terminal tail of each of the three neurofilament protein subunits. For more detailed information on microtubules and intermediate filaments in neurons, see Sect. V.D.2.

Occasionally large numbers of neurofilaments and microtubules can appear densely packed in a discrete region of the neuronal perikaryon. Large, spherical or elliptical, strongly argyrophilic bodies consisting of densely packed neurofilaments and microtubules have been found within the ganglion neurons of the nervus terminalis in some species of teleosts (Rossi and Palombi 1969). Whereas these bodies are present within neurons under apparently normal conditions, masses of densely packed neurofilaments have been observed within the spinal ganglion neurons of the lizard (Fig. V.9) in the course of axonal regeneration and neuronal hypertrophy (Pannese 1963a, b). These aggregations of neurofilaments are devoid of a limiting membrane. Bodies filled with filaments and delimited by a membrane are considered in the section on cytoplasmic inclusions (see Sect. V.A.9).

8. Centrioles and Cilia

Centrioles within the neuronal perikaryon were first reported toward the end of the nineteenth century by Lenhossék (1895) in the spinal ganglion neurons of the frog. Later, they were found by other authors (for a list, see Hatai 1901) within the neurons of both vertebrates (mammals included) and invertebrates. In electron micrographs, each neuronal centriole, much as those in other cell types, appears as a short cylinder whose wall consists of nine longitudinally oriented, evenly spaced triplets of microtubules.

Unlike centrioles, cilia were observed in neurons only after the advent of electron microscopy. Neuronal cilia are solitary primary appendages that extend from the basal body, an organelle with a centriole-like structure. Cross-striated bundles of thin filaments may radiate from the wall of the basal body into the surrounding cytoplasmic matrix (Dahl 1963; Allen 1965). Sometimes, a centriole can be found near the basal body; in such cases, the long axes of the two organelles form a right angle. The shaft of the cilium contains nine peripheral doublets of microtubules but lacks a central pair of microtubules; short and thin connections may sometimes be found between members of neighboring doublets, but arms, radial extensions, and other accessory structures seem to be missing. Cilia arising from neurons have been described in most regions of the nervous system (for a list, see Bishop et al. 2007). Neuronal cilia are thought to lack dyneins and be nonmotile. In any event, the shaft of a neuronal cilium may be enclosed in a deep narrow channel of the extracellular space formed by an invagination of the neuronal plasma membrane or may be surrounded by closely interwoven neuronal and neuroglial processes so that its motion in any direction is definitely hindered by lack of space. Recent studies have shown that specific receptors and signaling molecules are localized on the primary cilia

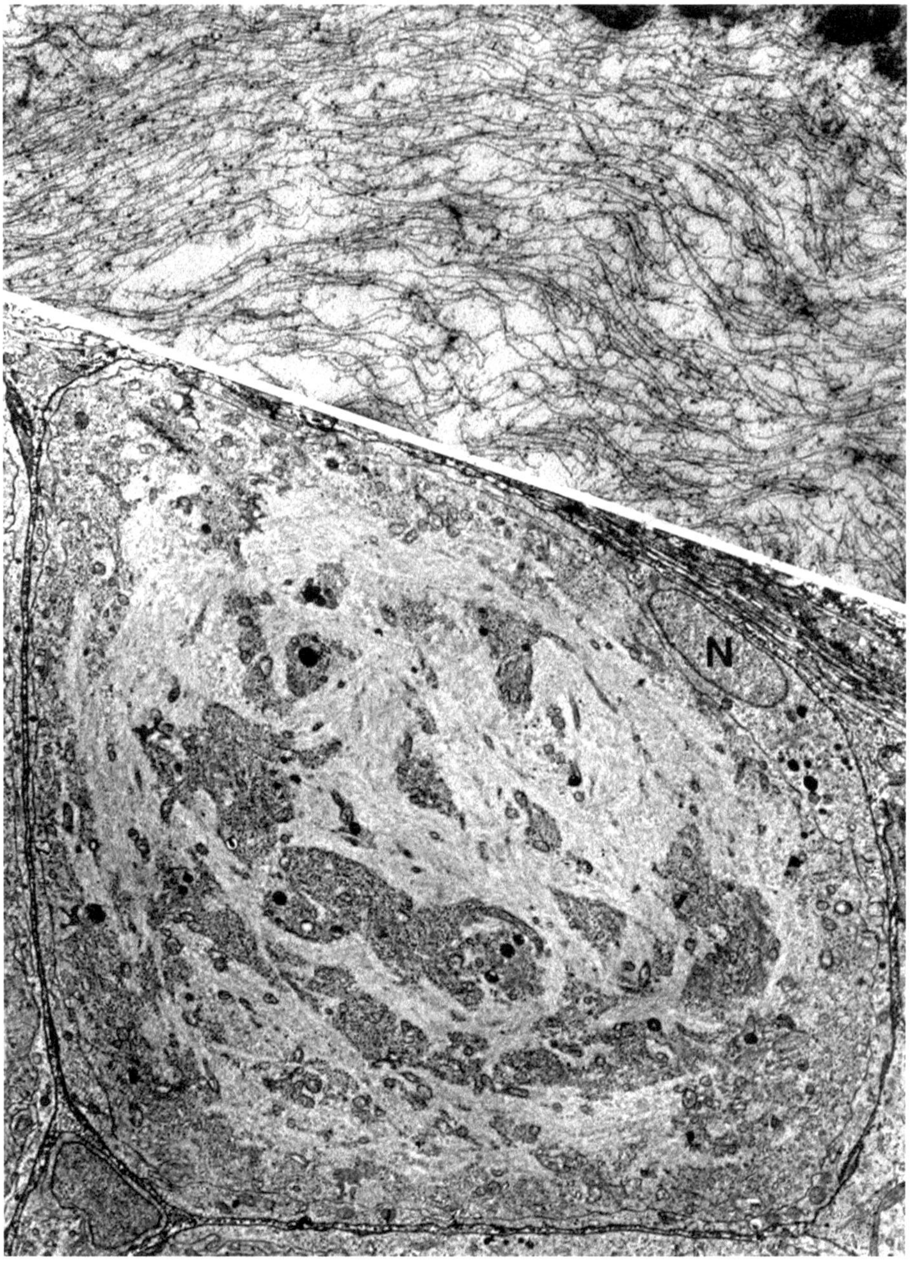

Fig. V.9 Masses of neurofilaments within the neuronal perikaryon. The perikaryon of a sensory neuron can be seen in the *lower picture*; the neuron is undergoing hypertrophy as a consequence of the enlargement of its peripheral field of innervation. The perikaryon contains masses of densely packed neurofilaments. The latter can be seen at higher enlargement in the *upper picture*. **N** nucleus of a satellite cell. Lizard spinal ganglion; *lower picture*, ×4,500; *upper picture*, ×30,000

of certain neurons (for a review, see Pazour and Witman 2003).

Various hypotheses for the function of these cilia have been advanced. They could simply be vestigial organelles inherited from the ancestors of neurons, i.e., from the ciliated cells that line the embryonic neural tube. Another suggestion is that these cilia are sensory organelles. This hypothesis rests on the observation that cilia lacking the central pair of microtubules occur in numerous sensory organs, for example, the vertebrate retina (Sjöstrand 1953a; De Robertis 1956; Carasso 1958; Tokuyasu and Yamada 1959), the locust (Gray 1960a) and cicada (Young 1973) auditory organs, the locust femoral chordotonal organs (Moran et al. 1975; Slifer and Sekhon 1975), the campaniform sensilla of cockroach legs (Moran et al. 1971), and the ocellus of ascidian tadpoles (Eakin and Kuda 1971). Primary cilia might give rise to signals caused, for example, by changes in the concentration of particular molecules or ions in the extracellular environment surrounding the cilia. It has also been suggested that cilia have some influence on neuronal maturation.

9. Cytoplasmic Inclusions

Spherical or ovoid basophilic bodies less than 1 μm in diameter have been found within the perikarya of a variety of neurons in the CNS and PNS (for a list, see Santolaya 1973). Cytochemical studies have suggested that these bodies, called nematosomes, contain nonhistone proteins and RNA (Grillo 1970). The nematosome, which lacks a limiting membrane, appears to consist of a convoluted network of dense strands (Fig. V.12), each of which is made up of tightly packed thin filaments or particles (Grillo 1970). Its structure resembles that of the fibrous component of the nucleolus. Inclusions different from those described above have been observed within the neuronal perikaryon in the CNS and PNS (Rosenbluth 1962a, and subsequently many others). These inclusions are devoid of a limiting membrane and exhibit a granulo-fibrillar texture. Several terms have been used to refer to these structures (for a list, see Shinoda et al. 1993).

Laminar bodies up to 5 μm in diameter have been detected within the neuronal perikaryon and dendrites of the lateral geniculate body (Morales et al. 1964; Smith et al. 1964; Peters and Palay 1966; Barron et al. 1967; Doolin et al. 1967), the thalamus (Herman and Ralston 1970), the cerebellar cortex (Morales and Duncan 1966; Mugnaini 1972), and the cerebellar central nuclei of the cat (Sotelo and Angaut 1973) as well as in the striate cortex of the monkey (Kruger and Maxwell 1969). Each laminar body consists of sheets of parallel tubules arranged at regular intervals which alternate with layers of a rather dense, granular material. These individual tubules, like microtubules, measure about 25 nm in diameter, but their wall, unlike that of microtubules, exhibits a trilaminar structure. Sometimes, the tubules of these bodies appear to be continuous with the endoplasmic reticulum.

Systems consisting of cisternae stacked closely together (Fig. V.10) or concentrically arranged have been found within the neuronal perikaryon (rat cerebral cortex, Rosenbluth 1962b; rat and cat substantia nigra, LeBeux 1972b; substantia gelatinosa of the spinal cord of the rat and monkey, Nathaniel and Nathaniel 1966; Inomata et al. 1983). The lumen of each cisterna is very narrow, except at its lateral edges where the cisternae may appear continuous with one another or in continuity with the granular endoplasmic reticulum. A rather dense, granular or filamentous material is evident in the cytoplasmic matrix between the cisternae. Only the external surfaces of the outermost cisternae in the system are studded with ribosomes. The cisternal systems which can also be found within dendrites (see Sect. V.C.1) bear some resemblance to the spine apparatus (see Sect. V.C.2).

Complexes composed of smooth-surfaced cisternae arranged in parallel arrays and of glycogen particles aligned in single rows between adjacent cisternae (Fig. V.11) have been observed within the neuronal perikaryon in the spinal ganglia of the guinea pig and rabbit (Pannese 1969a). Small aggregates of glycogen granules without any apparent relationship to cytoplasmic organelles can also be seen (Figs. V.7 and V.11); glycogen is

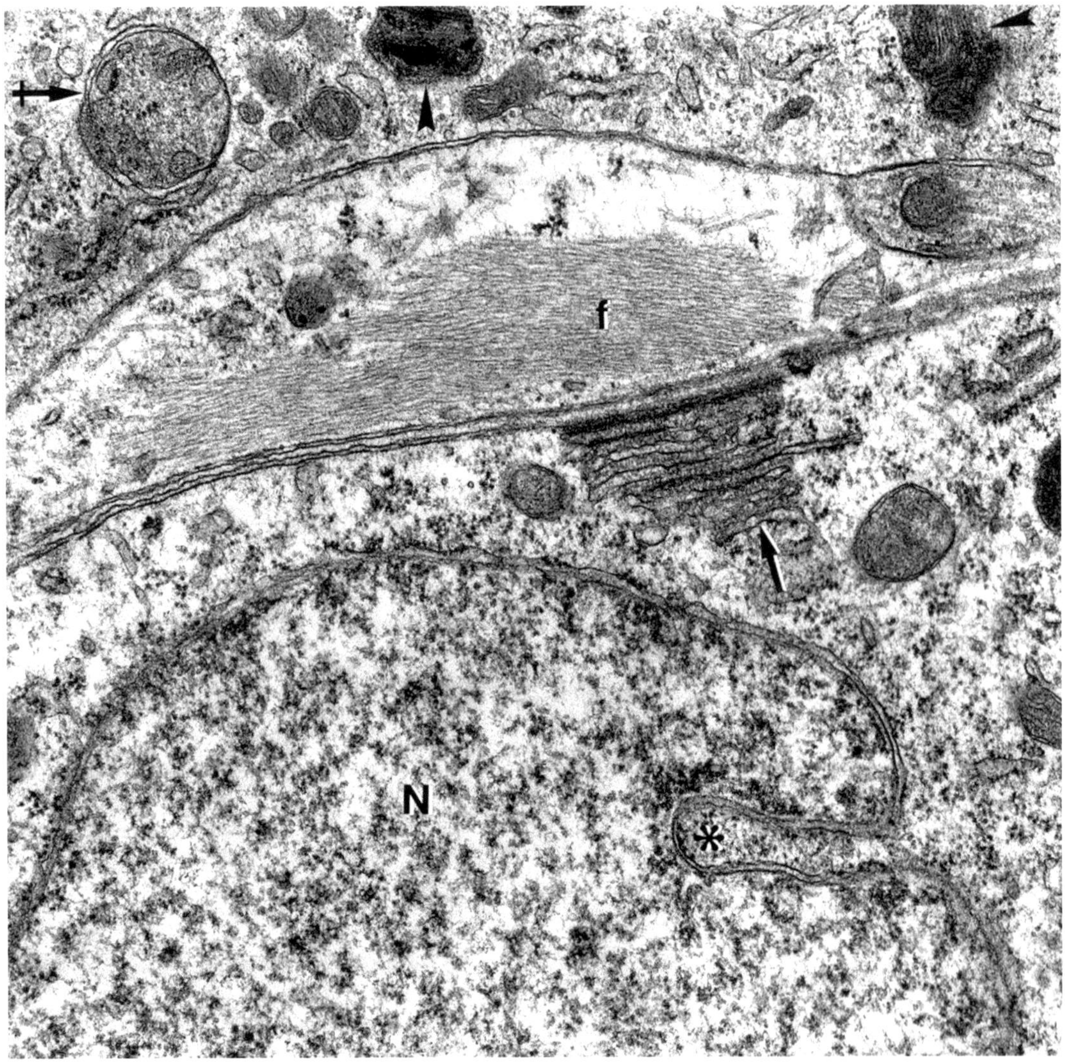

Fig. V.10 Cytoplasmic inclusion and autophagic vacuole in the neuronal perikaryon. Portions of two neuronal perikarya can be seen in the *uppermost quarter* and in the *lower half* of the field. A stack of flattened cisternae (*arrow*) lies close to the plasma membrane of the lower neuron; a rather dense, granular material intervenes between these cisternae. The nucleus (**N**) of this neuron displays a deep invagination (*) in its surface. An autophagic vacuole (*crossed arrow*) and lipofuscin granules (*arrowheads*) can be seen in the upper neuron. In contact with this neuron is the process of a fibrous astrocyte which contains a bundle of filaments (**f**) and some microtubules. Dorsal horn of the rat spinal cord, ×35,000

more abundant in cold-blooded vertebrates, especially during winter dormancy.

Bodies delimited by a wall and filled with filaments morphologically identical to neurofilaments elsewhere have been found within rat neurons (lateral vestibular nucleus, Sotelo and Palay 1968; substantia nigra, LeBeux 1972a). Aggregations of neurofilaments devoid of a limiting membrane are considered in Sect. V.A.7. The roles and significance of the above-described cytoplasmic inclusions are unclear.

10. Age-Related Changes

These changes can differ and also begin at different times depending on species, strain, and

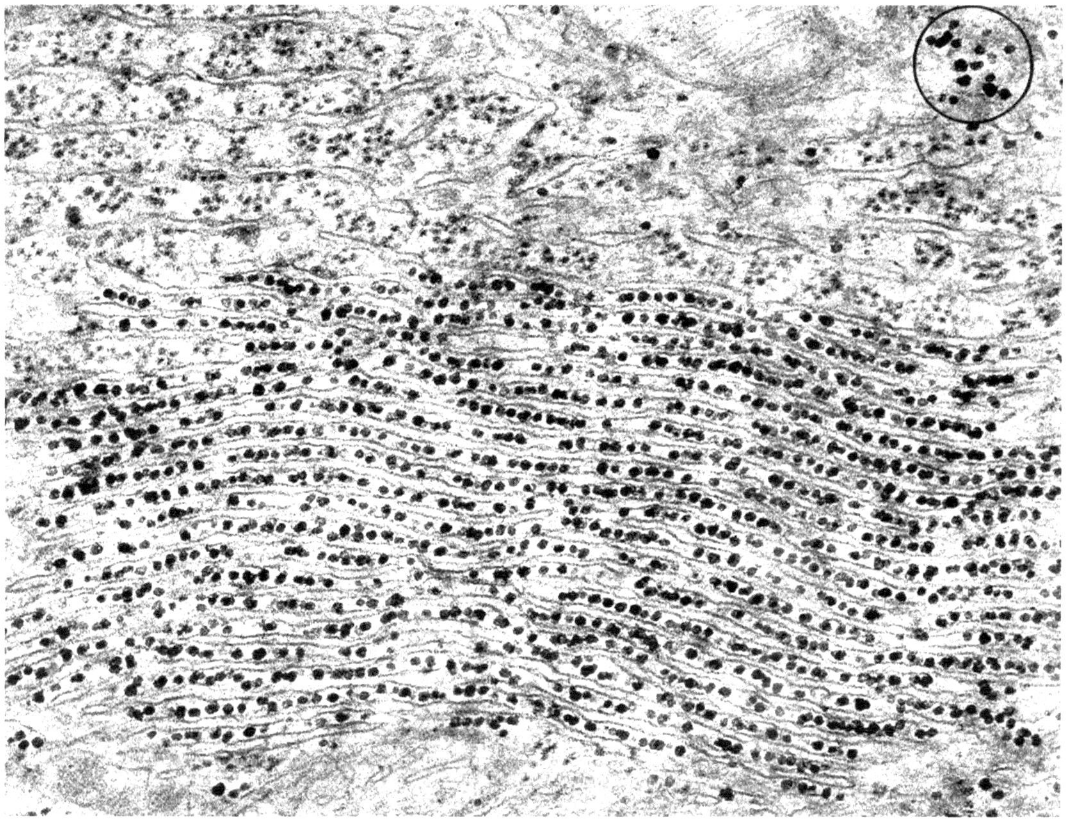

Fig. V.11 Cytoplasmic inclusion in the neuronal perikaryon. Most of the field is occupied by a complex composed of smooth-surfaced cisternae arranged in a parallel array with glycogen particles aligned in single rows between adjacent cisternae. Glycogen granules without any apparent relationship to cytoplasmic organelles are *encircled* in the *upper right corner* of the picture. Glycogen granules can readily be distinguished from ribosomes, which are smaller and less dense. Guinea pig spinal ganglion, ×60,000

region of the nervous system. The most commonly encountered changes are described in what follows.

The accumulation of lipofuscin is one of the most common and striking age-related changes occurring in neurons. It was observed already in the nineteenth century (Schulz 1883; Koneff 1887; Hodge 1894; Pilcz 1895) and later confirmed by quantitative investigations (Zeglio 1936; Brody 1960; Samorajski et al. 1968; Brizzee et al. 1969, 1976). There are differences in the time of appearance of lipofuscin pigment in different regions of the nervous system. For example, in humans it appears during fetal development in the superior cervical ganglion, during the first months of postnatal life in the inferior cervical ganglion, and by the seventh postnatal year in the inferior olive (Zeglio 1936). In rats and mice, lipofuscin accumulation begins earlier in the hippocampus than in the cerebral and cerebellar cortices (Reichel et al. 1968). In the monkey, the earliest appearance of lipofuscin occurs in the inferior olive during the third postnatal month (Brizzee et al. 1974).

Different regions of the nervous system accumulate pigment with age at different rates. Because of this and the fact that the time of appearance of lipofuscin also varies in different regions of the nervous system, for a given species and a given age, marked differences in the pigment distribution exist between the various regions, as the following examples show. In rats of 26–29 months and in mice of 28–29 months, the hippocampus contains a greater amount of lipofuscin per section volume than the cerebral and cerebellar cortices (Reichel et al. 1968); in 10-year-old monkeys the rank order of regional lipofuscin accumulation from lowest to highest is cerebellum, cerebral cortex, pons, midbrain, hippocampus, and medulla (Brizzee et al. 1974). Furthermore, the age-related accumulation of lipofuscin in a given region of the nervous system differs from species to species. For example, in spinal ganglia the percentage of perikaryal volume occupied by lipofuscin pigment is 12 % in old mice (Samorajski et al. 1968), 25 % in old rats (van den Bosch de Aguilar and Vanneste 1980), and only 2.5 % in old rabbits (Ledda et al. 1999).

The age-related lipofuscin accumulation seems to have little or no effect on neuronal metabolism, as the following findings show. The large amount of lipofuscin in spinal ganglion neurons of aged rats does not significantly modify H^3-lysine uptake (van den Bosch de Aguilar and Vanneste 1980). In the mouse brain during aging, no relationships between the regional decrease in metabolic rate and the regional accumulation of lipofuscin were found (Ferrendelli et al. 1971). The RNA content and nucleolar size (considered as indicators of the rate of protein synthesis) of several types of human nerve cell decrease in a similar way with age, despite the widely differing extents of lipofuscin accumulation (Mann et al. 1978).

The granular endoplasmic reticulum changes little in old age. In old rats, it may contain granular dense deposits (van den Bosch de Aguilar et al. 1992). The fine structure of the Golgi apparatus does not differ between young and old animals (rat spinal ganglia, Glees and Gopinath 1973, van den Bosch de Aguilar and Vanneste 1981; rabbit spinal ganglia, Ledda et al. 2001, 2006). Fragmentation and peripheral displacement of the Golgi apparatus have been reported in the spinal ganglia of old rabbits (Sosa and De Zorrilla 1966), but more recent studies have failed to confirm this result (Ledda et al. 2001, 2006). In neurons of rabbit spinal ganglia, the total volume of the Golgi apparatus remains essentially constant with increasing age; since, however, the perikaryal volume of these neurons increases during aging, the percentage of the perikaryal volume occupied by the Golgi apparatus decreases significantly in senescence (Ledda et al. 2001, 2006).

Particular attention has been paid to the mitochondrial changes that occur in old animals since evidence that these organelles are important in the aging process, particularly in postmitotic cells, has accumulated in recent years. The age-related changes to mitochondria may differ from species to species and with region of the nervous system. Structural alterations in these organelles have been reported in neurons of old rats; these mitochondria can contain filamentous or glycogen inclusions in dilated cristae or dense material in the matrix (Vanneste and van den Bosch de Aguilar 1981). By contrast, the structure of mitochondria in neurons of rabbit spinal ganglia does not appear to change even in extremely advanced age (Martinelli et al. 2006b). A significant increase in the mean mitochondrial size with age has been described in some regions of the nervous system. In neurons of rabbit spinal ganglia, the total volume of mitochondria does not change significantly during life (Martinelli et al. 2006b). This indicates that the mitochondria of these neurons are able to divide at a sufficient rate to completely compensate for mitochondrial degradation. However, since the perikaryal volume of rabbit spinal ganglion neurons increases during aging, the percentage of the perikaryal volume occupied by mitochondria decreases significantly. Thus, in old animals, mitochondria seem unable to adapt their total volume to the increasing cytoplasmic volume of these neurons. A similar finding was obtained for neurons of the rat vestibular ganglion (Lyon and King 1997). These results are consistent with the observation that the neurons of old animals have a reduced ability to respond to high energy demands (e.g., Sylvia and Rosenthal 1979). However, conflicting data exist in the literature. For example, the percentage of cytoplasmic volume occupied by mitochondria remains stable with age in pyramidal cells of the rat auditory cortex (Peters and Vaughan 1981) and even increases in the olfactory bulb mitral cells of old rats (Hinds and McNelly 1979).

Structures that are expression of cytoskeletal alterations can also be found in the neuronal perikaryon of old animals. Filamentous inclusions forming a regular lattice and bodies composed of closely packed microtubules have been reported in aged spinal ganglion neurons (van den Bosch de Aguilar and Vanneste 1981). In normal elderly individuals and also in some pathological conditions (e.g., Alzheimer's disease and postencephalitic parkinsonism), intraneuronal argyrophilic fibrous masses have been observed under the light microscope. In electron micrographs, these masses appear mainly composed of pairs of helically intertwined filaments (Kidd 1963), each filament being greater than 10 nm in diameter. These paired helical filaments, which have a periodicity of about 70 nm, are composed primarily of abnormally phosphorylated tau protein (Grundke-Iqbal et al. 1988; Lee et al. 1991; Appelt and Balin 1993).

Eosinophilic rod-like inclusions can be found under the light microscope in preparations of nerve tissue from normal aged individuals. In electron micrographs, these inclusions appear to consist of highly organized arrays of interlacing filaments. These inclusions were first described by Hirano (1965) in patients with amyotrophic lateral sclerosis and parkinsonism-dementia complex and later were referred to as Hirano bodies. They are also present in other pathological conditions (e.g., see Ogata et al. 1972).

Finally, the volume ratio between nucleus and perikaryal cytoplasm has been observed to decrease with age in various types of neuron (Silverman and Sladek 1991; Peinado et al. 1993; Ledda et al. 2000). This decrease is not the consequence of the perikaryal enlargement due to lipofuscin accumulation. The causes of this decrease are at present unknown.

The intriguing question of neuronal loss with aging and the extent of this loss has been much debated (for a review, see Pannese 2011). Several studies, which generally relied on measures of neuronal packing density, suggested that a substantial neuronal loss occurs with aging (e.g., human cerebellum, Hodge 1894; human spinal ganglia, Gardner 1940; human cerebral cortex, Brody 1955; Henderson et al. 1980; monkey cerebral cortex, Brizzee 1973; rat subcortical regions, Sabel and Stein 1981). Brody (1955) reported that in certain areas of the human cerebral cortex from very old subjects, neuronal loss could reach 50 %. These estimates captured the imagination of the public since it was thought that the extensive and substantial loss of neurons could explain the cognitive decline that often accompanies normal aging. It should be noted, however, that other investigations found no decrease in the number of neurons in senescence (e.g., Delorenzi 1931; Cragg 1975). Later it was shown that the embedding procedure results in a shrinkage of the nervous tissue, that is greater in younger than older individuals (Haug et al. 1984). Hence, microscopic sections of nervous tissue from younger individuals show higher neuronal density than those of older individuals (Haug 1985). Studies that took this fact into consideration and used newly developed methods for counting neurons reached the conclusion that during normal aging there is only a slight or even no significant loss of neurons (e.g., cat spinal ganglia, Aldskogius and Risling 1989; rat spinal ganglia, Mohammed and Santer 2001; mouse cerebral cortex, Curcio and Coleman 1982; rat cerebral cortex, Merrill et al. 2001; human cerebral cortex, Haug 1985, Terry et al. 1987, Leuba and Kraftsik 1994, Pakkenberg and Gundersen 1997; monkey cerebral cortex, Peters and Sethares 1993, Morrison and Hof 1997, Peters et al. 1998a,

Giannaris and Rosene 2012; tree shrew hippocampus, Keuker et al. 2004; monkey dorsal lateral geniculate nucleus, Ahmad and Spear 1993; rat suprachiasmatic nucleus, Madeira et al. 1995; monkey hypothalamus, Roberts et al. 2012). These results suggested that the underlying causes of the cognitive decline that often occurs during normal aging should be sought in more subtle cellular alterations (see Sects. V.C.4, V.D.3, VI.G, and VIII.F.3).

B. The Nucleus

1. General Features

In general, the nerve cell has only one nucleus. However, binucleate neurons are not rare, for example, in the autonomic ganglia (Schwalbe 1868; Apolant 1896; Carpenter and Conel 1914, and subsequently many others), and have been also found in other regions of the nervous system (see, among others, Marinesco 1909). The nucleus is often spherical or ellipsoidal, but it may also be completely irregular in shape, as can be seen, for example, in the large neurons of fish. In certain mammalian neurons (e.g., Purkinje cells, giant cells of Meynert in the visual cortex of primates, pyramidal cells in the cerebral cortex, supraoptic neurons), the nuclear surface is deeply indented (Fig. V.10) or has a wrinkled portion. As a rule, the nucleus lies in a central position within the nerve cell body (Fig. V.1), but this is not always the case; for instance, in supraoptic neurons the nucleus is in general eccentrically located. The nuclear volume is usually proportional to the neuronal size; hence, in large neurons the nucleus looks bigger than that of most cells in the same organism. However, since the nuclear/cytoplasmic ratio lessens as the neuronal size increases, the relative nuclear size decreases going from smaller to larger neurons. The nuclear/cytoplasmic ratio in large neurons can therefore be much lower than that of most cells from the same organism. In certain regions of the rat CNS, neurons exhibit nuclear sizes significantly lower in females than in males.

2. The Nuclear Envelope

As in other cell types (Sjöstrand and Rhodin 1953), the nuclear envelope of the neuron (Fig. V.12) consists of two membranes (outer and inner nuclear membranes) with a clear space between them (Hartmann 1953). Each membrane is about 7 nm in thickness. The outer one may appear in continuity at one or more sites with the membranes of the endoplasmic reticulum, so that the clear space between the outer and inner nuclear membranes can be continuous with the lumen of the endoplasmic reticulum cisternae. Thus, the nuclear envelope is regarded as a modified portion of the endoplasmic reticulum (Watson 1955), more precisely as a perinuclear cisterna. Polysomes can be seen on the cytoplasmic surface of the nuclear envelope (Fig. V.12), but usually they are less numerous in neurons than in other cell types.

The outer and inner nuclear membranes are continuous with each other at the periphery of circular apertures, called nuclear pores. In tangential thin sections of the nuclear envelope (Fig. V.13) and in freeze-fracture preparations (Fig. V.14), the pores appear arranged in an ordered array in certain neurons, but scattered at random in others. Pore density (number per unit surface area) may vary greatly from one neuron to another. Investigations carried out on non-neuronal cells have suggested that the pore density may increase in nuclei actively engaged in RNA synthesis. Hence, the differences in pore density observed in various neurons could be related to different rates of nuclear metabolism at the time of fixation. Note, however, that the pore density can also be affected by the preparation method employed (for a review, see Franke 1974). The structure of nuclear pores does not seem to have been specifically studied in neurons.

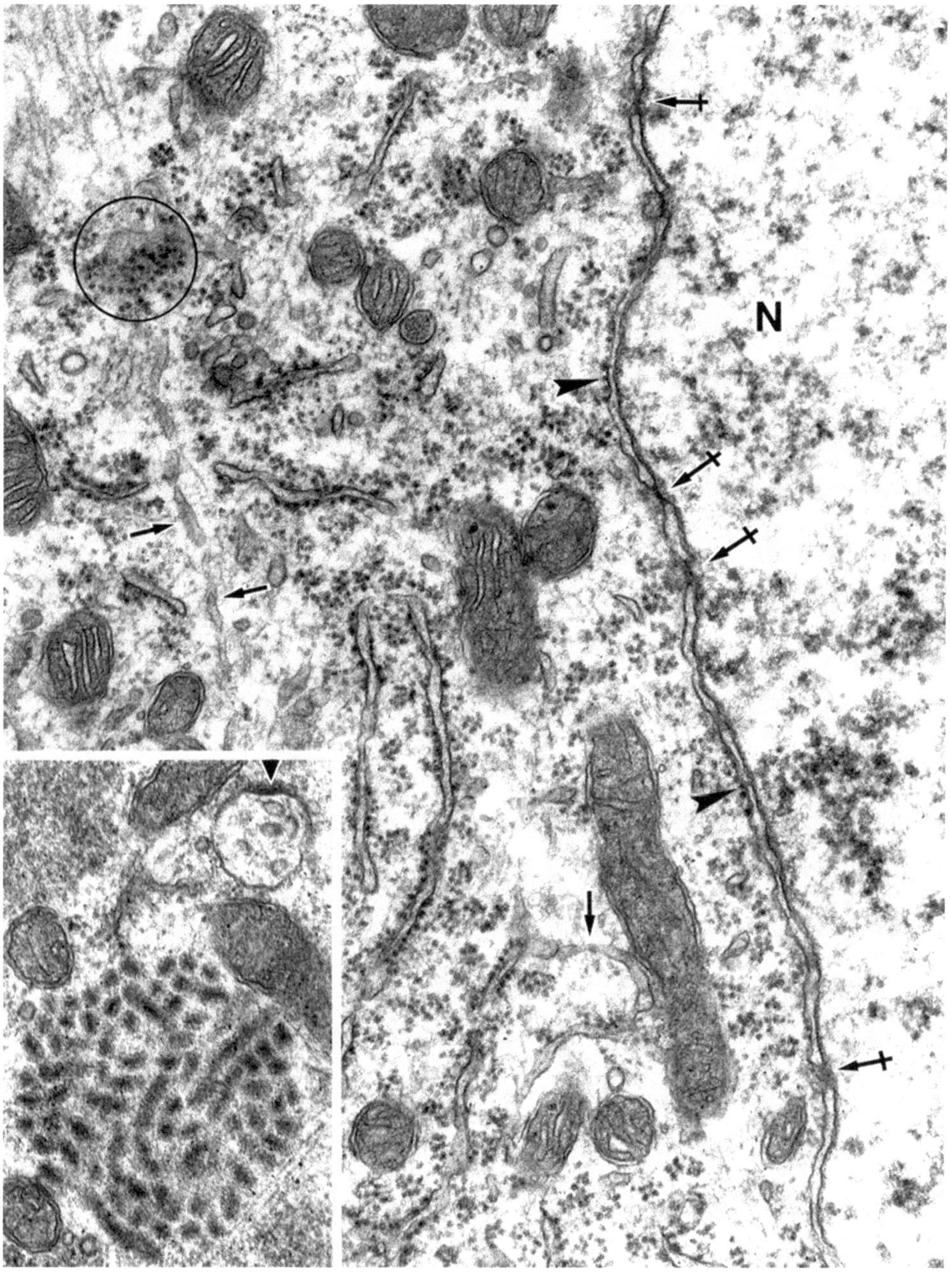

Fig. V.12 Nuclear envelope and some cytoplasmic organelles of a sensory neuron. The nucleus (**N**) is bounded by an envelope consisting of two membranes with a clear space between them. At each nuclear pore (*crossed arrows*), the outer and inner nuclear membranes are continuous with each other; each pore appears bridged by a diaphragm. Here and there, ribosomes (*arrowheads*) can be seen attached to the cytoplasmic surface of the nuclear envelope. Mitochondria, profiles of the granular and agranular (*arrows*) endoplasmic reticulum, microtubules, and neurofilaments are visible in the cytoplasm. Within the *circle* a small area of the membrane limiting a cisterna of the granular endoplasmic reticulum is grazed, so that the arrangement of the attached polysomes can be seen. A nematosome, consisting of a convoluted network of dense strands, and a multivesicular body are shown in the *inset*. A patch of the membrane bounding the multivesicular body appears flattened and lined by a fuzzy coat (▼). Rabbit spinal ganglion, ×50,000

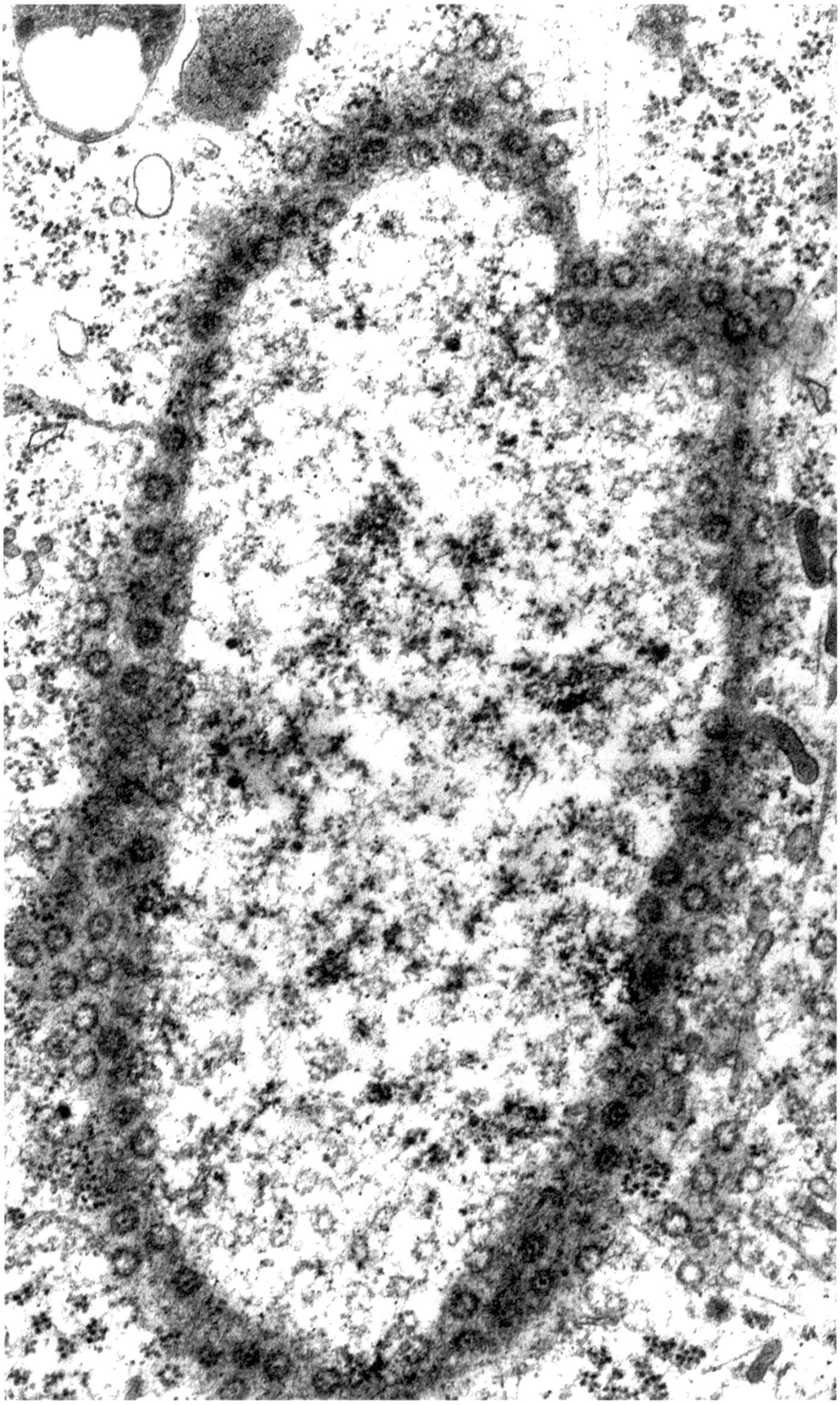

Fig. V.13 Nuclear envelope of a neuron. In this tangential section of the nuclear envelope, nuclear pores appear as dark circles. A small, dense spot can be seen in the center of some of the pores. Ventral horn of the rat spinal cord, ×44,000

Fig. V.14 Nuclear envelope of a neuron in a freeze-fractured preparation. A large area of the nuclear envelope can be seen in the *lower half* of the field. The cleavage plane shifts from the outer (**O**) to the inner (**I**) nuclear membrane and reveals numerous nuclear pores. Rabbit spinal ganglion, ×17,500 (Courtesy of E. Reale and L. Luciano)

However, the pore structure in neurons is likely to be similar to that observed in other cell types. Negatively stained preparations of nuclear envelopes isolated from nonneuronal cells show globular units, 15–20 nm in diameter, which are arranged in radial symmetry around the edge of the pore. Thin, radially arranged filaments project inward from these globules toward a central granule. These structures all together are known as the pore complex. In sections perpendicular to the nuclear envelope, each pore seems to be closed by a diaphragm (Fig. V.12). However, since there is constant traffic of molecules in both directions through the pores, it is unlikely that they are closed. The appearance of a diaphragm that seems to close the pore is probably the result of a superimposition of the images of the structures that constitute the pore complex. The structures of the pore complex partially obturate the aperture. Therefore, although the pore diameter is about 90 nm, the effective openings are about 10 nm. Ions and molecules less than 10 nm in size pass through the pores by passive diffusion. Larger molecules are thought to pass through by active transport.

A thin layer of very fine, interwoven filaments can be seen in most cell types interposed between the inner nuclear membrane and the peripheral clumps of chromatin (Fawcett 1966; Patrizi and Poger 1967). This fibrous lamina, which consists of polypeptides called lamins, is well developed and easily observed in certain invertebrate neurons, for example, leech ganglion cells (Gray and Guillery 1963a; Fawcett 1966), whereas it is lacking except around the edges of nuclear pores in mammalian neurons. The fibrous lamina probably has a role in the support of the nuclear envelope and provides sites of attachment for components of the karyoplasm.

3. The Karyoplasm

In large- and medium-sized neurons, the chromatin is widely and homogeneously dispersed so that the nucleus appears as a pale vesicle under the light microscope (Fig. V.1) even after staining with basic dyes. In many small neurons (e.g., granule cells in the cerebellar cortex), the chromatin is instead condensed into clumps which stain deeply with basic dyes. A substantial technical improvement in the structural study of the neuronal karyoplasm was obtained after the introduction of aldehydes as fixatives for electron microscopy (Sabatini et al. 1963). Notwithstanding this achievement, the neuronal karyoplasm has proved more difficult to preserve satisfactorily than its counterpart in other cell types. The nerve cell nucleus displays most of its chromatin complement in extended form, conceivably consisting of thin filaments helically coiled into 20-nm-thick fibers. In large neurons, these chromatin fibers appear condensed into small clumps only around the nucleolus and against the nuclear envelope, while they are extensively dispersed elsewhere (Fig. VII.1) so that the nucleus has a generally clear appearance. In the nucleus of small neurons, however, a great quantity of chromatin fibers is condensed into irregular clumps (Fig. V.4), some of which lie adjacent to the nuclear envelope, others around the nucleolus, and still others in the center of the nucleus. The chromatin fibers are suspended in a cottony matrix, in which there are also dense granules of various sizes.

4. The Nucleolus

The nucleus of large- and medium-sized neurons contains a large, spherical nucleolus which stains deeply with basic dyes (Fig. V.1). In electron micrographs (Fig. VII.1), the nucleolus was found to consist of dense granules (15–20 nm diameter) and 5-nm filaments. The latter, although less dense than granules, are often so closely arranged that the regions consisting mainly of filaments may appear denser than those predominantly made up of granules. In many cell types, granules and filaments are segregated into concentric regions, the granular surrounding the filamentous component more or less completely. Within the nucleolus of large- and medium-sized neurons, both components are intermingled into an irregular, three-dimensional network, in whose meshes karyoplasmic portions may be entrapped. Such portions are usually small, although occasionally large enough to appear like vacuoles under the light microscope. Both granular and filamentous components consist largely of RNA and proteins; however, the filamentous component also contains DNA. In small neurons, the nucleoli are fragmented and dispersed among chromatin clumps. The nucleolus is the site of ribosomal RNA synthesis and initial ribosomal assembly.

A mass of chromatin, about 1 μm in diameter, was found attached to the nucleolus in approximately 90 % of neurons in females and in less than 10 % of neurons in males of several species of carnivores and artiodactyls (Barr et al. 1950; Moore and Barr 1953). This sexual dimorphism was subsequently observed in neurons and other cell types (including neuroglial cells) of many other mammalian species, but not in amphibians, reptiles, or birds. Later still, it was shown that this mass of chromatin, called sex chromatin, represents one of the two X chromosomes. Sex chromatin is not, however, associated with the nucleolus but is closely applied to the nuclear envelope in neurons and other cell types of certain primates, including humans.

The nucleolar size in rat hypothalamic neurons has been found to be significantly smaller in female than in male animals and appears to be influenced by several hormones. Diurnal rhythmic variations in nucleolar size have also been reported (Pebusque et al. 1981).

5. DNA Content

Cytophotometric studies on sections of nervous tissue stained by the Feulgen technique had previously suggested that in several mammalian species certain large neurons (e.g., Purkinje cells of the cerebellar cortex, pyramidal cells of the hippocampus, Betz cells of the cerebral cortex, spinal cord motoneurons, certain neurons of the olfactory bulb) are tetraploid (Lapham 1965, 1968; Sandritter et al. 1967; Herman and Lapham 1969; Swartz and Bhatnagar 1981). However, contrasting results were obtained by several authors (Cohen et al. 1973; Mann and Yates 1973; Fujita 1974; Manuelidis and Manuelidis 1974; Mareš et al. 1974; Hobi et al. 1984). Later studies showed that not all neurons had the same amount of DNA (Rehen et al. 2001; Westra et al. 2010) and that in the normal adult human CNS only 0.2–0.5 % of neurons are tetraploid (Mosch et al. 2007). True polyploidy has been observed in some invertebrate neurons: the nuclear DNA content in certain neurons of the cerebral ganglion of the insect *Drosophila virilis* (Swift 1962) and in certain giant neurons of the marine mollusc *Aplysia californica* (Coggeshall et al. 1970; Lasek and Dower 1971) is many times greater than that of a diploid cell of the same species.

6. Nuclear Inclusions

Rod-like or crescent-shaped bodies were discovered within the nucleus of nerve cells at the end of the nineteenth century. Mentions of such inclusions can be found in papers by Hodge (1892) and Mann (1894). The latter interpreted them as homologues of the centrosomes. Soon afterward, they were carefully described by Roncoroni (1895), who hypothesized that they were crystalloid structures; electron microscope investigations have demonstrated this to be the case for many such inclusions. Intranuclear bodies have also been recorded in nerve cells by Lenhossék (1897), Prenant (1897), Holmgren (1899), and other authors. These bodies stain deeply with iron hematoxylin, erythrosin, safranin, and methyl blue and are impregnated by Cajal's reduced silver technique.

Nuclear inclusions from a broad range of species (see Table V.1) may appear in electron micrographs as rod-like or laminar structures. The rod-like bodies are composed of closely packed, parallel-oriented filaments, each about 7 nm thick; sometimes, microtubules are associated with them. Histochemical studies have suggested that these rod-like intranuclear bodies consist mainly of proteins and do not contain either DNA or RNA (Kim et al. 1970; Seïte et al. 1971). These inclusions were found with greater frequency in the neurons of the parietal cortex of aged than young mice (Field and Peat 1971). The laminar inclusions are composed of several stacked layers, each consisting of 7-nm-thick filaments; the filaments run in a parallel array within each layer, but are oriented at an angle with respect to those in adjacent layers, thus forming a paracrystalline lattice. Microtubules may also appear associated with laminar inclusions. These nuclear inclusions are not confined to neurons but have also been found within ependymal, choroidal, and neuroglial cells, as well as within many other cell types.

Another nuclear inclusion was first identified by Ramón y Cajal (1903; see also 1910) in many mammalian neurons and soon afterwards in nerve cells of reptiles and birds (Tello 1904). This structure was subsequently called Cajal body in honor of its discoverer. Two or more Cajal bodies may be present in a single nucleus. However, these inclusions are not confined to neurons but have also been reported in nonneuronal cells of insects, amphibians, and mammals as well as in plant cells. In vivo the Cajal body moves within the karyoplasm (Platani et al. 2000) and in fixed cells may be found either free in the karyoplasm or adjacent to the nucleolus. Because it may be associated with the nucleolus, Ramón y Cajal called it the nucleolar accessory body. Under the light microscope, it appears as a spherical body, 0.2–1 μm in diameter, stains with Cajal's reduced silver technique, is Feulgen negative, and lacks basophilic properties. This body is distinct from sex chromatin. Under the transmission electron microscope, the Cajal body appears as an aggregate of coiled threads (Monneron and Bernhard 1969) and hence has also been called a coiled body. The Cajal body contains small ribonucleoproteins, the survival motor neuron protein, and p80 coilin; the latter protein is often used as a marker for the body. Notwithstanding recent progress in the molecular characterization of this inclusion, its function is still intensively debated (for reviews, see Gall 2000; Lafarga et al. 2009).

In 1902, Marinesco described yet another type of intranuclear body, this time in pigmented neurons of the substantia nigra and locus coeruleus of normal humans. These bodies, subsequently called Marinesco bodies, generally occur singly within the nucleus, but sometimes there may be more than one. They may be as large as or a little smaller than the nucleolus, but unlike the latter, stain with acid dyes. In electron micrographs, the Marinesco body appears devoid of a limiting membrane and to consist of an aggregate of fine granules (10–20 nm in diameter) which are sometimes associated with filaments (Okamoto and Hirai 1981; Ono et al. 1989). Marinesco bodies, whose significance is obscure, seem to increase in number with advancing age.

7. Age-Related Changes

In some regions of the nervous system, the nucleus of neurons does not show morphological changes in old age (e.g., the rat occipital cortex, Diamond et al. 1977; layer II of the rat auditory cortex, Vaughan and Vincent 1979; rat dentate fascia, Fifková et al. 1987). In other regions, the nucleus shows only slight changes (e.g., in layer V of the rat auditory cortex, the mean sectional area of the nucleus decreases in old age, Vaughan and Vincent 1979). As noted previously (see Sect. V.A.10), the volume ratio

Table V.1 A partial list of intranuclear inclusions found in vertebrate neurons (electron microscopy)

Site	Family	Species	Common name	Author(s)
Sphenopalatine ganglion	Canidae	*Canis familiaris*	Dog	Costa and Paula-Barbosa (1979)
Sympathetic chain ganglia	Phasianidae	*Gallus domesticus*	Domestic fowl	Masurovsky et al. (1970)
	Leporidae	*Oryctolagus cuniculus*	Rabbit	Dixon (1970)
	Felidae	*Felis catus*	Cat	Seïte (1970)
Celiac ganglion	Felidae	*Felis catus*	Cat	Seïte et al. (1977b)
Ganglia of the intestinal nerve (nerve of Remak)	Phasianidae	*Gallus domesticus*	Domestic fowl	Young (1990)
Myenteric plexus	Leporidae	*Oryctolagus cuniculus*	Rabbit	Komuro et al. (1982)
Cuneate nuclei	Muridae	*Rattus norvegicus*	Rat	David and Nathaniel (1978)
Ventral cochlear nucleus	Muridae	*Rattus norvegicus*	Rat	Feldman and Peters (1972)
Lateral vestibular nucleus	Muridae	*Rattus norvegicus*	Rat	Sotelo and Palay (1968)
Dorsal motor vagal nucleus	Leporidae	*Oryctolagus cuniculus*	Rabbit	Aldskogius (1978)
	Felidae	*Felis catus*	Cat	McLean and Hopkins (1981)
Midbrain	Felidae	*Felis catus*	Cat	Liu and Hamilton (1976)
Thalamus and hippocampus	Muridae	*Mus musculus, Rattus norvegicus*	Mouse, rat	Chandler and Willis (1966)
Hypothalamus	Muridae	*Rattus norvegicus*	Rat	Lafarga and Palacios (1977)
	Leporidae	*Oryctolagus cuniculus*	Rabbit	Clattenburg et al. (1972)
	Cricetidae	[a]	Hamster	Fernandez et al. (1984)
Lateral geniculate nucleus	Muridae	*Rattus norvegicus*	Rat	Karlsson (1966b)
Retina	Tupaiidae	*Tupaia glis*	Tree shrew	Hajdu et al. (1982)
	Muridae	*Rattus norvegicus*	Rat	Magalhães (1967)
	Leporidae	*Oryctolagus cuniculus*	Rabbit	Magalhães (1967)
Caudate nucleus	Muridae	*Rattus norvegicus*	Rat	Seïte et al. (1977a)
Olfactory system	Felidae	*Felis catus*	Cat	Willey and Schultz (1971)
	Leporidae	*Oryctolagus cuniculus*	Rabbit	Siegesmund et al. (1964)
	Cebidae	*Saimiri sciureus*	Squirrel monkey	Siegesmund et al. (1964)

(continued)

Table V.1 (continued)

Site	Family	Species	Common name	Author(s)
Cerebellar cortex	Muridae	*Rattus norvegicus*	Rat	Palay and Chan-Palay (1974)
	Leporidae	*Oryctolagus cuniculus*	Rabbit	Siegesmund et al. (1964)
Cerebral cortex	Phasianidae	*Gallus domesticus*	Domestic fowl	Dahl (1970)
	Muridae	*Mus musculus*	Mouse	Gambetti and Gonatas (1967)
	Muridae	*Rattus norvegicus*	Rat	Gambetti and Gonatas (1967); Dahl (1970)
	Leporidae	*Oryctolagus cuniculus*	Rabbit	Siegesmund et al. (1964)
	Cercopithecidae	*Macaca mulatta*	Rhesus monkey	Dahl (1970)
	Hominidae	*Homo sapiens*	Man	Gambetti and Gonatas (1967); Cragg (1976)
Cerebral hemispheres	Myxinidae	*Myxine glutinosa*	Atlantic hagfish	Mugnaini (1967)

N.B.: In the cases where intranuclear inclusions in a given site of a particular species have been mentioned on various occasions by the same or different authors, only the first paper has been cited

[a]Species not indicated in the original paper

between nucleus and perikaryal cytoplasm has been observed to decrease with age in various types of neuron.

C. Dendrites

Some features of dendrites have already been described in Chap. III.

1. General Features

The following can be observed under the light microscope. In Golgi preparations and in the material stained by the Ehrlich methylene blue technique, small dendrites sometimes exhibit a beaded appearance due to the presence of bulbous dilatations which occur at rather regular intervals along their course. As already reported by Koelliker (1896) and Ramón y Cajal (1896a), such dilatations are artifactual. In silver-impregnated preparations, small dendritic branches can hardly be distinguished from the background. In material stained according to the usual techniques, only the largest Nissl bodies lying within the large dendrites can be detected. As a consequence, in material embedded in paraffin and processed according to the usual staining techniques, only the initial portion of the dendritic tree, containing large Nissl bodies, can be recognized (Fig. V.1); the greater part of the dendritic tree, which contains Nissl bodies decreasing in size with increasing distance from the perikaryon, cannot be appreciated. In semithin (1-μm) sections of material fixed with aldehydes and osmium, embedded in resin, and stained with toluidine blue or methylene blue, dendrites can be followed for long distances beyond their proximal segments since in these preparations they appear colorless against a blue background.

In electron micrographs, the primary dendrites appear as extensions of the neuronal perikaryon, containing the same organelles found in it. Just beyond the dendrite origin, however, all organelles assume a longitudinal orientation (Figs. V.15 and V.16). Nissl bodies (Fig. V.15) can be found throughout the dendritic tree, but they become progressively less frequent, smaller, and more simply organized with increasing distance from the perikaryon. Each Nissl body in the large dendrites, like those within the perikaryon, consists of a number of granular endoplasmic reticulum cisternae and of free polysomes lying in the cytoplasmic matrix between the cisternae. Near the

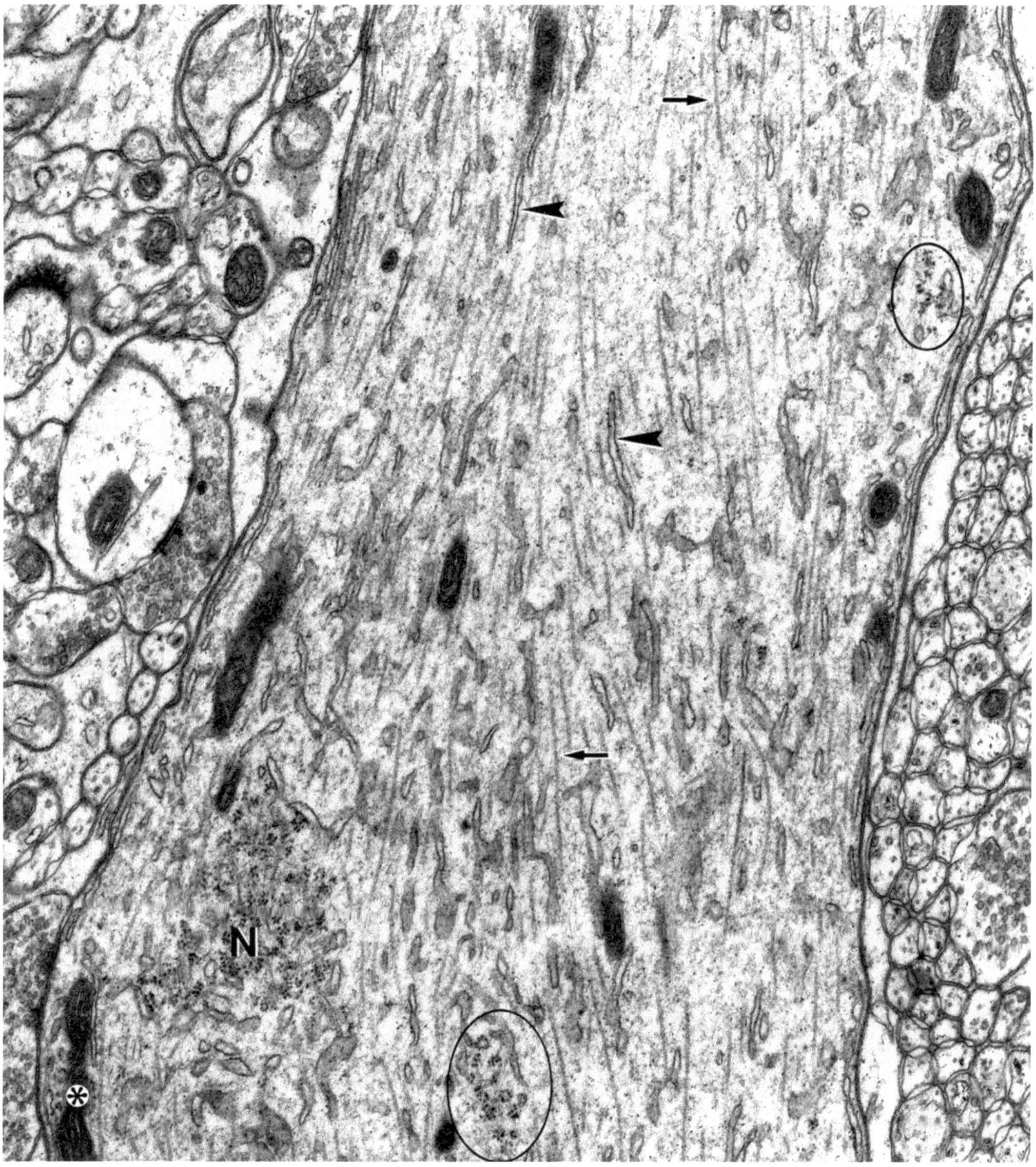

Fig. V.15 Dendritic organelles. A longitudinally sectioned dendrite of a Purkinje cell occupies most of the field. Mitochondria, tubules of the agranular endoplasmic reticulum (some indicated by *arrowheads*), and microtubules (some *arrowed*) are arranged parallel to the long axis of the dendrite. The mitochondrion marked * has longitudinally oriented cristae. Note that numerous cisternae lie close to and parallel with the plasma membrane of the process; these are called subsurface cisternae. The dendrite also contains a Nissl body (**N**) and groups of free polysomes (within the *ovals*). An axonal end-bulb synapsing with the dendrite can be seen in the *lower left corner*. Rat cerebellar cortex, ×24,000

perikaryon, such Nissl bodies can lie either in the core of the dendrite or more peripherally, while a little more distally they are confined to outpocketings and branching points of dendrites, as already observed by light microscopists (see, among others, Ramón y Cajal 1896c). In the distal segments of the dendritic tree, Nissl bodies

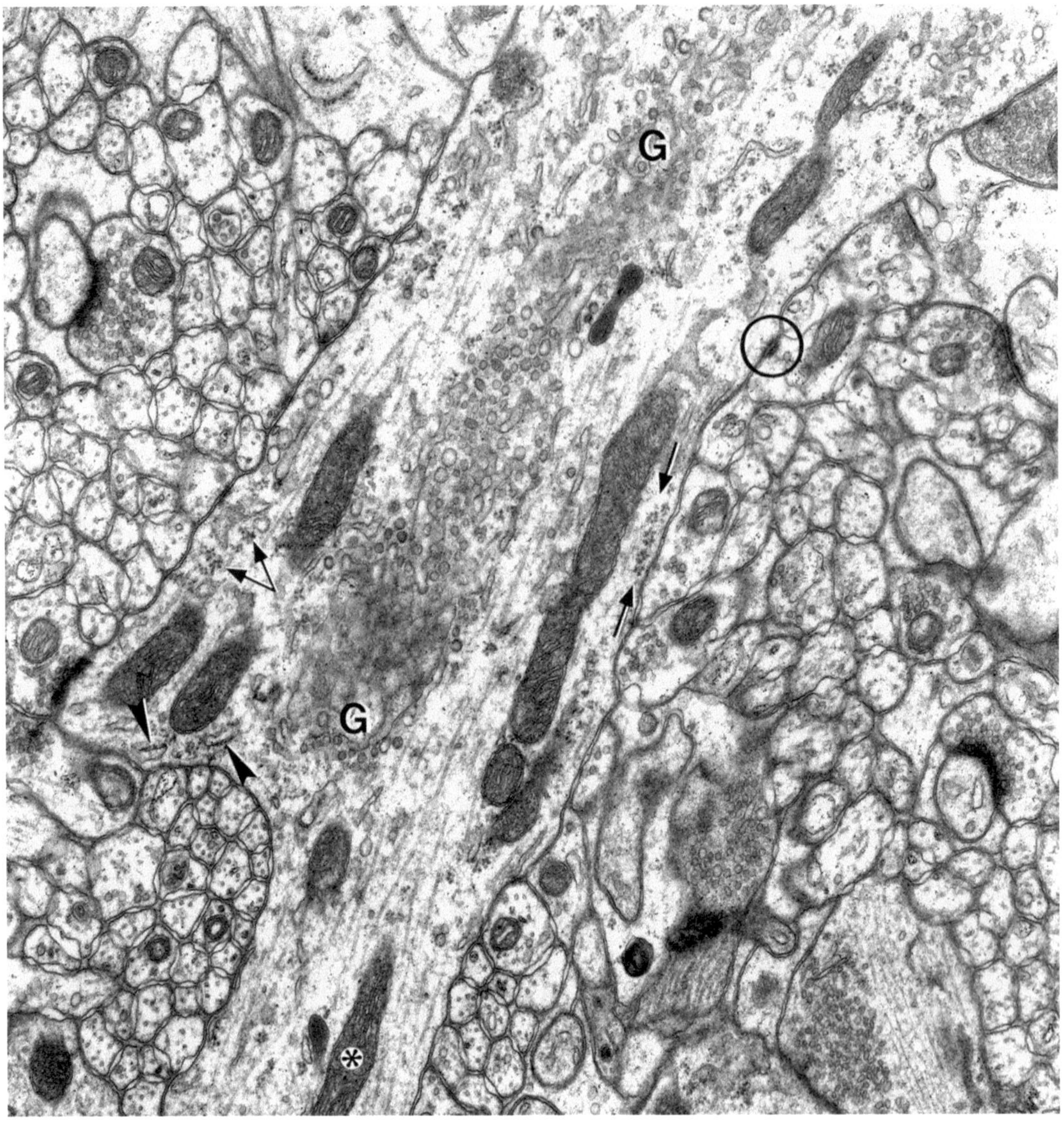

Fig. V.16 Dendritic organelles. A longitudinally sectioned dendrite passes obliquely across the field. A Golgi complex (**G**) is evident in the core of this process, which also contains tubules of the granular endoplasmic reticulum (some indicated by *arrowheads*) and groups of free polysomes (some *arrowed*). Note that the mitochondrion marked * has longitudinally oriented cristae. A small interneuronal adherent junction is *encircled*. An axonal end-bulb synapsing with the dendrite can be seen in the *upper right corner*. Rat cerebellar cortex, ×22,000

consist of isolated cisternae or tubules with a few polysomes. This gradual reduction in the size and frequency of Nissl bodies with increasing distance from the perikaryon explains why only the proximal portion of the dendritic tree can be stained with basic dyes; within the distal segments of the dendritic tree, both attached and free polysomes do not retain sufficient dye to enable them to be distinguished under the light microscope (Peters et al. 1991). In addition to polysomes, dendrites are equipped with all the other components of the translational apparatus necessary for the synthesis of proteins. In fact selected proteins are synthesized within dendrites (Tiedge and Brosius 1996).

The agranular reticulum (Figs. V.15 and VI.1) is a system of membrane-bounded tubules and cisternae which forms a continuous network throughout the dendritic tree and is in continuity with the endoplasmic reticulum lying within the perikaryon. This organelle is not obvious within primary dendrites, where the Nissl substance is well developed; it becomes more prominent in the distal segments of the dendritic tree, where the amount of Nissl substance is lower. Occasionally, a fenestrated cisterna of the agranular reticulum is arranged perpendicular to the major axis of the dendrite and is crossed by microtubules running longitudinally through its perforations. Subsurface cisternae (Rosenbluth 1962b) and stacked cisternae (Gray and Guillery 1963b; Kerns and Peters 1974), similar to those occurring in the perikaryon (see, respectively, Sects. V.A.2 and V.A.9), can be found also within dendrites. The subsurface cisternae vary in frequency according to the different types of neuron; for example, in the Purkinje cells (Fig. V.15) they are present throughout the dendritic tree (see also Sect. V.A.2). For functions of the dendritic agranular reticulum, see Sect. V.A.2.

The Golgi apparatus (Fig. V.16) can often be observed in primary dendrites with its cisternae arranged parallel to the long axis of the process. Among dendritic organelles, the Golgi apparatus is the first to disappear with increasing distance from the perikaryon (Peters et al. 1991).

Mitochondria (Figs. V.15 and V.16) of various shapes are seen within dendrites; some are slender and elongated, others small and spheroidal, and others again branched. They usually range from 0.1 to 0.5 μm in width and from 1 to 10 μm in length; slender mitochondria, up to 20 μm in length, are present in terminal dendritic branchlets as well as within the preterminal segments of axons (see Sect. V.D.2). The number of mitochondrial profiles per unit of cross-sectional area appears higher in the small branches than elsewhere in the dendritic tree (Palay 1964).

Microtubules, the most conspicuous organelles in dendrites, usually run parallel to the long axis of the process (Figs. V.15 and VI.1). In cross-sectioned large dendrites, microtubules appear rather uniformly arranged (Fig. VI.1) except where other organelles, particularly mitochondria, intervene among them. A network of fine threads (Fig. VI.1) is present among microtubules and is attached to them, to neurofilaments, and to other neighboring organelles (Weiss 1967; Wuerker and Palay 1969; Wuerker and Kirkpatrick 1972). This network seems to consist mainly of microtubule-associated proteins (see Sect. V.D.2). For the polarity orientation of microtubules in dendrites, see Sect. V.D.2.

Neurofilaments usually occur in small numbers (Fig. VI.1) and are scattered as single units within the dendrites. However, in certain types of neuron (e.g., large motoneurons of the rat spinal cord, Wuerker and Palay 1969; Betz cells in the cerebral cortex of the cat, Kaiserman-Abramof and Peters 1972; large neurons of the Clarke's nucleus in the cat, Smith 1972; giant Meynert cells of the monkey visual cortex, Chan-Palay et al. 1974), dendrites contain numerous neurofilaments, often arranged in loosely organized bundles; nonetheless, microtubules remain preponderant. Neurofilaments run parallel to the long axis of the process (Fig. VI.1) and are surrounded by the network of fine threads referred to above.

Dendrites are as a rule unmyelinated, so that the presence of a myelin sheath around a neuronal process is usually considered diagnostic for the axon. However, dendrites enveloped by myelin sheaths have been detected in several species of fish (e.g., mormyrid electrosensory lobe, Meek et al. 2001), amphibians (e.g., toad cerebellum, Rosenbluth 1966; frog spinal cord, Taugner 1980), and mammals [e.g., olfactory bulb of mouse (Burd 1980), hedgehog (Valverde and

Lopez-Mascaraque 1991), cat (Willey 1973; Remahl and Hildebrand 1985), nonhuman primates (Pinching 1971; Tigges and Tigges 1980), and humans (Braak et al. 1977)]. The myelin sheaths enveloping dendrites are much thinner and shorter than those around axons of a similar cross-sectional area. The structural specializations typically present in axons at nodes of Ranvier are absent in dendrites, both where isolated dendritic myelin sheaths terminate and between successive sheaths (Remahl and Hildebrand 1985). The functional role of the myelin sheath surrounding dendrites is at present unknown.

2. Dendritic Spines

In Golgi preparations and in the material stained by Ehrlich methylene blue technique, the surface of the dendrites of certain neurons appears covered with minute protrusions which are called dendritic spines (Ramón y Cajal 1888, 1891, 1896a; Retzius 1891; Schaffer 1892; Edinger 1893). Where these protrusions are very numerous, the dendritic tree appears covered by moss.

According to Jacobson (1993), dendritic spines were discovered by Philip Owsjannikov who in 1864 described a dense covering of short, fine hairs on the processes of Purkinje cells. For a long time, dendritic spines were interpreted as artifactual metallic precipitates (Meyer 1895; Koelliker 1896; Heidenhain 1911; Levi 1946) but were later demonstrated to be real structures. Spines have also been observed on the neuronal perikaryon (Mugnaini et al. 1967b) and on the initial segment of the axon (Westrum 1970), but here the spine density, i.e., the number of spines per unit surface area, is always much lower than that on the dendritic tree.

Use of fluorescent techniques in combination with laser scanning confocal microscopy (or two-photon laser scanning microscopy) has made it possible to study a single portion of a dendrite at the surface of the brain in a living animal over long periods of time. These studies have revealed that dendritic spines are dynamic structures. Individual spines may undergo changes in shape and size and may appear and disappear over time even in the mature brain (e.g., see Lendvai et al. 2000; Trachtenberg et al. 2002). These data should be considered when assessing the results of studies carried out on static images.

The shape of a dendritic spine can be recognized both under the light microscope (in Golgi preparations or after neuronal injection with a marker, e.g., biocytin or Lucifer yellow) and in electron microscope. Since spines may undergo changes in shape, they exhibit a wide range of forms. Most spines consist of an ovoid or spherical end-bulb (usually called head) and a narrow stalk (or neck) attached to the parent dendrite (Fig. V.17). Other spines appear as digit-like projections of constant width throughout their length; others again are mushroom-shaped with a short, thick neck and a particularly large head. Numerous intermediate forms are also found.

The spine size can be determined with higher precision in electron micrographs than in Golgi preparations, where the accumulation of silver chromate precipitates on the surface of the spines can lead to an overestimation of their size (Chan-Palay and Palay 1972). Both the transverse diameter of the head and the width of the neck were found to be significantly greater in the adult than newborn guinea pig (Schüz 1986). Systematic differences in mean spine size have been found in different cortical areas of the same species, and in the same area of different species. In the mouse, the spines of the temporal cortex are larger and longer than those of the occipital cortex; temporal cortex spines are larger and longer in humans than the mouse (Benavides-Piccione et al. 2002). The dendritic spines of the stratum radiatum neurons of the rat hippocampus measure 0.2–0.8 μm in length and have a head with a transverse diameter of some 0.25 μm (Westrum and Blackstad 1962); those of the Purkinje cells

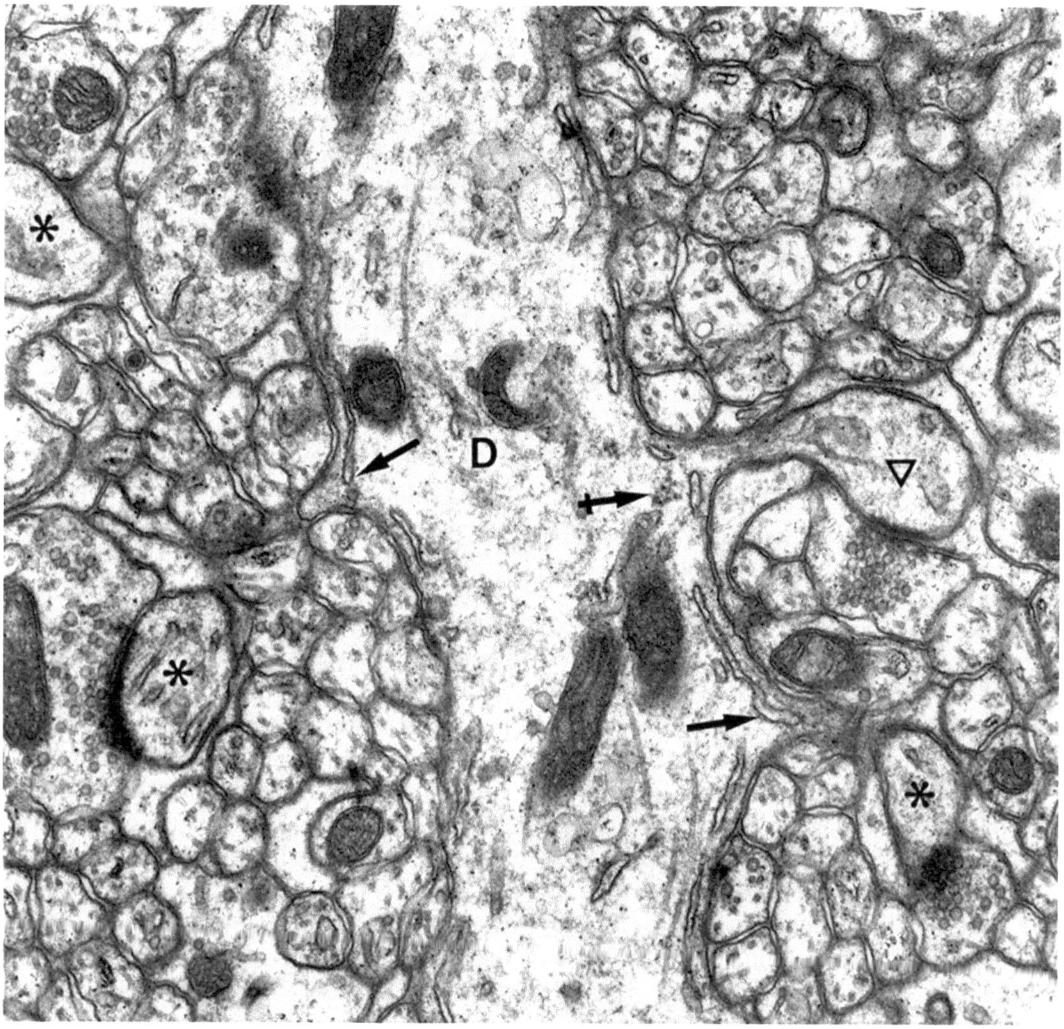

Fig. V.17 Dendritic spines. A Purkinje cell dendrite (**D**) runs almost vertically across the field. A spine (▽) emerges from this dendrite and forms an asymmetrical synapse with the axon terminal of a parallel fiber. This spine consists of an ovoid end-bulb with a narrow stalk by which it is attached to the parent dendrite. Note a polysome (*crossed arrow*) under the base of the stalk. The stems (*arrows*) of two other spines protrude from the dendrite; a tubule of the agranular endoplasmic reticulum extends from the dendrite into the right stem. Other spines (*) in the surrounding neuropil appear as isolated profiles and can be identified on the basis of their structural features: the presence of a diffuse cottony material and of profiles of the agranular endoplasmic reticulum, the absence of microtubules and neurofilaments, and the presence of a synapse. Rat cerebellar cortex, ×37,500

of the rat cerebellum are 1–2 μm long, their necks are 0.2–0.3 μm in diameter, and their heads are 0.4–0.5 μm across (Palay and Chan-Palay 1974). The maximum diameter of the head of the dendritic spines from pyramidal neurons of mouse primary visual cortex averages 0.42 μm (Konur et al. 2003). Since spines may undergo changes in size, the maximum and minimum values reported above give an idea of the limits within which spine size may vary. These morphological characteristics correlate with the physiological properties of spines. For example, the size of the spine head correlates with the

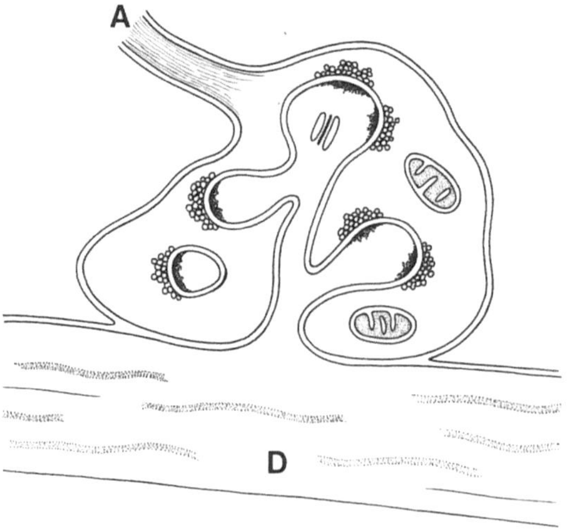

Fig. V.18 Dendritic spines. A branched spine emerges from a dendrite (**D**) of a hippocampus pyramidal neuron. This spine forms several synapses with the ending of an unmyelinated axon (**A**)

area of the postsynaptic density and with the number of presynaptic docked vesicles. The area of the postsynaptic density in turn correlates with the number of receptors, and the number of docked vesicles correlates with the readily releasable pool of neurotransmitter. To simplify somewhat, the size of the spine head is an indicator of the synaptic strength (Schikorski and Stevens 1997). Since spine size correlates with synaptic strength, changes in synaptic size indicate that changes in synaptic strength may occur throughout life. Changes in spine shape and size, in which actin and myosin are involved (for reviews, see Halpain 2000; Matus 2000), are more frequent and rapid in young animals but also occur in adult life (for a review, see Alvarez and Sabatini 2007).

Spines of greater length and more complex shape (Fig. V.18) have been observed at various sites. For example, the dendrites of the hippocampus pyramidal cells bear long, branched spines, each of which comes into intimate contact with the expanded ending of a mossy fiber where numerous synaptic junctions are formed (Blackstad and Kjaerheim 1961; Hamlyn 1962; Chicurel and Harris 1992; Gonzales et al. 2001). Spine equipment varies according to species and neuron type. Dendrites of cerebral cortex pyramidal cells and of cerebellar cortex Purkinje cells are particularly rich in spines. Spines are also found on the dendrites of the spinal cord motoneurons, vestibular nucleus neurons, thalamic neurons, olfactory bulb neurons, neurons of the substantia gelatinosa of the spinal cord, etc.

Detailed investigations have been concerned with establishing the number of spines in several types of neuron or in different portions of the dendritic tree of a given type of neuron. However, as shown by Feldman and Peters (1979), these studies have usually only quantified those spines extending laterally beyond the edges of the dendritic shafts. Those spines obscured by the opacity of the impregnated dendritic shafts have been ignored because they could not be seen. Such spine counts are, therefore, serious underestimates of the total spine populations. Furthermore, even in the adult, a fraction of spines appear and disappear over time (e.g., Trachtenberg et al. 2002; Zuo et al. 2005). Therefore, the results reported below should be considered as indicative only. In Golgi preparations of the cat somatic sensory cortex, dendritic spines were counted in three different portions of the dendritic tree of pyramidal neurons (Mungai 1967). The number of spines per 100 μm^2 of dendritic surface area varied between 0 and 2 in the proximal portion, between 5 and 25 in the intermediate portion, and between 10 and 37 in the distal portion. In this type of neuron, the spine density, therefore, increases with increasing distance from the perikaryon. The total number of spines for a whole neuron was estimated to be between 1,500 and 4,000; these spines would contribute 25–43 % of the total surface area of the cell body and dendrites. Dendrites of the long-axon pyramidal cells of the rabbit visual cortex studied in Golgi preparations always appear spine-free in the initial segments arising directly from the perikaryon; beyond these segments, spine populations average 0.75 spines/μm on the shaft of the apical dendrite, 0.40 spines/μm on the oblique and terminal branches of the apical dendrite, and 0.30 spines/μm along the basal dendrites (Globus and Scheibel 1967c). The total number of spines was estimated to be between 154,000 and 175,000 in rat Purkinje cells (Napper and Harvey 1988) and about 36,000 in a Meynert cell (giant pyramidal neuron) of the monkey visual cortex (Chan-Palay et al. 1974). In this latter cell, the initial segments of all dendrites are spine-free, whereas the rather short basal dendrites, which ramify mainly into cortical layer VI, bear over 77 % of the spines. Segments rich in spines and segments poor in spines alternate along the length of the apical dendrite. Specifically, proceeding in a proximal-distal direction, a spine-free initial segment is encountered, followed by the segment traversing cortical layer V which bears about 10 % of the spines, then by a segment passing through cortical layers IV and III which accounts for less than 2.5 % of the spines, and finally by the bouquet of the terminal branches expanding into cortical layers II and I which has 8–13 % of the spines. It has been suggested that this differential distribution of spines is related to the differing afferent inputs that a Meynert cell receives in the various cortical layers (Chan-Palay et al. 1974). Later, spines have been counted following intracellular injection of a marker. In the CA1 region of the rat hippocampus, the total number of spines per pyramidal neuron was estimated at about 23,000 in cells injected with the fluorescent marker Lucifer yellow (Trommald et al. 1995) and about 30,000 in biocytin-injected cells (Bannister and Larkman 1995). These estimates are much higher than those from Golgi-impregnated neurons (for a comparison, see, e.g., Englisch et al. 1974). It is emphasized that the disappearance of preexisting spines and the formation of new spines entail synapse elimination and synapse formation, respectively, and consequently the remodeling of neural circuits (see Sect. VI.F).

A large variety of molecules is concentrated within spines (for a review, see Tashiro and Yuste 2003): β-actin (the actin concentration is much greater in spines than in dendritic shafts), actin-associated proteins (myosin IIB, myosin V, drebrin, α-actinin, fodrin, spectrin, gelsolin, adducin, cofilin, spinophilin, synaptopodin, and cortactin), the intermediate filament protein α-internexin, MAP2, glutamate receptors, neurotrophin receptors (p75 and TrkB), molecules related to Ca^{2+} homeostasis, cell adhesion molecules [N-cadherin, E-cadherin, cadherin-related neuronal receptor (CNR), integrins, neural cell adhesion molecule (NCAM), densin-180, neuroligin 1, and syndecan-2], etc. There is also evidence that other receptors and various ion channels are present in the plasma membrane of spines.

With regard to the structure, the spine has a cytoskeleton consisting mainly of actin microfilaments (Fifková and Delay 1982; Matus et al. 1982). It was long thought that spines were wholly devoid of microtubules. It was later shown that microtubules may enter spines transiently (Gu et al. 2008; Hu et al. 2008). It has been hypothesized that these microtubules play a role in the formation and plasticity of spines (Hu et al. 2008). Occasionally small vesicles and a few ribosomes are found in the cytoplasmic matrix of the spine. Neurofilaments do not enter the dendritic spines; a tubular profile of the agranular endoplasmic reticulum (Fig. V.17) very often extends from the dendrite into the neck and head of spines (Peters and Kaiserman-Abramof 1970; Špaček 1985a). Many spines contain an organelle, known as the spine apparatus, which in single sections appears to consist of a stack of two to four small cisternae alternating with plates of dense material (Gray 1959). Reconstructions from serial sections have shown that the cisternae of the spine apparatus are not separate entities but really diverticula of a single complex cisterna, which is often in continuity with the agranular endoplasmic reticulum of the dendritic trunk via a thin, tubular connection (Peters and Kaiserman-Abramof 1970; Špaček 1985a, b). Not all dendritic spines contain a spine apparatus. Synaptopodin has been found in close association with the spine apparatus (Deller et al. 2000); however, the functional relevance of this association is still unclear. Organelles bearing some resemblance to the spine apparatus (Fig. V.10) have also been observed within the perikaryon (see Sect. V.A.9), dendritic trunks (see Sect. V.C.1), and the initial segment of the axon (see Sect. V.D.1). Like other compartments of the agranular endoplasmic reticulum (see Sect. V.A.2), the spine apparatus may store Ca^{2+} and hence may contribute to the local regulation of Ca^{2+} signaling and homeostasis. Polysomes in the dendritic shaft, at the base of spines (Fig. V.17), were first observed in granule cells of the rat dentate gyrus (Steward and Levy 1982), and later in a wide range of neuronal types (Steward 1983). Proteins related to the synaptic site are probably synthesized on these polysomes (Steward and Schuman 2001). The location of these polysomes suggests that the synthesis of spine proteins may be regulated locally by electrical and chemical signals from the synapse on the spine (see later, this section). In single electron micrographs, spines can only rarely be seen in continuity with their parent dendrite; often they appear as isolated profiles (Figs. V.17 and X.1) which can be identified on the basis of their structural features: the presence of a cytoskeletal network and of profiles of the agranular endoplasmic reticulum, the absence of microtubules and neurofilaments, and the presence of a synapse (see later, this section). Identification of isolated profiles is further confirmed when a spine apparatus is present.

Just a few years after the discovery of spines, at a time when many authors considered them artifacts, Berkley (1895) hypothesized that their function was "to receive nerve-impulses from the endings of … nerve-fibers." The demonstration that spines are sites of synaptic contact came much later from an electron microscope study by Gray (1959). More precisely, spines are almost always the postsynaptic elements of axo-dendritic synapses (Fig. V.17). However, the dendritic spines (called gemmules) of granule cells in the mammalian olfactory bulb (Fig. IV.10) participate in reciprocal synapses (see Sect. VI.B.6) with mitral cell dendrites; hence, they may be presynaptic (Rall et al. 1966). Notwithstanding the considerable research effort in this area, the precise functions of dendritic spines are still not well understood. It does not appear likely that spines are simply devices to increase the surface area available for synaptic contact since the dendritic trunks bearing spines often have extensive areas between spines which are free of synapses. It is more likely that spines are synaptic sites allotted exclusively for afferent axons derived from particular sources, while axons in direct contact with the dendritic shafts are derived from different sources. An arrangement of this nature has in fact been observed in the Purkinje cells of the cerebellar cortex. Climbing fibers and parallel fibers form excitatory synapses on the spines of Purkinje cells (on the spines of large dendrites and on the spines of the distal dendritic branches, respectively), whereas the axons of basket cells form inhibitory synapses directly on the dendritic shafts of Purkinje cells (Llinás and Hillman 1969). It has also been suggested that the narrow spine neck in some way serves to isolate the synapse of the spine from the electrical influences of other synapses in the same neuron (Diamond et al. 1970; Rall 1978). More recently, it has been proposed that the spine-neck resistance is diffusional rather than electrical. According to this suggestion, the spine neck restricts the flow of signaling ions, Ca^{2+} in particular, and molecules between the spine head and the parent dendrite such that the spine head is a localized environment in which reactions specific to a particular synapse can occur (Holmes 1990; Koch and Zador 1993). In summary, dendritic spines seem to be relatively isolated synaptic sites assigned to particular afferents. Recent findings suggest that dendritic spines underpin some forms of learning and memory. For reviews on spine functions, see Shepherd (1996) and Bhatt et al. (2009).

Experimental studies have shown that dendritic spines are highly sensitive to changes in their surrounding environment. One particular change to which they are responsive is afferent axon degeneration. Following interruption of the visual projection in newborn rabbits, about 30 % of spines disappear in the deafferented region, i.e., along the central third of the apical dendrite of pyramidal cells in the visual cortex (Globus and Scheibel 1967a). Similarly, following section of the corpus callosum in newborn rabbits, spine number decreases by about a third along the oblique branches of the apical dendrite of pyramidal cells in the parietal cortex (Globus and Scheibel 1967b). Spine loss on Purkinje cell dendrites also occurs after sectioning of the parallel fibers in adult cats (Mouren-Mathieu and Colonnier 1969). In hippocampal neurons of adult rats, spine number decreases on distal dendritic portions of granule cells in the dentate gyrus following lesioning of the entorhinal cortex (Parnavelas et al. 1974), and spine number decreases in the rabbit stratum oriens (CA3) following commissural lesions (Frotscher 1983).

While the above experiments show that dendritic spines disappear after the degeneration of their presynaptic terminals and thus emphasize the importance of afferent axons in the maintenance of dendritic spines in newborn and adult animals, the result of the following study is at variance with this conclusion. An increase (instead of a decrease) in the number of spines emerging from the thick dendritic trunks of Purkinje cells was observed in adult rats following the degeneration of climbing fibers produced by 3-acetylpyridine (Sotelo et al. 1975a). The dendritic spines which increased in number were just those which in the normal animal are in synaptic contact with the climbing fibers. However, this increase could be explained as follows: after the degeneration of the climbing fibers, the postsynaptic space left free might become occupied by collaterals from neighboring axons (probably parallel fibers), which would then form synapses on the deafferented spines and might also stimulate the formation of new spines.

In newborn and adult animals, the formation and maintenance of dendritic spines seem to depend on the presence of afferent axons, whereas it appears that these have little to do with the formation of spines during embryonic development. In the cerebellar cortex of the "weaver" mutant mice, there are no parallel fibers; however, the spines on the distal dendritic branches of Purkinje cells, which in normal animals are the postsynaptic mates of parallel fibers, are present in large numbers (Hirano and Dembitzer 1973; Rakic and Sidman 1973a,b). These spines, none of which appear to be contacted by afferent axons, are either contiguous with each other or are enveloped by glial processes. These observations are compatible with the assumption that the spines on the distal dendritic branches of Purkinje cells may develop independently of the direct influences of afferent axons. Hormones also may affect dendritic spines. Gonadal steroid hormones influence spine density in particular hippocampal pyramidal cells both in vivo (Gould et al. 1990; Woolley and McEwen 1993; Leranth et al. 2004) and in vitro (Murphy and Segal 1996).

A number of experiments have revealed that spines are highly sensitive to synaptic activity. Rat hippocampal neurons cultured in vitro develop spines when innervated by axonal afferents via a synaptic activity-dependent mechanism; the number of developing spines is reduced when tetrodotoxin or a cocktail of three glutamate receptor blockers is applied (Kossel et al. 1997). The formation of dendritic spines in rat striatal neurons cultured in vitro depends on the activity of excitatory afferent axons (Segal et al. 2003). In mice raised in darkness from birth to 22–25 days, the portion of the apical dendrite of layer V pyramidal cells traversing layer IV of the visual cortex shows a significant reduction in spine density compared to control animals (Valverde 1967). In mice raised in darkness from birth until the 20th postnatal day but subsequently placed under normal lighting conditions for 1–30 days, some of the above-described apical dendrites reach a nearly normal spine density, while others maintain a lower than normal spine density (Valverde 1971). A decrease in spine density along the segment of the apical dendrite of pyramidal cells traversing layer IV of the visual cortex occurs in young rats after eyelid suturing (Fifková 1968) and after intraocular injection of tetrodotoxin, which prevents optic impulses from being generated in the retina (Riccio and Matthews 1985). A synaptic activity-dependent increase in spine density has been observed not only during development but also in the mature brain (hippocampal dentate gyrus of adult rats, O'Malley et al. 1998; primary olfactory cortex of adult rats, Knafo et al. 2001).

Abnormalities in dendritic spines (changes in number, shape, and length) have been reported in some forms of mental retardation (13–15 trisomy or Patau syndrome, Marin-Padilla 1972, 1974; 21 trisomy or Down syndrome, Marin-Padilla 1976, Suetsugu and Mehraein 1980, Ferrer and Gullotta 1990; fragile X syndrome, Rudelli et al. 1985, Irwin et al. 2001; mental retardation of unknown etiology, Purpura 1974). Abnormalities in dendritic spines have also been observed in epilepsy (Scheibel et al. 1974; Multani et al. 1994; Jiang et al. 1998). However, it is unclear whether dendritic spine abnormalities are causes or occurrences accompanying neurological disorders (for a review, see Fiala et al. 2002).

3. Plasticity of Dendrites

As shown in Sect. V.C.2, dendritic spines exhibit a high degree of plasticity. It is possible, however, that this is just one aspect of the plasticity of the entire dendritic tree. Repeated direct observations of the same dendritic tree over periods of weeks or months in living animals have revealed that the dendritic tree undergoes substantial remodeling during adult life; some dendritic branches are lost, whereas others lengthen, and still others arise de novo (Purves and Voyvodic 1987). The capacity for remodeling of the dendritic tree seems to vary according to the neuron type. The remodeling tends to mainly involve the high-order dendritic branches and not the primary shafts. Note that the remodeling of the dendritic tree entails the remodeling of neural circuits.

Furthermore, the extent and pattern of the dendritic tree may change in response to changing environmental conditions (see also Chap. III). For example, the dendrites of stellate and pyramidal neurons in the visual cortex are less developed in animals housed in an isolated environment and more developed in animals housed in a complex environment compared with those of animals kept under standard conditions (Holloway 1966; Coleman and Riesen 1968; Uylings et al. 1978; Green et al. 1983). In rats trained on certain tasks, total dendritic length, number of oblique branches, and length of terminal dendritic segments are all increased in the apical tree of layer V pyramidal cells of the sensory motor cortex (Greenough et al. 1985).

The dendritic tree of certain neurons undergoes substantial changes under the influence of gonadal steroid hormones both in vitro and in vivo, as the following observations illustrate. Mouse preoptic area neurons cultured in vitro are supplied with significantly more dendrites in estradiol-treated cultures than in untreated cultures (Toran-Allerand et al. 1983). Treatment of adult male rats with testosterone significantly reduces the extent of the neuronal dendritic tree in the hypothalamic arcuate nucleus (Danzer et al. 2001). Hormonally induced seasonal changes occur in the extent of the dendritic tree of some vocal control neurons in certain birds (for a review, see Arnold 1985). The dendritic extent of certain motoneurons in adult mammals changes seasonally with cyclical variations in the levels of circulating gonadal steroid hormones (Forger and Breedlove 1987). The dendritic trees of oxytocin and vasopressin neurons in the rat supraoptic nucleus undergo significant changes during lactation (Stern and Armstrong 1998). For other examples of the effects of gonadal steroid hormones on the morphology of dendrites in adult mammals, see Cooke and Woolley (2005).

4. Age-Related Changes

With increasing age, the dendritic tree undergoes a progressive regression: dendritic shafts decrease in number, and those which remain become shorter and less branched. These changes, which vary according to species, region of the nervous system, and type of neuron, have been quantitatively assessed, for example, in neurons of the rat superior cervical ganglion (Andrews et al. 1994), in preganglionic neurons of the rat spinal cord (Dering et al. 1996), in rat hypothalamic neurons (Leal et al. 1998), in rat hippocampal neurons (Geinisman et al. 1978a), in neurons of the mouse striatum (Rafols et al. 1989), in pyramidal neurons of the cerebral cortex (human, Scheibel et al. 1975, 1976, 1977; Nakamura et al. 1985; Jacobs et al. 1997; de Brabander et al. 1998; monkey, Uemura 1985b; Peters et al. 1998b; rat, Vaughan 1977; Wong et al. 2000; Grill and Riddle 2002; Wang et al. 2009; mouse, Leuba 1983), and in mitral cells of the olfactory bulb (rat, Hinds and McNelly 1977). These changes are frequently accompanied by a reduction in spine number, as has been observed in the cerebral cortex (rat, Feldman and Dowd 1975, Wong et al. 2000, Wang et al. 2009; mouse, Leuba 1983; human, Scheibel 1978, Jacobs et al. 1997; monkey, Uemura 1985a, Peters et al. 1998b, Duan et al. 2003). In the monkey cerebral cortex, the reduction in spine number ranges from 25 to 50 % according to the area considered (Peters et al. 1998b; Duan et al. 2003; Kabaso et al. 2009; Dumitriu et al. 2010). It seems that not all spines are affected to the same extent during aging. In some areas of the monkey cerebral cortex, it is mainly the thin spines that are lost (Dumitriu et al. 2010). These spines are highly motile and plastic and are thought to be linked to learning. Since dendrites and spines undergo continuous remodeling (specifically growth and regression) throughout life (see Sects. V.C.2 and V.C.3), the reduction in number of dendrites and spines often observed in old animals can be attributed to greater loss than new formation of these structures. In fact dendrite growth has been detected even in old age (e.g., Buell and Coleman 1981; Coleman and Flood 1986). Finally, the decreased expression of a number of receptors that has been reported in old age (e.g., Gazzaley et al. 1996; Hof et al. 2002) is probably related to the regression of dendrites and spines.

D. The Axon

Some features of the axon have been described in Chap. III.

1. The Axon Hillock and Axon Initial Segment

Under the light microscope, the axon hillock appears as a conical region of the perikaryon which, particularly in large neurons, fails to stain with basic dyes. In electron micrographs, free polysomes, mitochondria, microtubules, neurofilaments, and a few cisternae of the granular endoplasmic reticulum can be observed. From the axon hillock, microtubules and neurofilaments funnel into the initial segment without interruption.

As described by many authors (Kohno 1964; Palay 1964; Conradi 1966, 1969b; Palay et al. 1968; Peters et al. 1968; Jones and Powell 1969; Sotelo 1969; Westrum 1970; Mugnaini 1972, and others), the following structural features characterize the initial segment both in myelinated and unmyelinated axons: (a) free polysomes, (b) fascicles of microtubules, and (c) a 15–25-nm-thick layer of dense material undercoating the axolemma. In this axonal region, the polysomes are not associated with the endoplasmic reticulum, and their number varies greatly from one axon to another. The fasciculation of microtubules begins in the narrowest portion of the axon hillock; in the initial segment, the fascicles run parallel to the long axis of the axon. In transverse sections, it can be seen that within individual fascicles the microtubules are interconnected by thin, crossbridges. The number of microtubule fascicles is roughly proportional to the transverse diameter of the initial segment, so that large initial segments may contain several such fascicles, while small segments may have only one. The layer of dense material undercoating the axolemma (see Sect. V.E) mainly consists of actin microfilaments.

A variety of proteins is concentrated under the axolemma at the level of the axon initial segment. These proteins include the cell adhesion molecules (CAMs) neurofascin-186 (NF-186) and neuronal cell adhesion molecule (NrCAM) (Davis et al. 1996), the cytoskeletal proteins ankyrin$_G$ (Kordeli et al. 1995) and βIV-spectrin (Berghs et al. 2000), and an amphiphysin-II-like protein (Butler et al. 1997). Densely clustered Na^+ channels are a prominent feature of the axon initial segment; their opening triggers the generation of action potentials. Even K^+ channels are present at high density in the axon initial segment. It has been suggested that ankyrin$_G$ and βIV-spectrin play major roles in anchoring Na^+ and K^+ channels to the subaxolemmal network of actin microfilaments (Berghs et al. 2000; Lacas-Gervais et al. 2004).

An axolemmal undercoating and densely clustered ion channels are also present at the nodes of Ranvier (see Sects. V.E and VII.B.4g). The Na^+ channels present here are responsible for the propagation of action potentials. In addition to densely clustered ion channels and the axolemmal undercoating, axon initial segments and nodes of Ranvier are also characterized by the presence of brevican, an extracellular chondroitin sulfate proteoglycan (see also Sect. VII.B.4g). An organelle similar to the spine apparatus (see Sect. V.C.2) has occasionally been observed in the initial segment of the axon (Peters et al. 1968; Jones and Powell 1969; Westrum 1970). In some neurons lacking an axon (see Chap. IV), action potentials are generated in a dendritic process in which Na^+ channels are densely clustered (Wu et al. 2011). Where an axon ending forms a synapse with the initial segment of the axon (see Sect. VI.B.5), the dense undercoating is absent beneath the axolemma in contact with the presynaptic terminal; at this site, the usual postsynaptic density can be observed (see Sect. VI.B.3).

The distal end of the axon initial segment is easily recognized in myelinated fibers since it corresponds to the site where the axon acquires its myelin sheath; at about this level, both the fasciculation of microtubules and the dense material undercoating the plasma membrane terminate. Analogously, for unmyelinated fibers the distal end of the initial segment of the axon may be identified as corresponding to the site where both the fasciculation of microtubules and the plasma membrane undercoating come to an end.

The length of the axon initial segment is highly variable, being 23–38 μm in cat spinal cord motoneurons (Conradi 1969b), 23–54 μm in monkey cerebral cortex neurons (Sloper and Powell 1973), about 40 μm in rat cortical pyramidal neurons (Palmer and Stuart 2006), about 17 μm in cat and rat cerebellar cortex Purkinje cells (Somogyi and Hámori 1976), and about 28 μm in a single neuron of the cat dorsal spinocerebellar tract (Saito 1972). Although in the monkey cerebral cortex a positive correlation has been found between the mean diameter of a nerve cell body and the transverse diameter of its initial segment, no correlation has been detected between the mean diameter of the nerve cell body and the length of its initial segment (Sloper and Powell 1973).

In cases where an axon originates from a dendrite, the fasciculation of the microtubules which pass into the initial segment of the axon begins a short distance prior to its origin; however, this fasciculation is evident only on the dendritic side from which the axon arises, while on the opposite side the microtubules remain separate from one another (Peters et al. 1968). Spines similar in shape, size, and structure to dendritic spines (see Sect. V.C.2) may occur on the axon initial segment (Westrum 1970); synaptic contacts have been seen on these axonal spines. Some neurons in the goldfish olfactory bulb have an axon initial segment which exhibits unusual features (Kosaka 1980). Both the undercoating of the axolemma and fascicles of microtubules can be seen only in the portion of the initial segment close to the perikaryon; the undercoating of the axolemma (but no microtubule fascicles) is present in the successive portion, while neither of these features can be observed in the portion of the initial segment most distant from the perikaryon. In the long initial, unmyelinated portion of the single process of the large neurons in the rat spinal ganglia, the structural features which characterize the axon initial segment cannot be detected close to the axon hillock but are visible just proximally to the onset of the myelin sheath (Zenker and Högl 1976). Patches of a dense undercoating of the plasma membrane are sometimes observed only in some regions of the initial portion of the single process in frog spinal ganglion neurons (Matsumoto and Rosenbluth 1985).

2. The Axon Beyond the Initial Segment

Mainly mitochondria, agranular endoplasmic reticulum, microtubules, and neurofilaments can be observed within the axon beyond the initial segment (Figs. V.19, VII.9, VII.10, and VII.14), whereas granular endoplasmic reticulum and ribosomes are very rarely detected (Jones and Powell 1969; Zelená 1970; for more details, see Pannese and Ledda 1991); concerning RNA in axons, see the last paragraph of this section. Multivesicular bodies in axons are relatively rare compared with perikarya and dendrites.

Axonal mitochondria are usually elongated, seldom branched organelles which are longitudinally oriented; their cristae may be oriented lengthwise. Slender mitochondria, up to 20 μm in length, can be observed within the preterminal segments of axons as well as within terminal dendritic branchlets (see Sect. V.C.1). Within presynaptic endings (see Sect. VI.B.3), however, the mitochondria are small and spheroidal. For the axonal transport of mitochondria, see Sect. V.D.4.

The agranular endoplasmic reticulum (Fig. VII.14) consists of membrane-bounded tubules and cisternae, which are generally arranged parallel to the long axis of the axon. Results obtained studying thick sections under the high-voltage electron microscope (Droz et al. 1975; Tsukita and Ishikawa 1976) and serial thin sections under the conventional transmission electron microscope (LaVail et al. 1980) show that tubules and cisternae of the agranular reticulum form a continuous system throughout the length of the axon. Agranular endoplasmic reticulum cisternae are occasionally observed arranged perpendicular to the long axis of the axon, thus forming partial septa lying at rather regular intervals across it. These cisternae are interconnected by longitudinally oriented tubules of the agranular reticulum and give the axon a cross-striated appearance when viewed in longitudinal section. These septa are incomplete and do not, therefore, obstruct the axonal transport (see Sect. V.D.4). This arrangement has been observed in lobster giant axons (de Lorenzo et al. 1968), in the crayfish ventral nerve cord (Peracchia 1970), in fish olfactory nerves (Holz and Weber 1970), in the olfactory bulb of mouse embryos (Hinds 1972), in the fish hypothalamus (Follenius 1970), and in the frog brain (Lieberman 1971). Within

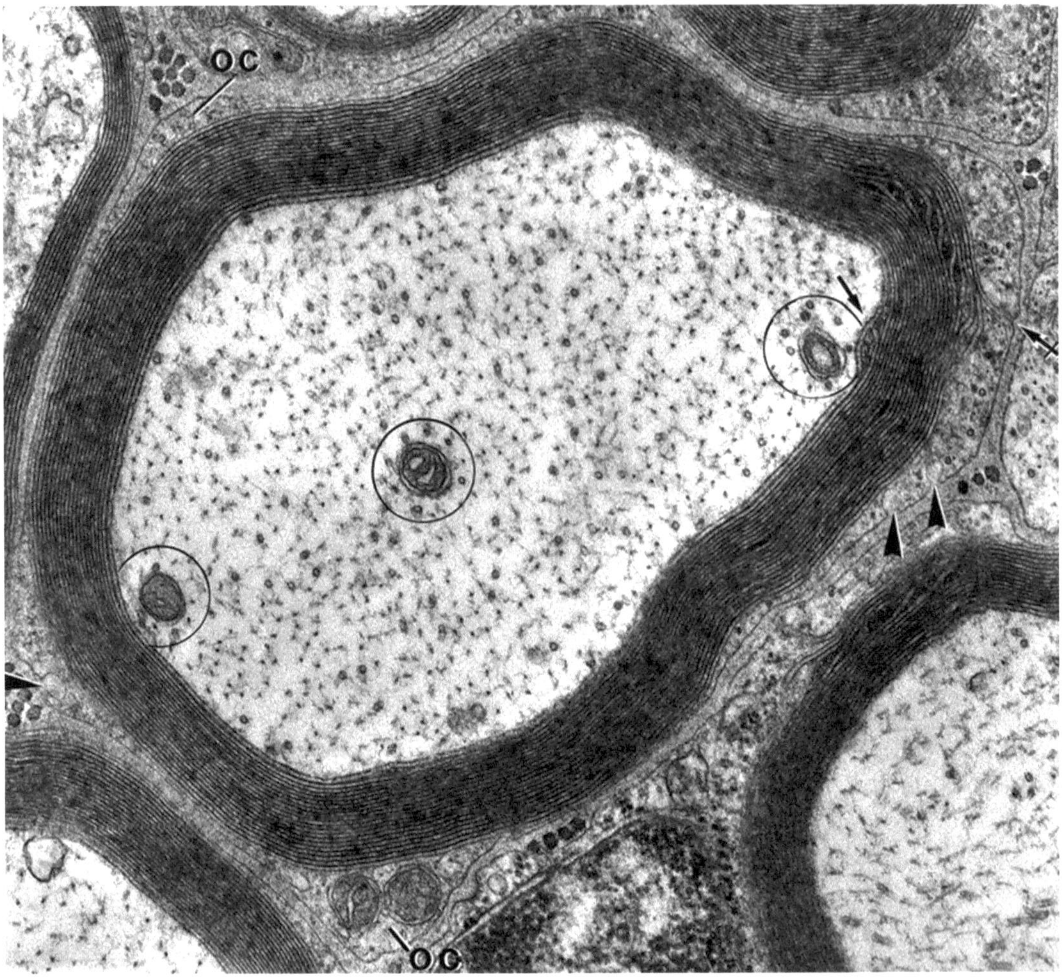

Fig. V.19 Relationships between microtubules and mitochondria in the axon. In this transverse section of a myelinated axon, three mitochondria (*encircled*) can be seen; each of these is surrounded by several microtubules. The inner compartment of the Schwann cell cytoplasm is absent; the outer compartment (**oc**) is confined to narrow bands. *Arrow* points to the inner mesaxon, *crossed arrow* points to the outer mesaxon. *Arrowheads* point to caveolae along the plasma membrane of the Schwann cell. Lizard dorsal root, ×51,000

axons of the CNS, assemblies of parallel, closely packed tubules of the agranular endoplasmic reticulum have been observed; they have been interpreted as the result of axonal degenerative processes which take place in apparently normal adult animals (rat cerebellum, Andres 1965b; cat cerebellum, Morales and Duncan 1966, Hámori and Szentágothai 1968; hamster cerebellum, Takahashi 1969; lateral vestibular nucleus of the rat, Sotelo and Palay 1971).

Microtubules (see also Sect. V.A.7) are ever present components of the axonal cytoskeleton; axons without microtubules have never been described. In electron micrographs of axon longitudinal sections, microtubules are seen to extend along considerable distances. In the past, it was suggested that each microtubule was a continuous structure extending from the perikaryon to an axon ending (Weiss and Mayr 1971). However, when the number of microtubules in a given unbranched axon was determined on cross sections cut at different levels, it was found that the microtubule number varied from one level to the next (crayfish, Nadelhaft

1974; mouse, Tsukita and Ishikawa 1981). These findings raised serious doubts about the validity of this supposition. By tracing individual microtubules through consecutive serial sections of transversely cut axons, it was shown that they do not extend the entire length of the axon. For example, the microtubule length averages 100 μm in the axons of spinal ganglia of fetal rats (Bray and Bunge 1981) and varies from 300 to 800 μm in myelinated axons of the mouse saphenous nerve (Tsukita and Ishikawa 1981). The length of microtubules in frog olfactory axons was estimated by Burton (1987) to be about 400 μm in the proximal portion (i.e., that close to the olfactory epithelium) and about 100 μm in the distal portion (i.e., that close to the olfactory lobe of the brain). Microtubules in the axons of a nematode varied in length from 1.2 to 27 μm; their length never exceeded 5 % of the total axonal length (Chalfie and Thomson 1979).

Usually, microtubules appear unevenly distributed throughout the cross section of the axon. Often they are more concentrated around the mitochondria (Raine et al. 1971; Smith 1973; Pannese et al. 1981, 1986; Fig. V.19) and just beneath the axolemma (Endo et al. 1979). Mitochondria appear connected to microtubules by fine crossbridges (Hirokawa 1982). In a few cases only, the distribution of microtubules has been found to be uniform throughout the axonal cross section (Nadelhaft 1974, in the crayfish abdominal nerve cord).

The average microtubule density (average number of microtubules per unit of axoplasmic area) appears to be related to the axonal size; more precisely, it is inversely proportional to the cross-sectional area of the axon (Friede and Samorajski 1970; Smith 1973; Zenker et al. 1973, 1975; Ohnishi et al. 1976; Brimijoin et al. 1979; Pannese et al. 1981, 1984b, c). Microtubules are therefore more closely packed within small than within large axons. In the smallest axons, neurofilaments are rarely found, and microtubules represent the major cytoskeletal component (Fig. V.20). Since the same condition pertains in the smallest dendrites, it may be difficult to distinguish between the two types of process in sections of the neuropil. The following points may serve to distinguish between the two: (a) usually, small axons are regular in outline, while small dendrites are rather irregular; (b) in the neuropil, small axons are often arranged in bundles, while small dendrites appear as individual entities; (c) small dendrites may bear spines and usually contain ribosomes (Peters et al. 1991). The presence of presynaptic or postsynaptic specializations is not a reliable criterion for the identification of a process since, although axodendritic synapses are the most common type, axoaxonal and dendrodendritic synapses have also been described (see Sect. VI.B.5). In myelinated axons, the microtubule density is much higher at the nodes of Ranvier than in the internodal segments as a result of the fact that at the nodes microtubules are not reduced in number, while the axon diameter is generally lower there than in the internodal segments (Landon and Hall 1976).

It is often held that the microtubule density is higher in unmyelinated than in myelinated axons, but it has not been found to be significantly different between unmyelinated and myelinated axons of similar cross-sectional area (Pannese et al. 1984a). This seems to indicate that the higher microtubular density often observed in unmyelinated axons with respect to myelinated ones correlates with axon size rather than with the absence of the myelin sheath; as a matter of fact, in vertebrates unmyelinated axons are often smaller than myelinated ones. The findings of Samorajski et al. (1970) in the "quaking" mouse point to the same conclusion; in this mutant, many axons are found which have much thinner myelin sheaths but similar microtubule densities with respect to normal axons of comparable size. The findings of Nadelhaft (1974) showing that in crayfish the microtubule density is higher in the axons of motoneurons than in those of interneurons might lead to the conclusion that a correlation exists between microtubule density and the functional role of the axon. However, no significant difference in the microtubule density

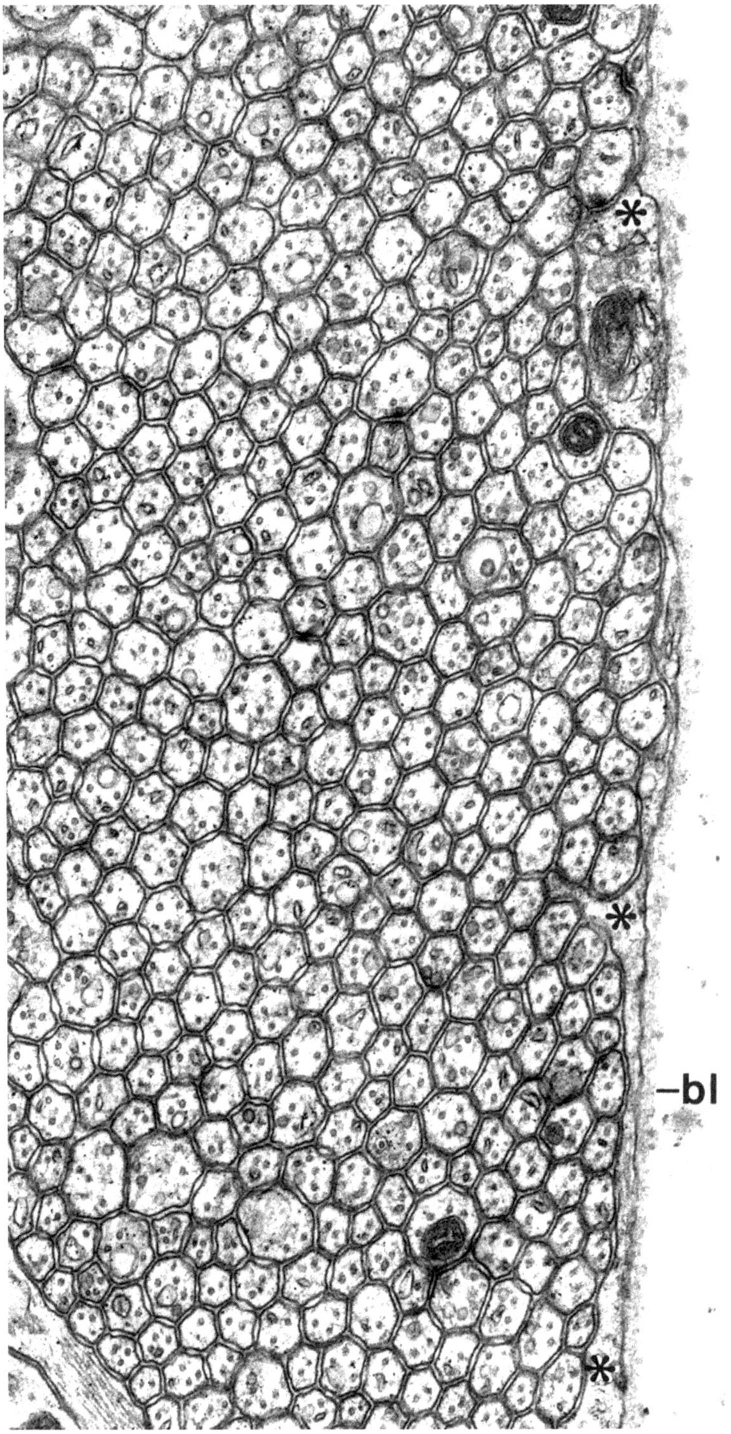

Fig. V.20 Small unmyelinated axons of the CNS. The granule cell axons, which appear transversely sectioned in this picture, all run parallel to each other and are packed close together without intervening neuroglia. Each axonal profile contains two to four microtubules, one or two tubules of the agranular endoplasmic reticulum, and occasionally a neurofilament. A mitochondrion can be seen in three axonal profiles. On the *right* is a particularly thin neuroglial limiting membrane (*), whose surface facing the pia mater is covered by a basal lamina (**bl**). Molecular layer of the rat cerebellar cortex, ×36,000

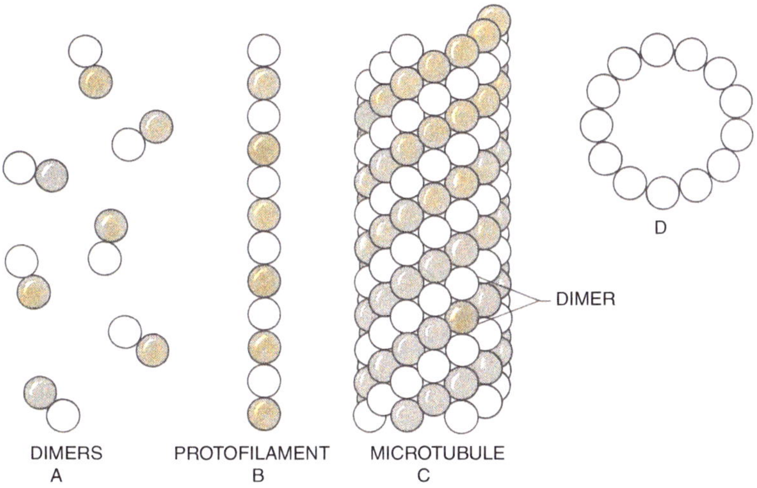

Fig. V.21 The molecular organization of microtubules. (**A**) Microtubules are formed from tubulin dimers, each of which consists of two globular subunits. These subunits are slightly different polypeptides, called α-tubulin and β-tubulin. (**B**) In the microtubule, dimers are aligned into rows called protofilaments, each protofilament being made of alternating α and β subunits. (**C**) In side view of a short segment of microtubule, protofilaments are packed together and helically arranged. (**D**) In cross section, the wall of the microtubule is made up of 13 protofilaments arranged side by side

has been detected between motor and sensory axons of comparable size in two species of vertebrates (South African toad, Smith 1973; lizard, Pannese et al. 1984c).

Microtubules from a wide variety of sources display the same structural features. When examined in cross section at high resolution, the wall of the microtubule appears to be made up of 13 globular subunits (Fig. V.21), each 4–5 nm in diameter (Ledbetter and Porter 1964; Ringo 1967; Tilney et al. 1973; Warner and Satir 1973). When microtubules are examined longitudinally in negatively stained preparations, their wall appears to consist of 13 long, helically arranged protofilaments, each 4–5 nm in diameter. Each protofilament seems to be built as a string of globular subunits (Fig. V.21). Neuronal microtubules share this kind of organization (Erickson 1974; Cohen et al. 1975). Microtubules have polarity; the end where subunit addition occurs (see later, this section) is designated plus, the other minus. Within the axon, microtubules are uniformly oriented (Fig. V.22) with their plus ends directed away from the cell body (Burton and Paige 1981; Filliatreau and Di Giamberardino 1981; Heidemann et al. 1981; Viancour and Forman 1987; Baas et al. 1988). By contrast, in dendrites (Fig. V.22), microtubules are oriented in both directions (Baas et al. 1988; Burton 1988).

Microtubules are composed of an acidic protein, called tubulin, which has a molecular weight of 110–120 kD and a 6S sedimentation coefficient. The native 6S protein has proved to be a dimer (Fig. V.21) composed of two globular polypeptides, designated α- and β-tubulin, which are slightly different in both molecular weight (between 55 and 60 kD) and amino acid composition. Multiple isoforms of tubulin have been demonstrated to occur in a single neuron. Tubulin tightly binds a guanine nucleotide and interacts specifically with the plant alkaloids colchicine and vinblastine, which are known to arrest mitosis at metaphase. Neurons contain proteins that promote microtubule assembly. These include microtubule-associated proteins (MAPs) such as tau, MAP1, MAP2, and others. These proteins form the very slender lateral projections which are usually attached to microtubules (Dentler et al. 1975). The microtubule composition varies in the different compartments of a single neuron (see later in this section, where the

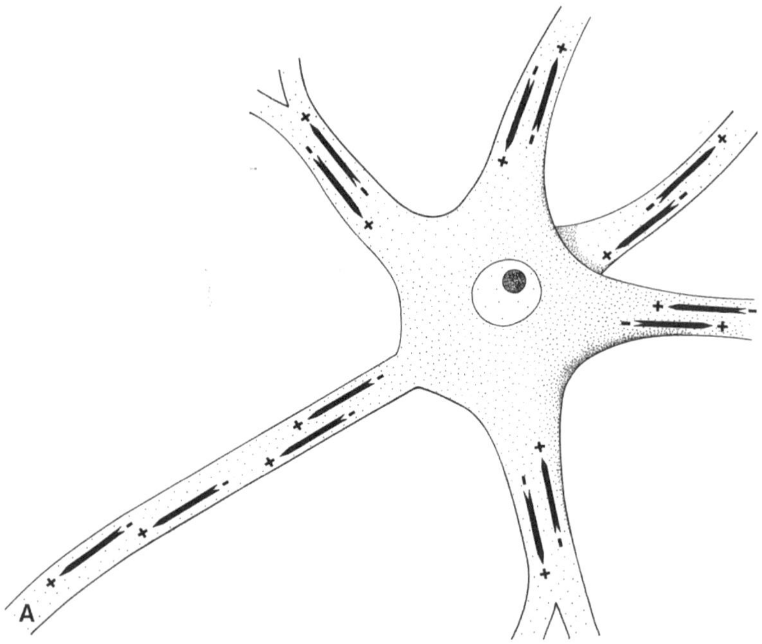

Fig. V.22 Polarity orientation of microtubules in the axon and dendrites. In the axon (*A*) microtubules are uniformly oriented with their minus ends directed towards the cell body, whereas in dendrites microtubules are oriented in both directions

regional heterogeneity of cytoskeletal proteins is discussed).

A number of findings have suggested that (a) a pool of tubulin subunits is usually present in the cytoplasmic matrix of the neuron and other cell types and (b) that these subunits exist in a dynamic equilibrium with the assembled microtubules (dynamic equilibrium theory). This theory explains the rapid reassembly of microtubules in the presence of protein synthesis inhibitors (e.g., puromycin and cycloheximide). The equilibrium between tubulin subunits and assembled microtubules may be upset by numerous factors (e.g., temperature, hydrostatic pressure, metal ions, and plant alkaloids). Microtubules can disassemble and reassemble rapidly as these variables change and hence are highly sensitive to environmental changes. In other words, they are rather labile structures. The preservation of microtubules in electron microscope preparations largely depends on the characteristics (chemical nature, ionic composition, pH, and temperature) of the fixative. There is evidence that not only tubulin subunits (Hemminki 1973) but also microtubule-associated proteins (Safaei and Fischer 1990) are subjected to turnover. Not all microtubules respond in the same way to environmental changes; a wide spectrum of susceptibility is seen, extending from the most stable microtubules of the outer doublets of cilia and flagella to the most labile microtubules of the mitotic spindle. Neuronal microtubules also show a quite variable susceptibility to changes in environmental conditions; for example, the microtubules of certain neurons disassemble in the cold, whereas those of other neurons are well preserved at low temperatures (for further details, see Pannese et al. 1982). It is not yet known whether variations in microtubule stability are due to tubulin-inherent chemical differences, to differences in MAPs or the intraneuronal environment, or to other factors. Finally, microtubules are the most rigid cytoskeletal components.

Functions attributed to neuronal microtubules include providing skeletal support for axons and dendrites, and involvement in axonal transport. The importance of microtubules in the maintenance of anisometric cell processes is shown by the following experiment on heliozoans, protozoans with long axopodia containing a well-organized microtubule bundle. The disruption of microtubules by cold or by hydrostatic pressure results in a retraction of the axopodia, which do not reform until the microtubules are reassembled (Tilney et al. 1966; Tilney and Porter 1967). It has been suggested that within the nervous system microtubules provide skeletal support for the axon during the early stages of its development. In this connection, it may be recalled that many microtubules but very few neurofilaments occur within young axons (Peters and Vaughn 1967) and that in immature nerve cells cultured in vitro colchicine disrupts the microtubules and causes the cessation of axon elongation and subsequently its retraction (Yamada et al. 1970). It is not known, however, whether microtubules are necessary for the maintenance of axon shape in the nervous system of the adult, where other structures (e.g., connective fibers in the PNS) or other conditions (e.g., tight packing of CNS cells) may provide sufficient support. The role of microtubules in axonal transport is considered in Sect. V.D.4.

Filaments within the axon were observed under the electron microscope by Schmitt (1950) and Schmitt and Geren (1950); the structures described by these authors most likely correspond to those elements later termed neurofilaments (see also Sect. V.A.7). Neurofilaments, a class of distinct intermediate filaments (see Table V.2) unique to neurons, are the major intermediate filaments in mature nerve cells. In axonal cross sections, neurofilaments usually occur scattered individually and, unlike microtubules, often appear uniformly dispersed throughout the axoplasm (Fig. VII.10). At the node, the number of neurofilaments is markedly lower than that in the internodal segment (Tsukita and Ishikawa 1981); given that at the node the axon diameter is generally lower than in the internodal segments, the average number

Table V.2 Intermediate filament proteins

Class	Common name	Distribution
I	Acidic keratins	Epithelial cells
II	Basic keratins	Epithelial cells
III	Vimentin	Cells of mesenchymal origin, precursor cells of both neuronal and glial lineages, certain mature neurons, certain astrocytes, radial glia, ependyma, endothelia
	Desmin	Muscle cells, endothelia
	Glial fibrillary acidic protein (GFAP)	Astrocytes, radial glia
	Peripherin	Many immature neurons, a subset of mature neurons of the PNS, mature neurons of the CNS with axons projecting outside the CNS
	Plasticin	A subset of neurons during axonal outgrowth
IV	Light, middle, and heavy subunits of neurofilaments	Mature neurons
	Nestin	Stem cells of the CNS, developing skeletal muscle cells
	α-Internexin	A subset of neurons, particularly during development
V	Lamins	Eukaryotes
VI	Filensin	Developing lens epithelial cells

Note: Intermediate filaments derive their name from their diameter (about 10 nm), which is intermediate between that of microfilaments (about 6 nm) and that of microtubules (20–26 nm). They are all built according to a common plan. They possess three domains: a central α-helical rod of ~310 (or ~350 for the lamins) amino acid residues, an amino-terminal head, and a carboxyl-terminal tail. The central α-helical rod domain is rather well conserved through the classes of intermediate filaments, while the sizes and sequences of the amino- and carboxyl-terminal domains are highly variable. The characteristics of these terminal domains determine the specific properties of each intermediate filament (for reviews, see Steinert and Roop 1988; Fuchs and Weber 1994; Lee and Cleveland 1996)

of neurofilaments per unit of axoplasmic area, unlike the microtubule density, is about the same at the nodes of Ranvier and in the internodal segments. In the mature neuron, large axons contain great numbers of neurofilaments and also many microtubules, whereas in the smallest axons neurofilaments are rarely found, and microtubules are the major component (Fig. V.20). Axons without microtubules are unknown, while neurofilaments have never been observed by electron microscopy in the axons of arthropods (Fernandez et al. 1971; Phillips et al. 1983; Lasek et al. 1985), although biochemical and immunological techniques have revealed the presence of neurofilament-like proteins in an arthropod species (Halls et al. 1987).

Mammalian neurofilaments are heteropolymers. According to most authors, they consist of three subunit proteins (usually called triplet proteins) whose molecular weights are about 70, 150, and 200 kD (Hoffman and Lasek 1975; Liem et al. 1978; Schlaepfer and Freeman 1978). However, according to other authors, peripherin and α-internexin are a fourth component of neurofilaments in the PNS and CNS, respectively (see later, this section). The three "classical" subunits are referred to as light, middle, and heavy, respectively. Each of these subunits is encoded by a separate gene (Lewis and Cowan 1985) and is synthesized on free neuronal polysomes (Czosnek et al. 1980). The smaller subunits therefore do not arise from degradation of a larger precursor. Three domains (a central α-helical rod, an amino-terminal head, and a carboxyl-terminal tail) are present in each subunit. The heavy and middle subunits have carboxyl-terminal tail domains longer than the light subunit. The rod domains of all three subunits form the core of the 10-nm neurofilament and the carboxyl-terminal tails of the heavy and middle subunits form sidearm projections that extend from the filament core (Hirokawa et al. 1984; Hisanaga and Hirokawa 1988), thus forming crossbridge structures. The sidearm projections are highly phosphorylated (Julien and Mushynski 1983; Carden et al. 1985), and it has been suggested that phosphorylation influences the ability of neurofilaments to interact with one another and with other cytoskeletal components (e.g., microtubules). Most phosphate groups are probably added to the tail domains of the heavy and middle subunits only after neurofilaments have entered the axon (Sternberger and Sternberger 1983). Finally, while neurofilaments were once considered to be mainly static structures, there is now evidence that they are dynamic, undergoing subunit exchange with a pool of free subunits (Angelides et al. 1989; Okabe et al. 1993; Takeda et al. 1994). Some data suggest that in axons, myosin Va associates with neurofilaments and binds selectively to the light subunit (Rao et al. 2002).

Neurofilaments are a major determinant of axon diameter (see Chap. III), as suggested by the finding that axons without neurofilaments are much smaller than those with a normal complement of neurofilaments (Yamasaki et al. 1991; Ohara et al. 1993). Phosphorylation of the sidearm projections is thought to regulate the spacing between adjacent neurofilaments, thereby contributing to the modulation of axon diameter (de Waegh et al. 1992; Nixon et al. 1994). For the axonal transport of neurofilaments, see Sect. V.D.4. As noted previously, in early developmental stages, many microtubules and very few neurofilaments occur within axons (Peters and Vaughn 1967). During axonal maturation, neurofilaments increase progressively in number until they greatly outnumber microtubules within the mature axon.

Initially, neurofilament triplet proteins were considered to be the only intermediate filament proteins specific to neurons. Successively, the neuronal intermediate filament proteins peripherin and α-internexin were discovered. These are present exclusively in cells of the neuronal lineage. Peripherin (Portier et al. 1984; Parysek and Goldman 1988) appears in many cells of the neuronal lineage after their final division. Subsequently it disappears in most of these cells but persists throughout life (Troy et al. 1990) in well-defined sets of neurons (e.g., many neurons of the PNS and neurons of the CNS with axons projecting outside the CNS). α-Internexin (Pachter and Liem 1985) is also expressed early in development and persists in certain neurons of the adult CNS. Some authors (e.g., Chien et al. 1996) maintain that peripherin forms an independent filament system in the mature PNS, as α-internexin does in the mature CNS. By contrast, other authors (Yuan et al. 2006, 2012) maintain that peripherin and α-internexin form a fourth neurofilament subunit (with neurofilament triplet proteins) in the mature PNS and mature CNS, respectively (see above, this section).

Proliferating neuronal precursor cells may transiently express intermediate filament proteins, which are not specific to neurons (Tapscott et al. 1981). These proteins are vimentin (found primarily in cells of mesenchymal origin) and nestin (present in muscle cells). When neuronal precursor cells differentiate, vimentin (Tapscott et al. 1981; Cochard and Paulin 1984) and nestin (Lendahl et al. 1990) disappear and are gradually replaced by the neurofilament triplet proteins. For a brief period during embryonic development, vimentin and neurofilament proteins may coexist in some young neurons (Tapscott et al. 1981; Bignami et al. 1982). For information on the sequential expression of various intermediate filament proteins in cells of the neuronal lineage, see the review by Nixon and Shea (1992). There is evidence that vimentin is present in certain neurons of adult animals, such as the olfactory neurons of the rat (Schwob et al. 1986) and the axonless horizontal cells of the mouse retina (Dräger 1983). These neurons are also unique in other respects: olfactory nerve cells are the only neurons in adult mammals to be continuously replaced through the proliferation of precursor cells; the axonless horizontal cells of the retina have such unusual features that they were at one time suggested to be intermediate between nerve and neuroglial cells. Finally, another intermediate filament protein has been detected in certain neurons during the early stages of axonal outgrowth. This protein has been called plasticin because it is thought to be concerned with axonal growth, plasticity, and regeneration (Glasgow et al. 1992; Canger et al. 1998).

Actin microfilaments have been identified within neurons (LeBeux and Willemot 1975; Metuzals and Tasaki 1978; Kuczmarski and Rosenbaum 1979; Baumgold et al. 1981). A dense network of actin microfilaments is present just beneath the axolemma (Metuzals and Tasaki 1978; Baumgold et al. 1981). Actin has also been reported within the peripheral region of the perikaryon (Pannese et al. 1996a), in dendritic spines (see Sect. V.C.2), and within the dense postsynaptic material (see Sect. VI.B.3). Of the three main forms of actin (alpha, beta, and gamma), it is the beta form that is mainly expressed in mammalian neurons (Letourneau 2009). Various myosins are also expressed in vertebrate neurons (for a review, see Bridgman 2009). Functions of actin and myosin in neurons include shape changes of dendritic spines (see Sect. V.C.2), short-range transport of organelles (see Sect. V.D.4), endocytosis (see Sect. V.E), and docking of integral membrane proteins (see Sects. V.D.1, V.E, and VII.B.4g). Microtubules, neurofilaments, and actin microfilaments form a cytoskeletal network within the axon. The predominant view is that this network is largely stationary, and its components undergo very slow turnover and replacement by precursors moving by slow transport (Nixon 1998; Yuan et al. 2009).

A number of studies have led to the concept that there is a regional heterogeneity of the cytoskeleton within the neuron (Cumming and Burgoyne 1983; Matus et al. 1983). The composition of microtubules varies between the axon and dendrites in that the relative amounts of tubulin isoforms and the pattern of MAPs are different. The MAP2 is abundant in the perikaryon and dendrites but is either absent or present at very low levels in the axon of mature neurons. The distribution of tau protein is the inverse of that of MAP2 (De Camilli et al. 1984). Neurofilaments are unphosphorylated or poorly phosphorylated in the perikaryon, dendrites, and axon initial segment but are heavily phosphorylated in the distal axon (Sternberger and Sternberger 1983). Further, the spectrin isoform present in the axon is different from that found in the perikaryon and dendrites. This compartmentalization of neuronal cytoskeletal proteins might have a number of important consequences. For example, axonal microtubules might be more stable than dendritic ones, and the neurofilaments of the various neuronal compartments might have differing abilities to interact with one another and with other cell constituents. As stated above, the cytoskeleton of the axon also differs from that of the dendrites with regard to the polarity orientation of microtubules (Fig. V.22).

Some comments on the evolution of neuronal cytoskeletal proteins are appropriate at this point. Microtubules and actin microfilaments consist of proteins which have undergone very few changes during evolution (Luduena and Woodward 1975). By contrast, neurofilament proteins are much less conservative. For example, the proteins which comprise invertebrate neurofilaments clearly differ in their composition from those of mammalian neurofilaments (Gilbert et al. 1975; Lasek et al. 1979; Roslansky et al. 1980; Fuchs and Weber 1994; Lee and Cleveland 1996). The same is true for the neurofilament proteins of many lower vertebrates (Lasek et al. 1985; Pleasure et al. 1989; Mencarelli et al. 1991; Jacobs et al. 1995).

A traditional view is that there is no extramitochondrial protein synthesis within axons and that all axonal proteins are synthesized in the nerve cell body and then transported to the axon. In accord with this view, ribosomes are very rarely detected in electron micrographs of vertebrate axons beyond their initial segments (see beginning of this section). However, in serial sections (Pannese and Ledda 1991) or in whole mounts isolated from myelinated nerve fibers (Koenig et al. 2000), ribosomes can be found in vertebrate axons. Moreover, nonmitochondrial RNA has been detected within the axoplasm by means of microchemical techniques both in invertebrates (Andersson et al. 1970; Lasek 1970; Lasek et al. 1973) and vertebrates (Edström 1964; Koenig 1965; Gambetti et al. 1973; for reviews, see Mohr and Richter 2000; Piper and Holt 2004). RNA is much less concentrated in the axon than the perikaryon; however, given that in certain neurons the axon accounts for most of the cell volume (see Chap. III), the total amount of RNA may be greater in the axon than in the nerve cell body. Furthermore, the local synthesis of some proteins has been shown to occur not only in axons of invertebrates but also in vertebrate axons (Eng et al. 1999; Zheng et al. 2001; Verma et al. 2005). Taken together, the above observations suggest that some proteins are synthesized within the vertebrate axon. Comprehensive reviews of the evidence for a local protein synthesis in the axon can be found in Alvarez et al. (2000) and Giuditta et al. (2008).

3. Age-Related Changes

Changes in the axon, like those in the perikaryon and dendrites, vary according to species and region of the nervous system. Structural changes occurring in axons of aged animals and humans include degenerated mitochondria, accumulations of filaments, and glycogen inclusions (e.g., Thomas et al. 1980; Grover-Johnson and Spencer 1981; Moore et al. 1982; Tigges et al. 1992; van den Bosch de Aguilar et al. 1992; Zhang et al. 1998). These changes may culminate in axon degeneration. Degenerated or degenerating axons and reductions in the number of axons have been reported in aged subjects (e.g., Ochoa and Mair 1969; Gilmore 1972;

Kawamura et al. 1977; Cotard-Bartley et al. 1981; Casey and Feldman 1985; Johnson et al. 1987; Fujisawa 1992; Tigges et al. 1992; van den Bosch de Aguilar et al. 1992; Karlsson and Hildebrand 1996; Ceballos et al. 1999; Sandell and Peters 2001, 2003; Marner et al. 2003; Bowley et al. 2010; Peters et al. 2010). The extent of age-related axonal loss has been estimated with fair precision in some well-circumscribed fiber bundles. For example, in the anterior commissure, genu of the corpus callosum, fornix, and optic nerve of monkeys losses are about 20, 25, 40, and 45 %, respectively (Sandell and Peters 2001, 2003; Bowley et al. 2010; Peters et al. 2010). However, other studies have not found evidence of a significant loss of axons in senescence (e.g., rat, Sharma et al. 1980, Rao and Krinke 1983; cat, Aldskogius and Risling 1989).

It can be presumed that at least some of the axons that are lost from the white matter of the cerebral hemispheres in old age are long projecting axons originating from pyramidal cells in the cortex. Since there is no substantial loss of cortical neurons in senescence (see Sect. V.A.10), these axons must degenerate without loss of their parent cell bodies. It has been proposed that the latter do not degenerate because sustained by a significant number of lateral axonal branches that remain connected to target neurons (Peters 2002). The loss of substantial numbers of axons results in some disconnection between groups of neurons, and this may be one of the factors responsible for the cognitive decline that often occurs during normal aging (Peters 2002). Finally, we note that although axon regeneration following injury of peripheral nerves is retained throughout life, it is delayed (Tanaka and Webster 1991; Vaughan 1992) and proceeds at a slower rate (Pestronk et al. 1980) in older animals than younger ones.

Age-related changes in axonal transport are discussed in Sect. V.D.4 and those in myelin structure are considered in Sects. VII.B.4k and VIII.F.3.

4. Axonal Transport

Beginning from the mid-nineteenth century, many observations have shown that the maintenance of the axon depends on its connection with the cell body (Waller 1850, 1852a; Ranvier 1872; and subsequently many others). On the basis of his axon sectioning experiments, Waller (1852b) advanced the hypothesis that some influence passes from the cell body into the axon to maintain its viability. The first experimental verification of this hypothesis appeared in the 1950s when nerves subjected to a local constriction were studied. It was seen that the axonal segment just proximal to the constriction widens up and keeps swelling until the constriction was removed, whereupon it gradually shrinks back to normal (Weiss 1943; Weiss and Hiscoe 1948). Observations of this type led to the view that the axoplasmic column moves as a whole from the cell body to the axon terminals. This movement was originally referred to as axoplasmic flow, and its velocity was estimated to be of 1–2 mm/day (Weiss and Hiscoe 1948). The use of radioactive tracers (see, among others, Samuels et al. 1951; Koenig 1958; Ochs and Burger 1958; Miani 1960, 1962; Droz and Leblond 1962, 1963) provided results which led to a revision of the concept of axoplasmic flow. With this technique, it was soon established that the phenomenon is not one of simple downflow of the axoplasm as a whole but consists in the movement of only selected components within the axon. Accordingly, the term axoplasmic flow was replaced by the more correct axonal transport. Many observations made it clear that the axonal transport occurs not only from the cell body to the axon terminals but also in the opposite direction (for a review, see Schwab and Thoenen 1983). The transport in the somatofugal direction was called anterograde and that occurring in the somatopetal direction was called retrograde.

Much information on anterograde transport has been obtained using radioactive tracers. When a labeled amino acid is injected near a group of nerve cell bodies, it is taken up and incorporated into the proteins synthesized within these bodies. Subsequently, the newly synthesized, radioactively labeled proteins enter the axons, along which they migrate toward the terminals. When a nerve is constricted at a point, the labeled proteins moving along it accumulate on the proximal side of the constriction; often the accumulated proteins can be collected and identified. By examining at intervals the position reached by the radioactive peak along the axons, the transport velocity of such material can be determined. By employing suitably labeled precursors, it is also possible to study axonal transport of substances other than proteins.

Radioactive tracer studies have shown that the materials that migrate anterogradely along the axon do so at different velocities (Lasek 1966; Grafstein 1967), and a distinction is made between fast and slow transport. Fast transport takes place at rates of 100–400 mm/day in birds and mammals and at rates of 10–50 mm/day in fish and amphibians; the corresponding rates for slow transport are 0.2–8 and less than 1 mm/day. The rapidly transported materials are usually carried in vesicles. Among proteins conveyed by fast transport are enzymes (Anderson and McClure 1973) needed for local synthesis and turnover of neurotransmitters within axonal endings (see Sects. V.A.3 and VI.E). Also lipids, other small molecular weight materials (e.g., amino acids, sugars, and nucleosides), Ca^{2+}, and mitochondria are carried by fast transport.

Slow transport carries more than half of the total proteins moving along the axon. These include cytoskeletal proteins (e.g., tubulin, actin, neurofilament triplet, spectrin, and tau), proteins involved in vesicle dynamics (e.g., clathrin and synapsin), regulatory proteins (e.g., calmodulin), metabolic enzymes (e.g., creatine kinase, aldolase, pyruvate kinase, lactic dehydrogenase, and enolase), and motor proteins (e.g., dynein and myosins; for a review, see Brown 2003). These slowly moving proteins are not enclosed by a membrane. Short segments of microtubules are also carried by slow transport, whereas the longer microtubules are essentially stationary. For a long time, the difference in the rate of fast and slow axonal transport was assumed to indicate that membranous and nonmembranous cargoes move by fundamentally different mechanisms. However, direct observations of the movement of these cargoes in living neurons has shown that the rate of movement does not differ substantially in fast and slow transport. Many membranous organelles move along axons unidirectionally and continuously with high instantaneous velocities. Cytoskeletal proteins also move at rates approaching those of membranous organelles; however, their anterograde movements are interrupted by long pauses and also brief reversals (Roy et al. 2000; Wang et al. 2000; Wang and Brown 2002). Although, therefore, the instantaneous velocity of cytoskeletal proteins is high, their overall velocity is low because of these long pauses.

Anterograde transport provides the materials necessary for the continuous renewal of structural elements of the axon as well as the substances required for synaptic function. This role is of fundamental importance since the spatially separated component parts of a neuron show some degree of functional specialization (see Chap. II). In most other cells, most substances are utilized close to their production sites. In the neuron, however, the synthetic machinery of many compounds is mainly confined to the cell body, while these substances are used mainly in the axon and presynaptic terminals. Thus, the sites of main utilization of many substances are often distant from the site of synthesis; this requires a highly efficient system for a long-range transport. The importance of this system is underlined by observations that the block of transport (e.g., with colchicine) leads to an impairment of synaptic transmission (Perišić and Cuénod 1972) and to structural changes in the axon terminals which are similar to those caused by axotomy (Cuénod et al. 1972). Anterograde axonal transport has other roles, too. A part of the material anterogradely transported along an axon is released into the surroundings of that axon. Such a release may occur both along the axon trunk and at the axon terminals. Autoradiographic studies have suggested that in myelinated fibers phospholipid molecules and bases such as choline or its derivatives, conveyed by axonal transport, are delivered to the enveloping cells and utilized for the renewal of myelin components (Droz et al. 1978, 1981; Haley and Ledeen 1979; Gould et al. 1982; see also Sect. VII.B.4d).

The material released at the terminals of an axon is predominantly transferred to neurons or muscle fibers which are synaptically linked to that axon (Korr et al. 1967; Grafstein 1971; Miani 1971; Korr and Appeltauer 1974; Schubert and Kreutzberg 1974). The specificity of transcellular transport suggests that the material transfer is con fined to the synaptic contact zone. However, the amount of the material transferred does not appear to correlate with the level of electrophysiological activity of the axon, so that its release does not seem to be linked with the release of the neurotransmitter (Grafstein and Forman 1980). At times, however, the material released at the axon terminals can be localized in neuroglial cells situated in close proximity to them, a fact indicating that the material may be released in a nonspecific way and may diffuse into the extracellular space. However, this seems to occur in growing systems rather than in mature ones. Among materials transcellularly transferred are amino acids, nucleosides, and probably also proteins. It is via this transcellular transport that neurons may transfer information to or may exert trophic influences on postsynaptic cells. Transynaptic degeneration which occurs after the arrest of axonal transport could be the consequence of the failure of these materials (usually called neurotrophic substances) to reach the postsynaptic cells. However, there are alternative explanations for transynaptic degeneration. Since after the block of axonal transport synaptic transmission is also impaired, the degenerative changes could be a consequence of the arrest of the functional activity of the postsynaptic cell or a consequence of the fact that this cell is deprived of the synaptic transmitter it normally receives. As discussed subsequently, transfer may also occur from the postsynaptic to the presynaptic site, i.e., in a direction opposite to that of synaptic transmission (Schwab and Thoenen 1977).

Retrograde transport has been studied mainly by the application of tracers to nerve terminals. The most widely used tracer in these studies is horseradish peroxidase, probably because of the sensitive cytochemical techniques available for its detection. This protein is taken up by pinocytosis at the axonal endings and subsequently transported inside vesicles, short tubular elements or multivesicular bodies to the neuronal perikaryon (Kristensson and Olsson 1971; LaVail and LaVail 1974); the agranular endoplasmic reticulum does not seem to be involved in the retrograde transport of this protein (LaVail et al. 1980). In addition to peroxidase, numerous other proteins may be carried by retrograde transport, along with glycine, glutamate, serotonin, adenosine, Ca^{2+}, and phospholipids. Since some exogenous proteins which are conveyed by the retrograde transport (e.g., peroxidase) end up within lysosomes in the neuronal perikaryon, it is supposed that one of the main functions of this kind of transport is to return materials to the lysosome-rich perikaryon for degradation. Some molecules may return to the perikaryon for recycling. Another major function of retrograde transport is to convey neurotrophic signals from axon terminals to the perikaryon. Trophic substances released by a tissue (e.g., skeletal muscle) innervated by a specific group of neurons may be transferred from that tissue to axon terminals and retrogradely transported along these axons to their perikarya (for a review, see Thoenen and Barde 1980). In this way, tissues may exert trophic influences on innervating neurons. One trophic substance carried by retrograde transport is nerve growth factor (Levi-Montalcini 1964) released by peripheral tissues; this protein is carried along the axon within vesicles (for further details, see Thoenen and Barde 1980). Deprivation of trophic substances as a consequence of, for example, axotomy could be a signal that starts the chromatolytic reaction in the nerve cell body.

The following findings are of particular pathological importance: (a) tetanus toxin and cholera toxin can be retrogradely transported along the axon to the nerve cell body; the transport of tetanus toxin along the axon occurs inside tubular elements; (b) several neurotropic viruses (herpes simplex virus, herpes virus suis, rabies virus, and probably poliomyelitis virus) are retrogradely transported along peripheral nerve axons. The calculated velocity of retrograde transport varies between 60 and 200 mm/day according to the experimental model employed.

Microtubules are directly involved in the mechanism of axonal transport. This view is supported by (a) the close spatial association of microtubules with transportable organelles (Fig. V.19) such as microvesicles (Smith et al. 1970) and mitochondria (Raine et al. 1971; Smith 1973; Smith et al. 1975; Pannese et al. 1986), (b) the observation that drugs or procedures which cause the disassembly of microtubules can inhibit or significantly reduce fast transport (Kreutzberg 1969; Fernandez et al. 1970; James et al. 1970; for a review, see Grafstein and Forman 1980), and (c) the observation that following the administration of β,β'-iminodipropionitrile most microtubules are segregated in the center of the axon and most of the transported proteins are located in the central region of the axon with microtubules (Papasozomenos et al. 1982). Note that certain organelles move preferentially toward a given end of a microtubule. For example, small vesicles appear to be specifically translocated toward the plus end of microtubules, whereas ribosomes appear to be specifically transported toward their minus end. Microtubules serve as tracks along which long-range transport occurs. However, some axonal compartments (e.g., presynaptic terminals) have very few or no microtubules, yet a transport of materials occurs. These compartments contain actin filaments that may serve as tracks for short-range transport.

Specialized motor proteins generate the force for the translocation of materials along tracks. For some time, it was thought that the motor proteins involved in slow transport differed from those involved in fast transport, but their identity remained elusive. The observation that cytoskeletal proteins in axons move as fast as membranous organelles (see above) suggests that both fast and slow transport are driven by similar or identical motors. The relevant motor proteins are dynein and members of the kinesin and myosin families (Lasek and Brady 1985; Vale et al. 1985; Schnapp and Reese 1989; for reviews, see Goldstein and Yang 2000; Bridgman 2004). Most evidence suggests that kinesins and dynein are involved in long-range transport along microtubules; in particular, kinesins mediate anterograde transport, while dynein is responsible for retrograde transport. Myosins may contribute to the short-range transport of organelles in neuronal compartments (e.g., dendritic spines and presynaptic terminals) devoid of microtubules and rich in actin filaments (for a review, see Bridgman 2004).

Both anterograde and retrograde axonal transport form the basis of new tracing techniques that are now widely used for visualizing the terminal axonal arborizations of a group of nerve cell bodies or for locating the cell bodies of axons ending in a given region of the CNS or the periphery. Previously, axon degeneration techniques were widely used to these purposes. A nervous pathway was surgically interrupted or its nucleus of origin destroyed; the lesion resulted in the degeneration of the corresponding terminals, which was detected either at the light microscope level using special staining methods or at the electron microscope level. Because of the difficulties in their use and in the interpretation of the results, degeneration techniques have been superseded by new procedures which do not involve destructive lesions but are based on physiological activities of neurons. A commonly used anterograde tracing technique is as follows. Radioactively labeled amino acids injected in the proximity of a group of nerve cell bodies are taken up by them and incorporated into proteins which are subsequently carried along the axons to accumulate at their terminals. The labeled terminals may be visualized by light or electron microscope autoradiographic techniques. Another technique takes advantage of retrograde axonal transport. Horseradish peroxidase is injected in a circumscribed zone of an organ and then localized by a cytochemical technique in the bodies of the neurons whose axons end within the injection site (for more details, see Jones and Hartman 1978; Mesulam 1982). Fluorescent labels are also employed as retrogradely transported markers. The availability of various retrograde tracers has allowed the use of two markers in combination to effect a double labeling. Thus, it is now possible to establish whether axonal terminals located at two different sites originate from the same population of neurons with branching axons or from different populations of neurons. For example, when injecting peroxidase at one site and a substance labeled with a fluorescent dye at another, the two labels accumulate within the same nerve cell bodies in the first case and within different nerve cell bodies in the second case (for further details, see Jones and Hartman 1978; Mesulam 1982). These tracers have not proved very effective in visualizing chains of synaptically connected neurons. Applied to the neuron at one end of a chain, these tracers usually label the successive neurons of the chain only weakly, probably because only a small amount of tracer is transferred from one neuron to the next. After it was shown that certain live neurotropic viruses (e.g., herpes and pseudorabies viruses) may be transferred from one neuron to the next by a transsynaptic mechanism (Kristensson et al. 1982), they began to be employed with success in transneuronal tracing (Martin and Dolivo 1983). These viruses produce strong transneuronal labeling since, unlike nonviral tracers, they are replicated in the recipient neurons after transfer. Virus antigen can be detected in neurons by means of immunocytochemical techniques (for further details, see Card 1998).

The effect of aging on axonal transport has been studied in several regions of the nervous system. In general, these studies have shown that the rate of fast transport decreases in old age under normal conditions both in the peripheral (e.g., Stromska and Ochs 1982; Viancour and Kreiter 1993; Frolkis et al. 1997) and central (e.g., Geinisman et al. 1977b; Cross et al. 2008) nervous systems. However, the rate of fast transport was not found changed in old age by Inestrosa and Alvarez (1988), Robertson et al. (1993), or Jacob (1995). The rate of slow transport decreases in old age (e.g., Komiya 1980; McQuarrie et al. 1989). This change may have important implications for the ability of axons to regenerate in old animals (see earlier, this section).

Somatofugal transport in dendrites was directly demonstrated by the intracellular injection of labeled protein precursors (Globus et al. 1968; Schubert et al. 1971). RNA and certain proteins are transported in dendrites along microtubules at a rate that is within the range of fast transport (Knowles et al. 1996). Kinesins are thought to be involved in this transport. Mitochondria move both anterogradely and retrogradely within dendrites.

E. The Plasma Membrane

Neurons, like all cells, are bounded by a plasma membrane which plays a key role in their functional activities. Specifically, neuron excitability is dependent on the difference in ion composition between the intracellular compartment and the extracellular space and consequently on the properties of the membrane which separates these two compartments; moreover, interactions of neurons with other cells (neuronal or not), one of their major functions, also take place via the plasma membrane. The study of the neuronal plasma membrane is beset with particular difficulties. While much information on the characteristics of the plasma membranes of other cell types has been obtained from the study of homogenous membrane fractions, this approach is not well suited to neurons owing to the great variety of cell types composing the nervous system and to the fact that the plasma membrane of a neuron does not show uniform properties over its whole surface. For these reasons, our knowledge of the organization of the neuronal plasma membrane is still fragmentary and mainly based on comparisons with the much better known plasma membranes of other cell types.

As the plasma membrane cannot be seen under the light microscope, its chemical, physical, and functional aspects were elucidated prior to its morphological ones. On the basis of the experimental evidence that lipid-soluble molecules penetrate cells more quickly than water-soluble compounds, Overton (1899) advanced the hypothesis that the plasma membrane has high lipid content. Gorter and Grendel (1925) extracted the lipids from the plasma membranes of a known number of erythrocytes and measured the total area covered by them when arranged in a monolayer. Since the total area of such preparations was twice that of the total surface area of the cells from which the lipids had been extracted, these authors concluded that in the plasma membrane of erythrocytes the lipids form a bilayer. As a matter of fact, the extraction procedure employed by Gorter and Grendel (1925) did not remove all the plasma membrane lipids, but this shortcoming was "compensated" by an underestimation of the surface area of the cells, so that their conclusion may still be regarded as valid. Impedance measurements on cells strengthened the idea that a substantial part of the plasma membrane is formed by lipids. Later the surface tension was measured in some types of cell, and values of 0.08–0.2 dyn/cm were obtained (Cole 1932; Cole and Michaelis 1932; Harvey and Marsland 1932; Harvey and Fankhauser 1933); these values are much lower than those (10–15 dyn/cm) obtained by measuring the surface tension at an oil-water interface. By analogy with what one observes in lipid-water systems, where the addition of proteins markedly lowers the surface tension at the oil-water interface, the low values measured at the cell surface led to the conclusion that the plasma membrane always contains proteins.

On the basis of all the data then known, Danielli and Davson (1935) proposed the model shown in Fig. V.23, which provided the first significant attempt to interpret the organization of the plasma membrane in molecular terms. Subsequent studies by polarization optical methods (Schmidt 1936; Schmitt et al. 1936) supported the view expressed by this model that the long axes of the lipid molecules are oriented perpendicular to the surface of the plasma membrane. Careful chemical analyses later showed that in addition to proteins and lipids, oligosaccharides are also present in the plasma membrane. Proteins form 60–70 % and lipids 30–40 % of the membrane components; only in the myelin sheath are there quantitatively more lipids than proteins.

The subsequent advent of the electron microscope made it possible to visualize the plasma membrane and to measure its thickness (7–8 nm). In low-power electron micrographs of thin sections cut perpendicular to the membrane surface, it appears as a thin, dense line, but at higher resolution this line can be resolved into three layers, each 2–3 nm thick; the two bordering layers appear dense and are separated by a lighter layer (Fig. V.24). After primary fixation with OsO_4 or after fixation with aldehydes followed by OsO_4, the inner (=cytoplasmic)

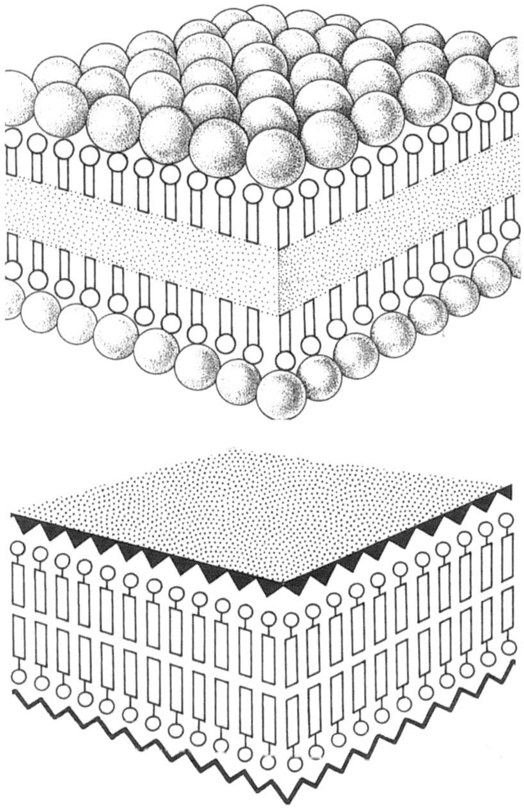

Fig. V.23 Plasma membrane. *Above*: The Danielli and Davson (1935) model of the molecular organization of the plasma membrane. A lipid bilayer is sandwiched between two continuous layers of protein. The axes of the lipid molecules are oriented perpendicularly to the surface of the membrane, and the polar heads (*white circles*) are in direct contact with the protein layers. The proteins have a globular configuration. The membrane has a symmetrical structure. *Below*: The membrane model proposed by Robertson (Robertson 1959a). While in the Danielli and Davson model the proteins have a globular configuration, in this model they are unfolded. The chemical asymmetry of the membrane is indicated by the different representation of the proteins on the outside surface and on the inside surface

layer appears denser than the outer one. On the basis of these observations, Robertson (1959a) proposed some modifications to the Danielli and Davson model. Taking into account the measured thickness of the membrane, Robertson (1959a) assumed that the membrane proteins were unfolded (Fig. V.23) rather than in a globular configuration, and on the basis of the density difference between the outer and inner layers, he introduced the idea of the chemical asymmetry of the membrane, according to which a chemical difference exists between the proteins in the outside surface and those on the inside (Fig. V.23). Originally, it was believed that the layers of the plasma membrane appearing dense under the electron microscope represented the two protein layers of the models shown in Fig. V.23, while the lighter intervening layer represented the lipid bilayer. Later, it was shown that the two dense layers corresponded to the polar heads (also called hydrophilic groups) of the lipid molecules in the bilayer and the intervening lighter layer corresponded to the fatty acid chains (also called hydrocarbon chains or hydrophobic tails) of the lipid molecules, which take up little osmium.

Important information on the plasma membrane organization was obtained subsequently thanks to the introduction of the freeze-fracture technique. This involves the splitting of the membrane along the plane of the lipid bilayer so that the interior of the membrane becomes accessible to morphological study. When examined under the electron microscope, the fractured surfaces thus obtained display a substantial number of globular particles (Fig. V.25), about 6–10 nm in diameter, whose presence is compatible neither with the Danielli and Davson model (1935) nor with the version proposed by Robertson (1959a). Information obtained from the freeze-fracture and other techniques led to the proposal of a new model for the plasma membrane (Fig. V.26), usually called the fluid mosaic model (Singer and Nicolson 1972). This differs from previous models principally with respect to the distribution of proteins which, instead of forming two continuous layers on both sides of the lipid bilayer, are partially embedded in it in the form of globular particles. Most protein molecules are amphipathic, that is, containing both hydrophilic and hydrophobic groups;

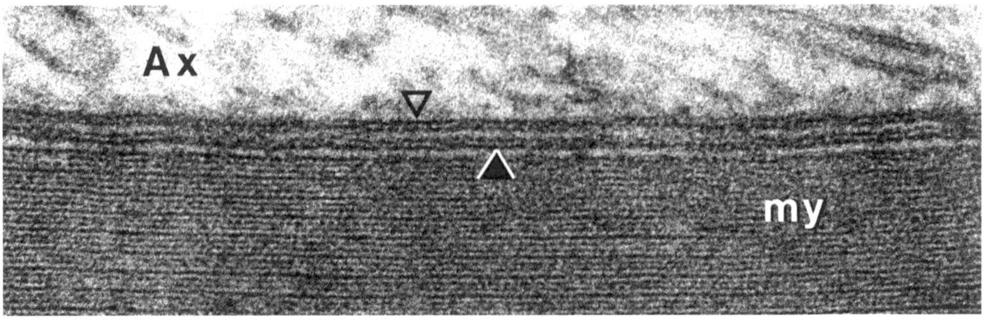

Fig. V.24 Plasma membrane. The plasma membrane of an axon (axolemma, ▽) and that of a Schwann cell (▲) sectioned perpendicular to their surfaces. Each membrane appears as a linear profile consisting of two dense layers separated by a lighter layer. **Ax** axoplasm, **my** myelin. Lizard ventral root, ×200,000

the hydrophilic ends of the amphipathic proteins (containing ionic amino acid residues) protrude from the lipid bilayer, whereas the hydrophobic regions (containing many of the nonpolar residues) are embedded in the hydrophobic interior of the lipid bilayer. Some protein particles penetrate the lipid bilayer partially and protrude either from the external or from the internal surface, whereas others span the whole thickness of the lipid bilayer and protrude from both the inner and outer surfaces. The latter are called transmembrane proteins. The chemical composition of the lipid bilayer is likely to vary from one region of the cell surface to another. Moreover, the chemical properties and physical state (see later, this section) of the lipids surrounding individual proteins may influence their functions. Oligosaccharides linked to proteins thus forming glycoproteins or linked to lipids thus forming glycolipids are placed on the external surface of the plasma membrane, where they form a layer called the glycocalyx (Fig. V.26).

The membrane proteins protruding into the cytoplasm can be linked to microfilaments (see later, this section). Among the proteins which protrude from the outer surface of the lipid bilayer, a number are endowed with a high affinity for certain ligands (e.g., hormones, certain plasma proteins, neurotransmitters, lectins, toxins, and viral proteins), thus functioning as surface receptors. They are also known as receptor molecules. Certain proteins which span the lipid bilayer and protrude from both its inner and outer surfaces mediate adhesion between neurons or between nerve and neuroglial cells, target recognition, certain cell-matrix interactions, and so on, thus contributing during development to the establishment of specific neuronal pathways and connections (for a review, see Rutishauser and Jessell 1988). Proteins responsible for cell adhesion in the nervous system include cadherins, integrins, immunoglobulin superfamily members, neurexins, and neuroligins. These proteins are called cell adhesion molecules (CAMs). It has been suggested that the cytoplasmic segment of a cell adhesion molecule may interact with the cytoskeleton, thus regulating certain cytoskeletal functions. More generally, cell adhesion molecules transduce information across the plasma membrane, generating signals which may regulate intracellular ion concentrations or influence certain cell activities (e.g., cell proliferation or enzyme activity).

A large fraction of plasma membrane lipids are unsaturated, thus lowering the fusion point of the bilayer to below that of the body temperature of the organism concerned. The lipid bilayer would therefore be in a fluid state (except in the myelin sheath, however). Lipid domains of differing fluidity would coexist in the plasma membrane of a

Fig. V.25 Plasma membrane. Plasma membranes of a neuron (**Nc**) and of a satellite cell (**Sc**) in a freeze-fractured preparation. The fractured surfaces of both membranes display numerous globular particles. A gap junction (*encircled*) is present in the plasma membrane of the satellite cell. Rabbit spinal ganglion, ×23,000 (Courtesy of E. Reale and L. Luciano)

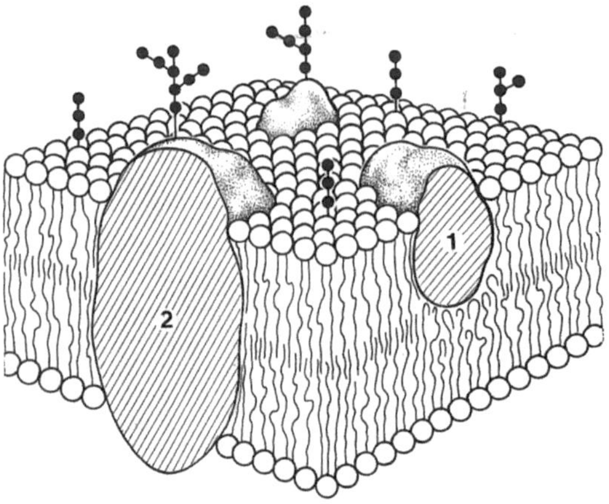

Fig. V.26 Plasma membrane. The fluid mosaic model of the plasma membrane proposed by Singer and Nicolson (1972). The bodies indicated by **1** and **2** represent proteins. Note that, while in previous models the proteins formed two continuous layers on both sides of the lipid bilayer, in this model they are partially embedded in it. Protein **1** penetrates the lipid bilayer partially and protrudes only from the outer surface, whereas protein **2** spans the whole thickness of the lipid bilayer and protrudes from both surfaces. Straight or branched chains of *joined black circles* represent oligosaccharides. These are linked to proteins or lipids on the outer surface of the membrane

given cell. The fluidity of the lipid bilayer implies that the protein molecules embedded in it are free to move in the plane of the membrane maintaining, however, their usual orientation.

The constituents of the plasma membrane are distributed asymmetrically; for example, in an erythrocyte the outer leaflet of the lipid bilayer contains mainly phosphatidylcholine and sphingomyelin, whereas the inner leaflet contains mainly phosphatidylethanolamine and phosphatidylserine. At the level of the synaptic membranes, phosphatidylcholine is thought to be likewise primarily located in the outer half of the lipid bilayer, while phosphatidylethanolamine and phosphatidylserine would be primarily located in the inner half. This chemical asymmetry, due to the nonuniform distribution of phospholipids across the membrane, is further accentuated by the arrangement of oligosaccharides, which are present only on the outer surface of the membrane.

The fluid mosaic model endows the membrane with more flexibility than previous models and provides a more convincing explanation for some of its properties. It is compatible with the differences in chemical composition observed in the plasma membranes studied so far and with the changes in structure and composition which may take place in the plasma membrane during the life of a cell. Moreover, this model makes it possible to explain the transport not only of water- and lipid-soluble molecules across the membrane but also ion transport along channels localized between the subunits of the proteins which span the whole thickness of the plasma membrane. Finally, it clarifies the localization of recognition and immune reaction sites at the cell surface.

While the basic characteristics of the neuronal plasma membrane appear similar to those of the plasma membrane of other cell types, there are certain differences in detail. Of relevance is the greater abundance of gangliosides, a class of glycolipids. Moreover, the neuronal plasma membrane exhibits regional differences in its physiological and chemical properties. Electrophysiological experiments have revealed that both Na^+ and K^+ channels are unevenly distributed in the plasma membrane of the soma, axon, and dendrites. There are differences in electrical properties between the dendritic membrane and the axolemma, between the perikaryal membrane and the membrane of the axon initial segment, between the postsynaptic and the nonjunctional areas of the dendritic membrane, as well as between the membrane of the node of Ranvier and that of the internodal axon. Neurochemical techniques, on the other hand, have shown that the receptor molecules are not uniformly distributed over the neuronal surface and that different regions of the surface of a single neuron may contain different receptor molecules and therefore be sensitive to different neurotransmitters.

The demonstration that the properties of the neuronal plasma membrane vary over the cell surface has prompted morphological studies aimed at defining the structural bases of such differences. These investigations have yielded some particularly illuminating results as regards the axolemma of myelinated axons, where a high degree of regional specialization has been evidenced. (a) At the level of the axon initial segment (Conradi 1966) and of the node of Ranvier (Elfvin 1961), a layer of dense material (Fig. VII.18) beneath the axolemma (see Sects. V.D.1 and VII.B.4g) as well as particularly thick and dense layer of oligosaccharides disposed on its outer surface are found. (b) The concentration of the large particles revealed by means of the freeze-fracture technique in the E face of the axolemma (Figs. VII.15 and VII.16) of myelinated axons (see Sect. VII.B.4g) is much higher in the nodal than in the internodal axolemma (Rosenbluth 1976). (c) In the paranodal region of myelinated nerve fibers (see Sect. VII.B.4f), the axolemma and the plasma membrane of the enveloping cell form junctional complexes (Livingston et al. 1973; Dermietzel 1974b; Schnapp and Mugnaini 1975; Fig. VII.17). Finally, (d) the concentration of intramembranous particles is higher within the presynaptic membrane than in the nonjunctional portion of the axolemma (see Sect. VI.B.3). Since Na^+ channels are much more concentrated in the nodal than in the internodal axolemma and since the density (number per unit surface area) of these channels at the node of Ranvier (see, among others, Conti et al. 1976) correlates well with that of the E face particles in freeze-fractured preparations (Rosenbluth 1976), it has been suggested that each of these particles corresponds to one Na^+ channel (see Sect. VII.B.4g). The distribution of these intramembranous particles within the axolemma would be influenced by neuroglial cells (for a review, see Rosenbluth 1988). The particles of the nodal axolemma are thought to be anchored to the subaxolemmal network of actin microfilaments (Fig. VII.18; see also Sect. V.D.1). The existence of the structural characteristics quoted above, whatever their precise functional significance may be, indicates that the various regions of the neuronal plasma membrane have a high degree of structural specialization.

Investigations carried out on erythrocytes led to a model in which a network of microfilaments, the major constituents of which are spectrin and actin, is located on the cytoplasmic surface of the plasma membrane. This network, which is linked to integral membrane proteins through other proteins such as ankyrin, appears to regulate the segregation of proteins to specific domains of the plasma membrane and to control the movements of membrane proteins. Since a spectrin isoform (also called fodrin) and ankyrin isoforms have been found in the CNS, this model has been also applied to nervous system cells. The control of the segregation and lateral mobility of integral membrane proteins, which include ion channels and cell adhesion molecules, seems important because redistribution of these proteins may have profound consequences for certain membrane functions (see Sects. V.D.1 and VII.B.4g).

Shallow depressions of the plasma membrane, flask-shaped invaginations connected with the plasma membrane by a narrow neck, and closed vesicles can be seen along the neuronal surface. These specializations are probably involved in endocytosis. As in most cell types, two categories of pits and vesicles can be found at or near the neuronal plasma membrane: those with a smooth surface and those surrounded by a protein coat. The protein in question is clathrin (Pearse 1976). Smooth-surfaced vesicles are involved in the nonselective pinocytosis of fluid and solutes. Coated vesicles, on the other hand, are concerned with the pinocytosis of specific proteins (e.g., hormones and other regulatory molecules). The latter process is highly selective and its selectivity depends on the presence of specific receptors clustered in the portion of the plasma membrane which undergoes invagination. It is thought that both the plasma membrane pinched off in endocytosis and the receptors contained in it are later recycled. Finally, myosin is thought to play a role in endocytosis (for a review, see Bridgman 2004).

Various ion species are not equally distributed on both sides of the neuronal plasma membrane. For a nerve cell at rest, of the four major ion species, K^+ and the large organic anions (e.g., amino acids and polypeptides) are more concentrated inside the cell, whereas Na^+ and Cl^- are more concentrated on the outside. The neuronal plasma membrane separating the endo- and extracellular compartments contains channels, located in membrane proteins, which are permeable to K^+, Na^+, and Cl^-, but impermeable to large organic anions. Despite the fact that these channels are permeable to K^+, Na^+, and Cl^-, the endo- and extracellular concentrations of these ions do not reach equilibrium since an ion transport mechanism, called the Na-K pump, is operating in the membrane. This pump, which is located in integral membrane proteins, actively extrudes Na^+ from the neuron while taking in K^+ against the transmembrane gradients. The energy required for the functioning of this pump derives from the hydrolysis of adenosine triphosphate.

As a consequence of the different endo- and extracellular concentrations of positive and negative ions, a nerve cell at rest is positively charged on the outside and negatively charged on the inside; between the two sides of the membrane there is therefore a potential difference, called the resting membrane potential. This potential difference, which can be recorded by placing a microelectrode on the outside of the neuron, inserting a second microelectrode into the neuron, and connecting both to a cathode ray oscilloscope, is about 70 mV with the inside of the cell negative with respect to the outside. According to a widely accepted hypothesis, at the point of stimulation, the Na^+ channels in the neuronal plasma membrane open, with the consequent rapid influx of Na^+ into the cell under the influence of concentration and electrical gradients. As a consequence, the resting potential between the two sides of the plasma membrane drops quickly to zero and soon reverses as a net excess of positive charges accumulates inside the cell and a net excess of negative charges remains on the outside. The resulting potential difference is about 20 mV with the inside of the cell positive with respect to the outside. This sudden reversal of polarity, usually called depolarization, constitutes the nerve signal. Soon after, the Na^+ channels close, and the K^+ channels open, whereby K^+ diffuses out of the neuron under the influence of the new concentration and electrical gradients. In this way, the original resting membrane potential of about 70 mV is reestablished. The Na-K pump then brings the Na^+ and K^+ concentrations in the endo- and extracellular compartments to their resting values. The manifestation of the excitation is the so-called action potential, a large depolarizing signal which propagates actively and with great speed along the neuronal plasma membrane without decreasing in amplitude (see Sect. VI.E).

VI. Intercellular Junctions Involving Neurons

A. Interneuronal Adherent Junctions

The adhesion between neighboring neurons is sometimes reinforced by adherent junctions. These junctions are characterized, with few exceptions, by the presence of symmetrical plaques of dense material applied to the cytoplasmic face of the two apposed plasma membranes and of fine filaments converging upon these plaques. Along the midline of the intercellular space, which at the junction is slightly wider than elsewhere, a thin dense line parallel to the two apposed plasma membranes can sometimes be seen. Interneuronal adherent junctions are generally very small (Fig. V.16) and are hence referred to as puncta adhaerentia. At these junctions, cell cohesion is maintained by adhesion molecules belonging mainly to the cadherin family.

Any portion of a neuron can be linked to any portion of another neuron by means of adherent junctions; thus, they may occur between dendrites, between a dendrite and a nerve cell body, between a dendrite and an axon, between nerve cell bodies, etc. These attachment devices have also been found at many synaptic interfaces (see Sect. VI.B.3), where they can be distinguished from synaptic junctions by the occurrence of symmetrical densities and a lack of synaptic vesicles. These small attachment devices have also been observed between nerve and neuroglial cells.

B. Interneuronal Chemical Synapses

1. General Features

As noted previously (see Chap. I), the interneuronal chemical synapse is an intercellular junction specialized for the transmission of signals from one neuron to the next. In most synapses, the presynaptic element consists of an axon which makes contact with a dendrite or the body of another neuron; these synapses, which were the first to be recognized, are called axodendritic and axosomatic synapses, respectively. Later it was discovered that synaptic contacts may occur between any portion of a neuron (dendrite, soma, axon) and any portion of another neuron (see Sect. VI.B.5).

Where the axon is presynaptic, it may form synapses at the tips of its branches or along its course. In the first case, the presynaptic element is called a terminal bouton, a synaptic bouton, an end-bulb, an end foot, or a presynaptic ending (all these terms are synonymous); in the second case, it is called a bouton "en passant." In no case is the axonal portion forming the presynaptic element covered by a myelin sheath; myelinated axons lose their myelin sheaths before giving rise to end-bulbs and may possess boutons "en passant" only at the nodes of Ranvier.

E. Pannese, *Neurocytology: Fine Structure of Neurons, Nerve Processes, and Neuroglial Cells*, 2nd Edition,
DOI 10.1007/978-3-319-06856-5_6, © Springer International Publishing Switzerland 2015

2. Number and Density

Quantitative studies using the light microscope have shown that the total number of synapses on a neuron may vary from a few dozen (e.g., on a granule cell of the cerebellum) to several hundred thousand (e.g., on a Purkinje cell of the cerebellum). In the lateral vestibular nucleus, some 16 presynaptic endings per 100 μm^2 of cell surface have been counted in the cat (David 1957) and 15–20 in the rabbit (Hydén and Pigon 1960). This would imply that on these neurons there are up to about 3,000 synapses in the cat and 10,000 in the rabbit. The synapse density would be of the same order in the spinal cord: 10–20 (Barr 1939), 15–20 (Wyckoff and Young 1956), and 17–22 (Conradi 1969a) presynaptic endings per 100 μm^2 of the cell surface have been counted on the anterior horn motoneurons of the cat. A mean density of 20 synapses per 100 μm^2 of the cell surface has been calculated by Gelfan and Rapisarda (1964) for all types of neuron in the gray matter of the dog spinal cord. On a motoneuron with a total surface area of 10,000 μm^2, there would be about 2,000 terminal boutons (Wyckoff and Young 1956). As to the cerebral cortex, 25 presynaptic endings per 100 μm^2 of the cell surface, corresponding to a total of more than 8,000 synapses per cell, have been found on a large neuron in the cat sensory-motor cortex (Armstrong and Young 1957). The total number of synapses on a neuron of the visual cortex averages 7,000 in the mouse and 5,600 in the monkey; for a motor cortex neuron, the mean values are 13,000 in the mouse and 60,000 in the monkey (Cragg 1967). For many neurons, over 95 % of the presynaptic endings make contact with dendrites (Rall and Segev 1987).

3. Structure

In the past, the morphology of synapses was studied by light microscopy on material stained with a variety of techniques, particularly involving metallic impregnation. With these procedures the presynaptic ending appeared as a small swelling of a nerve fiber closely applied to the surface of the postsynaptic neuron (Fig. I.4). If stained appropriately, the terminal bouton appeared to be filled with mitochondria. The advent of electron microscopy made it possible to study the structural organization of synapses in much greater detail and to clarify the relationships between neurons at the synaptic level (Robertson 1953, 1955; Sjöstrand 1953b; Palade and Palay 1954; De Robertis and Bennett 1955; Palay 1956, 1958a).

At every synapse there is a discontinuity between the pre- and postsynaptic elements, each element being bounded by its own plasma membrane. The two apposed membranes are separated by the synaptic cleft (Fig. VI.1). The pre- and postsynaptic elements have different structures, so that the synapse possesses a structural asymmetry corresponding to its functional polarity, which manifests itself in the capacity to transmit nervous signals in one direction only.

The following occur commonly in the presynaptic element: synaptic vesicles, mitochondria, tubules of the agranular endoplasmic reticulum, and a dense material underlying a portion of the presynaptic membrane. The portion of presynaptic membrane coated with dense material has a protein composition that differs distinctly from that of the surrounding regions.

Synaptic vesicles, 35–50 nm in diameter, are in part closely apposed to the presynaptic membrane (so-called docked vesicles), while others occur at varying distances from it (Figs. VI.1 and VI.3). These vesicles have clear centers, and in a given presynaptic ending are quite uniform in size, although their mean diameter may vary somewhat in different synapses. They contain nonpeptide neurotransmitters (e.g., acetylcholine, glutamate, γ-aminobutyric acid [GABA], and glycine). Evidence that the neurotransmitter is stored within these vesicles comes chiefly from

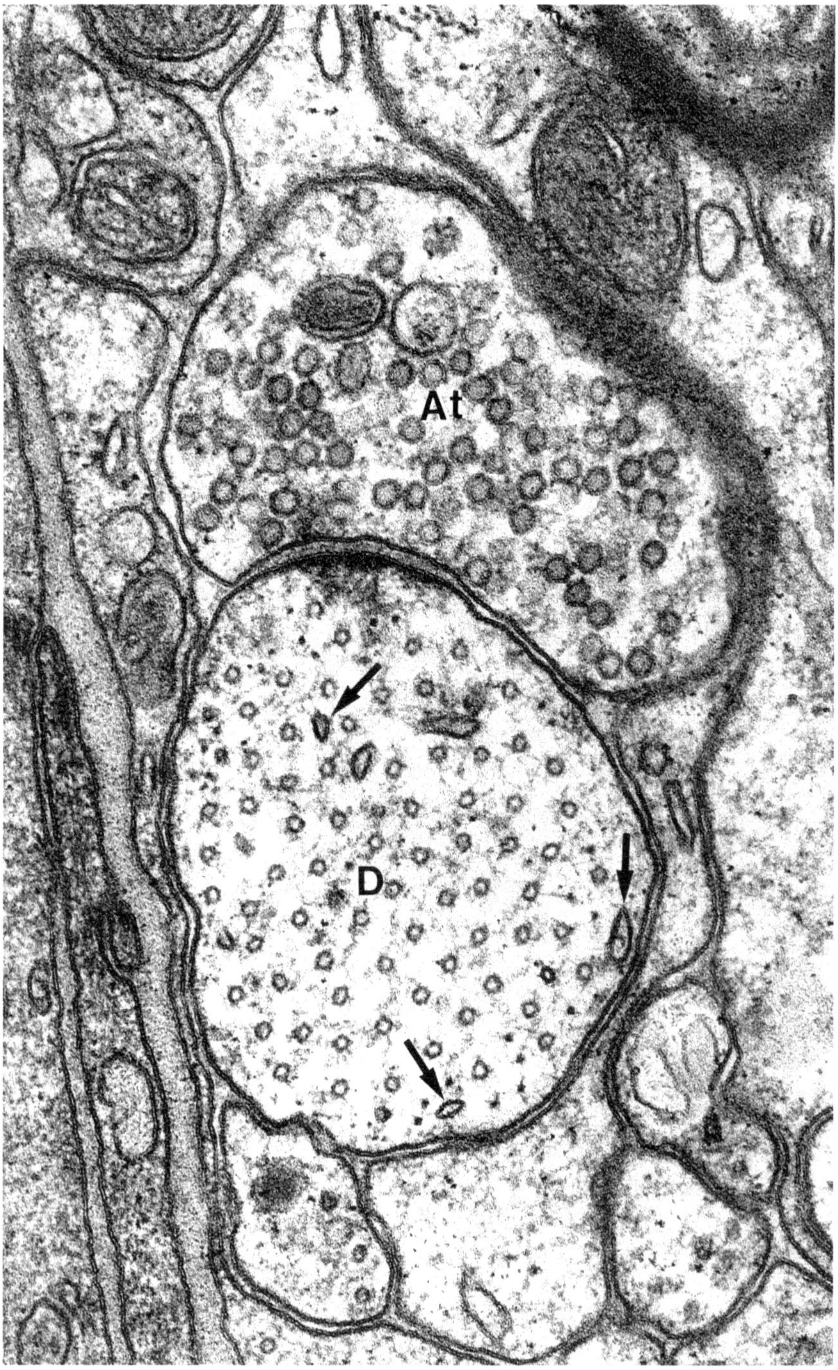

Fig. VI.1 Axodendritic synapse. An axon terminal (**At**) forms an asymmetric synapse with a dendrite (**D**), which is transversely sectioned. The axon terminal contains spherical vesicles. At the junction, the presynaptic and postsynaptic membranes are separated by a cleft containing intercellular material. The cytoplasmic face of the postsynaptic membrane is coated with a fuzzy layer of dense material (postsynaptic density). Tubules of the agranular endoplasmic reticulum (some *arrowed*), numerous microtubules, and a number of neurofilaments can be seen within the dendrite. Note the uniform arrangement of microtubules. Both microtubules and neurofilaments run parallel to the long axis of the process, so that they appear in cross section. Note that a diffuse cottony material is dispersed among the microtubules and neurofilaments. Ventral horn of the rat spinal cord, ×90,000

biochemical analyses of subcellular fractions of nerve tissue (De Robertis et al. 1963; Whittaker et al. 1964; Israel et al. 1968; Iversen and Bloom 1972). The part of the presynaptic element characterized by dense material underlying the presynaptic membrane and by its associated synaptic vesicles is called the active zone. The number of active zones per presynaptic element ranges from 1 in many synapses of the CNS to about 600 in the calyx of Held. Active zones may also vary in size. Numerous proteins are associated with the active zone. They include Ca^{2+} channels, cytoskeletal proteins (e.g., actin, tubulin, myosin, and spectrin that are also present in other compartments of the neuron), protein families that play specific roles in the synaptic vesicle cycle (e.g., Munc13s, RIMs, ELKS proteins, Piccolo, and Bassoon; for a review, see Schoch and Gundelfinger 2006), and syntaxin and synaptosome-associated protein 25 (also known as SNAP-25) which play a key role in neurotransmitter release.

Synaptic vesicles, 40–60 nm in diameter, containing catecholamines (noradrenaline, adrenaline, or dopamine) may present a dense core, which is especially evident after permanganate fixation. Large, dense-cored vesicles about 100 nm in diameter are usually present within presynaptic terminals. These vesicles, which are much more heterogeneous in size and much less numerous than clear vesicles, are usually located some distance from the presynaptic membrane. They are also present in extrasynaptic regions of axons, dendrites, and perikarya. Large, dense-cored vesicles contain neuropeptides, neurotrophic factors, monoamines, and other modulatory substances.

Clear synaptic vesicles appear spherical after primary fixation with OsO_4 or $KMnO_4$ as well as in freeze-fractured unfixed material, but after primary fixation with aldehydes, certain presynaptic boutons (Fig. VI.2) are seen to contain spherical vesicles, while others have a high percentage of flattened vesicles (Uchizono 1965; Walberg 1965a, 1966; Bodian 1966; Hirata 1966; Lund and Westrum 1966). Numerous observations suggest that, in the living state, clear synaptic vesicles are spherical (e.g., see Tatsuoka and Reese 1989) and that this shape is preserved by primary fixation with OsO_4, which is a powerful membrane stabilizer. Conversely, the native shape of vesicles is not always preserved by primary fixation with aldehydes, probably because they do not stabilize membranes quite as well. Vesicle flattening does not, however, depend primarily upon aldehydes, but rather upon buffer osmolarity (Valdivia 1971); osmic postfixation would then stabilize the already modified vesicle shape. The flat shape of the majority of vesicles observed within certain presynaptic boutons under given fixation conditions is, therefore, an artifact. Nonetheless, since the majority of vesicles exhibit a characteristic and constant shape for each presynaptic bouton with a particular fixation techniquc, such an artifact can be regarded as useful for distinguishing presynaptic boutons belonging to different nerve cells. For example, in the aldehyde-fixed cerebellar cortex of the rat, the terminal boutons of the mossy and parallel axons contain spherical vesicles, whereas those of the basket cell and Golgi cell axons contain a high percentage of flattened vesicles (Valdivia 1971). Several types of presynaptic endings have been observed on the dendrites and perikaryon of the Mauthner cells of the goldfish. After primary fixation with aldehydes, both the large club endings of the myelinated fibers of the ipsilateral eighth nerve and the endings of the spiral fibers from the medial longitudinal fasciculus contain spherical vesicles, whereas the club endings of the collateral branches of the contralateral eighth nerve fibers mainly contain flattened vesicles (Nakajima 1974). The relationships between vesicle shape and function of the synapse are discussed in Sect. VI.B.4.

Some protein components of synaptic vesicles have been isolated and characterized; among these are synaptophysin, synapsins, synaptotagmin, and synaptobrevin.

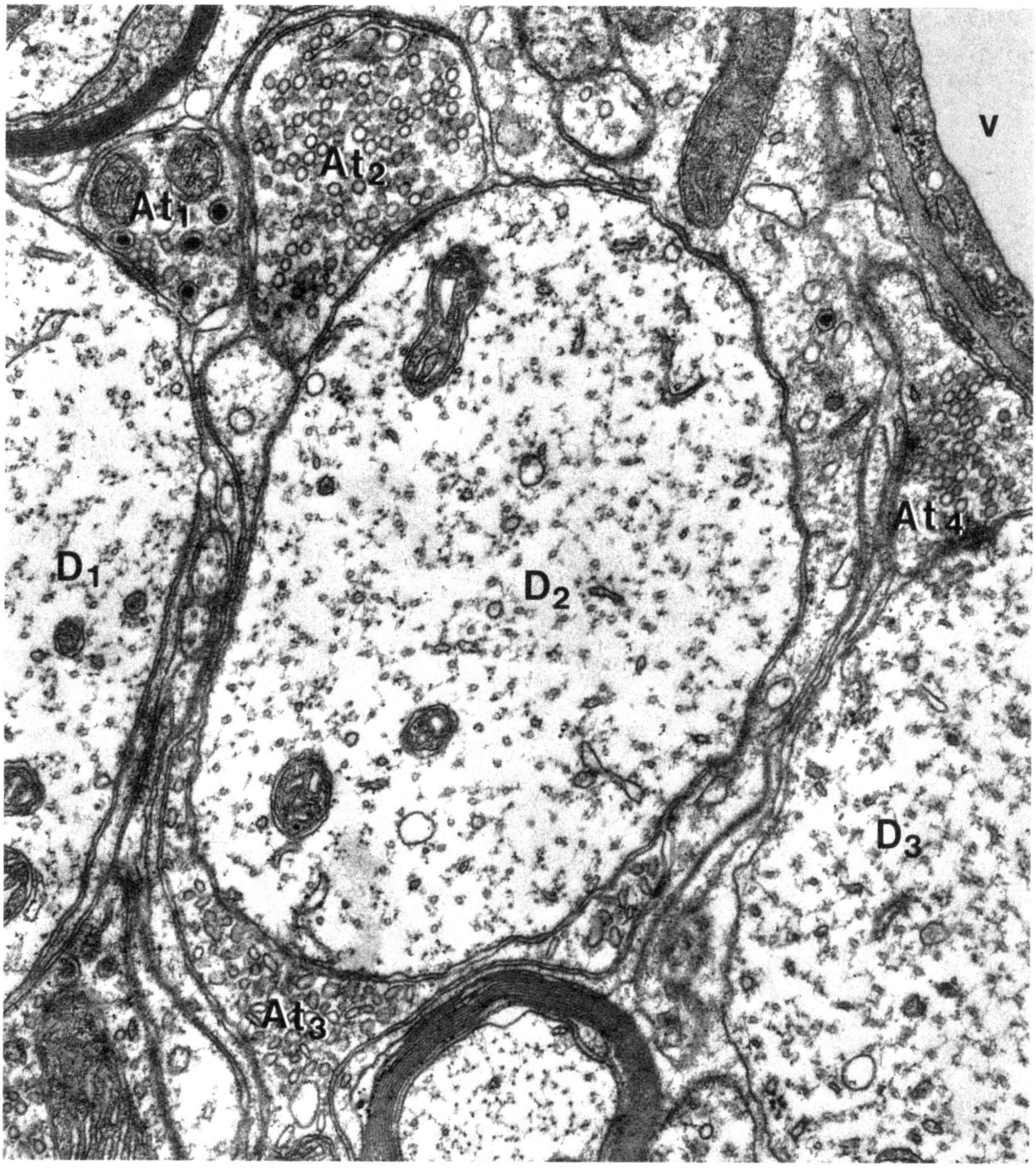

Fig. VI.2 Features of synaptic vesicles. In this field three transversely sectioned dendrites (**D_1–D_3**) can be seen; synapsing with these are four axon terminals (**At_1–At_4**). The terminals **At_2–At_4** contain only vesicles with clear centers, but terminal **At_1** also contains a number of large, dense-cored vesicles. While in terminals **At_2** and **At_4** the synaptic vesicles are spherical, in terminal **At_3** a high proportion are flattened. **v** blood vessel. Ventral horn of the rat spinal cord, ×44,000

Synaptophysin, a glycoprotein, is the most abundant integral membrane protein of synaptic vesicles (Wiedenmann and Franke 1985). Synapsins are a family of phosphoproteins associated with the cytoplasmic surface of the synaptic vesicle (De Camilli et al. 1983); synapsins may interact with various cytoskeletal components, and a role in neurotransmitter release has also been proposed. Synaptotagmin is a Ca^{2+}-binding protein that functions in the regulation of synaptic vesicle exocytosis. Synaptobrevin (also known as VAMP, vesicle-associated membrane protein) plays a key role in exocytosis.

Numerous small and spheroidal mitochondria, often located away from the presynaptic membrane, usually occur within the presynaptic bouton. The abundance of these organelles probably reflects the high energy requirement at this site. As noted previously (see Sect. V.A.2), the tubules of the agranular endoplasmic reticulum may store and release Ca^{2+}. Ca^{2+} release regulates a number of presynaptic functions.

The structural organization of the dense material has been studied in greater detail using block-staining with phosphotungstic acid or bismuth-iodide (Gray 1963; Akert and Pfenninger 1969). Discrete, dense projections, which extend from the presynaptic membrane into the cytoplasm, can be seen in sections perpendicular to the presynaptic membrane; they are approximately 60 nm in diameter and spaced about 80 nm apart. In tangential sections, these dense projections appear interconnected by fine strands and are arranged in an hexagonal pattern, constituting the so-called presynaptic grid. It is thought that docked synaptic vesicles usually lie within the holes of the grid. Occasionally, dense projections can also be seen in material that had not been stained before embedding.

At certain sites (e.g., in the retina and pineal gland, Sjöstrand 1958; Kidd 1962; Villegas 1964; Dowling and Boycott 1967; Raviola and Raviola 1967; Kappers 1971; Oksche 1971; Vollrath and Huss 1973), the presynaptic ending contains a dense lamella of large surface area, called the synaptic ribbon, which is closely associated and often perpendicular to the presynaptic membrane. The surface of the ribbon is studded with small particles; hundreds of synaptic vesicles are linked to these particles via fine filaments. Similar ribbons are also present in certain junctions between a neuron and a sensory cell (e.g., in the inner ear). Some molecular components of the ribbon have been recently identified (e.g., see Sterling and Matthews 2005; tom Dieck et al. 2005). For the function of the ribbon synapses, see Sect. VI.E.

Some interesting characteristics of the presynaptic membrane have been revealed with the freeze-fracture technique. Firstly, the concentration of intramembranous particles is higher within the presynaptic membrane than in the surrounding nonjunctional regions. Some of the particles concentrated in the presynaptic membrane are probably receptors, while others, distinguishable by their larger size, probably correspond to Ca^{2+} channels (Pumplin et al. 1981). Furthermore, on the P face of the presynaptic membrane, corresponding to the inner leaflet of the split membrane viewed from outside the presynaptic ending, small craters can be seen, while tiny protuberances are apparent on the E face, corresponding to the outer leaflet of the split membrane viewed from inside the presynaptic terminal (Pfenninger et al. 1972). The small craters on the P face are thought to be membranous necks which connected synaptic vesicles with the presynaptic membrane at the time the tissue was prepared for freeze-fracturing, whereas the tiny protuberances on the E face are thought to correspond to the convexities of the synaptic vesicles attached to the presynaptic membrane.

Autoradiographic techniques and fluorescent microscopy have shown that neurotransmitter receptors are concentrated within the postsynaptic membrane. These receptors may correspond to the particle clusters revealed by freeze-fracturing. The inner aspect of the postsynaptic membrane is coated with a fuzzy layer of dense material (Figs. VI.1 and VI.3); this postsynaptic density is exactly aligned with, and often more evident than, the dense presynaptic material. Fine filaments are sometimes associated with the postsynaptic density and extend for a short distance into the adjacent cytoplasm. The postsynaptic dense material may form either a number of discrete patches or a

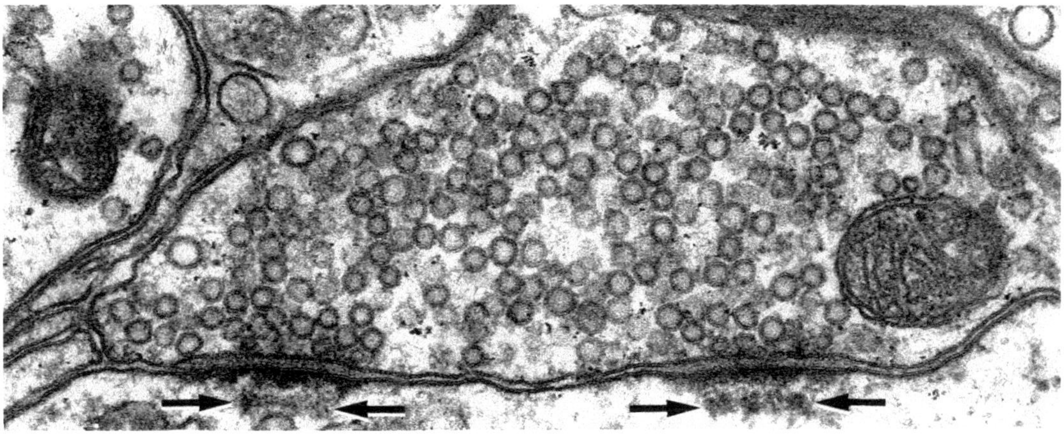

Fig. VI.3 Subsynaptic dense bodies. The bodies (*arrows*) shown in this picture are situated beneath the postsynaptic densities of axodendritic junctions. Ventral horn of the rat spinal cord, ×82,000

single plaque having one or more perforations. Postsynaptic densities can be isolated from brain homogenates by means of detergents, thus becoming accessible to biochemical analysis. Among the proteins that have been identified within postsynaptic densities, are actin, spectrin, tubulin, myosin Va, protein of postsynaptic density 95 (PSD-95), calmodulin, and fodrin (for a review, see Sheng 2001). It has been hypothesized that these proteins anchor receptors in the postsynaptic density, mediate receptor clustering, modulate receptor activity, control receptor internalization and turnover, have signaling roles, and link the postsynaptic membrane to the cytoskeleton.

Accumulations of dense material subjacent to, and separated from, the postsynaptic density can be observed in certain synapses (Fig. VI.3). These entities, which have been given various names (subsynaptic dense bodies, subsynaptic particles, subjunctional bodies, subsynaptic apparatus), may consist of one or more dense bars, a plaque, or a row of dense, spherical bodies. Since they may persist for a long period after the experimentally induced degeneration of presynaptic terminals (Sotelo 1968; Lund 1969), it has been suggested that they are not dependent upon the integrity of the presynaptic endings. The observations made by Rakic and Sidman (1973b) on the cerebellar cortex in "weaver" mutant mice are also consistent with this hypothesis. Another entity which can be observed within the postsynaptic element is the spine apparatus (see Sect. V.C.2).

The synaptic cleft (Fig. VI.1), which is about 20 nm across, is continuous with the narrow intercellular space of the nervous tissue (see Chap. X). Presynaptic and postsynaptic membranes adhere to each other so tightly that they remain attached under acute mechanical stress (e.g., when pre- and postsynaptic neurons are distorted by shrinkage during fixation). Even when the brain is fragmented by homogenization, the postsynaptic membrane clings to the presynaptic one, and they can be separated only by means of strong proteolytic agents. Cell adhesion molecules, e.g., syndecans and cadherins, are implicated in the cohesion between pre- and postsynaptic membranes (see Yamagata et al. 2003 for a review). Puncta adhaerentia, which contain the cadherin-catenin cell-adhesion complex, can also contribute to this cohesion. In certain synapses, many fine filaments bridge the synaptic cleft (De Robertis 1962; Van der Loos 1963; Ichimura and Hashimoto 1988), and in others, a thin layer of dense material is arranged parallel to the pre- and postsynaptic membranes. The material contained within the synaptic cleft does

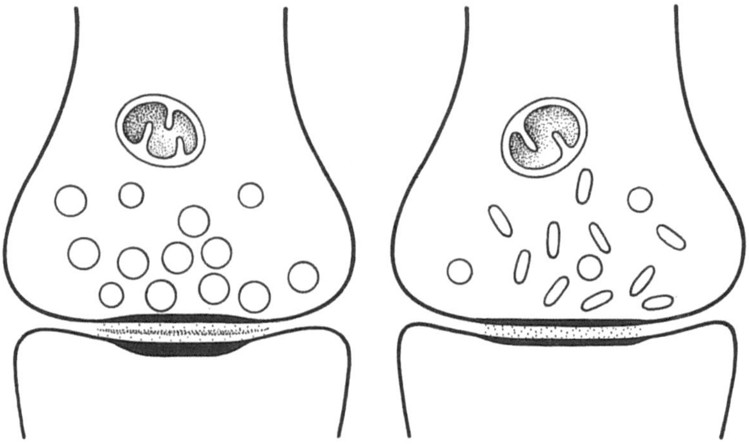

Fig. VI.4 Asymmetrical (*left*) **and symmetrical** (*right*) **synapses**; the main features after primary fixation with aldehydes. Compared with the symmetrical synapse, the asymmetrical synapse has a more conspicuous postsynaptic density and a wider synaptic cleft, in which an intermediate layer of dense material is present. The vesicles of the asymmetrical synapse are spherical, whereas many of the vesicles in the symmetrical synapse are flattened

not prevent the diffusion of transmitter molecules released into it, which are thus able to reach the postsynaptic membrane (see Sect. VI.E). Note that, although synapses possess a remarkable mechanical stability (see above), they do nevertheless undergo remodeling in normal adult animals (see Sect. VI.F). This suggests the presence of mechanisms that block the action of the adhesion molecules at these sites.

4. Correlation Between Structure and Function

Two types of interneuronal chemical synapses have been distinguished (Fig. VI.4) on the basis of their structural characteristics (Gray 1959; Colonnier 1968). The type 1 junction has a conspicuous postsynaptic density and a wide synaptic cleft containing an intermediate layer of dense material. Since the postsynaptic dense material is much more abundant than that found presynaptically, these junctions are often referred to as asymmetrical. The type 2 junction is characterized by a less conspicuous postsynaptic density and by a narrower cleft than the type 1 junction. Since in type 2 junctions the pre- and postsynaptic densities are roughly of the same thickness, they are often called symmetrical junctions.

Comparisons of results of electron microscopy and electrophysiology studies on the cerebellar cortex have shown that the synapses between the parallel fibers and the dendritic spines of Purkinje cells as well as those between the climbing fibers and the dendritic spines of Purkinje cells, which are both excitatory, are morphologically type 1, whereas the synapses between the basket cell axons and the dendritic shafts of Purkinje cells, which are inhibitory, are morphologically type 2. Similar relationships have also been discovered in other regions of the CNS. Since, however, synapse forms intermediate between type 1 (asymmetrical) and type 2 (symmetrical) have been described, it is not possible in all neuronal systems to characterize a synapse as excitatory or inhibitory on the basis of its structure alone.

Excitatory and inhibitory synapses have been compared in freeze-fractured preparations of the cerebellar cortex and the olfactory bulb. In these regions, prominent collections of particles were found associated with the postsynaptic membrane at excitatory synapses, whereas this specialization was absent at inhibitory synapses (Landis and Reese 1974a; Landis et al. 1974). It has been proposed that these particles are receptor molecules; therefore, the above observations suggest that in excitatory synapses receptors are clustered in groups, while in inhibitory synapses they are dispersed. However, further observations in other neuronal systems are necessary before the correlation can be accepted as a general rule.

Attempts to relate the structure of synapses to their function have also considered differences in the size and shape of the synaptic vesicles. The synaptic vesicles within excitatory synapses were found to be significantly larger than those within inhibitory synapses in the following regions of the nervous system: ventral cochlear nucleus of the rat, cat, and chinchilla (Lenn and Reese 1966); mouse cerebellum (Larramendi et al. 1967); spinal cord of the cat and monkey (Akert et al. 1972); and crayfish stretch receptor organs (Nakajima an Reese 1983).

Since, as described above (see Sect. VI.B.3), after primary aldehyde fixation certain synapses contain spherical vesicles, whereas others contain a high percentage of flattened vesicles (Fig. VI.2), attempts have been made to establish whether any correlation exists between vesicle shape and synaptic function. In the cat and mouse cerebellar cortex, synapses electrophysiologically identified as excitatory exhibit spherical vesicles, whereas synapses electrophysiologically identified as inhibitory contain a high percentage of flattened vesicles (Uchizono 1965; Larramendi and Victor 1967). Likewise, in the crayfish stretch receptor organs, excitatory synapses have spherical vesicles, while flattened vesicles predominate within inhibitory synapses (Tisdale and Nakajima 1976). Consistent with the above results are those reported in the hippocampus and the dentate gyrus of the rat and cat (Gottlieb and Cowan 1972). However, it is unclear whether the size and shape of synaptic vesicles are in all cases reliable indicators of the excitatory or inhibitory function of a synapse (e.g., see Hámori et al. 1990). A more reliable way to determine the excitatory vs. inhibitory function of a synapse is to identify (e.g., by immunocytochemistry) the neurotransmitter(s) present at the presynaptic ending and the receptor molecules present within the postsynaptic element. Unfortunately, methods of identification are still not available for all neurotransmitters and receptors.

5. Types of Synaptic Relations

Axodendritic and axosomatic contacts have been known since the end of the nineteenth century (see Sect. VI.B.1); later other types of synaptic relations were demonstrated. In this regard, it should be noted that any portion of a neuron (dendrite, soma, axon) may form a synapse with any portion of another neuron in every conceivable combination (Fig. VI.5).

Axodendritic synapses are the most common and may be type 1 or type 2, i.e., asymmetrical or symmetrical. They may occur at the dendritic shafts (Figs. VI.1 and VI.2) and at the dendritic spines (Fig. V.17). When spines are present, they are the more common sites of axodendritic synapses.

Axosomatic synapses may also be type 1 or type 2, the latter being much more common.

In some axoaxonal synapses, the terminal bouton of an axon is in contact with the endbulb of a second axon, which in turn is the presynaptic ending of an axodendritic synapse. Junctions having these features are sometimes referred to as serial synapses (Fig. VI.5). They have been described in the rat and cat spinal cord (Gray 1962, 1963; Ralston 1965), in the monkey lateral geniculate nucleus (Colonnier and Guillery 1964), in the cat cuneate nucleus (Walberg 1965b), in the cat gracile nucleus (Rustioni and Sotelo 1974), and at many other sites in the nervous system. According to Gray (1962, 1963)

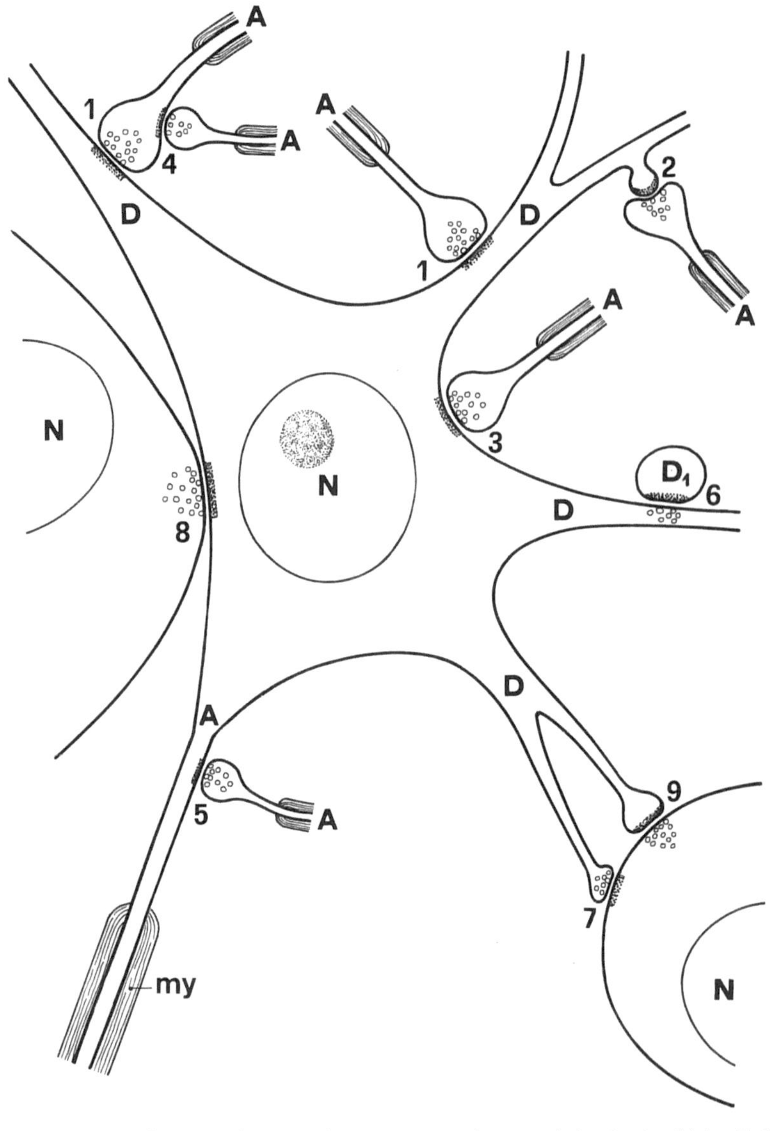

Fig. VI.5 Various types of synaptic relations: **1** axodendritic at a dendritic shaft, **2** axodendritic at a dendritic spine, **3** axosomatic, **4** axoaxonal between two axonal end-bulbs, **5** axoaxonal at the initial segment of axon, **6** dendrodendritic, **7** dendrosomatic, **8** somatosomatic, **9** somatodendritic. **A** axon, **D** dendrite, $\mathbf{D_1}$ transversely sectioned dendrite, **my** myelin, **N** neuronal nucleus

these junctions form the morphological basis for a type of presynaptic inhibition.

At other axoaxonal junctions, an axon ending synapses upon the initial segment of another axon. Junctions with these features have been found on the axon initial segment of the following neurons: Purkinje cells in the cerebellar cortex (Palay 1964; Hámori and Szentágothai 1965; Gobel 1971; Palay and Chan-Palay 1974), pyramidal cells of the cerebral cortex (Westrum 1966; Peters et al. 1968), giant cells of Deiters of the lateral vestibular nucleus (Sotelo and Palay 1970), Mauthner neurons of the brainstem (Nakajima 1974), neurons of the spinal dorsal horn (Réthelyi and Lozsádi 1990), sympathetic preganglionic neurons of the spinal cord (Llewellyn-Smith et al. 1995), and neurons of the thalamic reticular nucleus (Pinault et al. 1997). It has been suggested that these synapses upon the axon initial segment play an inhibitory role. Axoaxonal synapses have also been observed in invertebrates (Atwood and Morin 1970).

Axoaxonal relations in which membrane specializations are lacking and the junctional interspace is much wider than that found in typical synapses have been described in the CNS. These are known as synapses at distance. In the rat and cat cerebellar cortex, the terminal branches of the axons of a number of basket cells form a tangle around the initial segment of a Purkinje cell axon. Only a few of the basket cell axon terminals make true chemical synapses with the initial segment of the Purkinje cell axon, while the great majority end freely some distance away (Palay 1964; Hámori and Szentágothai 1965; Gobel 1971; Palay and Chan-Palay 1974). Physiological evidence shows that this junction at distance exerts an inhibitory influence on the Purkinje cell. In the goldfish, many thin, unmyelinated axons form a spiral plexus around, but at some distance from, the initial segment of the Mauthner cell axon, while some of these spiral axons end on this initial segment (Robertson et al. 1963; Kohno 1970; Nakajima 1974). The spiral axons form symmetrical synapses between themselves. This spiral plexus, also called the axonal cap, is covered by neuroglial cells whose delicate processes insert themselves between the spiral axons to end very close to the initial segment of the Mauthner cell axon. Stimulation of the spiral plexus axons was reported to produce an inhibition of the Mauthner cell.

Dendrodendritic synapses were first demonstrated both morphologically and physiologically between granule cells and mitral cells in the olfactory bulb (Hirata 1964; Andres 1965a; Rall et al. 1966). Later, they were identified between mitral and tufted cells of the salamander olfactory bulb (Allen and Hamilton 2000); in the primate retina (Dowling and Boycott 1967); in the lateral geniculate body of the cat (Famiglietti 1970), monkey (Wong 1970), mouse (Rafols and Valverde 1973), and rat (Lieberman 1973); in the medial geniculate body of the cat (Morest 1971); in the thalamus of the cat (Ralston and Herman 1969), monkey (Harding 1971), and rat (Pinault et al. 1997); in the monkey cerebral cortex (Sloper 1971); in the suprachiasmatic nucleus of the rat hypothalamus (Güldner and Wolff 1974); in the cat spinal cord (Gobel et al. 1980); in the rat superior cervical ganglion (Kawai and Senba 1995); and in other regions of the nervous system.

Dendrosomatic synapses have been found in the primate retina (Dowling and Boycott 1967), in the ventral thalamus of the cat (Ralston and Herman 1969), in the rat olfactory bulb (Pinching 1970), and in the suprachiasmatic nucleus of the rat hypothalamus (Güldner 1976).

Somatosomatic synapses have been observed in the rat olfactory bulb (Pinching 1970) and in the cat retina (Fisher 1972), whereas somatodendritic synapses have been found in the frog optic tectum (Sétáló and Székely 1967), in the lateral geniculate body of the monkey (Wong 1970) and rat (Lieberman 1973), and in the suprachiasmatic nucleus of the rat hypothalamus (Güldner 1976).

Finally, the small cells containing intensely fluorescent granules in autonomic ganglia have been observed to synapse upon axonal processes (see, among others, Yamauchi et al. 1975). If these cells are regarded as interneurons, then the junctions should be interpreted as somatoaxonic synapses.

6. Reciprocal Synapses

This term refers to adjacent areas of a single synaptic interface which are polarized in opposite directions. Synapses of this type were first detected in the olfactory bulb (Hirata 1964; Andres 1965a).

In most cases observed to date, reciprocal synapses occur between two dendrites. Reciprocal dendrodendritic junctions have been found in the olfactory bulb of the rat (Rall et al. 1966) and cat (Willey 1973), in the lateral geniculate body of the cat (Famiglietti 1970), in the ventral thalamus of the monkey (Harding 1971), and in the suprachiasmatic nucleus of the rat hypothalamus (Güldner and Wolff 1974). However, reciprocal synapses have also been found between two axons (Atwood and Kwan 1979; Elekes and Florey 1987; Pearce and Govind 1993, in invertebrates), between an axon and a dendrite (Dowling and Boycott 1967, in the primate retina; Witkovsky and Stell 1973, in the retina of a dogfish), and between a dendrite and a nerve cell body (Price and Powell 1970, in the rat olfactory bulb). Reciprocal synapses also occur in the PNS. They have, for example, been identified in the human organ of Corti between outer hair cells and afferent axons (Nadol 1988).

Rall et al. (1966) have correlated their morphological observations with electrophysiological probes and concluded that in the reciprocal synapses of the olfactory bulb, the component oriented in a mitral-to-granule direction is excitatory and that oriented in a granule-to-mitral direction is inhibitory.

7. Synaptic Glomeruli

The term glomerulus was introduced at the end of the nineteenth century to designate clusters of intertwined nerve cell processes connected by numerous synapses and localized in circumscribed, cell body-free areas of the olfactory bulb and the cerebellar cortex (see, among others, Held 1897b). These formations can be distinguished in the light microscope at low magnification and are often partially encircled by glial processes.

Within the olfactory glomeruli, afferent olfactory axons make asymmetric synapses with the dendritic tufts of mitral, tufted, or periglomerular cells; moreover, the dendrites of mitral or tufted cells make synaptic contacts with the dendrites of periglomerular cells (Andres 1970; Reese and Brightman 1970; Pinching and Powell 1971b; White 1972). Certain dendrodendritic synapses between mitral and periglomerular cells are of the reciprocal type (see Sect. VI.B.6). The synapse in which the mitral cell dendrite is presynaptic to the periglomerular cell dendrite is thought to be asymmetrical, whereas that polarized in the opposite direction would be symmetrical (White 1972). Each glomerulus is partially encircled by astrocytic sheet processes. A summary of the neurotransmitters identified in neurons of olfactory glomeruli can be found in Shipley and Ennis (1996).

The cerebellar glomeruli (Fig. VI.6) are located in the cortical granular layer. While the first accurate description of the synapses between mossy fibers and granule cell dendrites was provided by Ramón y Cajal (1896c), many details of the structure of these glomeruli were later provided by electron microscope studies (Gray 1961; Palay 1964; Hámori and Szentágothai 1966; and subsequently other workers). The center of the glomerulus is occupied by the large ending of a mossy fiber, which contains synaptic vesicles and mitochondria. This axonal ending makes numerous small asymmetrical synapses with the granule cell dendrites indenting its surface, and often it synapses also upon a Golgi cell dendrite. In the peripheral zone of the glomerulus, axons of the Golgi cells make symmetrical axodendritic synapses with the granule cell dendrites. After primary aldehyde fixation, the synaptic vesicles appear spherical within the ending of the mossy fiber and mostly flattened within the end-bulbs of the Golgi cell axons. The glomerulus is partially separated from the surrounding neuropil by astrocytic processes.

Concentrations of numerous synaptic junctions within small discrete areas were later detected under the electron microscope in other regions of the nervous system (e.g., dorsal lateral geniculate nucleus, substantia gelatinosa of the spinal cord, main sensory trigeminal nucleus, and inferior olivary nucleus). Though not recognizable under the light microscope because of their small size, these areas have characteristics comparable with those of the above-described glomeruli and hence are termed as such.

In each synaptic glomerulus of the dorsal lateral geniculate nucleus, there is a centrally

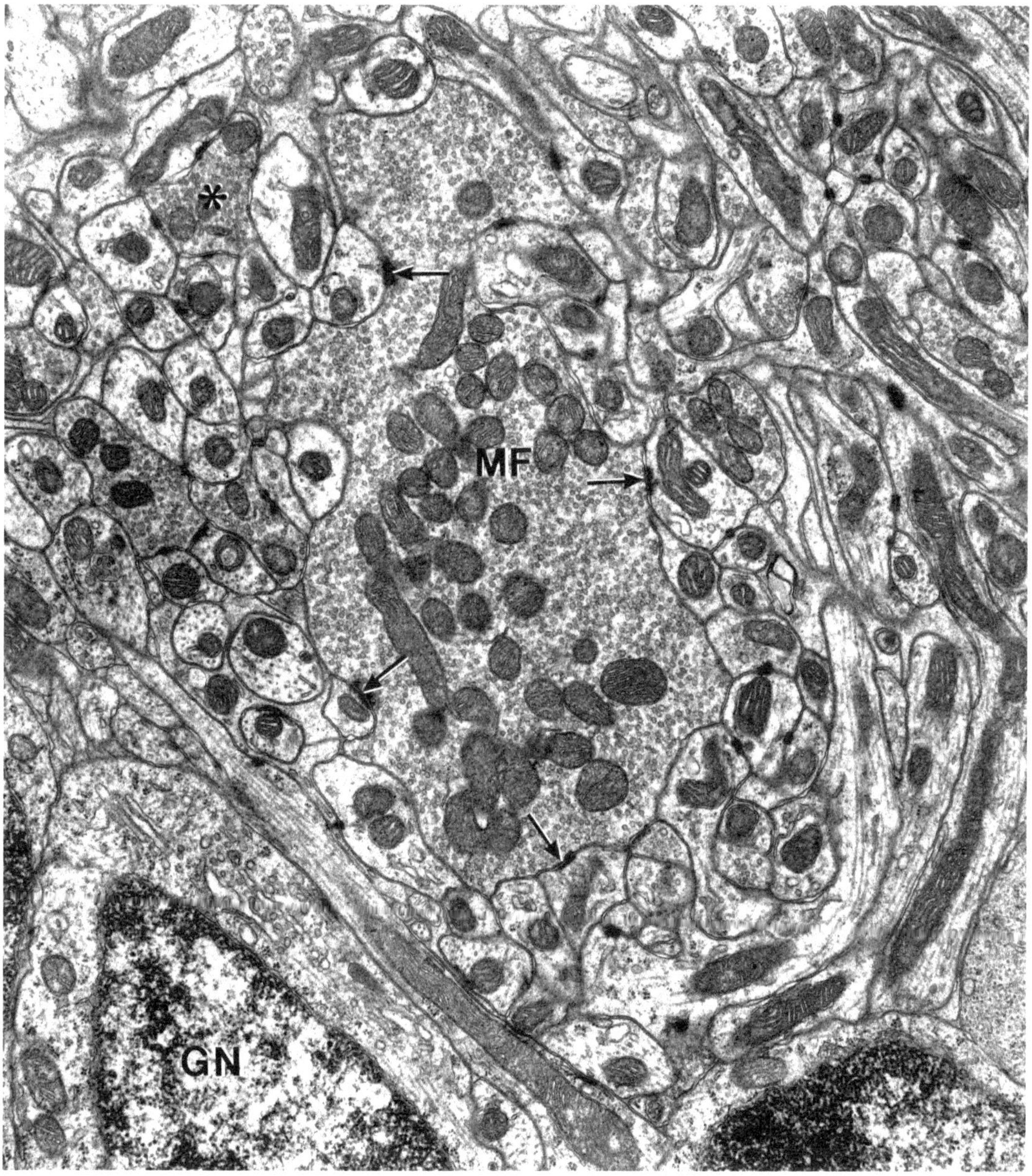

Fig. VI.6 Cerebellar glomerulus. The center of the glomerulus is occupied by the large ending of a mossy fiber (**MF**), which contains many synaptic vesicles and mitochondria. This axonal ending makes numerous small synapses (some *arrowed*) with the granule cell dendrites. In the peripheral zone of the glomerulus, axons of the Golgi cells (one of which is marked *) make axodendritic synapses with the granule cell dendrites. **GN** granule cell nucleus. Granular layer of the rat cerebellar cortex, ×18,000

placed axon terminal of a retinal ganglion cell. This retinogeniculate axon terminal, which is filled with spherical synaptic vesicles, is presynaptic to two types of dendritic elements: (a) spinous dendritic branches which lack synaptic vesicles and belong to geniculocortical neurons and (b) dendrites which contain a high percentage of flattened vesicles and probably arise from local circuit neurons (Guillery 1969; Guillery and

Colonnier 1970; Szentágothai 1970; Le Vay 1971; Famiglietti and Peters 1972; Rafols and Valverde 1973; Rapisardi and Miles 1984). The latter dendritic components of the glomerulus are GABAergic and in their turn are presynaptic to the dendrites of geniculocortical neurons. In this glomerulus, therefore, the synaptic connections are arranged in a triad. Both dendritic elements of the triad are postsynaptic to the retinogeniculate axon terminal, and the dendritic branch of a geniculocortical neuron is also postsynaptic to the dendrite containing flattened vesicles. It is thought that the retinogeniculate axon terminal exerts excitatory influences on both dendritic elements of the triad and that the GABAergic dendrite containing flattened vesicles transmits inhibitory signals to the dendritic branch of the geniculocortical neuron. In this glomerulus, which is partially encapsulated by neuroglial processes, other axon terminals whose origins have not been clarified are also present.

In the dorsal horn of the spinal cord are found glomeruli whose central core consists of a large end-bulb of the central axon of a primary sensory neuron. This axonal end-bulb, which contains spherical synaptic vesicles, is surrounded by numerous dendritic spines, some dendritic shafts, and some axonal terminal boutons, the latter containing a high percentage of flattened vesicles. The large, centrally located end-bulb of the sensory axon is presynaptic to the surrounding dendritic spines and shafts, with which it forms asymmetric synapses, and is postsynaptic to the axonal boutons, with which it forms symmetric synapses (Réthelyi and Szentágothai 1969; Szentágothai 1970; Maxwell and Réthelyi 1987).

At the center of each synaptic glomerulus of the main sensory trigeminal nucleus is a trigeminal primary afferent axon which is surrounded by small axonal terminal boutons and by small dendrites. Both the central axon and the peripheral axonal boutons are presynaptic to adjacent dendrites; presumably, the peripheral axonal boutons also synapse upon the central axon (Gobel and Dubner 1969).

While the central cores of the synaptic glomeruli described above consist of one or more axons, the central cores of the glomeruli in the inferior olivary nucleus consist of dendrites, which arise in all likelihood from neurons of that nucleus (Němeček and Wolff 1969; Bowman and King 1973; Sotelo et al. 1974). Gap junctions frequently occur between the dendrites forming the central core (Sotelo et al. 1974). The central core is surrounded by many axonal end-bulbs, several containing spherical vesicles and others mainly flattened vesicles after primary aldehyde fixation. These axonal end-bulbs (whose origin is still unknown) are in synaptic contact with the central dendrites. The glomerulus is almost completely surrounded by thin astrocytic laminar expansions.

C. Autapses

As noted at the beginning of Sect. VI.B, the synapse is a junction specialized for the transmission of signals from one neuron to another. However, neurons may also form synapses with themselves. In fact it has been observed that axon collaterals of a neuron may synapse with a dendrite or with the soma of the same neuron. These junctions were called autapses by Van der Loos and Glaser (1972).

It seems that Held (1897a) was the first to describe autapses. He observed them in Purkinje cells of the human cerebellar cortex. Initially these contacts were detected with the light microscope in Golgi preparations. Subsequently their existence was confirmed by light or electron microscope examination of neurons injected with intracellular markers (e.g., horseradish peroxidase or biocytin). Autaptic contacts have been observed in various regions of the nervous system of different species: e.g., in the cerebral cortex of rat (Peters and Proskauer 1980; Lübke et al. 1996), rabbit (Van der Loos and Glaser 1972), and cat (Tamás et al.

1997) and in the substantia nigra of cat (Karabelas and Purpura 1980). A high proportion of pyramidal neurons in the rat neocortex contain autapses, although the number of autapses per cell is small: on average two per neuron (Lübke et al. 1996).

Several investigators have considered autapses as aberrant or rudimentary nonfunctional structures. However, it has been shown that the autapses of certain neurons are functional (e.g., Bacci et al. 2003), and it has been suggested that autaptic contacts modulate the integrative properties and firing rhythm of the neuron (Tamás et al. 1997) or provide a form of self-inhibition (Bacci et al. 2003).

D. The Neuromuscular Junction

The junction between one of the terminal axonal branches of a motoneuron and a striated muscle fiber is called the neuromuscular junction or the motor end plate. The discovery of the neuromuscular junction is attributed by certain authors to Doyère (1840), who, however, studied an invertebrate endowed with smooth muscle cells only. The earliest descriptions of neuromuscular junctions were probably those provided in 1862 by W. F. Kühne [1837–1900] and in 1863 by W. Krause [1833–1910].

As may be seen under the light microscope, in the immediate vicinity of the muscle fiber, one of the terminal branches of a motor axon loses its myelin sheath and ramifies to form a circumscribed arborization. Schwann cells, also called here terminal or perisynaptic Schwann cells, cover the branchlets of this arborization and separate them from the connective tissue but do not come between them and the muscle fiber; thus, the branchlets of the terminal arborization and the muscle fiber face each other directly. These Schwann cells possess purinergic and muscarinic receptors. For the role of these cells, see Sect. VII.D.7. The nervous arborization with its related Schwann cells forms a slight mound on the surface of the muscle fiber (Rouget 1862). This protuberance has a smooth, elliptic or circular outline in mammals and a scalloped outline in reptiles. Its size, which ranges from 20 to 80 μm in length and from 10 to 30 μm in width, varies with the muscle and the species and in general is greater in reptiles than mammals (Kühne 1886). See Ogata (1988) for a comparative review of motor end plates in many vertebrate species.

The branchlets of the nerve arborization lie in synaptic troughs, also called primary synaptic grooves, which indent the surface of the muscle fiber. By means of certain techniques (e.g., staining with ferric hematoxylin, supravital staining with Janus green, cytochemical technique for cholinesterase), it is possible to distinguish numerous ribbon-shaped lamellae attached by one edge to the convex outline of the synaptic troughs and projecting into the underlying sarcoplasm (Couteaux 1944). The structure consisting of the primary synaptic grooves and the lamellae was termed by Couteaux the subneural apparatus. The sarcoplasm which underlies the nerve arborization is rich in nuclei and mitochondria and depleted in myofibrils; it is often referred to as the sole of the motor end plate. An accumulation of nuclei is present at the level of this junction (Rouget 1862). Of these nuclei, some belong to the Schwann cells and are known as arborization nuclei; others, which are larger and less stainable than the former, belong to the muscle fiber and are called sole nuclei.

Electron microscopy has shown that the nerve endings associated with the muscle fiber (Fig. VI.7) contain synaptic vesicles, mitochondria, and tubules of the agranular endoplasmic reticulum (Palade and Palay 1954; Reger 1955; Robertson 1956; Andersson-Cedergren 1959). A synaptic cleft about 60 nm wide separates the plasma membrane of the nerve ending (presynaptic membrane) from that of the muscle fiber (postsynaptic membrane). The former membrane is relatively smooth, whereas the latter forms a series of narrow infoldings. A single basal lamina occupies the synaptic cleft and

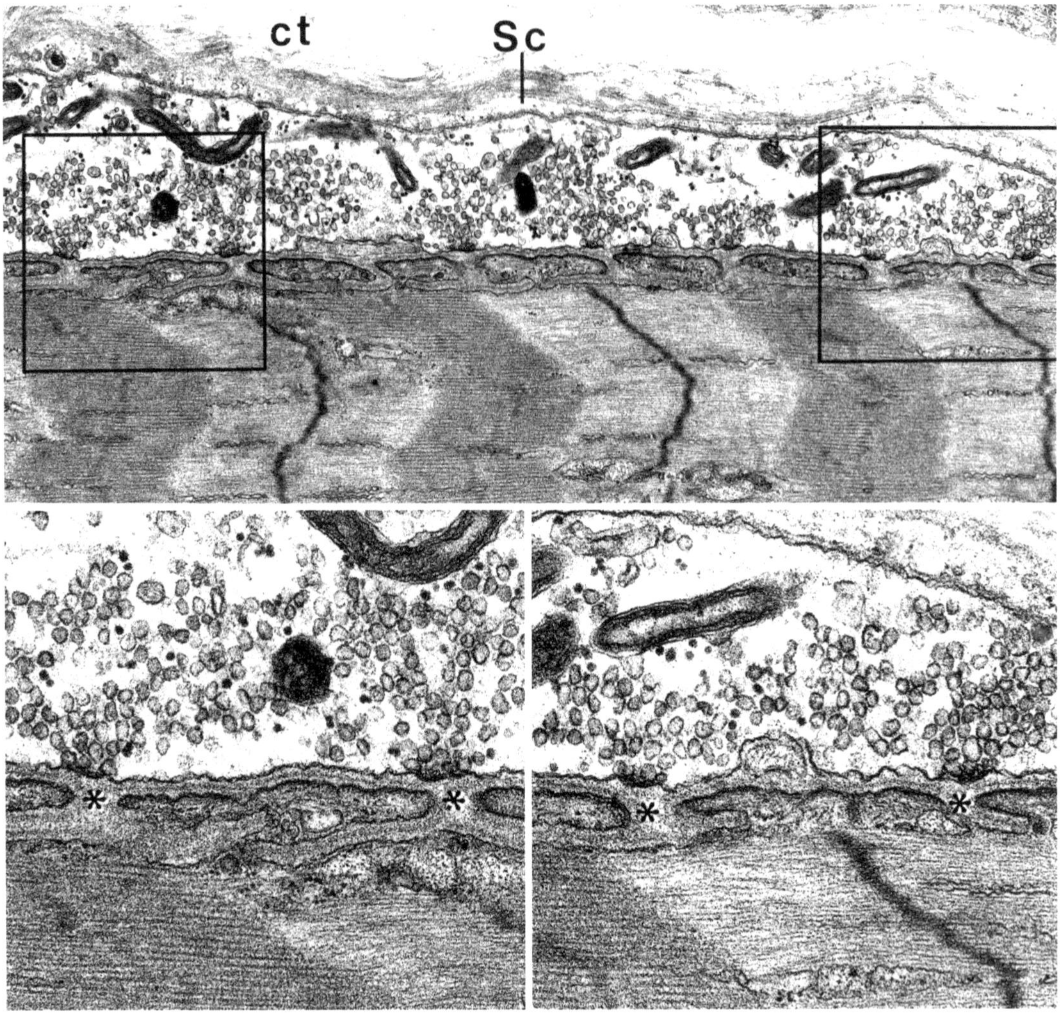

Fig. VI.7 Neuromuscular Junction. The nerve ending, which contains synaptic vesicles and mitochondria, is separated from the surface of the muscle fiber by a large cleft which contains the basal lamina. While the plasma membrane of the nerve ending (presynaptic membrane) is relatively smooth, the plasma membrane of the muscle fiber (postsynaptic membrane) forms a series of narrow postsynaptic folds, often called secondary synaptic clefts, into which the basal lamina extends. Synaptic vesicles appear clustered around tufts of dense material, which face the openings (*) of the postsynaptic folds. The *boxed areas* in the *upper figure* are shown enlarged in the corresponding *lower figures*. **ct** connective tissue, **Sc** Schwann cell. Cutaneous pectoris muscle of the frog; *upper figure*, ×15,500; *lower figures*, ×33,000 (Courtesy of N. Iezzi and B. Ceccarelli)

extends into the clefts formed by the infoldings of the postsynaptic membrane (Fig. VI.7). This basal lamina contains a variety of proteins, including acetylcholinesterase, agrin, and a laminin isoform. The postsynaptic folds (also called secondary synaptic clefts) can vary considerably in number, depth, and orientation according to the muscle and the species (Couteaux 1961). Although very numerous in mammals, these folds may be almost completely absent in fish; in the frog, fast muscle fibers have numerous postsynaptic folds, whereas slow muscle fibers have few folds. The secondary synaptic clefts correspond to the lamellae of the subneural apparatus visible under the light microscope (see above). Both primary synaptic grooves and secondary synaptic

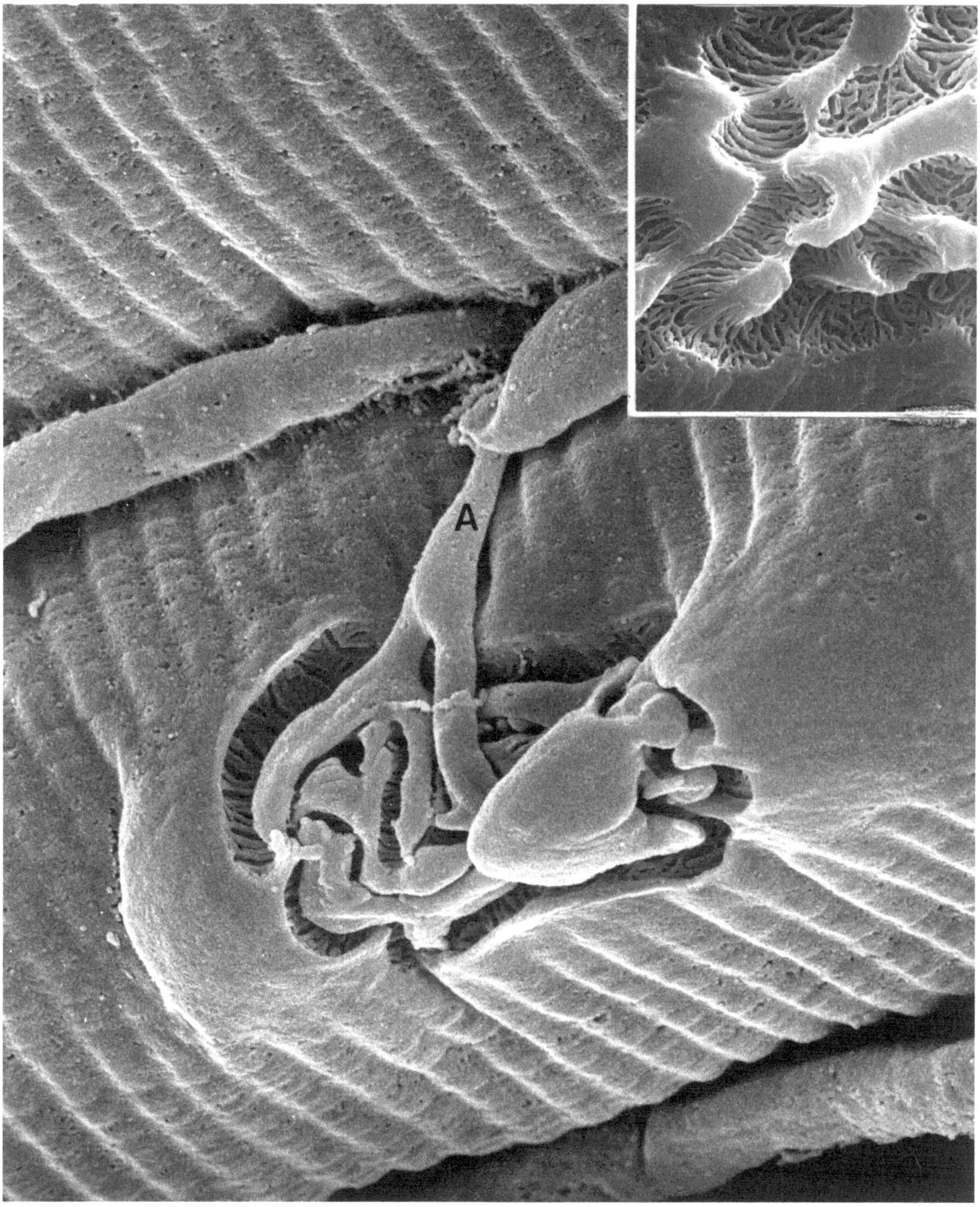

Fig. VI.8 Neuromuscular Junction. This scanning electron micrograph shows a terminal branch of a motor axon (**A**) which forms a circumscribed arborization on the surface of a muscle fiber. The branchlets of the nerve arborization lie in synaptic troughs (primary synaptic grooves). In the *inset*, both primary synaptic grooves and secondary synaptic clefts are revealed after removal of the terminal nerve arborization. Peroneus longus muscle of Chinese hamster, ×6,000 (Courtesy of Y. Matsuda)

clefts may be easily seen with the scanning electron microscope after removal of the terminal nerve arborization (Fig. VI.8).

As in the case of the membranes of the interneuronal chemical synapses (see Sect. VI.B.3), the pre- and postsynaptic membranes of the neuromuscular junction adhere strongly to each other; they can be separated by prolonged treatment with protease.

In thin sections oriented perpendicular to the pre- and postsynaptic membranes of a neuromuscular junction, synaptic vesicles often appear clustered around tufts of dense material (Fig. VI.7) which project from the presynaptic membrane into the axoplasm (Birks et al. 1960). These presynaptic densities have, in fact, the form of narrow bands running perpendicular to the long axis of the nerve ending and often face the secondary synaptic clefts (Fig. VI.7); each of these bands with its associated cluster of vesicles is called an active zone (see Sect. VI.B.3). Transmitter release occurs exactly at these sites. In freeze-fractured replicas, the abovementioned bands are easily identified since they are typically bordered by a double row of large particles located in the cytoplasmic leaflet of the split presynaptic membrane. These particles are probably Ca^{2+} channels (Robitaille et al. 1990). Clusters of particles have also been found in the cytoplasmic leaflet of the freeze-fractured postsynaptic membrane (Dreyer et al. 1973; Heuser et al. 1974).

The organization of the neuromuscular junction is similar to that of interneuronal chemical synapses. However, these two junctions exhibit a number of clear differences: (a) whereas both cells meeting at an interneuronal chemical synapse are derived from the ectoderm, the postsynaptic element of a neuromuscular junction is of mesodermal origin; (b) the synaptic cleft is wider in the motor end plate than in the interneuronal synapse; (c) in the motor end plate, the synaptic cleft is occupied by a basal lamina, which is missing in the interneuronal synapse; (d) well-defined postsynaptic densities are lacking in the neuromuscular junction; and (e) the interneuronal synapse lacks postsynaptic folds.

The efferent innervation of the smooth musculature is considered in Sect. VI.H.

E. Structural Aspects of Synaptic Activity

When an action potential arrives at the presynaptic bouton in a cholinergic synapse, the bouton releases acetylcholine, a neurotransmitter (Dale et al. 1936). The neurotransmitter molecules diffuse across the synaptic cleft and bind rapidly (in only a few microseconds) to specific receptors located on the surface of the postsynaptic membrane. Each receptor is a component (binding component) of a membrane protein; the other portion (ionophore component) of the same protein circumscribes a fine channel spanning the postsynaptic membrane. As a consequence of the interaction between the neurotransmitter and the receptor, the ionophore component is believed to change its conformation, resulting in the opening of the corresponding channel and thereby permitting the flow of certain ions through the postsynaptic membrane.

While acetylcholine directly opens ion channels located in the postsynaptic membrane, other neurotransmitters open ion channels indirectly, i.e., via a second messenger. In both cases, the binding of the neurotransmitter to the receptor appears to cause a transitory increase in the permeability of the postsynaptic membrane to certain ions.

In excitatory synapses, the Na^+ channels open, permitting a rapid and substantial flow of Na^+ into the postsynaptic element under the influence of concentration and electrical gradients (see Sect. V.E). In spite of the fact that K^+ channels also open under these conditions, permitting a smaller quantity of K^+ to flow out of the postsynaptic element under the influence of the concentration gradient, the membrane potential of the postsynaptic neuron, which in the resting state is −70 mV (see Sect. V.E), becomes less negative. If this change in the membrane potential reaches the threshold value, a depolarization of the postsynaptic membrane is triggered; in other words, a nervous signal, which manifests itself as an action potential, arises in the postsynaptic neuron. In most neurons, the action potential initiates at the initial segment of the axon, which is the region with the lowest threshold, and propagates both along the axon and, retrogradely, along the cell body and dendrites.

At inhibitory synapses, the interaction between neurotransmitter and receptor on the postsynaptic membrane would cause the opening of K^+ channels, so that K^+ flow out of the postsynaptic element under the influence of the concentration gradient. Because the Na^+ channels remain closed, the membrane potential of the postsynaptic element changes from −70 to −75 or −80 mV, thus becoming more negative than at rest. In these synapses, neurotransmitter release results, therefore, in the hyperpolarization of the postsynaptic membrane, i.e., a change in membrane potential of opposite sign to that described above for excitatory synapses. Under such conditions, excitatory synapses on the same postsynaptic neuron will find it more difficult to raise the membrane potential to the threshold value and therefore trigger depolarization. In brief, the activity of both excitatory and inhibitory synapses causes an increase in the ionic permeability of the postsynaptic membrane. However, in excitatory synapses the Na^+ channels open so that Na^+ flow into the postsynaptic neuron, whereas in inhibitory synapses the opening of K^+ channels takes place, and K^+ flow out of the postsynaptic neuron, thus triggering reverse effects on the polarization of the postsynaptic membrane.

It is generally accepted that in cholinergic presynaptic endings acetylcholine exists in two main pools. About 75 % of this transmitter would be stored within the synaptic vesicles, and about 25 % would be free in the cytoplasmic matrix (Weiler et al. 1982).

While recording at the neuromuscular junction under resting conditions, Fatt and Katz (1952) observed random sequences of brief, minute, spontaneously occurring local waves of depolarization and called them spontaneous miniature end plate potentials. On the basis of this and other observations, del Castillo and Katz (1955) advanced an explanation of the mechanism of chemical transmission consisting of the following points: (a) each miniature potential arises from the release of a multimolecular packet of neurotransmitter, (b) the number of transmitter molecules necessary to produce a miniature potential is remarkably constant in size and hence is called a quantum, (c) a quantum consists of a few thousand (probably less than 10,000) molecules of the neurotransmitter, (d) each quantum is released independently of all others, and (e) each quantum corresponds to the content of a single synaptic vesicle. This theory is known as the vesicular release theory or vesicular hypothesis. Phenomena analogous to those originally observed by Katz and collaborators at the neuromuscular junction were subsequently observed at ganglionic and central chemical synapses, and it was therefore assumed that the vesicular release theory is applicable to all chemical synapses and to all neurotransmitters.

When the presynaptic terminals release small quantities of neurotransmitter, i.e., the contents of several synaptic vesicles, the changes in the postsynaptic membrane potential do not reach the threshold value, and no signal is transferred from one cell to the next. When a nerve impulse arrives at the presynaptic terminal, the contents of several hundred synaptic vesicles are discharged in less than 1 ms, i.e., nearly simultaneously. Under such conditions, the change in the postsynaptic membrane potential reaches the threshold value, and a propagating electrical signal arises in the postsynaptic cell. According to this theory, the discharge of transmitter molecules stored within a single synaptic vesicle, i.e., the release of a quantum of neurotransmitter, is the basic event of synaptic activity both in the resting state and when a nerve impulse arrives at the presynaptic terminal. The structural counterpart of this basic event would be the coalescence of a vesicle with the presynaptic membrane and the subsequent opening of the vesicle into the synaptic cleft. What distinguishes resting state phenomena from the events that occur when a nerve impulse arrives at the presynaptic terminal seems essentially to be the number of basic events occurring simultaneously or nearly so. Hence, release of the neurotransmitter occurs essentially by a secretory mechanism. Release of the neurotransmitter stored in synaptic vesicles is preceded by the entry of Ca^{2+} into the presynaptic ending and is probably triggered by the combination of Ca^{2+} with calmodulin.

The entire sequence of events involved in the transmission process (influx of Ca^{2+} into the presynaptic ending, exocytosis of the contents of synaptic vesicles, diffusion of the transmitter across the synaptic cleft, and transmitter action on the postsynaptic membrane) delays the postsynaptic potential with respect to the presynaptic potential. This synaptic delay is 0.5–0.9 ms for mammalian CNS synapses but is longer for the synapses in sympathetic ganglia.

The regulated release of neurotransmitters stored in synaptic vesicles occurs at the active zone (see Sects. VI.B.3 and VI.D). Since docked vesicles are believed to be released immediately after stimulation, they are usually referred to as the readily releasable pool of synaptic vesicles. As docked vesicles undergo exocytosis, it is thought that they are replaced by vesicles located at varying distances from the active zone; these latter vesicles constitute the reserve pool. By contrast, the release of the content of large, dense-cored vesicles occurs at parts of the presynaptic membrane distant from the active zones. It is widely held that synaptic strength (i.e., the efficacy of synaptic transmission) is related to the number and size of active zones (in which docked vesicles are concentrated) present in the presynaptic element and to the size of the postsynaptic density (in which neurotransmitter receptors are clustered).

The synaptic ribbon (see Sect. VI.B.3) occurs in synapses that release neurotransmitter at a high rate for relatively long periods. A single ribbon synapse can release several hundred synaptic vesicles per second. This release, which occurs at a much higher rate than in conventional synapses, seems to be made possible by the fact that the ribbon tethers a large pool of vesicles close to the active zone, so that they can rapidly replace released vesicles.

To confirm the vesicular release theory, much work has been carried out to determine whether the number of synaptic vesicles significantly diminished following stimulation. Work was done both on the PNS (e.g., frog, rat, and cat superior cervical ganglion, frog motor end plates), and on the CNS (e.g., lamprey spinal cord). While many studies showed significant reduction in the number of vesicles in presynaptic endings after prolonged and intense stimulation (Hubbard and Kwanbunbumpen 1968; Ceccarelli et al. 1972; Perri et al. 1972; Heuser and Reese 1973; Birks 1974; Pysh and Wiley 1974; Wickelgren et al. 1985; Koenig et al. 1989), divergent results were also obtained. It is probable that this lack of uniformity of results is attributable to the slowness of chemical fixation, which may allow time for the reformation of vesicles (see later, this section). After long periods of intense stimulation, an increase was observed in the surface area of the presynaptic membrane, which also became more convoluted (Heuser and Reese 1973; Pysh and Wiley 1974; Wickelgren et al. 1985). This result is consistent with the theory that transmitter release takes place by exocytosis of the contents of synaptic vesicles and with the idea that vesicle discharge involves the incorporation of the vesicle membrane into the presynaptic membrane.

Results more clearly supporting the vesicular release theory were obtained with the quick-freezing technique, by which a nerve-muscle preparation is frozen within a few milliseconds of nerve stimulation, making it possible to observe under the electron microscope the structural changes taking place at a

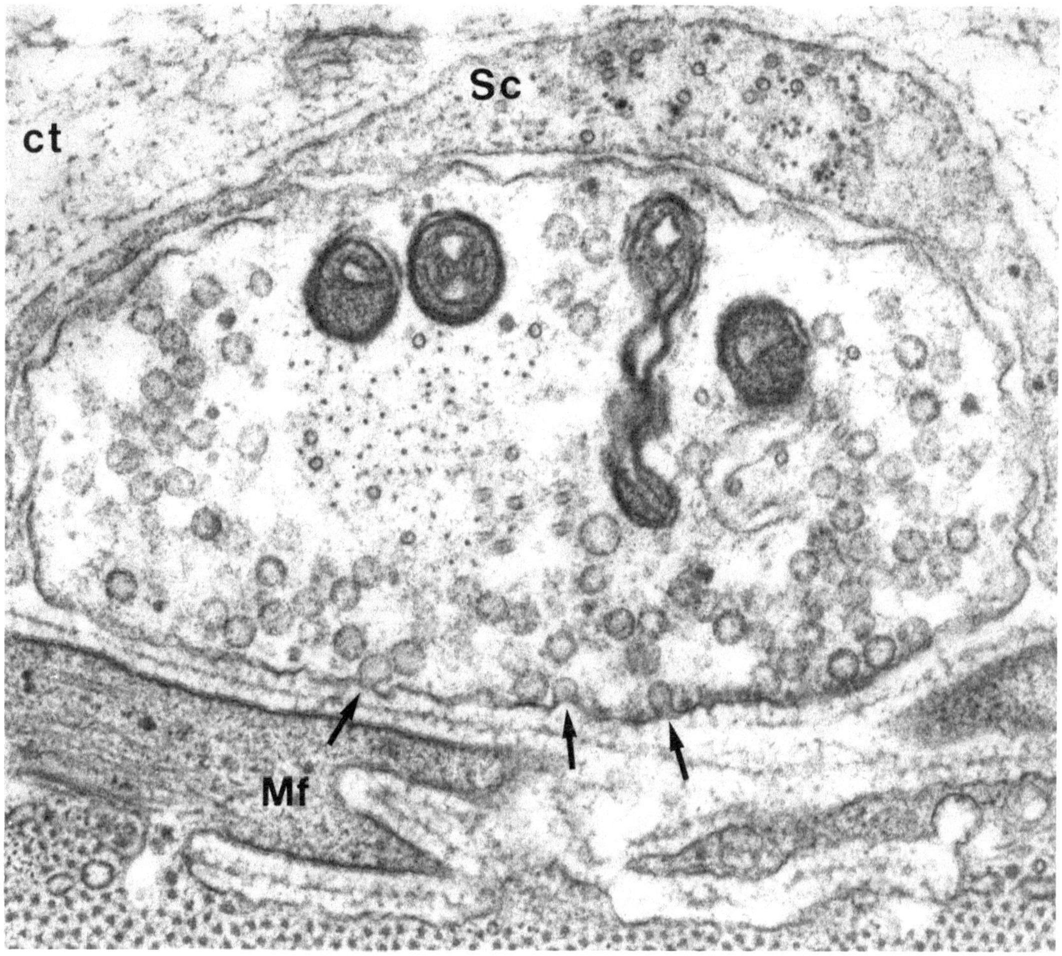

Fig. VI.9 Neuromuscular Junction. Electron micrograph from a nerve-muscle preparation quick-frozen 4 ms after application of a single electric stimulus in 1 mM 4-aminopyridine and cryosubstituted. Different degrees of association between vesicles and the plasma membrane of the nerve ending (presynaptic membrane) can be seen. *Arrows* indicate vesicles which have opened into the synaptic cleft. **ct** connective tissue, **Mf** muscle fiber, **Sc** Schwann cell. Cutaneous pectoris muscle of the frog, ×68,000 (Courtesy of F. Torri-Tarelli, F. Grohovaz, R. Fesce, and B. Ceccarelli)

motor end plate during transmitter release. As shown in Fig. VI.9, in frog motor end plates 3 or 4 ms after nerve stimulation, many synaptic vesicles were observed fused with the presynaptic membrane, their lumina being continuous with the synaptic cleft (Heuser and Reese 1981). There is strong evidence that the synaptic vesicle exocytosis into the synaptic cleft and the neurotransmitter release occur at nearly the same time and that the exocytotic vesicles are nearly as abundant as the number of quanta released (Heuser et al. 1979; Haimann et al. 1985; Torri-Tarelli et al. 1985).

After vesicles have discharged their contents into the synaptic cleft, they become part of the presynaptic membrane. Successively the vesicle membrane is retrieved and reused to form new synaptic vesicles. A number of observations suggest that vesicle membrane is retrieved from the presynaptic membrane by endocytosis. For example, when extracellular tracers (e.g., lanthanum ions or peroxidase) are applied, the stimulation of frog motor end plates is followed by the uptake of the tracer into a large proportion of vesicles. Two models of synaptic vesicle retrieval have been proposed. According to one model, endocytosis occurs near the active zones via the formation of uncoated vesicles (Ceccarelli et al. 1972, 1973); this mechanism would be clathrin independent. In the second model, endocytosis occurs away from the active zones; it involves the formation of coated vesicles (Heuser and Reese 1973; Miller and Heuser 1984) and is mediated by clathrin. For some time these were rival models. However, recent evidence suggests that both models may be valid and that vesicles budding near the active zones supply the readily releasable pool of synaptic vesicles, whereas vesicles arising away from the active zones enter the reserve pool. It is commonly held that vesicle membrane can be used several times in successive generations of vesicles, though direct validation of this hypothesis is still lacking.

The cholinergic synapses studied to date do not become depleted of vesicles even after prolonged and intense nerve stimulation. This is in keeping with the fact that the store of neurotransmitter in cholinergic synapses is not depleted by prolonged stimulation as acetylcholine is normally synthesized fast enough to replace the amount released. There is evidence that acetylcholine is normally synthesized in the presynaptic terminals of cholinergic synapses; in the process, the choline produced in the synaptic cleft by the acetylcholine hydrolysis effected by acetylcholinesterase is reused. The liberated choline is taken up by the presynaptic terminal through an active membrane mechanism. The enzymes involved in acetylcholine synthesis, however, are produced at the granular endoplasmic reticulum in the neuronal perikaryon and are then conveyed to the presynaptic terminals (see Sect. V.D.4). The enzymes involved in the synthesis of noradrenaline are also produced in the neuronal perikaryon to be then transported to the presynaptic endings (Thoenen et al. 1973).

Since the results of a number of experiments do not seem to be consistent with the vesicular release theory, some authors have rejected it, notwithstanding the wealth of evidence in its favor. According to Marchbanks (1978), when an action potential arrives at the presynaptic ending, acetylcholine is released directly from the cytoplasmic pool through membrane channels without the intervention of synaptic vesicles; these would provide a reserve pool of acetylcholine to replenish the cytoplasmic pool as required.

Taking into account the wealth of data available, it may be noted that the hypothesis that the neurotransmitter is released in response to nerve impulses without the intervention of synaptic vesicles rests on very weak evidence. However, a steady, diffuse leakage of transmitter molecules across the presynaptic membrane, i.e., a non-quantal release of the neurotransmitter, probably occurs under resting conditions (Katz and Miledi 1977), but usually it does not produce obvious electrical changes in the postjunctional element.

Until about 50 years ago, the molecules known to be involved in neurotransmission included acetylcholine, the catecholamines (noradrenaline, adrenaline, and dopamine), 5-hydroxytryptamine (or serotonin), histamine, and certain amino acids such as γ-aminobutyric acid and glutamic acid. Subsequently, several dozens of small protein molecules which may act as transmitters have been identified in neurons, mainly by immunocytochemical techniques. These protein molecules are usually referred to as neuropeptides. Unlike classical neurotransmitters, which are synthesized locally within the presynaptic terminals, neuropeptides (or their precursors) are produced within the neuronal perikaryon and subsequently transported to their sites of release. Note that neuropeptides may be released not only from axons but also from dendrites at regions showing no synaptic specializations (for a review, see Ludwig and Pittman 2003).

As a rule, two or more transmitters occur together in a single neuron (for a review, see Hökfelt 1987). Various combinations of neuropeptides with classical neurotransmitters have been described. This coexistence of multiple neuroactive substances in a single neuron has been observed in the CNS and PNS of mammals and some other vertebrates, as well as in invertebrates, including the coelenterates. It seems, therefore, that the phenomenon is phylogenetically ancient. The functional consequences of the coexistence in a single neuron of multiple neuroactive substances remain to be clearly established. It was initially suggested that neuropeptides played a transmitter role early in evolution but were replaced by more efficient, small molecules and yet remained within neurons with no functional role (Hökfelt 1987). However, a large number of receptors for neuropeptides have been identified and localized in the nervous system. These and other observations indicate that at least some neuropeptides are still physiologically active as transmitters. As a general rule, classical neurotransmitters induce rapid responses of short duration, whereas neuropeptides mediate slower onset and longer lasting responses. With regard to the interactions of classical neurotransmitters and neuropeptides, a number of possibilities have been advanced. For example, coexisting classical neurotransmitters and neuropeptides may be coreleased and interact producing synergistic or antagonistic effects. In other cases, coexisting classical neurotransmitters and neuropeptides may be released independently and have distinct physiological effects. Finally, note that neuropeptides have other roles in addition to neural signaling.

F. Synaptic Structural Plasticity

The term synaptic structural plasticity, as used here, encompasses loss and formation of synapses, as well as changes in synapse size, shape, and structure. All these modifications have likely functional consequences.

It has long been known that in vertebrates synaptic plasticity is conspicuous (a) during development both under normal circumstances and following experimental manipulation and (b) in the adult when it is a response to experimental manipulation. The issue of synaptic plasticity in adult vertebrates under normal conditions has been taken up in earnest more recently; for reviews, see Cotman et al. (1981) and Holtmaat and Svoboda (2009). Initially much of the work in this area concentrated on the PNS because of the relative simplicity of neuron-neuron and neuron-target cell relationships there. The skeletal muscle innervation has been a preferred object of these studies. Examination of fixed material suggested that changes in the shape and size of nerve endings could occur in the neuromuscular junctions of adult vertebrates under normal conditions (for a review, see Wernig and Herrera 1986). This impression was confirmed by means of repeated direct observations of a single neuromuscular junction over periods of weeks or months in living animals (see, among others, Purves 1988; Herrera et al. 1990). Furthermore, seasonal plasticity of neuromuscular junctions has been reported in adult mammals (e.g., see Hegstrom and Breedlove 1999). Synaptic endings on the surface of identified autonomic ganglion neurons of adult animals have also been shown to undergo remodeling with time under normal conditions (see, among others, Purves 1988).

Subsequently it was observed that synaptic remodeling also occurs normally in the CNS of adult animals. For example, a cyclic change in the number of axosomatic synapses has been observed in the hypothalamic arcuate nucleus of the rat; this change is linked to hormonal variations that are part of the estrous cycle (Olmos et al. 1989; Csakvari et al. 2007). Dendritic spines undergo changes in the mature brain (see Sect. V.C.2). Since in the brain many synapses are on dendritic spines, changes in these spines imply changes in the corresponding synapses. Moreover, there are findings suggesting that degeneration and new growth of axonal branches take place in the CNS of normal adult animals (Sotelo and Palay 1971). These axonal changes also imply some reorganization of synaptic connectivity and are a further indication that the synapses of the CNS retain their plasticity under normal circumstances in the adult. Long-term in vivo imaging studies have confirmed synaptic turnover. Finally, synaptic structural modifications occur in association with long-term potentiation (e.g., see Desmond and Levy 1986; Geinisman et al. 1996) and learning (e.g., see Moser et al. 1994; Kleim et al. 1997).

G. Age-Related Changes

Although some studies have reported no significant decline in the number of synapses with aging (e.g., Cragg 1975; Scheff et al. 2001), most studies report a mild to severe synapse loss with normal aging in various species of mammals (rat, cat, monkey, and human) and in different regions of the nervous system (Bondareff and Geinisman 1976; Geinisman et al. 1977a; Geinisman 1979; Hinds and McNelly 1979; Uemura 1980, 1985a; Adams and Jones 1982; Matsumoto et al. 1982; Gibson 1983; Rogers et al. 1984; Markus and Petit 1987; Levine et al. 1988; Matsumoto 1998; Wong et al. 1998; Chen and Hillman 1999; Helfert et al. 1999; Poe et al. 2001; Peters et al. 2008a). This decline varies from 15 % to about 50 % according to the species, the region of the nervous system considered, and the technical procedures employed. Furthermore, synaptic structural plasticity (see Sect. VI.F) decreases in old age. It is conceivable that the decrease both in the number of synaptic contacts and in the synaptic structural plasticity is in part responsible for the behavioral changes and cognitive decline that often occur during normal aging (e.g., see Morrison and Baxter 2012). The age-related decrease in the number of synapses is associated with the loss of both pre- and postsynaptic structures. This is supported by findings that while there is no substantial neuron loss with age (see Sect. V.A.10), the numbers of dendritic shafts, spines (see Sect. V.C.4), and axons (see Sect. V.D.3) all decline with increasing age.

The structure of most synapses does not change in old age (Markus et al. 1987; Peters et al. 2008a). However, neuromuscular junctions of old animals may have varicose nerve terminals, and acetylcholine receptors fragmented into small islands (e.g., see Courtney and Steinbach 1981). Finally, a reduction in markers for a number of neurotransmitters has been recorded in old age (e.g., Beal et al. 1991). This reduction could result from the loss of synapses that occurs during aging.

H. Relationship Between Axons of the Autonomic Nervous System and Effector Cells

Numerous varicosities, arranged in a more or less regular sequence and totally or partially free of their Schwann cell envelope, occur along the unmyelinated terminal portion of the axons which innervate the smooth musculature, myocardium, and glandular tissues. These varicosities, which are often fusiform and measure 1–3 μm in diameter and up to 4 μm in length, are generally interpreted as sites of transmitter release (for a review, see Burnstock and Costa 1975). Each neuron usually has a large number of these varicosities; it has been estimated that about 26,000 of them occur along the terminal axonal branches of a single adrenergic neuron of the cat cervical sympathetic ganglion (Dahlström and Häggendal 1966).

Where the varicosities are free of their Schwann cell envelope, they face the extracellular space and are directly covered by the basal lamina. These varicosities contain vesicles and mitochondria (Fig. VI.10), but in contrast to neuromuscular junctions and interneuronal chemical synapses, vesicles are infrequently clustered close to the varicosity plasma membrane. Moreover, only a small proportion of axonal varicosities bear the dense material which is present in presynaptic endings (see Sect. VI.B.3). The structural features of the effector cell in the region of the junction do not differ from those in other regions of the cell; in particular, the effector cell commonly lacks both the junctional folds typical of the neuromuscular junction (see Sect. VI.D) and the membrane specializations which are present in the postjunctional element of interneuronal chemical synapses (see Sect. VI.B.3). Dense material coating the cytoplasmic surfaces of the junctional membranes in both the axon and the effector cell has been described only rarely (e.g., adrenal medulla: Coupland 1965; pancreatic islets: Klein 1971).

The distance between axonal varicosities and effector cells varies according to the organ and species. In the smooth musculature of blood vessels, the ureter, the uterus, and the gastrointestinal tract (Fig. VI.10), the distance is often 40–100 nm (see Table VI.1). In these smooth muscles, the transmitter is released into the extracellular space through which it may diffuse over a long distance to affect several muscle cells. Close contacts (with a gap ≤30 nm between axon and smooth muscle cell) are common, for example, in the ductus deferens and intrinsic eye muscles (see Table VI.1).

In smooth muscles, the pattern of innervation may vary from situations in which every muscle cell is likely to be directly innervated to those in which only a few cells in a bundle are innervated and the excitation propagates from one cell to the next via gap junctions (Dewey and Barr 1964; Evans and Evans 1964; Oosaki and Ishii 1964; Richardson 1964; Cobb and Bennett 1969).

In cardiac muscle, the distance between the axonal varicosities and muscle cells varies considerably depending on the species and heart region (ordinary cardiac muscle cells: see, among others, Thaemert 1969; Yamauchi and Chiba 1973; conducting system: see, among others, Yamauchi and Burnstock 1968; Thaemert 1970; Jensen et al. 1978). However, gap junctions between cardiac muscle cells are extremely numerous, so that there is a high degree of electrotonic coupling between the cells of this tissue.

Autonomic axons located in proximity (within about 20 nm) to secretory cells have been described frequently in glands, although a much greater distance between axonal varicosities and secretory cells has also been found (see Tables VI.2 and VI.3).

Three types of vesicles can be found within the varicosities (Fig. VI.10) and the end-bulbs of autonomic axons: small vesicles with clear centers (35–50 nm in diameter), small dense-cored vesicles (40–60 nm in diameter), and large dense-cored vesicles (about 100 nm in diameter). Based on evidence obtained with several techniques (e.g., 5-hydroxydopamine incubation; electron microscope autoradiography; a triple fixation with aldehyde, dichromate, and osmium; immunocytochemical techniques; etc.) it is generally accepted that the small vesicles with clear centers contain acetylcholine, the small dense-cored vesicles contain catecholamines, and the large dense-cored vesicles store mainly neuropeptides.

I. Electrotonic and Mixed Junctions

As noted in Chap. I, intracellular recording techniques have shown that the information transfer between neurons is not always mediated by the release of a chemical transmitter; at certain sites it is electrically mediated, i.e., occurs via the direct flow of ions (largely K^+ ions) from neuron to neuron. This mode of transmission, called electrotonic, is distinguished from chemically mediated transmission mainly by its greater velocity and by its basically bidirectional nature. Electrotonic transmission, first demonstrated in the crayfish giant fiber system (Furshpan and Potter 1957, 1959), was later recognized at numerous

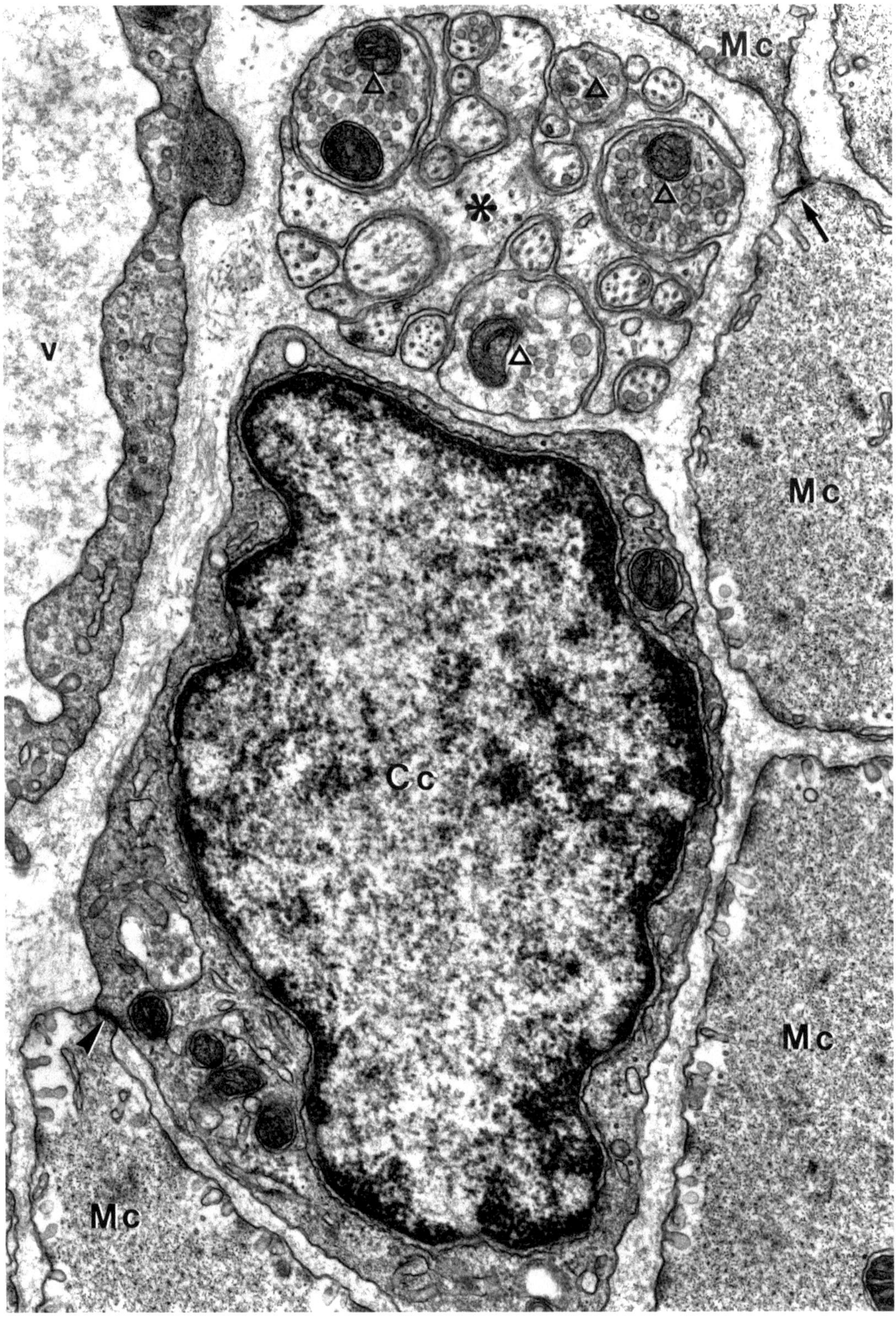
Mc
V
Mc
Cc
Mc
Mc

other sites. It is fairly common in invertebrates and lower vertebrates, whereas in mammals instances of electrotonic transmission are much less common than those of chemical transmission.

Electron microscope studies have been carried out to establish whether electrotonic transmission takes place at morphologically distinct junctions. It was established that the electrotonic transmission is mediated by gap junctions. The latter are macular junctions characterized by (a) a gap of about 2 nm between apposed plasma membranes, (b) the scarcity of dense material underlying the cytoplasmic aspects of these membranes, and (c) the absence of synaptic vesicles. In freeze-fractured preparations, it can be seen that globular particles of about 8 nm in diameter are contained within the apposed plasma membranes forming the gap junction (Fig. VI.11). Often, the particles are arranged in an orderly fashion with a center-to-center spacing of about 9 nm, but in other instances they form a less regular pattern; in any event, the particles are arranged in register, so that each particle within a membrane is matched with a particle in the confronted membrane. Each particle protrudes into the extracellular space and contacts its opposite one in the facing membrane forming a unit called a connexon, which bridges the extracellular space. Each particle is perforated by a hydrophilic channel, about 1.2 nm in diameter; the alignment of the channels belonging to each pair of particles allows direct communication between the two adjacent cells (Fig. VI.12). Connexons are composed of a family of closely related proteins collectively called connexins. Numerous connexins have been reported in neurons of normal adult animals; however, so far only connexin36 has been detected in neuronal gap junctions identified under the electron microscope (for a review, see Nagy et al. 2004).

Evidence for cell-to-cell communication through gap junctions has been obtained by the intracellular injection of low molecular weight substances (e.g., fluorescein, Lucifer yellow, Procion yellow, neutral red, sucrose, biocytin, neurobiotin, and microperoxidase). The substance injected into a single cell connected through gap junctions to a contiguous cell appears rapidly in the latter, whereas it fails to enter or does so only very slowly when injected extracellularly.

Many of the characteristics of electrotonic transmission may be explained on the basis of the structure of the gap junctions mediating it. The close approximation of the two plasma membranes forming the junction and the existence of channels allowing direct passage of ions from neuron to neuron reduce current leakage during signal transmission to a minimum. Since the release of transmitter molecules and their diffusion across the synaptic cleft do not occur at the electrotonic junction, the transmission of signals is faster here than at the chemical synapse (junctional delay is about 50 μs

Fig. VI.10 Relationship between axons of the autonomic nervous system and muscle cells. The picture shows a nerve bundle (*), several muscle cells (**Mc**), and an interstitial cell of Cajal (**Cc**). The intramuscular nerve bundle is made up of about 20 axons, some of which are sectioned through an expanded portion (varicosity, marked Δ) which contains vesicles and mitochondria. While some axons are enveloped singly by the Schwann cell, groups of two to four axons can be seen enclosed in the same trough. Note the considerable distance between axonal varicosities and muscle cells. A gap junction (*arrow*) connects two muscle cells; another gap junction (*arrowhead*) connects the interstitial cell of Cajal with a muscle cell. **v** capillary. Circular muscle layer of the guinea pig ileum, ×29,000 (Courtesy of G. Gabella)

Table VI.1 Examples of relationships between autonomic axons and effector cells (smooth muscles)

Site	Family	Species	Common name	Close contacts (distance <30 nm)	Nonclose contacts (distance >30 nm)
Vascular muscles	Muridae	*Rattus norvegicus*	Rat		Appenzeller (1964), Dieterich (1974), Sandow et al. (1998)
	Caviidae	*Cavia aperea porcellus*	Guinea pig		Lever et al. (1965), Luff et al. (1987), Klemm et al. (1993)
	Felidae	*Felis catus*	Cat		Lever et al. (1968, 1971), Knight et al. (1981)
	Bovidae	*Ovis ammon aries*	Sheep		Simpson and Devine (1966)
Ductus deferens	Muridae	*Rattus norvegicus*	Rat	Richardson (1962), Taxi (1964), Furness and Iwayama (1971)	
	Muridae	*Mus musculus*	Mouse	Furness and Iwayama (1971)	
	Caviidae	*Cavia aperea porcellus*	Guinea pig	Furness and Iwayama (1971)	Merrillces et al. (1963), Merrillees (1968)
	Hominidae	*Homo sapiens*	Man		Baumgarten et al. (1971)
Epididymis	Hominidae	*Homo sapiens*	Man		Baumgarten et al. (1971)
Myometrium	Cercopithecidae	*Macaca fascicularis* (=*M.irus*)	Crab-eating macaque		Barbe and Taxi (1986)
Ureter	Muridae	*Rattus norvegicus*	Rat		Dixon and Gosling (1971)
	Caviidae	*Cavia aperea porcellus*	Guinea pig		Dixon and Gosling (1971)
	Leporidae	*Oryctolagus cuniculus*	Rabbit		Dixon and Gosling (1971)
Urinary bladder	Muridae	*Rattus norvegicus*	Rat		Gabella and Uvelius (1990), Gabella (1995)
Gizzard	Phasianidae	*Gallus domesticus*	Domestic fowl		Bennett and Cobb (1969)
	Columbidae	*Columba domestica*	Pigeon		Bennett and Cobb (1969)
Intestine	Caviidae	*Cavia aperea porcellus*	Guinea pig		Bennett and Rogers (1967), Gabella (1972a)
	Canidae	*Canis familiaris*	Dog		Kyösola and Rechardt (1975)
Intrinsic eye muscles	Muridae	*Rattus norvegicus*	Rat		Nilsson (1964)
	Leporidae	*Oryctolagus cuniculus*	Rabbit	Richardson (1964)	
	Felidae	*Felis catus*	Cat	Taxi (1964)	

Table VI.2 Examples of relationships between autonomic axons and effector cells (exocrine glands)

Site	Family	Species	Common name	Close contacts (distance <30 nm)	Nonclose contacts (distance >30 nm)
Parotid gland	Muridae	*Mus musculus*	Mouse	Bloom et al. (1977)	
	Muridae	*Rattus norvegicus*	Rat	Bloom et al. (1977)	
	Cricetidae	*Mesocricetus auratus*	Syrian hamster	Bloom et al. (1977)	
	Caviidae	*Cavia aperea porcellus*	Guinea pig	Bloom et al. (1977)	
	Leporidae	*Oryctolagus cuniculus*	Rabbit	Garrett (1977)	
Submandibular gland, tubules	Leporidae	*Oryctolagus cuniculus*	Rabbit	Garrett (1977)	
Submandibular gland, acini	Leporidae	*Oryctolagus cuniculus*	Rabbit		Garrett (1977)
Submandibular gland	Cercopithecidae	*Macaca cyclopis*	Taiwan macaque	Kagayama (1971)	
		Macaca fuscata yakui	Japanese macaque	Kagayama (1971)	
Labial salivary glands	Hominidae	*Homo sapiens*	Man	Tandler and Ross (1969)	
Pancreas	Muridae	*Rattus norvegicus*	Rat		Zelander et al. (1962)
Lacrimal glands	Bovidae	*Ovis ammon aries*	Sheep	Yamauchi and Burnstock (1967)	
	Cercopithecidae	*Macaca mulatta*	Rhesus monkey	Ruskell (1968)	
		Macaca fascicularis (=*M. irus*)	Crab-eating macaque	Ruskell (1968)	
Sweat glands	Hominidae	*Homo sapiens*	Man		Yamada and Miyake (1960)

Table VI.3 Examples of relationships between autonomic axons and effector cells (endocrine glands)

Site	Family	Species	Common name	Close contacts (distance <30 nm)	Nonclose contacts (distance >30 nm)
Anterior pituitary	Muridae	*Rattus norvegicus*	Rat	Liu et al. (1996)	
Pancreatic islets	Poeciliidae	*Xiphophorus helleri*	Sword-tail	Klein (1971)	
Adrenal cortex	Cricetidae	*Mesocricetus auratus*	Syrian hamster	Unsicker (1969)	
	Muridae	*Rattus norvegicus*	Rat	Unsicker (1971)	
	Suidae	*Sus scrofa domesticus*	Pig	Unsicker (1971)	
Adrenal medulla	Muridae	*Rattus norvegicus*	Rat	Coupland (1965)	
	Felidae	*Felis catus*	Cat	Prentice and Wood (1975)	
Ovary interstitial gland	Caviidae	*Cavia aperea porcellus*	Guinea pig	Sporrong et al. (1985)	

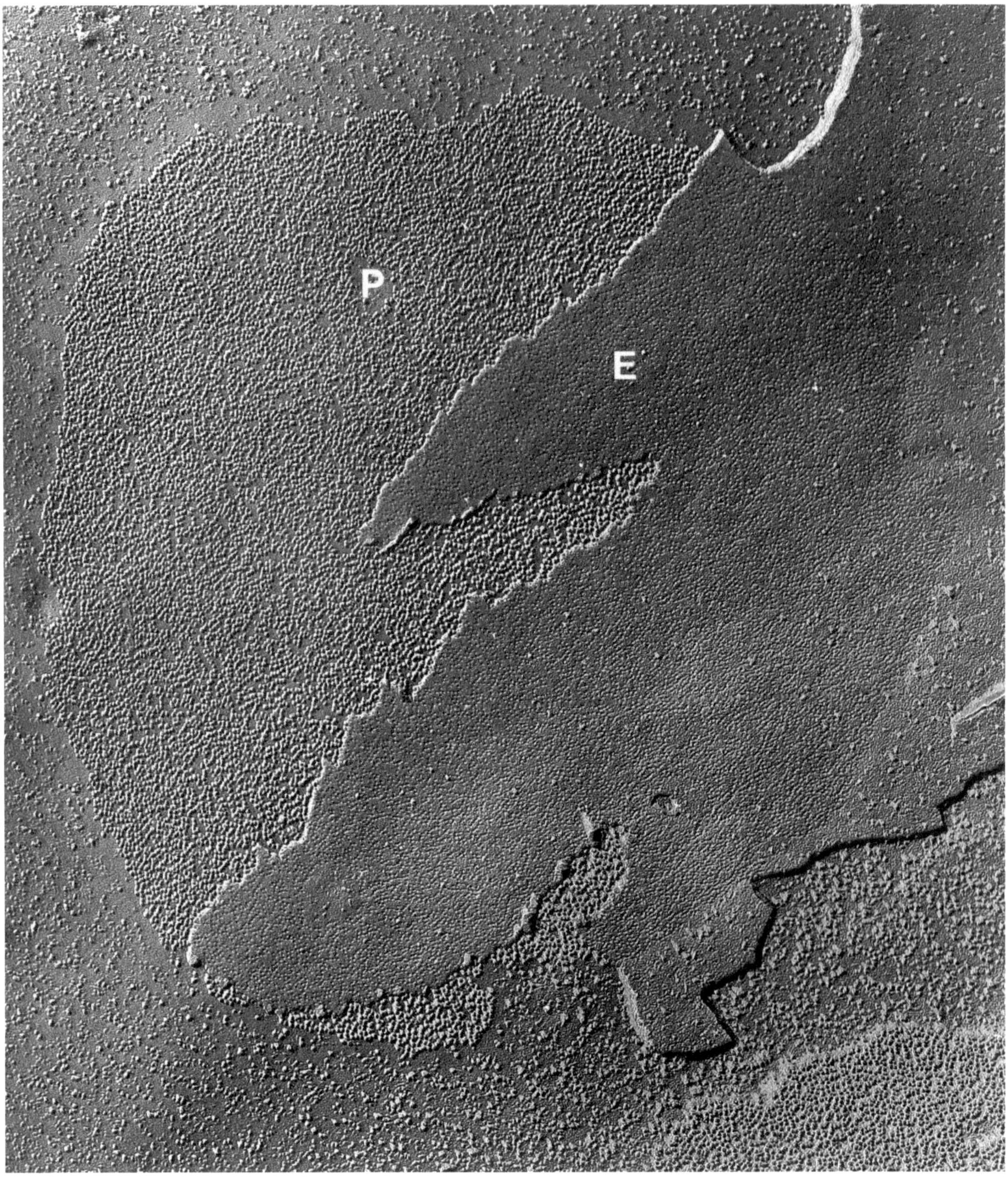

Fig. VI.11 Electrotonic junction. An extensive dendrodendritic gap junction between two horizontal cells of the retina in a freeze-fractured preparation. The cleavage plane shifts from the P face (**P**) to the E face (**E**) of the apposed plasma membranes and reveals particles on the P face and corresponding pits on the E face. Retina of turtle, ×59,000 (Courtesy of D. Cantino and M. Testa)

compared with the 0.5–0.9 ms at the chemical synapse).

Electrotonic junctions promote neuronal synchronization. A cluster of electrotonically coupled neurons can fire nearly simultaneously, as has been observed in the nuclei which innervate the electric organs of certain fish. Furthermore, electrotonic junctions are often found in the escape systems of both invertebrates and lower vertebrates;

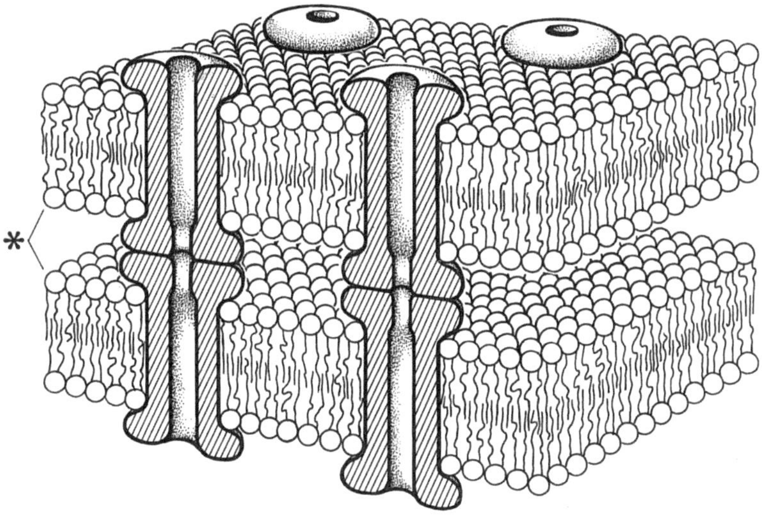

Fig. VI.12 Model of the molecular organization of an electrotonic junction. In the area of such a junction, the plasma membranes of the two adjacent neurons are separated by a gap (*) of about 2 nm. Particles perforated by a hydrophilic channel are arranged in register in the apposed membranes so that each particle within a membrane is matched with a particle in the confronted membrane. Each particle protrudes into the extracellular space and contacts its opposite one in the facing membrane forming a unit called a connexon, which bridges the extracellular space. The channels belonging to each pair of particles are in continuity, permitting direct communication between the two adjacent neurons

for example, they have been found between the motor giant axons and the segmental giant axons in the crayfish and at the level of Mauthner cells in fish. The speed of electrotonic transmission makes this kind of junction particularly suitable for mediating escape reactions.

A number of instances of electrotonic transmission identified both morphologically and physiologically are listed in Table VI.4 together with the corresponding interneuronal relation types (axodendritic, axosomatic, dendrodendritic, etc.). Electron microscope studies have brought to light interneuronal gap junctions at a number of sites in the CNS and PNS. These cases, however, still await physiological confirmation of electrotonic transmission. Some of these sites are listed in Table VI.5.

In certain regions of the nervous system, both chemical and electrotonic modes of transmission occur at the same junctional interface (e.g., chick ciliary ganglion, Martin and Pilar 1963; spinal cord of the sea lamprey, Rovainen 1974 and Martin and Ringham 1975). At these same sites, the electron microscope has revealed the presence of chemical synapses and gap junctions next to each other (chick ciliary ganglion, Takahashi and Hama 1965, de Lorenzo 1966, Brightman and Reese 1969, Cantino and Mugnaini 1975; spinal cord of the sea lamprey, Rovainen 1974). The term mixed junction has been proposed to indicate the presence of an electrotonic junction and a chemical synapse within a short distance of each other at the same junctional interface. Many of the electrotonic junctions described in this section coexist in fact with typical chemical synapses and may be considered therefore as components of mixed junctions. At a mixed junction, the stretch of membrane occupied by the chemical synapse is often much larger than that forming the adjacent gap junction. Future research should establish whether all morphologically mixed junctions are capable of a dual mode of transmission and elucidate the functional relevance of the coexistence of

Table VI.4 A partial list of the sites at which both electrotonic transmission and interneuronal gap junctions have been demonstrated

Site	Family	Species	Common name	Relation type	Author(s)
			(a) Cyclostomes		
Spinal cord (giant interneurons)	Petromyzontidae	*Petromyzon marinus*	Sea lamprey	Axodendritic	Rovainen (1974)
			(b) Fish		
Spinal electromotor nucleus	Mormyridae	*Gnathonemus sp.*	Elephant fish	Dendrodendritic	Bennett et al. (1967b),
	Apteronotidae	*Sternarchus (=Apteronotus) sp.*	Black ghost	Axosomatic	Waxman et al. (1972), Bennett et al. (1978)
	Electrophoridae	*Electrophorus electricus*	Electric eel	Axodendritic, axosomatic, dendrodendritic	Albe-Fessard and Martins-Ferreira (1953), Meszler et al. (1974)
Spinal giant electromotor neurons	Malapteruridae	*Malapterurus electricus*	Electric catfish	Axodendritic, axosomatic	Bennett et al. (1967c)
Medullary relay nucleus	Mormyridae	*Gnathonemus sp.*	Elephant fish	Dendrodendritic, dendrosomatic	Bennett et al. (1967b)
	Rhamphichthyidae	*Sternopygus sp.*	Knifefish	Axosomatic	Bennett et al. (1967d)
	Rhamphichthyidae	*Steatogenys sp.*	Knifefish	Axosomatic	Bennett et al. (1967d)
	Gymnotidae	*Gymnotus carapo*	Banded knifefish	Axosomatic	Bennett et al. (1967d)
Medullary pacemaker nucleus	Rhamphichthyidae	*Sternopygus sp.*	Knifefish	Axodendritic, axosomatic	Bennett et al. (1967d)
	Rhamphichthyidae	*Steatogenys sp.*	Knifefish	Axodendritic, axosomatic	Bennett et al. (1967d)
	Gymnotidae	*Gymnotus carapo*	Banded knifefish	Axodendritic, axosomatic	Bennett et al. (1967d)
Supramedullary neurons	Tetraodontidae	*Sphaeroides maculatus*	Puffer fish	Axoaxonal	Bennett et al. (1967a)
	Antennariidae	*Histrio sp.*	Sargassum fish	Axoaxonal	Bennett et al. (1967a)
Mauthner neurons	Cyprinidae	*Carassius auratus*	Goldfish	Axodendritic, axosomatic	Robertson et al. (1963), Furshpan (1964), Brightman and Reese (1969), Nakajima (1974)
Lateral vestibular nucleus	Batrachoididae	*Opsanus tau*	Toadfish	Axodendritic, axosomatic	Korn el al. (1977)
Magnocellular mesencephalic nucleus	Mormyridae	*Gnathonemus petersii*	Peter's elephantnose	Axosomatic	Szabo et al. (1975)
	Gymnotidae	*Gymnotus carapo*	Banded knifefish	Axosomatic	Sotelo et al. (1975b), Szabo et al. (1975)

Oculomotor nuclei	Tetraodontidae	*Sphaeroides maculatus*	Puffer fish	Axosomatic	Kriebel et al. (1969)
	Cottidae	*Myoxocephalus octodecemspinosus*	Long-horned sculpin	Axosomatic	Kriebel et al. (1969)
	Diodontidae	*Chilomycterus schoepfi*	Spiny boxfish	Axosomatic	Kriebel et al.(1969)
	Diodontidae	*Chilomycterus schoepfi*	Spiny boxfish	Axodendritic, axosomatic	Waxman and Pappas (1971)
Retina	Carcharhinidae	*Mustelus canis*	Smooth dogfish	Between horizontal cells	Kaneko (1971), Witkovsky and Stell (1973)
	Ictaluridae	[a]	Catfish	Between amacrine cells	Naka and Christensen (1981)
			(c) Amphibians		
Spinal cord (motoneurons)	Ranidae	*Rana pipiens*	Leopard frog	Dendrodendritic	Grinnel (1966), Erulkar and Soller (1980)
	Ranidae	*Rana temporaria*	Common frog	Dendrodendritic	Grinnel (1966)
	Ranidae	*Rana catesbeiana*	Bullfrog	Dendrodendritic	Grinnel (1966)
	Ranidae	*Rana esculenta*	Edible frog	Dendrodendritic	Sotelo and Taxi (1970), Sonnhof et al. (1977)
Retina	Bufonidae	[a]	Toad	Between "red" rods	Fain et al. (1976)
			(d) Birds		
Ciliary ganglion	Phasianidae	*Gallus domesticus*	Domestic fowl	Axosomatic	Martin and Pilar (1963), Takahashi and Hama (1965), de Lorenzo (1966), Brightman and Reese (1969), Cantino and Mugnaini (1975)
			(e) Mammals		
Lateral vestibular nucleus	Muridae	*Rattus norvegicus*	Rat	Axodendritic, axosomatic	Korn et al. (1973)
Trigeminal mesencephalic nucleus	Muridae	*Rattus norvegicus*	Rat	Somatosomatic	Hinrichsen (1970), Baker and Llinás (1971), Sotelo (1975)
Retina	Muridae	*Rattus norvegicus*	Rat	Dendrodendritic	Hidaka et al. (2004)
Cerebral cortex	Muridae	*Rattus norvegicus*	Rat	[b]	MacVicar and Dudek (1982)
	Muridae	*Rattus norvegicus*	Rat	Dendrodendritic, dendrosomatic	Kosaka (1983a, b)
	Muridae	*Rattus norvegicus*	Rat	Dendrodendritic, somatodendritic	Simon et al. (2005)
	Muridae	*Rattus norvegicus*	Rat	Dendrodendritic	Szabadics et al. (2001)

[a]Species not indicated in the original paper
[b]Relation type not indicated in the original paper

Table VI.5 A partial list of the sites at which interneuronal gap junctions have been demonstrated

Site	Family	Species	Common name	Relation type	Author(s)
			(a) Cyclostomes		
Vestibular nuclei	Petromyzontidae	*Lampetra planeri*	Brook lamprey	Axosomatic	Stefanelli and Caravita (1970)
			(b) Fish		
Spinal cord (ventral horn)	Cyprinidae	*Carassius auratus*	Goldfish	Axosomatic	Charlton and Gray (1966)
	Cyprinidae	*Carassius auratus*	Goldfish	Dendrodendritic	Schnitzlein and Brown (1975)
	Apteronotidae	*Sternarchus albifrons*	Blackghost	Axodendritic, axosomatic	Bennett et al. (1978)
Mauthner neurons	Cyprinidae	*Carassius auratus*	Goldfish	Axodendritic	Nakajima et al. (1987)
Tangential nucleus	Cyprinidae	*Carassius auratus*	Goldfish	Axosomatic	Hinojosa (1973)
Trigeminal mesencephalic nucleus	Rajidae	*Raja ocellata*	Eyed skate	Dendrodendritic, dendrosomatic, somatosomatic	Witkovsky and Roberts (1976)
Oculomotor nuclei	Salmonidae	*Salmo irideus*	Rainbow trout	Axosomatic	Schuster (1973)
	Cyprinidae	*Carassius auratus*	Goldfish	Axoaxonal, axodendritic, axosomatic	Sterling (1977)
Cerebellar cortex (glomeruli)	Mormyridae	*Gnathonemus petersii*	Peter's elephantnose	Axodendritic	Kaiserman-Abramof and Palay (1969)
	Gymnotidae	*Gymnotus carapo*	Banded knifefish	Axodendritic	Sotelo and Llinás (1972)
Preoptic nucleus	Balistidae	*Navodon modestus*	Modest filefish	Axodendritic, axosomatic	Uchiyama and Ito (1984)
Retina	Carcharhinidae	*Mustelus canis*	Smooth dogfish	Axoaxonal	Witkovsky and Stell (1973)
	Rajidae	a	Skate	Somatosomatic	Fain et al. (1976)
			(c) Amphibians		
Spinal cord (intermediate gray matter)	Ranidae	*Rana temporaria*	Common frog	Axodendritic, axosomatic	Charlton and Gray (1966)
	Ranidae	*Rana esculenta*	Edible frog	Axodendritic	Sotelo and Taxi (1970)
Oculomotor nuclei	Ranidae	*Rana pipiens*	Leopard frog	Axodendritic, axosomatic	Waxman and Pappas (1971)
Cerebellar cortex (glomeruli)	Ranidae	*Rana esculenta*	Edible frog	Axodendritic	Sotelo and Llinás (1972)
Retina	Ambystomatidae	*Ambystoma mexicanum*	Axolotl	Between photoreceptor cells	Custer (1973)
Cardiac parasympathetic ganglion	Proteidae	*Necturus maculosus*	Mudpuppy	Axosomatic, somatosomatic	McMahan and Purves (1976)

			(d) Reptiles		
Oculomotor nuclei	Iguanidae	*Anolis carolinensis*	Caroline anole	Axodendritic, axosomatic	Waxman and Pappas (1971)
	Varanidae	*Varanus exanthematicus*	Monitor lizard	Axoaxonal, axosomatic	Wouterlood et al. (1988)
Retina	Emydidae	*Pseudemys scripta elegans*	Pseudemid turtle	Between photoreceptor cells	Raviola and Gilula (1973)
			(e) Birds		
Tangential nucleus	Phasianidae	*Gallus domesticus*	Domestic fowl	Axosomatic	Hinojosa and Robertson (1967), Brightman and Reese (1969)
Cerebellar cortex (glomeruli)	Columbidae	*Columba domestica*	Pigeon	Axodendritic	Paula-Barbosa (1976)
Retina	Phasianidae	*Gallus domesticus*	Domestic fowl	Between photoreceptor cells	Fujisawa et al. (1976)
			(f) Mammals		
Spinal cord (ventral horn)	Muridae	*Rattus norvegicus*	Rat	Dendrodendritic	van der Want et al. (1998)
Spinal cord (dorsolateral nucleus)	Muridae	*Rattus norvegicus*	Rat	Dendrodendritic, somatodendritic somatosomatic	Matsumoto et al. (1989)
Inferior olivary nucleus (glomeruli)	Didelphidae	*Didelphis marsupialis*	Opossum	Dendrodendritic	King et al. (1975)
	Felidae	*Felis catus*	Cat	Dendrodendritic	Sotelo et al.(1974), Sotelo (1975)
	Cebidae	*Saimiri sciureus*	Squirrel monkey	Dendrodendritic	Rutherford and Gwyn (1977)
	Cercopithecidae	*Macaca mulatta*	Rhesus monkey	Dendrodendritic	Sotelo (1975)
	Muridae	*Rattus norvegicus*	Rat	Dendrodendritic	Sotelo et al. (1986)
Ventral cochlear nucleus	Muridae	*Rattus norvegicus*	Rat	Dendrosomatic, somatosomatic	Sotelo et al. (1976)
Lateral vestibular nucleus	Muridae	*Rattus norvegicus*	Rat	Axodendritic, axosomatic	Sotelo and Palay (1970)

(continued)

Table VI.5 (continued)

Site	Family	Species	Common name	Relation type	Author(s)
Descending vestibular nucleus	Muridae	*Rattus norvegicus*	Rat	Axosomatic	Sotelo (1975)
Trigeminal mesencephalic nucleus	Muridae	*Mus musculus*	Mouse	Axosomatic, somatosomatic	Hinrichsen and Larramendi (1970)
	Felidae	*Felis catus*	Cat	Axosomatic, somatosomatic	Hinrichsen and Larramendi (1970)
	Cricetidae	*Mesocricetus auratus*	Hamster	Axosomatic, somatosomatic	Alley (1974)
	Leporidae	*Oryctolagus cuniculus*	Rabbit	Axosomatic, somatosomatic	Bortolami and Manni (1981)
Cerebellar cortex (molecular layer)	Muridae	*Rattus norvegicus*	Rat	Dendrodendritic, dendrosomatic, somatosomatic	Sotelo and Llinás (1972)
	Felidae	*Felis catus*	Cat	Dendrodendritic, dendrosomatic, somatosomatic	Sotelo and Llinás (1972)
Olfactory bulb	Muridae	*Rattus norvegicus*	Rat	Dendrodendritic	Pinching and Powell (1971a)
	Muridae	*Mus musculus*	Mouse	Dendrodendritic	Landis et al. (1974), Kosaka and Kosaka (2005)
	Chinchillidae	*Chinchilla lanigera*	Chinchilla	Dendrodendritic	Landis et al. (1974)
	Leporidae	*Oryctolagus cuniculus*	Rabbit	Dendrodendritic	Landis et al. (1974)
	Muridae	*Rattus norvegicus*	Rat	Somatosomatic	Reyher et al. (1991)
	Cricetidae	*Gerbillus perpallidus*	Gerbil	Somatosomatic	Reyher et al. (1991)
	Callithricidae	*Callithrix jacchus*	Common marmoset	Somatosomatic	Reyher et al. (1991)

Cerebral cortex	Cercopithecidae	*Macaca mulatta*	Rhesus monkey	Dendrodendritic, dendrosomatic	Sloper (1972)
	Muridae	*Rattus norvegicus*	Rat	Dendrodendritic	Fukuda and Kosaka (2000, 2003)
	Felidae	*Felis catus*	Cat	Dendrodendritic	Fukuda et al. (2006)
	Muridae	*Rattus norvegicus*	Rat	Axodendritic, dendrodendritic	Hamzei-Sichani et al. (2012)
Retina	Cercopithecidae	*Macaca mulatta*	Rhesus monkey	Between photoreceptor cells	Raviola and Gilula (1973)
	Cercopithecidae	*Macaca arctoides*	Stump-tailed macaque	Between photoreceptor cells	Raviola and Gilula (1973)
	Leporidae	*Oryctolagus cuniculus*	Rabbit	Between photoreceptor cells	Raviola and Gilula (1973)
	Felidae	*Felis catus*	Cat	Between photoreceptor cells	Kolb (1977)
	Felidae	*Felis catus*	Cat	Between horizontal cells	Kolb (1977)
	Cercopithecidae	*Macaca mulatta*	Rhesus monkey	Between horizontal cells, between bipolar cells	Ravioia and Gilula (1975)
	Cercopithecidae	*Macaca arctoides*	Stump-tailed macaque	Between horizontal cells, between bipolar cells	Raviola and Gilula (1975)
	Felidae	*Felis catus*	Cat	Between amacrine cells	Kolb and Famiglietti (1974)
	Felidae	*Felis catus*	Cat	Between photoreceptor and amacrine cells	Kolb and Famiglietti (1974), Famiglietti and Kolb (1975)

[a]Species not indicated in the original paper

two types of junction at the same junctional interface.

Electrotonic transmission was formerly regarded as incapable of being modulated, and as a consequence was considered a less advanced form of intercellular communication than chemical transmission. However, recent studies have revealed that electrotonic transmission can be modulated by several mechanisms (for a review, see Bennett and Zukin 2004). For example, junctional conductance may be modified by connexin phosphorylation, changes in cytoplasmic pH, changes in the intracellular concentrations of Ca^{2+} and Mg^{2+}, and the influence of neurotransmitters and growth factors. At mixed junctions, the activity of the chemical synapse may modulate the electrotonic junction (Pereda et al. 2004).

Note, finally, that electrotonic junctions, whose channels are about 1.2 nm in diameter (see above, this section), not only allow the intercellular passage of ions but also permit the transfer of molecules of molecular weight up to about 1 kDa (e.g., signaling molecules, amino acids, and glucose).

J. Synapse-Like Junctions Involving Neuroglial Cells

Junctions with a structure similar to that of typical interneuronal synapses have also been found between nerve and neuroglial cells. In these entities, a vesicle-containing profile (which has usually been identified as an axon) contacts a cell (usually identified as an ependymal or macroglial cell), and a presynaptic density is usually present, whereas a postsynaptic density may be absent.

Certain of these junctions are transient structures present during early developmental stages, but not later (Henrikson and Vaughn 1974; Oppenheim et al. 1978; Singer et al. 1979; Wolff et al. 1979; Gorgels 1991). Some authors (Henrikson and Vaughn 1974; Soffer and Raine 1980) have suggested that these junctions may reflect developmental errors in the axonal recognition of targets, whereas other authors (Gorgels 1991) have hypothesized that these junctions influence neuroglial cell maturation.

Axoglial synapses have been observed during regeneration both in vitro (James and Tresman 1969) and in vivo (Nordlander and Singer 1978) as well as in normal adult animals (invertebrates, Colonnier et al. 1979, Roubos and Moorer-Van Delft 1979; vertebrates, Ebner and Colonnier 1975, Soffer and Raine 1980, Palacios-Prü et al. 1983). A vesicular release of glutamate occurs at the synapses between axons and NG2-expressing cells in some regions of the rat CNS (Bergles et al. 2000; Ziskin et al. 2007; see also Sect. VIII.G). These findings suggest that synaptic junctions mediate some form of neural influence over neuroglia. For junctions between axonal endings and the luminal surface of ependymal cells, see Sect. VIII.C.6.

There have been particularly abundant reports of synapses between axons and neuroglial cells in the neurosecretory system (Knowles and Vollrath 1965; Scharrer and Kater 1969; Wittkowski 1969; Nakai 1970; Zimmermann 1972; Güldner and Wolff 1973; Galabov and Schiebler 1978; Okado and Yokota 1982; Van Leeuwen et al. 1983). It has been suggested that a transfer of substances from axons to neuroglial cells may occur through these junctions (Wittkowski 1969).

While the neuroglial cell is usually the postsynaptic component of the junction, an example of a presynaptic astrocyte forming a synapse-like junction with a dendritic spine has been reported (Edmunds and Parnavelas 1983).

Some observations suggesting a gap junctional coupling between nerve and neuroglial cells of mammals have been reported in the last 20 years (for a review, see Haydon 2000). However, gap junctions between nerve and neuroglial cells have never been found in the mammalian CNS in studies that used strict criteria for identifying such junctions (Mugnaini 1986; Rash et al. 1997).

K. Other Types of Interneuronal Communication

It is possible that not all interneuronal communications are mediated by synapses or electrotonic junctions. Other possible modes of communication between neurons are described below.

When two axons lie in contact, the electrical activity in one may modify the excitability of the other. The term ephapse has been proposed to designate the site where two neurons electrically interact without forming a specialized junction (Arvanitaki 1942). Over time the term ephapse has also come to be used to indicate a point of contact (other than a synapse or an electrotonic junction) between two axons or dendrites across which ion exchange occurs.

Receptors have been found in membrane domains not faced by a synaptic bouton. It has been proposed that these extrasynaptic receptors interact with transmitters that diffuse within the extracellular space of the CNS. This type of interneuronal communication, which may occur at long distance, has been called volume transmission (for a review, see Agnati et al. 1995). Transmitters which diffuse within the extracellular space of the CNS include glutamate, glycine, γ-aminobutyric acid, and neuropeptides. The latter are often released from dendrites at regions showing no synaptic specializations (for a review, see Ludwig and Pittman 2003). Volume transmission may also be involved in neuron-glia and glia-glia communications within the CNS.

Finally, it has been shown that some small gaseous molecules (e.g., nitric oxide and carbon monoxide) play a role as intercellular messengers in the nervous system. It appears that the nitric oxide is an ancient intercellular messenger, its role dating back to the beginning of the evolution of the nervous system. Nitric oxide and carbon monoxide, which have short lifetimes, are released not only from axons but also from dendrites (for a review, see Ludwig and Pittman 2003) and diffuse into adjacent neurons and neuroglial cells to act directly on intracellular targets (for further information, see Barañano et al. 2001).

VII. The Neuroglia of the PNS

A. The Satellite Cells of Sensory and Autonomic Ganglia

1. Historical Note

A cell sheath enveloping the body of the neurons in sensory and autonomic ganglia was mentioned for the first time by Valentin (1836). In some illustrations of his paper, the nuclei of cells adjacent to the surface of the nerve cell body, both in the trigeminal ganglion and in the ganglia of the autonomic nervous system, were clearly shown even though they were misinterpreted as pigment granules (*Pigmentkörperchen*). Since Remak (1838) denied the existence of this perineuronal sheath, Valentin (1839) provided a more detailed description of it, illustrated with new drawings, the captions of which gave a correct interpretation of the satellite cell nuclei. There followed much debate as to whether or not the perineuronal sheath is a continuous envelope and whether it is built of discrete cells or has a syncytial structure (for more details, see Pannese 1981); this dispute was brought to an end only when it became possible to examine peripheral ganglia under the electron microscope.

2. Organization of the Perineuronal Sheath

In adult vertebrates, the basic organization of satellite cells around neurons is similar in sensory and autonomic ganglia. Satellite cells envelope each nerve cell body, its dendrites (when present), and the initial segment of the axon within a sheath which is in turn completely surrounded by the interstitial connective tissue of the ganglion. Thus, a discrete unit consisting of a single nerve cell body together with its satellite cell sheath is formed (see Pannese 1981, 2010), and each nerve cell body is completely isolated from every other (Fig. VII.1). However, clusters of two or three nerve cell bodies enclosed within a common connective envelope have also been found (Pannese et al. 1991). The percentage of nerve cell bodies arranged in clusters is highest in neonates and decreases progressively with age. In adult animals, the percentages of nerve cell bodies arranged in clusters range from 9.4 in the lizard and 5.6 in the rat to 4.3 in the rabbit (Pannese et al. 1991, 1993). Sometimes, the nerve cell bodies of a given cluster are in direct mutual contact; in other cases, they are separated from one another by a thin satellite cell sheet. Interestingly, chemical transmission can occur between the nerve cell bodies of a cluster even when the nerve cell bodies themselves are separated by a thin satellite cell sheet (Rozanski et al. 2012).

The satellite cell sheath may show a variously complicated pattern of organization. It may consist of one single layer of flattened cells or of several layers of cells which overlap in a complex fashion. Both the arrangement

E. Pannese, *Neurocytology: Fine Structure of Neurons, Nerve Processes, and Neuroglial Cells*, 2nd Edition, DOI 10.1007/978-3-319-06856-5_7, © Springer International Publishing Switzerland 2015

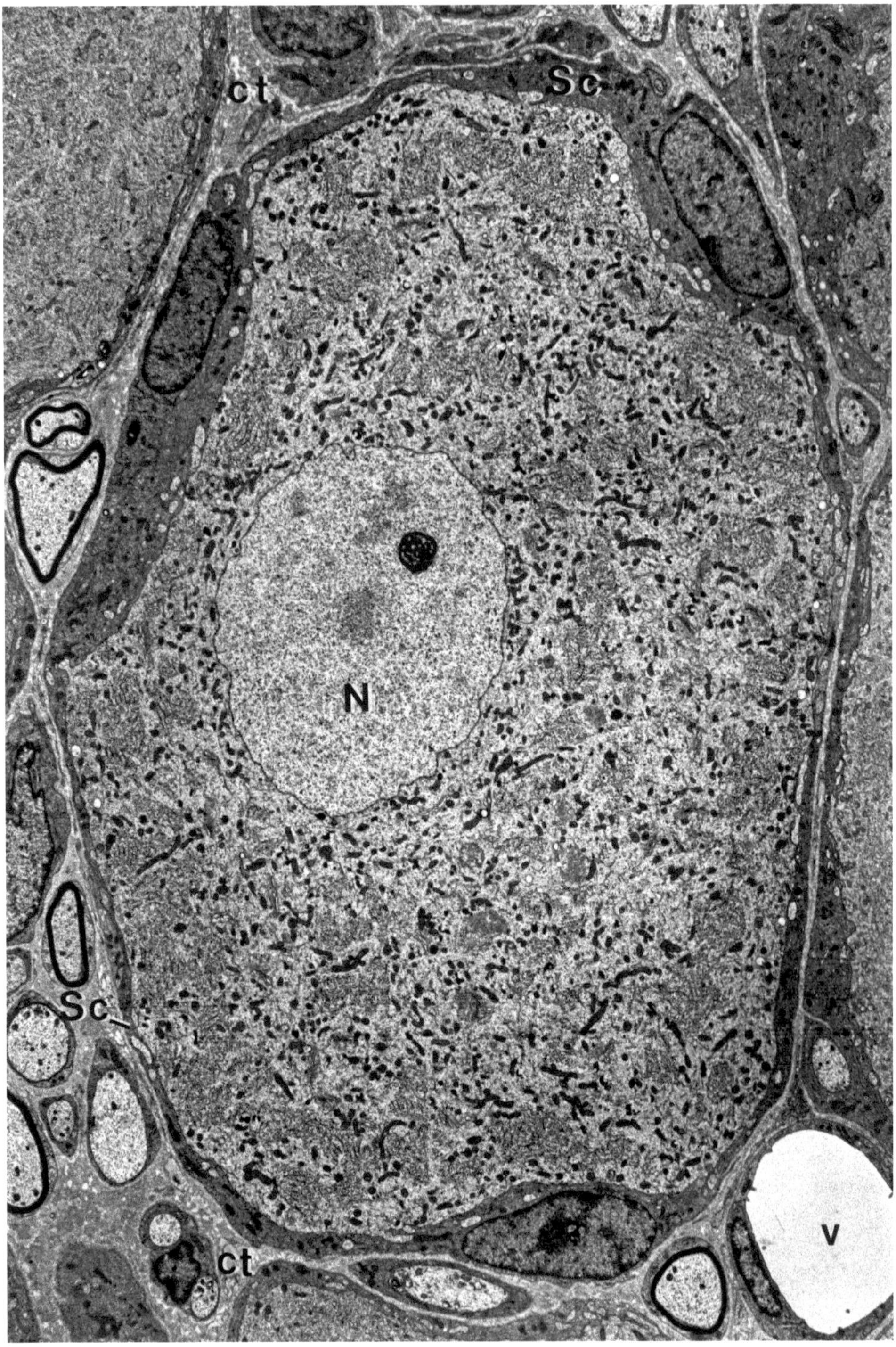

Fig. VII.1 Unit consisting of a nerve cell body and its enveloping satellite cell sheath (Sc). This unit is sharply separated from its neighbors by the connective tissue space (**ct**). Note that in the satellite cell nuclei the chromatin occurs in clumps located in the central region and in a peripheral band, whereas in the nucleus of the neuron (**N**) the chromatin is widely dispersed. **v** blood vessel. Rat spinal ganglion, ×6,000

of the satellite cells and the thickness of the sheath may vary from area to area within the same sheath. The sheath can in some places be as thin as a few dozen nanometers and, therefore, below the limit of resolution of the light microscope; this explains why the sheath may have appeared discontinuous to some optical microscopists. Although in electron microscope preparations the satellite cell sheath usually appears as a continuous structure, small gaps (from 0.06 to 0.75 μm in length in the rabbit spinal ganglia, Pannese et al. 1996b) have occasionally been observed; these gaps are less rare in autonomic than in sensory ganglia. In autonomic ganglia, satellite cells cover axon terminals that make synaptic contacts with the neuron.

3. Shape of Satellite Cells

As a rule, these cells are laminar and irregular in outline; the long, branching processes described in the past are likely the result of distortion of the cell shape due to technical procedures (Pannese 1960). Some of the satellite cells associated with the initial segment of the axon look like narrow sheets which wind spirally around it (named *espirocitos* by Río Hortega). Even after long-term in vitro maintenance of peripheral ganglia, satellite cells display a shape quite similar to that observed in vivo (Shimizu 1965; Bunge et al. 1967; Waxman et al. 1977); in cultures of dissociated ganglia, these cells often appear aggregated according to an epithelial-like pattern. In electron micrographs, satellite cells often exhibit fine lamellar expansions and microvilli, which greatly increase their surface area.

4. Structure of Satellite Cells

In general, the satellite cell is mononucleate; binucleate satellite cells have only occasionally been observed. The nucleus (Fig. VII.1) is often elliptic, less frequently round, kidney shaped, or nearly triangular in section; in the South African toad, it often shows one or more deep indentations. Chromatin usually occurs in centrally located clumps and in a peripheral band (Bunge et al. 1967; Pineda et al. 1967), which is separated from the inner surface of the nuclear envelope by a fibrous lamina. In the nuclei of neurons ensheathed by satellite cells (Fig. VII.1), in contrast, the chromatin is frequently widely dispersed (Hess 1955).

Free polysomes and cisternae of the granular endoplasmic reticulum (Fig. VII.2) are a common occurrence in the cytoplasm (Hess 1955; Wyburn 1958). As shown in Fig. VII.3, these cisternae may be arranged in stacks (Pannese 1960; Bunge et al. 1967). Tubules and vesicles of the agranular endoplasmic reticulum are less frequently observed, except in frog autonomic ganglia, where they are abundant. The Golgi apparatus and centrioles (Fig. VII.2), observed under the light microscope by Ramón y Cajal (1912) and Kubota and Hioki (1943) respectively, are usually localized in the juxtanuclear region. Mitochondria are scattered in the cytoplasm; autophagic vacuoles, lysosomes, lipofuscin granules, and peroxisomes (Citkowitz and Holtzman 1973) can also be observed.

Microtubules (20–30 nm in diameter) and intermediate filaments (about 10 nm in thickness) occur throughout the cytoplasm (Fig. VII.2), generally lying more or less parallel to both the outer and inner surfaces of the satellite cell sheath. They are particularly numerous around the axon hillock and the initial segment of the axon.

Solitary primary cilia (autonomic ganglia, Grillo and Palay 1963; sensory ganglia, Pannese 1964) have been observed in satellite cells. The cilium, which usually originates in the juxtanuclear region from a basal body (Fig. VII.2), projects into a narrow channel of the extracellular space formed by a deep invagination of the plasma membrane and may be directed toward the adjacent neuron or the connective tissue space. These

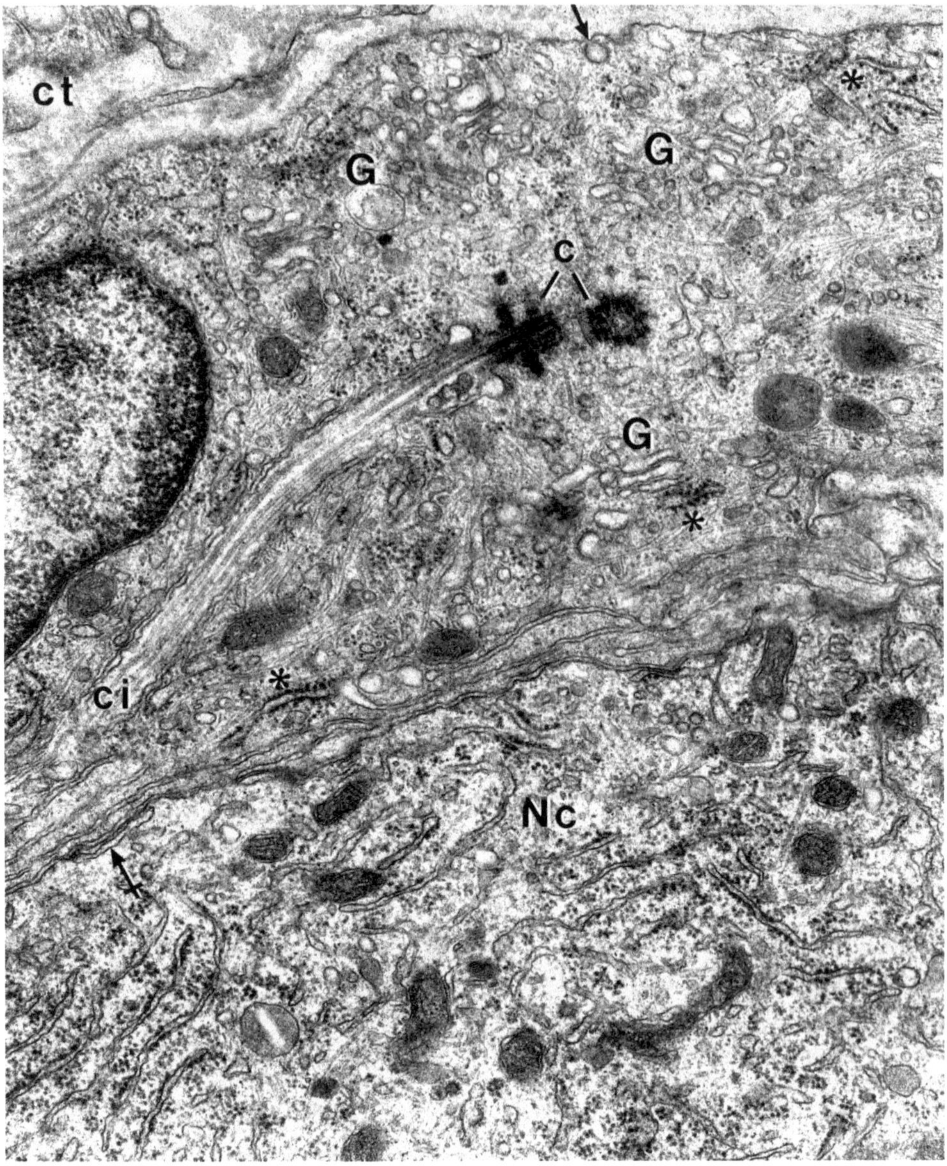

Fig. VII.2 Satellite cell. Mitochondria, dense bodies, cisternae of the granular endoplasmic reticulum (*), and Golgi complexes (**G**) can be seen in the cytoplasm. Microtubules and intermediate filaments are interposed between the other organelles. *Arrow* points to a pinocytotic vesicle along the satellite cell plasma membrane facing the connective tissue (**ct**). *Crossed arrow* points to a subsurface cisterna lying in the neuronal perikaryon, close to the plasma membrane. **c** centrioles, one of which serves as the basal body of a cilium (**ci**), **Nc** cell body of a sensory neuron. Rabbit spinal ganglion, ×30,000

Fig. VII.3 Satellite cells. *Above*: Satellite cell (**Sc**) undergoing mitosis. **Nc** neuronal perikaryon. Spinal ganglion of a normal adult rat, ×6,250. *Below*: Numerous cisternae of the granular endoplasmic reticulum arranged in a stack can be seen in the satellite cell cytoplasm. The outer contour of the satellite cell sheath facing the connective tissue (**ct**) is lined by a basal lamina (**bl**). Note that the cytoplasmic process of a connective tissue cell (*) is devoid of a basal lamina. Δ indicates a projection of the neuronal perikaryon (**Nc**) which invaginates the inner contour of the satellite cell sheath. Rat spinal ganglion, ×25,000

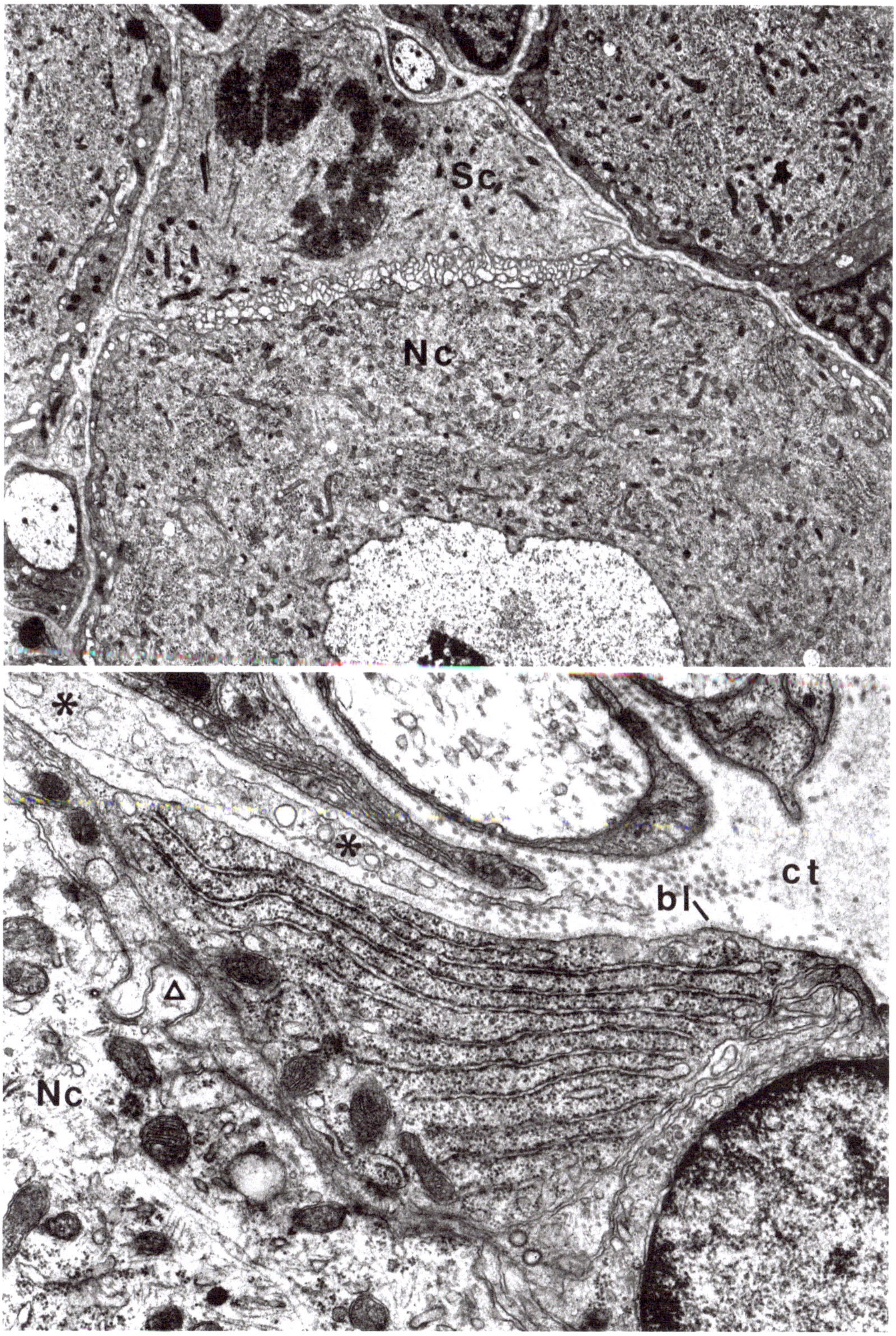
Sc
Nc
*
*
ct
bl
Δ
Nc

cilia lack the central pair of microtubules and only show the nine pairs of peripheral microtubules much as do those in neurons (see Sect. V.A.8), in Schwann cells (see Sect. VII.B.3b), and in astrocytes (see Sect. VIII.E.1). The finding of specific receptors on the solitary cilia of certain neurons supports the hypothesis that these cilia are sensory organelles (see Sect. V.A.8). The cilia of satellite cells may have a similar sensory function; however, there appear to be no data on the presence of specific receptors on the cilia of these cells.

Glycogen particles (25–30 nm in diameter) and lipid inclusions have been found in satellite cells. In mammals and birds, lipid inclusions are usually scarce and do not exceed 1 μm in diameter; in amphibians, by contrast, at certain periods of the year, they can be very numerous and so large (several micrometers in diameter) as to distort the outline of the nucleus and even that of the whole cell.

The plasma membrane of satellite cells, like that of neuroglial cells in the CNS, appears thinner and less dense than that of neurons. Assemblies of orthogonally arranged particles can be found in the freeze-fractured plasma membrane of satellite cells. The center-to-center spacing of these particles is about 7 nm. Such assemblies were found, for example, in the satellite cell plasma membrane of the chick (Pannese et al. 1977) and rat (Elfvin and Forsman 1978; Gotow et al. 1985) but not the frog (Matsumoto and Rosenbluth 1986). Similar specializations also are found in astrocytes (see Sect. VIII.E.1), ependymal cells (see Sect. VIII.C.2), and other cell types; however, the number per unit surface area of these assemblies is always lower in the plasma membrane of satellite cells than in that of astrocytes. For more information on these assemblies, see Sects. VIII.C.2 and VIII.E.1.

Another specialization can be seen in the freeze-fractured plasma membrane of satellite cells. This specialization consists of strands, which appear as ridges on the P face and as grooves on the E face. Since these strands show the same morphological characteristics as those composing tight junctions in other tissues, they could be interpreted as focal tight junctions (maculae occludentes) linking adjacent satellite cells.

5. Molecular Characteristics of Satellite Cells

Knowledge in this field is still fragmentary. Under normal conditions, glial fibrillary acidic protein (see Sect. VIII.E.1) is present at low levels in the perineuronal satellite cells of sensory ganglia (Woodham et al. 1989; Ajima et al. 2001; Jimenez-Andrade et al. 2006) and at higher levels in the satellite cells of autonomic ganglia (Elfvin et al. 1987; Alvarez et al. 1989). The expression of this protein increases markedly in both types of ganglia following axon injury (Elfvin et al. 1987; Woodham et al. 1989; Stephenson and Byers 1995; Vit et al. 2006) or compression of the ganglion (Zhang et al. 2009). The glial fibrillary acidic protein of satellite cells (and Schwann cells) has the same molecular weight as that of astrocytes, but their molecular identities differ slightly (Jessen et al. 1984). Satellite cells contain Ca^{2+}-binding S100 proteins (Vega et al. 1991; Sandelin et al. 2004), transporters for glutamate (Schon and Kelly 1974a; Hösli and Hösli 1978; Berger and Hediger 2000; Keast and Anderson 2000) and for γ-aminobutyric acid (Schon and Kelly 1974a, b), as well as the enzyme glutamine synthetase (Miller et al. 2002; Weick et al. 2003). This latter catalyzes the conversion of glutamate to glutamine. S100 proteins, the glutamate-aspartate transporter GLAST, and glutamine synthetase are often used as molecular markers of satellite cells. The enzyme NTPDase 2 (Braun et al. 2004), the transcription factor Erm (Hagedorn et al. 2000), and the NKCC1 cotransporter (Price et al. 2006) are expressed by perineuronal satellite cells. Immunoreactivity for the neuropeptide cholecystokinin has been demonstrated in satellite cells enveloping large neurons of the spinal ganglia of a cetacean (Bombardi et al. 2010).

The K^+ channels Kir4.1 and Sk3 (Hibino et al. 1999; Vit et al. 2006) and the acid-sensing ion channel 2 (Kawamata et al. 2006) have been identified in satellite cells. There is also evidence that these cells express purinergic receptors (P_2X, Zhang et al. 2005; P_2Y, Weick et al. 2003, Kobayashi et al. 2006) and aspartate receptors (Castillo et al. 2013). Furthermore, satellite cells express receptors to classical neurotransmitters (Bernardini et al. 1999), to neuropeptides (Li et al. 2008), to neurotrophins (Wetmore and Olson 1995; Pannese and Procacci 2002), to endothelins, to orexin-1, to bradykinin, to somatostatin, and to laminin-2 (Masaki et al. 2001). Laminin-2 receptors are involved in the adhesion of satellite cells to the basal lamina. Finally, cell adhesion molecules have been demonstrated in satellite cells (Mirsky et al. 1986). For a list of the molecules found in these cells, see Hanani (2005, 2010).

6. Relationships Between Satellite Cells

In certain areas, adjacent satellite cells contact simply, while in other areas within the same sheath their lamellar expansions overlap and interdigitate in a complex fashion. The gaps, from 15 to 20 nm in width, which intervene between the apposed plasma membranes of adjacent satellite cells open into the light space under the basal lamina at one end and into the narrow interval between the neuron and the sheath at the other. This potential pathway between the interstitial connective tissue and the neuronal surface usually follows a long and tortuous course, a condition which probably has some functional significance (see Sect. VII.D.1).

Satellite cells are linked to each other by adherent and gap junctions both in sensory (Pannese 1974; Pannese et al. 1977, 1978; Sakuma et al. 2001) and autonomic (Elfvin and Forsman 1978) ganglia. The presence of gap junctions between adjacent satellite cells has important functional consequences. The internal diameter (about 1.2 nm) of the gap junction channels not only allows the intercellular passage of ions but also permits the transfer of molecules with molecular weight up to about 1 kDa (e.g., signaling molecules, amino acids, and glucose). Dye coupling studies have shown that satellite cells enveloping a given nerve cell body are functionally coupled (Hanani et al. 2002; Pannese et al. 2003; Huang et al. 2005, 2006). Information on the connexins that form these junctions is relatively scarce. Connexin43 has been identified in the perineuronal satellite cells of rat trigeminal ganglion (Vit et al. 2006) and of mouse spinal ganglia (Procacci et al. 2008). However, it is not known whether the gap junctions linking perineuronal satellite cells consist exclusively of connexin43 or contain other connexins. Reflexive adherent and reflexive gap junctions, i.e., junctions linking different portions of the same satellite cell, have also been observed.

7. Perikaryal Myelin Sheaths

A myelin sheath enveloping nerve cell bodies in the PNS was described as long ago as the nineteenth century (for further details, see Pannese 1981); under polarized light, this perikaryal sheath presents the same characteristics shown by the myelin surrounding the axon. Myelinated nerve cell bodies occur frequently in the acoustic and vestibular ganglia in all classes of vertebrates and occasionally in the other sensory ganglia of bony fishes (for a review, see Scharf 1958). In the autonomic nervous system, they can be observed in the ciliary ganglion of birds.

Electron microscope studies by Rosenbluth and Palay (1961), Rosenbluth (1962a), and subsequently other authors showed that the perikaryal myelin may display either the compact structure exhibited by the myelin surrounding the axons (see Sect. VII.B.4d) or the characteristics of loose myelin. In the latter case, the perikaryal sheath is built of a varying number of lamellae (from 10 to 100 nm in thickness), each consisting of a layer of satellite cell cytoplasm bounded by its plasma membrane and separated from the nearby lamellae by a narrow space. Compact myelin has a highly regular pattern of organization (see Sect. VII.B.4d), whereas in loose myelin the number and thickness of the lamellae and the width of the interval between the lamellae in a given sheath may vary from one region to another. The main characteristics of perikaryal loose myelin are similar to those of the sheaths enveloping certain axons of invertebrates (see Sect. VII.B.2) and the developing myelinated nerve fibers of vertebrates. Perikaryal loose myelin differs from common multilayered perikaryal sheaths only in the thickness of the cytoplasmic layers and the location of the cell nuclei; in loose myelin, the cytoplasmic layers are thinner than in common multilayered perikaryal sheaths, and the satellite cell nuclei always lie in the outermost layer of the sheath.

Thus, a wide range of perikaryal sheaths can be found in the PNS, namely, a simple layer of satellite cells, multiple layers of satellite cells, loose myelin, loose and compact myelin coexisting in the same sheath, and finally compact myelin. Therefore, the perikaryal myelin sheaths are not basically different from the satellite cell envelopes usually found around the nerve cell bodies but rather represent variants of them. Myelin sheaths surrounding nerve cell bodies in the CNS are considered in Sect. VIII.F.2.

8. Boundaries of the Satellite Cell Sheath with the Neuron and Connective Tissue

The neuron-satellite cell boundary is usually complicated by the presence of many projections from both the neuron (Fig. III.1) and the satellite cells (Hess 1955; Palay 1957; Wyburn 1958; and subsequently other authors). In single sections, the neuronal projections, usually fingerlike, may sometimes appear continuous with the neuronal perikaryon (Fig. VII.3); more frequently, they arise from the neuron at another level, thus appearing as discrete entities completely surrounded by satellite cells (Fig. V.5) or located along the interval between the neuron and its satellite cell sheath. Even in these cases, however, they can be identified because their cytoplasmic matrix is less dense and their plasma membrane thicker and denser than their counterparts in the adjacent satellite cells.

Satellite cell projections, which occur less frequently than neuronal ones, invaginate the neuronal profile to variable depths; these projections possibly correspond to the trophospongium described by Holmgren (1901, 1902). The invaginations of the neuronal surface appear similar to those observed under the electron microscope in large neurons of invertebrates (see, among others, Rosenbluth 1963; Coggeshall 1967).

The interval between the satellite cell sheath and the enclosed neuron shows a rather uniform width (about 20 nm), except where numerous projections arise from the neuron; here it appears uneven and enlarged. Buttonlike, adherent junctions have occasionally been observed at the neuron-satellite cell boundary (Pannese 1969b; Tennyson 1970) but, as far as is known, tight or gap junctions have never been described at this interface.

In sensory ganglia, Ehrlich (1886) [P. Ehrlich 1854–1915], Ramón y Cajal (1890), and Dogiel (1896, 1897) described a plexus of unmyelinated fibers surrounding some nerve cell bodies. Some authors regarded these fibers as sympathetic fibers of extrinsic origin (coming from outside the ganglion), while others regarded them as unmyelinated branches of myelinated fibers from neurons situated within the ganglion (Dogiel type II cells). Although the presence of a perineuronal plexus of unmyelinated fibers in sensory ganglia was questioned by Van Gehuchten (1892) and Retzius (1894), the existence of this plexus has been repeatedly confirmed over the last 20 years. The perineuronal unmyelinated fibers are currently regarded as autonomic fibers of extrinsic origin (for more details, see Matsuda et al. 2005). In normal conditions, only a small proportion of ganglionic sensory neurons are surrounded by autonomic fibers. Sometimes these fibers are enclosed within the satellite cell sheath that envelopes the sensory nerve cell body, and very rarely these fibers are directly apposed to the plasma membrane of the sensory nerve cell body but never make synaptic contact with it. Since the number of autonomic fibers surrounding ganglionic sensory neurons increases notably after peripheral nerve injury, it has been suggested that this autonomic innervation of ganglionic sensory neurons may contribute to the generation and maintenance of some forms of chronic neuropathic pain.

It should be emphasized that electron microscope studies have never convincingly demonstrated the presence of synaptic contacts on nerve cell bodies or their axons in sensory ganglia, either in situ or in organotypic cultures. Only in dissociated cell cultures have spinal ganglion cells been shown to develop rare synaptic contacts (Miller et al. 1970; Lodin et al. 1973).

The outer surface of the satellite cell sheath facing the connective tissue is covered by a basal lamina (Fig. VII.3), which is continuous from one satellite cell to the next, i.e., bridges the 15–20-nm gaps between adjacent satellite cells. Notwithstanding the presence of evaginations and invaginations, the outer surface of the sheath is generally smoother than that facing the neuron. The evaginations of satellite cells appear as lamellar or finger-like cytoplasmic projections which extend into the connective tissue and are always surrounded by the basal lamina. In single sections, the profiles of these evaginations may sometimes be seen as discrete entities in the connective tissue since they arise from the satellite cell sheath at another level; in these cases, they may be distinguished from cytoplasmic processes of the connective tissue cells as the latter are devoid of a basal lamina (Fig. VII.3).

Invaginations, which again always appear lined by a basal lamina, enter the satellite cell sheath to variable depths (Fig. V.5). Such invaginations are generally deeper, more numerous, and more ramified around the initial segment of the axon than around the neuronal perikaryon; in the former location, they sometimes form a labyrinthine system which may nearly reach the axonal surface. In some amphibians and reptiles, the satellite cell sheath enveloping the initial segment of the axon is divided into two layers; the intervening cleft, which may contain collagen fibrils, communicates with the ganglionic connective tissue space.

9. Quantitative Relationships Between Nerve and Satellite Cells

Sensory ganglia are a particularly suitable material for studying the quantitative relationships between nerve and neuroglial tissues because both neuronal bodies and satellite cell sheaths have a rather regular shape. Moreover, unlike the situation in other regions of the nervous system, some connective tissue sharply separates the satellite cell sheaths belonging to adjacent neurons (Fig. VII.1), so that each nerve cell body and its related satellite cell sheath usually form a discrete structural unit (see Pannese 1981).

Light microscope studies showed that in spinal ganglia the number of satellite cells enveloping each nerve cell body is proportional to the volume of the latter (Pannese 1960, 1964). Studies carried out with the electron microscope have revealed that in spinal ganglia the volume of the satellite cell sheath is directly proportional to both the volume and surface area of the nerve cell body (Pannese et al. 1972, 1975). The surface of each nerve cell body is highly irregular (see Chap. III and Sect. VII.A.8), its area thus greatly exceeding that of a sphere of equal volume; this explains why the volume of the satellite cell sheath can be directly proportional to both the volume and the surface area of the nerve cell body. Note that the number of satellite cells associated with each neuron correlates highly with the neuronal volume also in autonomic ganglia (Gabella et al. 1988; Pomeroy et al. 1996).

As shown by the studies quoted above, a quantitative balance exists between nerve cell bodies and their associated satellite cell sheaths. The factors which maintain this balance are a matter of hypothesis. Neurons are known to differentiate earlier than satellite cells during ganglionic development. Moreover, since in experimentally induced enlargement of the nerve cell body there follows an increase in volume of the satellite cell sheath (Pannese 1963b, 1964; Humbertson et al. 1969), it is reasonable to suppose that the size and/or surface area of the nerve cell body influence and control the volume of the related satellite cell sheath. By comparing the quantitative results obtained from mammals and reptiles, it appears that, for a given size or surface area of the nerve cell body, the volume of the satellite cell sheath is greater in mammals than in reptiles (Pannese et al. 1972, 1975). It seems, therefore, that in addition to the size and surface area of the nerve cell body, other factors, for example, the metabolic rate of the ganglionic neuron, may influence the volume of the satellite cell sheath.

Note that, as is the case for satellite cell sheaths and nerve cell bodies in spinal ganglia, in both myelinated and unmyelinated nerve fibers of spinal roots the size of the Schwann cell sheath is directly proportional to the size of the related axon or axons (Pannese et al. 1987b, 1988b).

10. Mitotic Activity of Satellite Cells

Mitotic satellite cells and satellite cells labeled with tritiated thymidine have been observed both during body growth (Pannese 1969b) and in normal adult animals (sensory ganglia, Pannese 1960, Friede and Johnstone 1967, Kraus-Ruppert et al. 1975; autonomic ganglia, Dropp and Sodetz 1971). Under normal conditions, therefore, these cells retain their ability to divide also in adult life (Fig. VII.3). The half-life of spinal ganglion satellite cells has been estimated at about 600 days in normal adult mice (Elson et al. 2004). The question arises as to whether satellite cells arising by mitosis in the adult animal remain in situ and are added to the existing cell population or whether they replace degenerated or otherwise lost satellite cells. Since the size of the satellite cell population in a given ganglion does not change appreciably between young adult and adult animals, it seems that new cells replace lost satellite cells in these life periods and the satellite cell population is renewed, although at a very slow rate.

The rate of satellite cell proliferation in adult life increases under a variety of experimental conditions (e.g., after axonal injury, during neuronal hypertrophy, and following the administration of certain chemicals) as well as under some pathological conditions (for a more detailed account, see Pannese 1981).

In various in vitro systems, satellite cells of both sensory and sympathetic ganglia were reported to proliferate (Pomerat et al. 1967; Lodin et al. 1970; Varon et al. 1973; McCarthy and Partlow 1976). The addition of nerve growth factor to the culture medium increases the mitotic activity in these cells (Ebendal 1975; McCarthy and Partlow 1976), an effect which is probably indirect being mediated by neurons. Neurons and satellite cells, which have been initially separated in dissociated cell cultures, show a clear tendency with time to reassociate and to reestablish their usual in vivo relationships (Waxman et al. 1977). The mechanisms underlying this intercellular reaggregation may be surface recognition and/or attractive phenomena exerted by one cell type on the other. One can hypothesize that similar mechanisms are also at work during the in vivo development of ganglia.

There is evidence that in animals in which growth continues throughout life, neurons of the sensory ganglia increase in number even during adult life (Devor et al. 1985; Popken and Farel 1997). New neurons could derive from the late differentiation of committed nerve cells generated prenatally (Popken and Farel 1997; Ciaroni et al. 2000). However, the possibility cannot be ruled out that new neurons derive from precursors present in the ganglion, which proliferate during adult life. These precursors could be indistinguishable from perineuronal satellite cells. Therefore, some of the satellite cells that proliferate in adulthood could actually be neuronal precursors.

11. Phagocytic Activity of Satellite Cells

Fragments of degenerated nerve cells have been seen within perineuronal satellite cells both in the sensory (Pannese 1978) and autonomic ganglia (Matthews and Raisman 1972; Pilar and Landmesser 1976) of various species. These findings together with the identification of an engulfment receptor (Wu et al. 2009) on satellite cells suggest that in vivo these cells have phagocytic capacity. Satellite cells have been shown to be phagocytic when cultured in vitro (Wilkinson et al. 1999).

12. Plasticity of Satellite Cells

Satellite cells are endowed with marked plasticity. This property becomes much evident following injury of the axon of the ganglionic neuron with which they are associated. After such injury, satellite cells proliferate (see Sect. VII.A.10) and undergo modifications. Such modifications include the increase in the amount of glial fibrillary acidic protein (Woodham et al. 1989; Stephenson and Byers 1995; Takeda et al. 2007; Vit et al. 2006), the increase in one of the S100 proteins (Sandelin et al. 2004), and the expression of high levels of neurotrophins (Zhou et al. 1996). Further changes include the formation of bridges connecting previously separate perineuronal sheaths and a marked increase in number of the gap junctions between satellite cells (Pannese et al. 2003). While in normal conditions satellite cells are only coupled to other satellite cells within the same perineuronal sheath, following axon injury satellite cells become extensively coupled to satellite cells belonging to other perineuronal sheaths. This marked increase in number of gap junctions correlates with the increase in the incidence of dye coupling among satellite cells (Hanani et al. 2002). The increased coupling following axon injury could provide the basis for medium- and long-distance communication between satellite cells. Finally, both glial fibrillary acidic protein within satellite cells and coupling through gap junctions between these cells increase markedly following compression injury of the ganglion (Zhang et al. 2009).

13. Age-Related Changes

The structure of satellite cells changes little with age. The Golgi apparatus of these cells in rabbit spinal ganglia undergoes neither structural changes nor peripheral displacement with advancing age; both the total volume of the Golgi apparatus and the mean percentage of cytoplasmic volume occupied by this organelle decrease significantly in old animals compared to young adults (Ledda et al. 2003). The mitochondrial structure does not change either, whereas mitochondrial size increases progressively and significantly with advancing age. Both the mitochondrial mass and the mean percentage of cytoplasmic volume occupied by mitochondria decrease progressively and significantly with age (Martinelli et al. 2007). These findings suggest that the ability of satellite cells to produce energy decreases with age and that the reduced ability of sensory neurons to respond to high energy demands in old age (e.g., see Sylvia and Rosenthal 1979) may be in part due to the diminished contribution of perineuronal satellite cells. With age, satellite cells accumulate lipofuscin. Lipofuscin accumulation seems to have little or no effect on either the metabolism or functional activity of neurons (for a review, see Ledda et al. 1999). This is probably also the case for satellite cells.

In the sensory ganglia of adult vertebrates, each nerve cell body is usually enveloped by its own satellite cell sheath (for a review, see Pannese 1981). This particular organization does not change in old age. However, both the mean volume of the satellite cell sheaths and the mean value of the volume ratio between the satellite cell sheaths and the related nerve cell bodies are significantly lower in old animals than young adults (Pannese et al. 1996b; Martinelli et al. 2006a). Since satellite cells provide trophic and protective support for neurons, it is likely that the significant decrease in their number and the marked quantitative reduction in the volume ratio between the satellite cell sheaths and the related neurons have negative consequences for neuronal activity.

Furthermore, the gaps sometimes present in satellite cell sheaths (see Sect. VII.A.2) are significantly more numerous and extensive in aged animals than young adults. For example, in rabbit spinal ganglia, for comparable neuronal perimeters, these gaps occur more than twice as often in old animals than young ones. In addition, while in young adults these gaps are no longer than 0.75 μm, in old rabbits they may reach a length of 7.7 μm (Pannese et al. 1996b). Since the nerve cells in sensory ganglia lack the protection of a vascular barrier such as is present in the CNS, only the satellite cell sheath controls the traffic of substances from the blood to the ganglionic neurons (see Sect. VII.D.1). Therefore, the greater number and extent of these gaps in old animals may render the neurons more susceptible to damage by harmful substances from the circulation.

The gap junctions between perineuronal satellite cells increase in number with age in rabbit spinal ganglia, whereas the mean size of individual gap junctions remains constant (Martinelli et al. 2005). Both gap junction number and dye coupling between satellite cells increase (Huang et al. 2006), whereas connexin43 expression decreases (Procacci et al. 2008) in mouse spinal ganglia in senescence. A possible explanation for this discrepancy is that in these ganglia additional connexin types other than connexin43 contribute to the gap junctions between satellite cells (see Sect. VII.A.6).

B. Schwann Cells and the Myelin Sheath

1. Historical Note

It was Remak (1838) who for the first time mentioned those elements which ever since have been known as Schwann cells. This author stated that the unmyelinated nerve fibers of the sympathetic nervous system, termed by him "gray" or "organic" fibers and later called "Remak fibers," bear small, oval or round, and occasionally irregular, corpuscles (*Körperchen*) bulging into the overlying space. The following year, Schwann (1839) confirmed that these corpuscles are associated with the "gray" fibers, showed their occurrence also in the "white" fibers, and interpreted them correctly as cell nuclei.

The fact that Schwann cells are discrete elements was firmly stated by L. A. Ranvier [1835–1922]. Having discovered discontinuities occurring in the myelin sheath of peripheral nerve fibers (1871a), Ranvier further investigated the structure of these gaps in the living frog and in nerves of several mammals by various techniques. He came to the conclusion that they represent the regions of contact, which he called nodes, between successive Schwann cells (1878). In contrast to Ranvier's opinion, Nageotte (1922) regarded the Schwann sheath as syncytial in structure and continuous throughout the course of the nerve fiber.

The term myelin was probably introduced to the scientific literature by Virchow (1854b). Ranvier (1878) considered myelin as a fluid or nearly fluid fatty mass produced by and contained within the Schwann cell, which for this reason he likened to a fat cell. This same author attributed to myelin the function of mechanically protecting the axon and of isolating it, thereby increasing the efficiency of conduction. Other workers, including Ramón y Cajal (1909) and Nageotte (1922), on the other hand, regarded myelin as an extracellular material secreted by the axon and lying between it and the Schwann sheath. Koelliker (1886) held that in the PNS the outer protoplasmic layer of the nerve fiber became transformed into true myelin through the accumulation of fat under the influence of the Schwann sheath nuclei; conversely, in the nerve fibers of the CNS, which are without Schwann sheaths, he thought that the formation of myelin took place independently of external influences. The birefringence of myelin was discovered by Ehrenberg (1849); the results of careful polarized light studies carried out almost a century later (see, among others, Schmidt 1924, 1936, 1937; Schmitt and Bear 1937) led to the conception of a model of molecular organization of myelin, in which concentric layers of radially oriented, rod-like lipid components alternate with layers of tangentially arranged protein molecules. The first electron microscope images showing the radially repeating structure of myelin were published by Fernández-Morán (1950, 1952, 1954) and Sjöstrand (1950, 1953c). Sjöstrand (1953c) described both the major dense line and the intraperiod line and found a radially repeating period of about 12 nm. That PNS myelin is formed by the Schwann cell plasma membrane spirally wrapped around the axon was established by Geren (1954).

2. Evolutionary Aspects

In the primitive nerve nets of coelenterates, axons lack neuroglial ensheathments; only an intercellular gap, 20 nm in width, intervenes between the plasma membranes of adjacent axons. In all other invertebrates, axons are related to neuroglial cells. A very simple type of relationship between axons and neuroglial cells is one in which many small axons lie in direct contact with one another and the entire axonal bundle is ensheathed by a single neuroglial cell. This type of relationship can be observed in, e.g., the cockroach (Hess 1958), freshwater mussel (Nakajima 1961), and crab (Horridge and Chapman 1964); it is also present in the olfactory nerve of higher vertebrates.

Small axons are usually ensheathed as bundles by neuroglial cells. On the other hand, as the axonal diameter increases, each axon tends to be enveloped by an individual sheath, as is nearly always the case with large axons. In some of these (e.g., the giant axons from the squid and lobster), the sheath may consist of a single layer of neuroglial cells arranged as a mosaic around the axon (Geren and Schmitt 1954; Villegas and Villegas 1963). In other cases, the sheath enveloping an invertebrate large axon is multilayered and in some ways resembles vertebrate myelin. In arthropods (Trujillo-Cenóz 1962), the sheath enveloping a large axon consists of multiple, sometimes loosely spiralled neuroglial layers. Such a pattern was likewise reported in chordates (e.g., lancelet) by Peters (1963).

The sheath enveloping the giant axon in the earthworm is a mixture of compacted, noncompacted, and spirally arranged portions (Hama 1959; Günther 1976). In the interganglionic connectives of the ventral nerve cord of the shrimp, each axon is surrounded by the following structures: a periaxonal mantle of loosely interdigitated neuroglial cells, a large extracellular space, and finally a thick and multilayered sheath

(Hama 1966; Kusano 1966; Hsu and Terakawa 1996). This complex ensheathment very efficiently isolates the axon, whereby conduction takes place at a very fast rate (Kusano 1966; Hsu and Terakawa 1996). Finally, the sheaths enveloping certain axons in the prawn show a partial compaction of their constituent neuroglial processes (Heuser and Doggenweiler 1966). Each sheath consists of 10–50 neuroglial cell processes concentrically arranged around the axon; extracellular spaces 20 nm wide intervene between the innermost processes, whereas the outermost ones appear fused together with obliteration of the extracellular gaps. This sheath displays irregularly spaced interruptions. It should be stressed once more that the cells forming the periaxonal sheaths in invertebrates are often designated Schwann cells, but their exact nature has not been established.

Among vertebrates, myelin seems to be absent from hagfish and lampreys; in the other vertebrates, the myelin sheath has almost always a spiralled and compact structure and displays regularly spaced interruptions. These features are accompanied by a high efficiency in the propagation of action potentials (see Sect. VII.B.4h).

3. Unmyelinated Nerve Fibers

Schwann cells are usually confined to the PNS; however, their occurrence within the CNS under apparently normal conditions has occasionally been reported (Raine 1976; Büssow 1978; Jung et al. 1978). In the PNS of vertebrates, axons are surrounded for most of their length by Schwann cells. The unit consisting of one or more axons and the associated Schwann cells, which is surrounded by a basal lamina, is termed the nerve fiber. According to the type of the morphological relationship between the axons and Schwann cells, unmyelinated and myelinated nerve fibers can be distinguished. Under the light microscope, unmyelinated nerve fibers appear as long, narrow bands with elongated nuclei adhering to them. Electron microscope studies have shown that the characteristic feature of this type of nerve fiber is the rather simple morphological relationship between the axons and Schwann cells, the axons being enclosed in troughs which indent the surface of the Schwann cells.

3a. General Organization

Unmyelinated axons have diameters varying in mammals from 0.2 to 3.5 μm, but axons with diameters near the extreme values of this range are rarely seen. By studying transverse sections of unmyelinated nerve fibers under the electron microscope, it can be seen that each Schwann cell can accommodate one or more axons. Each axon may be enclosed in a single trough formed by its associated cell, the axon being more or less extensively surrounded by the Schwann cell. When the axon is only partially ensheathed by it, a portion of its plasma membrane is directly in contact with the basal lamina surrounding the Schwann cell (Fig. VII.4). When the axon is completely enclosed by the Schwann cell, the edges of the periaxonal collar come together and the two apposed portions of the Schwann cell plasma membrane run parallel to each other, separated by an interval of 10–15 nm (Fig. VII.5). By analogy with the mesentery enfolding the intestine, the two closely applied portions of Schwann cell plasma membrane have been called the mesaxon (Gasser 1955). The length and path of the mesaxon can vary greatly; it may be short and straight or long and sinuous (Fig. VII.5) or even developed into a loose spiral around the axon. Reflexive (see Sect. VII.A.6) adherent junctions have been observed across the mesaxon (Gamble and Gosset 1966; Dolapchieva and Lierse 1975). In other cases, a group of small axons is enclosed in the same trough (Fig. VI.10) so that axons lie in contact with one another.

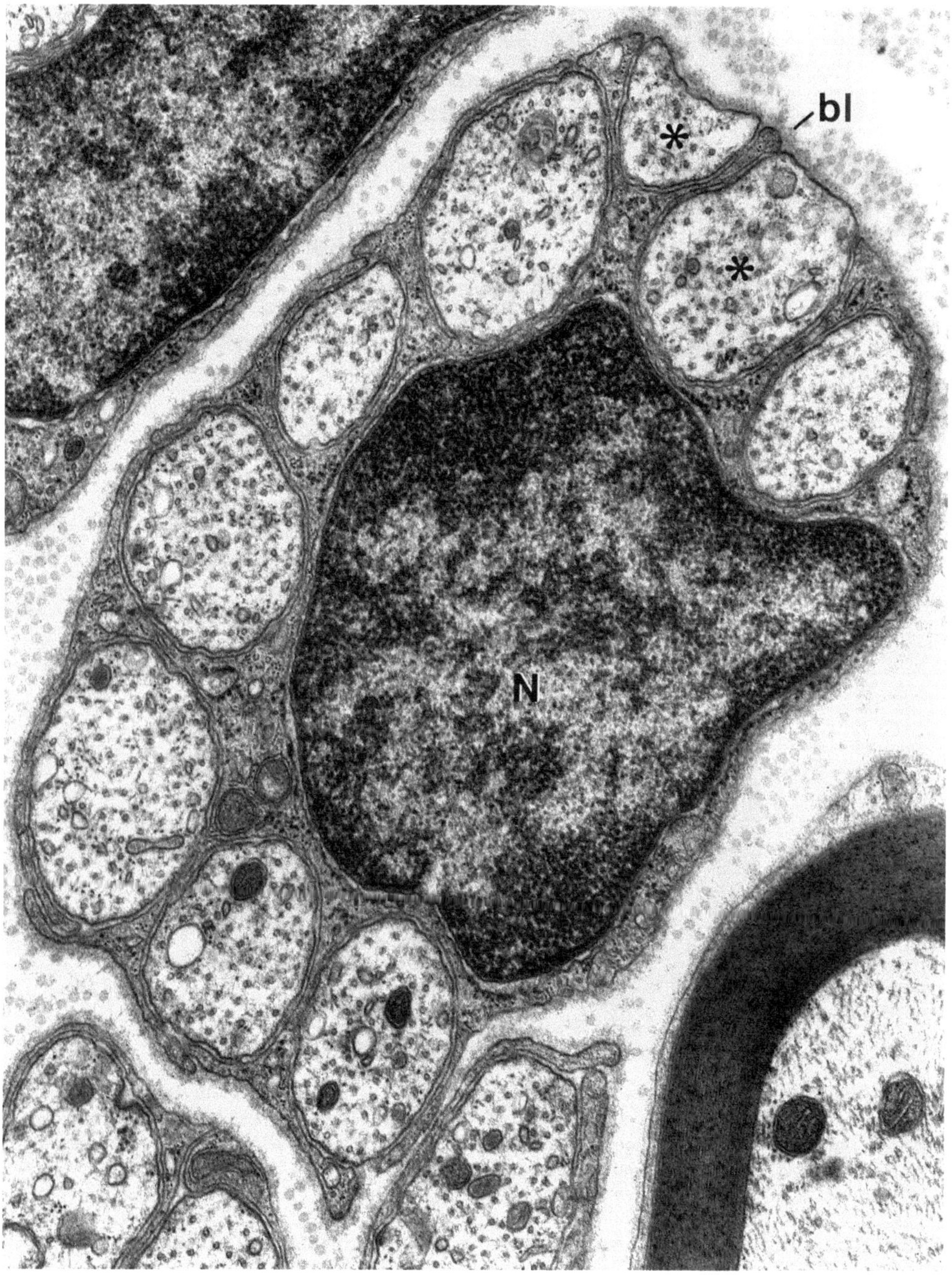

Fig. VII.4 Unmyelinated nerve fiber. The Schwann cell ensheathes nine transversely sectioned unmyelinated axons. Each axon is enclosed in a single trough. Note that the axons marked * are only partially surrounded by the Schwann cell so that portions of their plasma membranes are directly in contact with the basal lamina (**bl**). The nucleus (**N**) shows some shallow indentations; its chromatin occurs in centrally located clumps and in a peripheral band. Rat dorsal root, ×35,000

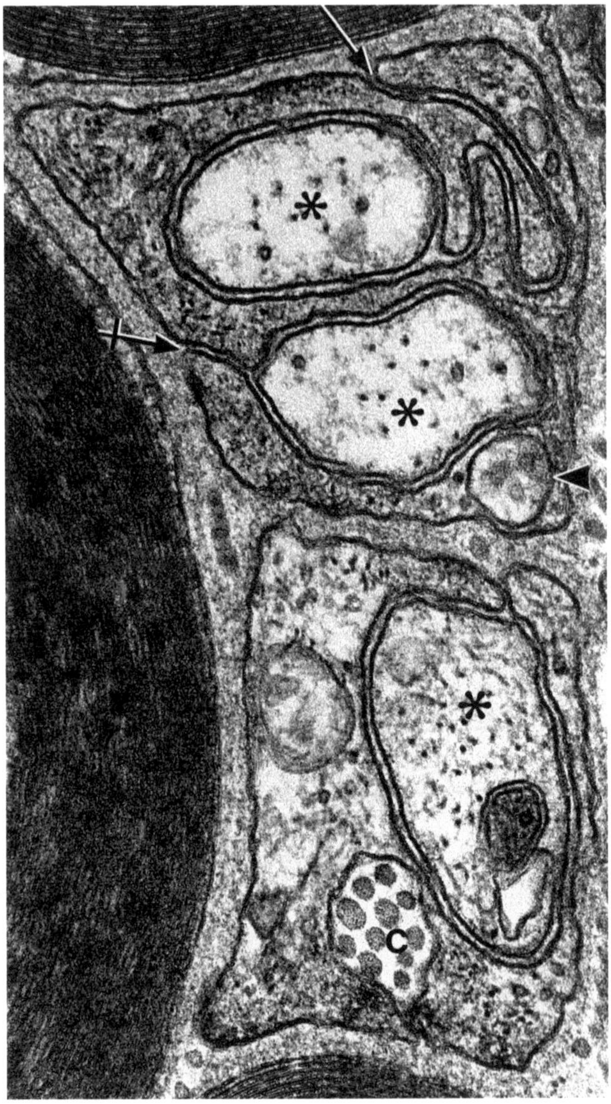

Fig. VII.5 Unmyelinated nerve fibers. The Schwann cell above ensheathes two transversely sectioned axons (*). Note that one mesaxon (*arrow*) is long and sinuous, while the other (*crossed arrow*) is short and fairly straight. *Arrowhead* points to a multivesicular body. The lower Schwann cell ensheathes one transversely sectioned axon (*) and also encloses a bundle of collagen fibrils (**c**). Lizard ventral root, ×75,000

In electron micrographs of a longitudinally sectioned nerve fiber, it can be seen that the Schwann cells are arranged in a chain along the entire length of the axons, each cell ensheathing a single axon or number of axons only for part of its (or their) length. Hence, contrary to Nageotte's (1922) belief, the Schwann cell sheath associated with unmyelinated axons is not a syncytium but consists of discrete cells (Elfvin 1958).

The morphological relationships between axons and Schwann cells can differ between species and in different nerves in the same species. In autonomic nerves and in deep-seated somatic nerves (cat, Gasser 1955; rat, Dixon 1963; human, Gamble and Eames 1964), each axon is enclosed in a single trough formed by its associated cell. In other cases, for example, in human cutaneous nerves (Gamble and Eames 1964), numerous laminar processes arise from the Schwann cell. Each of these processes envelopes one or rarely two axons, thus forming an entity bounded by a basal lamina; these entities are scattered singly within the connective tissue space of the nerve and appear connected with each other only in the proximity of the nucleated region of the Schwann cell.

In unmyelinated fibers, the Schwann cell length can vary greatly along a single nerve. In the rat sural nerve, for example, the lengths of Schwann cells can range between 20 and 310 μm, in the rat cervical sympathetic trunk from 20 to 250 μm (Peyronnard et al. 1973), and in the human sural nerve from 220 to somewhat more than 400 μm (Carlsen and Behse 1980).

3b. Structure of Schwann Cells

The nucleus is elongated with its major axis parallel to the long axis of the nerve fiber. It exhibits a rather irregular outline due to the presence of shallow indentations (Fig. VII.4) where neighboring axons are very close and, less frequently, to deep folds. As in the satellite cell nucleus (see Sect. VII.A.4), chromatin usually occurs in centrally located clumps and in a peripheral band (Fig. VII.4) separated from the inner membrane of the nuclear envelope by a fibrous lamina.

Cisternae of the granular endoplasmic reticulum (which usually occur singly, Pannese et al. 1988c), free polysomes, and mitochondria are scattered uniformly in the cytoplasm. The Golgi apparatus is usually localized in the perinuclear region, where one or two centrioles may also be found (Grillo and Palay 1963). Tubules and vesicles of the agranular endoplasmic reticulum, dense bodies, multivesicular bodies (Fig. VII.5), autophagic vacuoles, lipofuscin granules, and peroxisome-like bodies can also be observed. Solitary primary cilia have occasionally been noted in the Schwann cells associated with unmyelinated axons but never in those associated with myelinated ones (rat, Grillo and Palay 1963; Bunge et al. 1967; South African toad, Billings-Gagliardi et al. 1974; lizard, Pannese et al. 1988c). These cilia lack the central pair of microtubules and only have the nine pairs of peripheral microtubules, as do those in neurons (see Sect. V.A.8), in satellite cells of sensory and autonomic ganglia (see Sect. VII.A.4), and in astrocytes (see Sect. VIII.E.1). Microtubules and many bundles of intermediate filaments occur, most of these lying roughly parallel to the long axis of the nerve fiber. For the protein subunits of intermediate filaments, see Sect. VII.B.4c.

3c. Relationships Between Adjacent Schwann Cells

As shown in Fig. VII.6, adjacent Schwann cells along an unmyelinated fiber may contact by a simple abutment or by envelopment of the terminal part of one cell by that of its neighbor (Elfvin 1958) or by complex interdigitation of one cell with the next (Eames and Gamble 1970); adherent junctions may occur at contact zones between adjacent Schwann cells (Eames and Gamble 1970).

3d. Boundaries of Schwann Cells with the Axon and Connective Tissue

The axon-Schwann cell boundary is rather regular, with a uniform gap (15–20 nm) between the apposed plasma membranes (Fig. VII.5). Pinocytotic vesicles can be observed along both the axonal and Schwann cell plasma membranes. Small adherent junctions have occasionally been observed at the axon-Schwann cell boundary (Uga and Ikui 1968), as well as the neuron-satellite cell boundary in sensory and autonomic ganglia (see Sect. VII.A.8).

The contour of the Schwann cell sheath facing the connective tissue is encircled by a basal lamina (Fig. VII.4) whose major components are type IV collagen, laminin, entactin, and heparan sulfate proteoglycan. All these components are produced by the Schwann cells, but they are assembled to form the basal lamina only if the Schwann cells are in contact with axons (for a review, see Bunge et al. 1986). In addition to these components of the basal lamina, Schwann cells synthesize and secrete types I and III collagen. Sometimes, Schwann cell lamellar processes, consistently enveloped by a basal lamina (Eames and Gamble 1970), extend into the endoneurium.

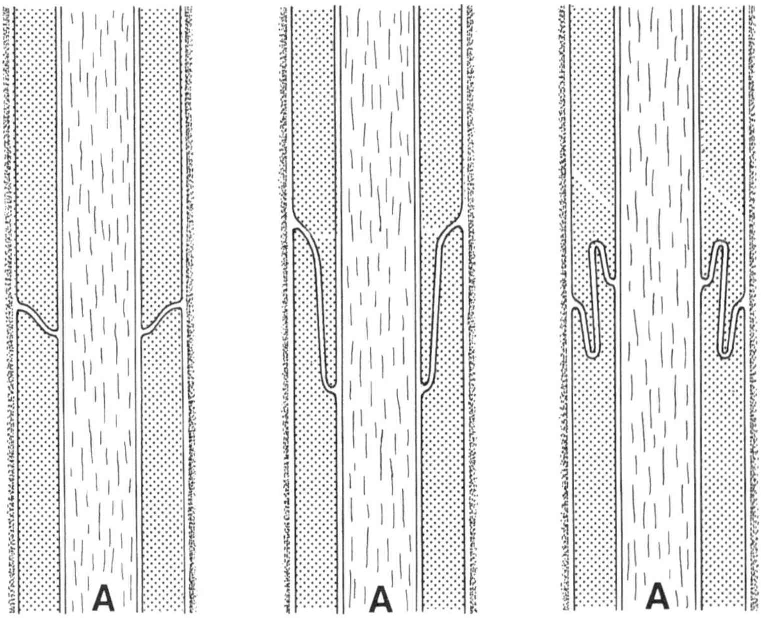

Fig. VII.6 Contact zones between adjacent Schwann cells along unmyelinated fibers. Schwann cells (*stippled*) may contact by a simple abutment (*left*), by envelopment of the terminal part of one cell by that of its neighbor (*middle*), and by interdigitation of one cell with the next (*right*). **A** axon

In single sections, profiles of individual processes may occasionally be seen as discrete entities within the endoneurium since they arise from the Schwann cell at another level; in these cases, they may be distinguished from cytoplasmic processes of connective tissue cells, the latter being devoid of a basal lamina.

A Schwann cell ensheathing unmyelinated axons can also enclose bundles of collagen fibrils (Fig. VII.5) with or without an intervening basal lamina (rat, Gamble 1964, Aguayo et al. 1972, Morris et al. 1972; human, Gamble and Eames 1964, Ochoa and Vial 1967). These authors regarded such "collagen pockets" as the result of the engulfment by Schwann cells of preexisting collagen. Similar relationships may become particularly frequent in some neuropathies; in these cases, however, they might be the result of the formation of new collagen fibrils which come to occupy the place of degenerated, unmyelinated axons.

4. Myelinated Nerve Fibers

Each fiber consists of a single axon invested by a chain of Schwann cells, which are arranged in longitudinal sequence and form the so-called myelin sheath around the axon. The minimum axon diameter for myelination is of about 1 μm in the PNS. To identify the factors which determine whether a Schwann cell will produce myelin or not, nerve grafts and cross-anastomosis between myelinated and unmyelinated nerves have been used (Weinberg and Spencer 1975, 1976; Aguayo et al. 1976). With these techniques, it has been shown that each Schwann cell is potentially able to produce myelin and that it is the axon which stimulates the Schwann cell to initiate myelinogenesis. The axon to be myelinated expresses high levels of type III neuregulin-1, an axonal surface molecule, which the Schwann cell recognizes as a signal to initiate myelination (Taveggia et al. 2005).

4a. General Organization

The axon appears as a thin cylinder, whose diameter in mammals varies from about 1 to about 15 μm. While the axon is continuous throughout the length of the fiber, the myelin sheath is interrupted (Ranvier 1871a) at regular intervals (Ranvier 1871b, 1878) by annular constrictions referred to as nodes of Ranvier (Fig. VII.7). The segment of myelin

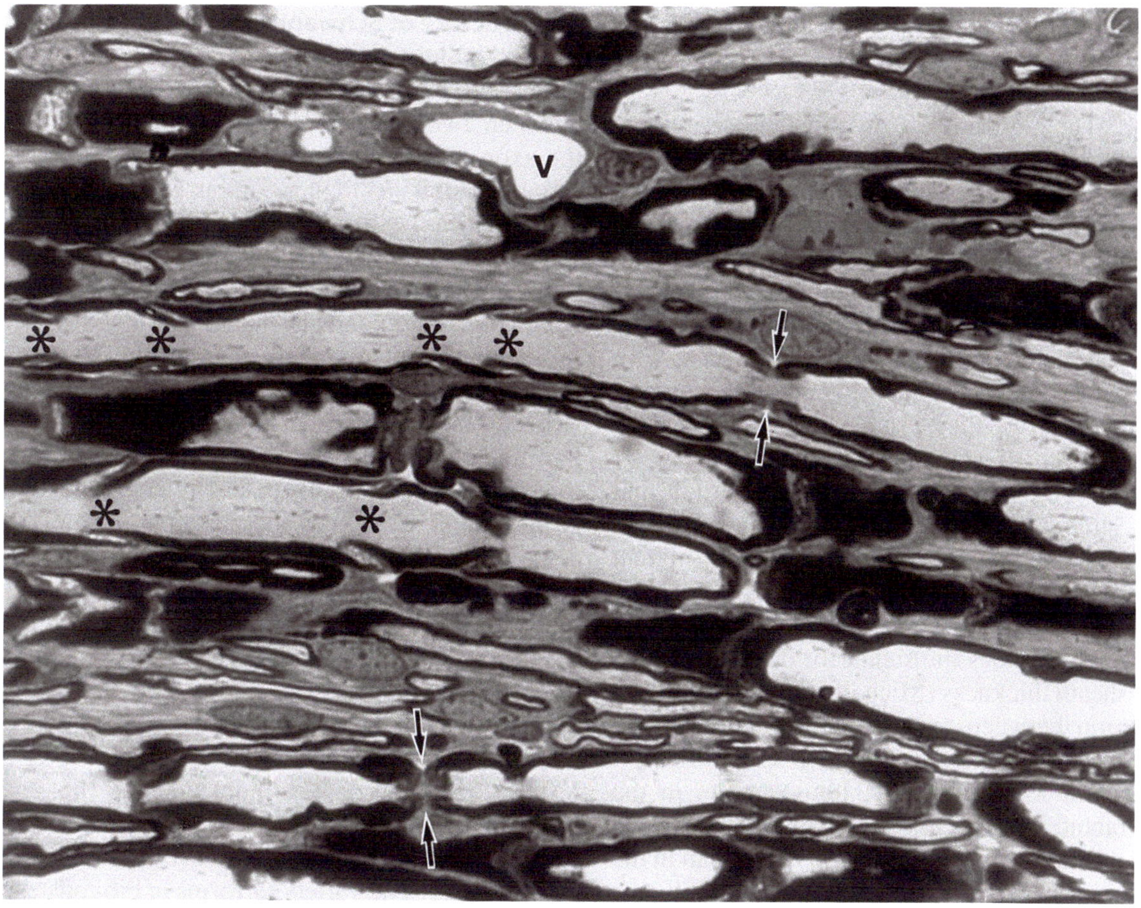

Fig. VII.7 Myelinated nerve fibers. Light micrograph showing nodes of Ranvier (*arrows*) and incisures of Schmidt-Lanterman (*) in longitudinally sectioned myelinated fibers of the PNS. **v** blood vessel. Rabbit dorsal root, 1-μm plastic section stained with toluidine blue, ×900

between two successive nodes is called an internode. Ranvier (1878), having observed that usually there is only one Schwann cell nucleus in each internode, advanced the hypothesis that each internode represents the territory of a single Schwann cell and that the node marks the boundary between two adjacent cells. However, some Schwann cells, although rarely, may have two nuclei (e.g., lizard spinal roots, Pannese et al. 1988a) or several nuclei (e.g., lateral line nerves and the vagus nerve of elasmobranchs and teleosts, Thomas and Young 1949).

The internodal length is roughly proportional to the diameter of the nerve fiber (Key and Retzius 1876; Ranvier 1878, and later many other authors) and is more rigorously proportional to the axonal diameter, as was subsequently demonstrated (Williams and Wendell-Smith 1971). In mammals, this length varies from about 200 μm in the thinnest fibers to more than 2 mm; the longest internodes found up to now measure 2.1 mm in the rat and 2.8 mm in the ox (Tuczinski and Friede 1984). In fish, the minimum internodal length is about 200 μm, while the maximum length may reach 8 mm (Thomas and Young 1949). The above correlation between internodal length and diameter of the fiber (or its axon) is thought to depend on the fact that, in a given nerve, myelination begins first around axons destined to reach the largest diameters. Once myelination is initiated in a given fiber, the total number of Schwann cells in that fiber does not increase, so that any subsequent increase in fiber length requires elongation of the existing internodes (Vizoso and Young 1948; Schlaepfer and Myers 1973).

The branching of myelinated nerve fibers always takes place at the nodes (Ranvier 1878). In fresh, unstained nerve fibers examined under the light microscope, the myelin sheath appears homogeneous and highly refractive; it stains gray with osmium tetroxide, pink with Sudan III, and blue or violet with hematoxylin lakes after treatment with potassium dichromate (Weigert technique). In general, there is a positive correlation between axonal diameter and thickness of the corresponding myelin sheath (Friede and Samorajski 1967; Williams and Wendell-Smith 1971). The thickness of the myelin sheath is determined by the axon, and the axonal surface molecule type III neuregulin-1 provides the signal for establishing the sheath thickness (Michailov et al. 2004). The ratio between axonal diameter and myelin sheath thickness, however, may be different in the various nerves; for example, in the rabbit, in axons of equal diameter the myelin sheath is thinner in the sural nerve than in the nervus gastrocnemius medialis (Williams and Wendell-Smith 1971). These differences suggest that specific local interactions between neurons and Schwann cells may influence some quantitative characteristics of the nerve fiber.

On both sides of the node of Ranvier, the internodes terminate as dilated bulbs known as paranodes. The region extending 10–15 μm from the paranode is called the juxtaparanode. In each internode, the myelin sheath appears obliquely traversed by bilateral clefts (Fig. VII.7); these clefts, called Schmidt-Lanterman incisures (see Sect. VII.B.4e), seem to subdivide each internode into a variable number of cylindrico-conical segments. In myelinated fibers, as in unmyelinated fibers, the contour of the Schwann cell facing the connective tissue is surrounded by a basal lamina. The major components of the basal lamina and the axon-Schwann cell interactions required for their assembly are considered in Sect. VII.B.3d.

4b. Structure of Schwann Cells

In electron micrographs of myelinated nerve fibers transversely sectioned midway between two successive nodes of Ranvier, the axon, which in freeze-substituted preparations exhibits a circular contour (Okamura and Tsukita 1986), appears as a rule surrounded by the following parts of the Schwann cell, moving from the axon outward (Fig. VII.8): the plasma membrane, the inner compartment of the cytoplasm, the myelin sheath with its concentric laminated structure (see Sect. VII.B.4d), the outer compartment of the cytoplasm, and finally the plasma membrane again, encircled by the basal lamina. The inner compartment of the Schwann cell cytoplasm (Figs. VII.8 and VII.10) is usually present in the largest fibers, but only rarely is it continuous around the whole circumference of the axon; it is often absent in the finest fibers. The outer compartment of the Schwann cell cytoplasm, which contains the nucleus (Figs. VII.8 and VII.9), may appear continuous along the entire circumference of the fiber or crescent shaped or even confined to narrow bands (Fig. VII.10). The variety of forms exhibited by the outer compartment of the Schwann cell cytoplasm in transverse sections of myelinated nerve fibers may depend on the fact that this compartment consists of a large portion containing the nucleus and of narrow, longitudinally arranged bands linked by cross or oblique bridges (Figs. VII.11) which form the extensive network observed under the light microscope by Nageotte (1910a), Nemiloff (1910), and Ramón y Cajal (1912) at the beginning of the twentieth century.

Except for its smoother contour, the characteristics of the nucleus (Figs. VII.8 and VII.9) are not substantially different from those of the nucleus of the Schwann cells ensheathing the unmyelinated axons. Free polysomes, cisternae of the granular endoplasmic reticulum, mitochondria, small dense bodies, the Golgi apparatus (Ramón y Cajal 1912), microtubules (20–30 nm in diameter), and intermediate filaments (about 10 nm in thickness) are usually present in

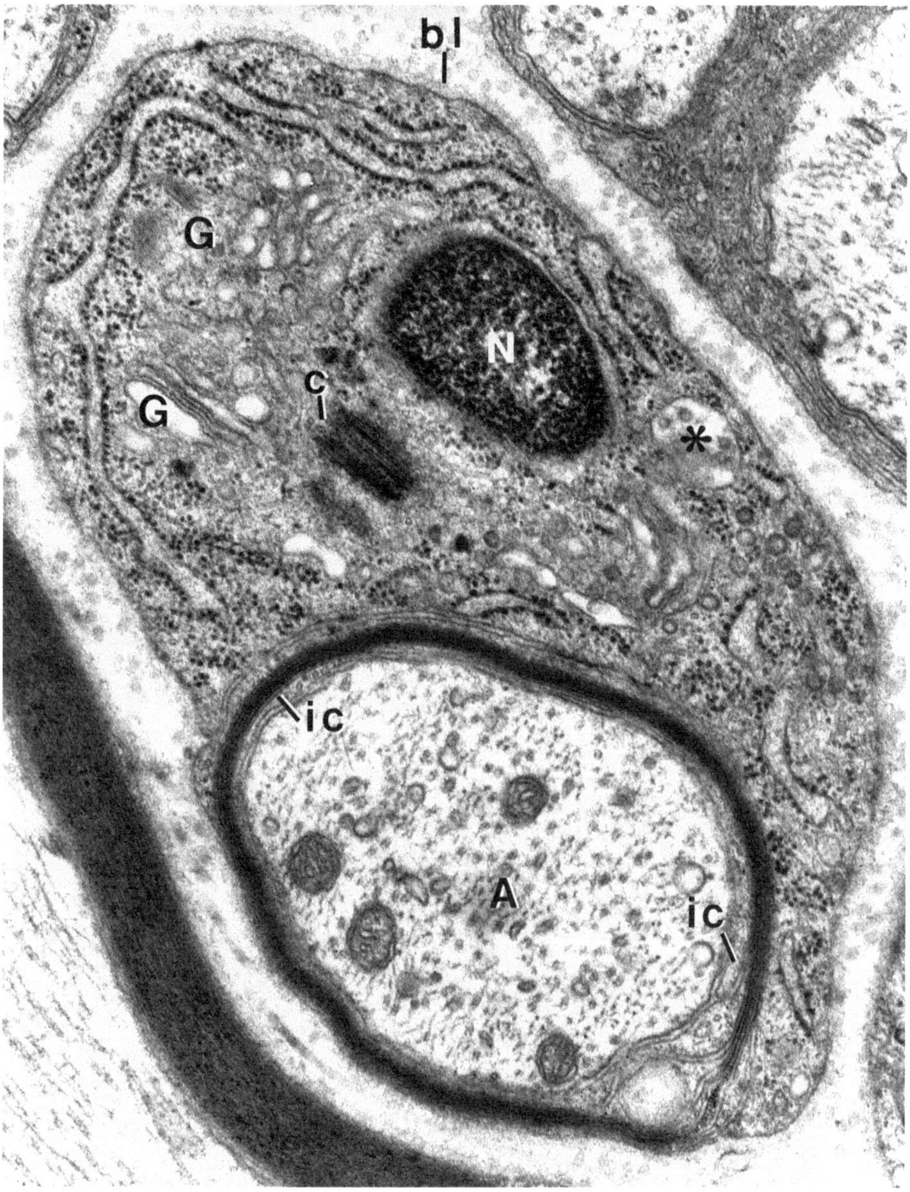

Fig. VII.8 Myelinated nerve fiber. The inner compartment (**ic**) of the Schwann cell cytoplasm can be seen immediately adjacent to the transversely sectioned axon (**A**). Cisternae of the granular endoplasmic reticulum, free polysomes, Golgi complexes (**G**), a centriole (**c**), and a multivesicular body (*) are evident in the outer compartment of the Schwann cell cytoplasm. **bl** basal lamina, **N** Schwann cell nucleus. Rat dorsal root, ×40,000

the perinuclear region (Fig. VII.8). As in satellite cells (see Sect. VII.A.4), the cisternae of the granular endoplasmic reticulum may be arranged in stacks (Coggeshall et al. 1974; Pannese et al. 1988a). Sometimes, lysosomes, glycogen particles, Reich's π granules (Reich 1903, 1907), and corpuscles of Elzholz (Elzholz 1898) can also be found in the same cytoplasmic region. Reich's π granules are ellipsoidal or rod-shaped bodies, 0.5–2 μm long, that stain metachromatically with thionin, toluidine blue, and methylene blue. When observed under the electron microscope, these granules appear

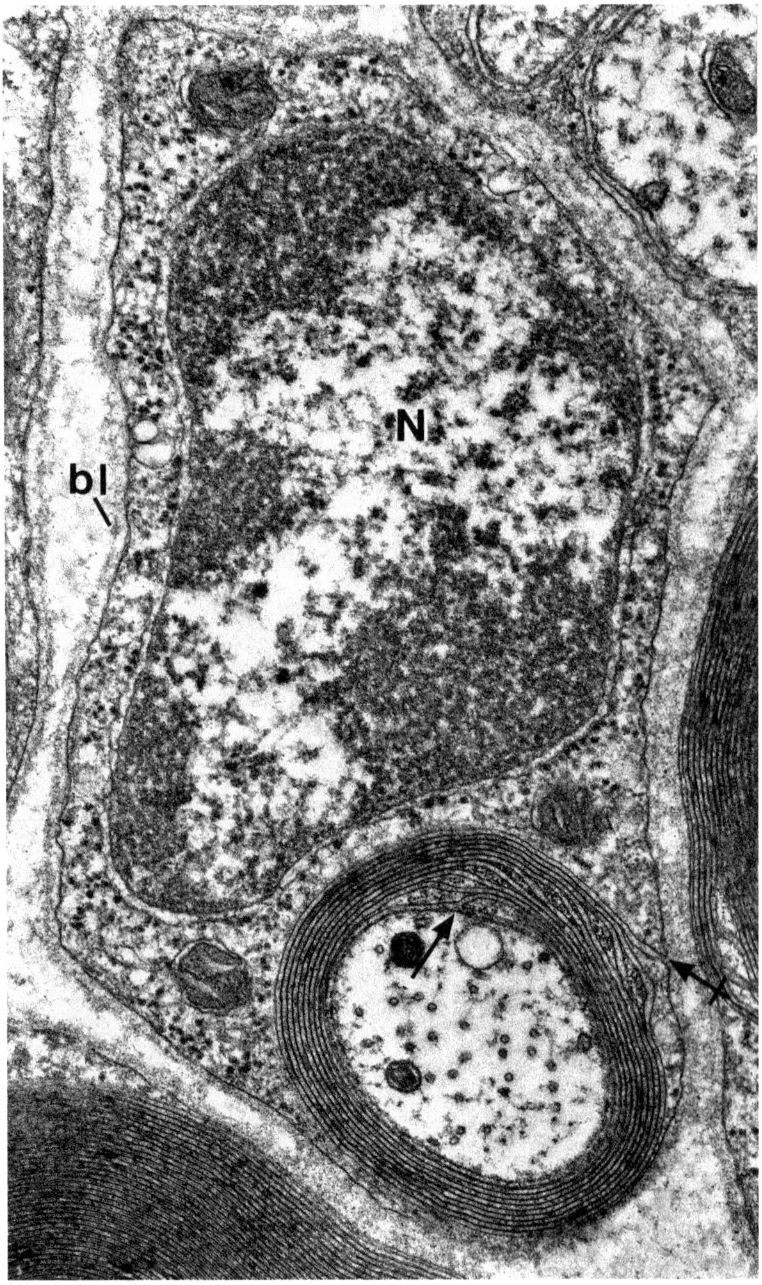

Fig. VII.9 Myelinated nerve fiber. This fiber is transversely sectioned through the nuclear region of a Schwann cell. *Arrow* points to the inner mesaxon, *crossed arrow* points to the outer mesaxon. **bl** basal lamina, **N** Schwann cell nucleus. Lizard dorsal root, ×44,000

as membrane-bounded bodies exhibiting a regular pattern of alternating dense and light layers (period of 4–5 nm). Since they have been observed only within the Schwann cells of myelinated fibers, they are likely concerned with myelin metabolism. They are particularly numerous in developing and regenerating nerve fibers but have also been

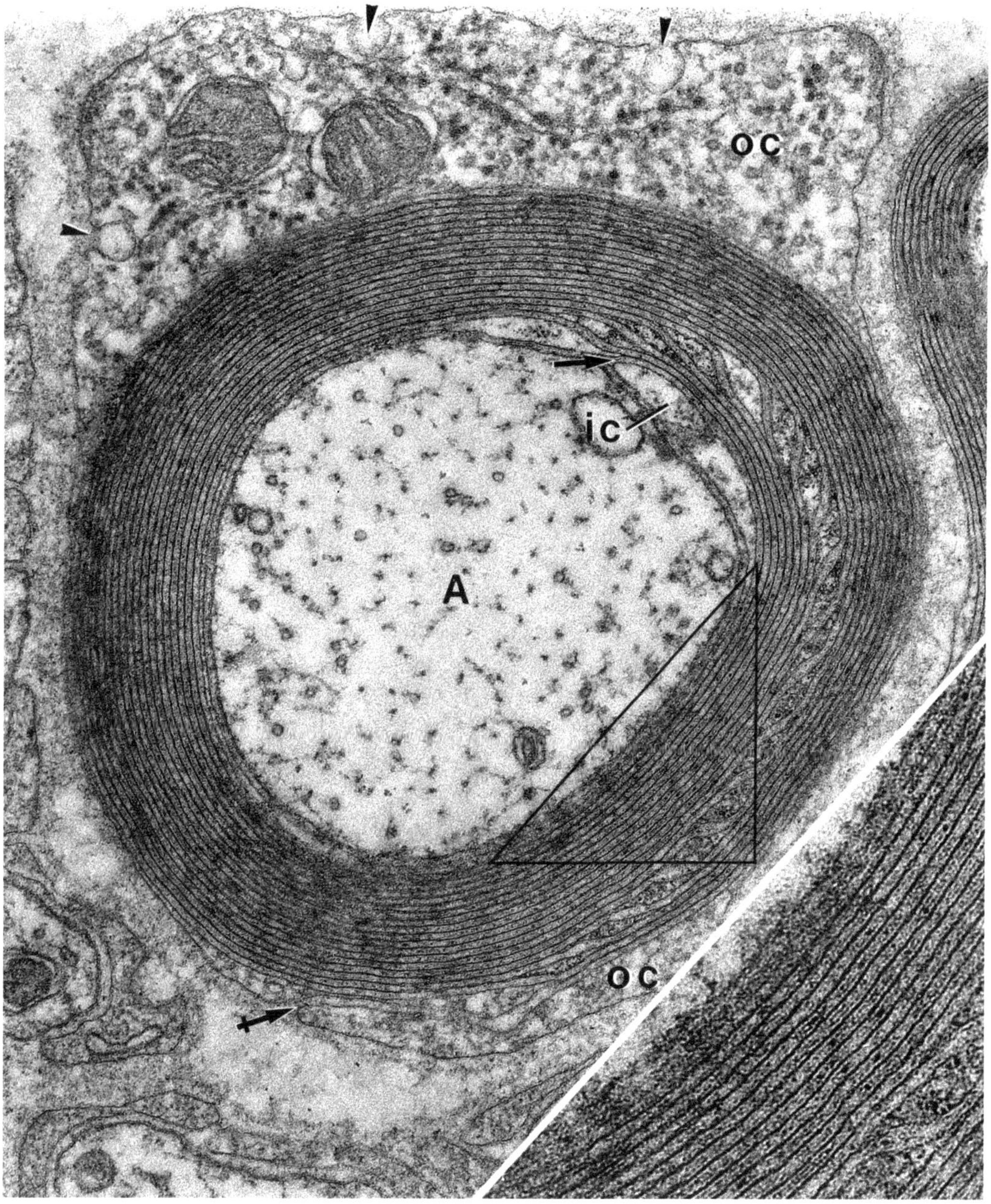

Fig. VII.10 Myelinated nerve fiber. The inner compartment (**ic**) of the Schwann cell cytoplasm is present only along a short portion of the circumference of the transversely sectioned axon (**A**). The outer compartment (**oc**), which in this section does not contain the nucleus, consists of two longitudinally arranged bands. Neurofilaments are fairly evenly distributed throughout the axoplasm. *Arrow* points to the inner mesaxon, *crossed arrow* points to the outer mesaxon. *Arrowheads* point to caveolae along the plasma membrane of the Schwann cell. The *outlined area* of the myelin sheath is enlarged in the *inset*; in the latter, the major dense lines and the intraperiod lines can be seen. Note that here and there the intraperiod line is split into two discrete lines. Lizard dorsal root, ×60,000; inset, ×120,000

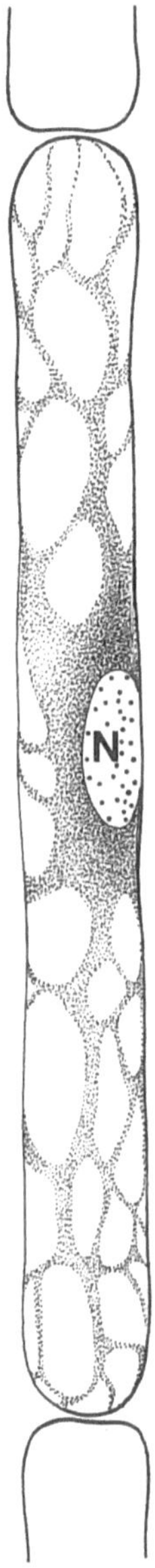

Fig. VII.11 The outer cytoplasmic compartment of a Schwann cell at the surface of a myelinated nerve fiber. This compartment extends from one node of Ranvier to the next and appears as a network consisting of a large portion containing the nucleus (**N**) and narrow, longitudinally arranged bands linked by cross or oblique bridges

observed in apparently normal myelinated fibers. Reich's π granules are often absent from the Schwann cells of common laboratory animals. The corpuscles of Elzholz, which are probably identical to Reich's μ granules, are bodies (0.5–2.0 μm diameter) which stain with osmic acid and with the Marchi technique (Elzholz 1898). In electron micrographs, they display a periodicity about 40 % less than that of myelin. These and other inclusion bodies can be absent from the Schwann cells of common laboratory animals and are most frequently observed when Schwann cells are engaged in myelin synthesis, regeneration, or during demyelination. Occasionally, microtubules and intermediate filaments and more rarely mitochondria and dense bodies can be observed within the bands of the outer cytoplasmic compartment of the Schwann cell. Large numbers of mitochondria are present in the cytoplasm of paranodal bulbs (see Sect. VII.B.4f).

4c. Molecular Characteristics of Schwann Cells

High levels of myelin-related proteins (see Sect. VII.B.4d) are expressed by the Schwann cells of myelinated fibers but not by the Schwann cells associated with unmyelinated axons. On the contrary, growth-associated protein 43 kD, the low-affinity neurotrophin receptor p75 ($p75^{NTR}$), the neural cell adhesion molecules N-CAM and L1, as well as some surface antigens (Jessen and Mirsky 1984) are expressed by the Schwann cells associated with unmyelinated axons but not by those of myelinated fibers. The main protein subunit of Schwann cell intermediate filaments is vimentin (Yen and Fields 1981; Autilio-Gambetti et al. 1982). Furthermore, glial fibrillary acidic protein (see Sect. VIII.E.1) has been detected immunohistochemically in Schwann cells surrounding unmyelinated axons (Yen and Fields 1981; Barber and Lindsay 1982; Dahl et al. 1982; Jessen et al. 1984), but it seems to be absent from Schwann cells surrounding myelinated axons. The glial fibrillary acidic protein of Schwann cells (and of perineuronal satellite cells) has the same molecular weight as that of astrocytes, but its molecular identity is slightly different (Jessen et al. 1984). The Schwann cells of myelinated fibers express high levels of ezrin, radixin, and moesin proteins, which are mainly present in nodal microvilli (Melendez-Vasquez et al. 2001; Scherer et al. 2001). Schwann cells express neurotrophic factors, glutamate transporters, and Na^+ channels (Chiu et al. 1984; Rende et al. 1992). In the Schwann cells associated with unmyelinated axons, Na^+ channels are present over an extended area of the cell membrane, while in the Schwann cells of myelinated fibers Na^+ channels are restricted to nodal microvilli. The role of these ion channels is unclear. It has been suggested that these channels are synthesized by Schwann cells and later transferred to the adjacent axolemma (Gray and Ritchie 1985); however, direct evidence for this "channel transfer hypothesis" is still lacking. Schwann cells also express K^+ channels (Chiu et al. 1984). Ca^{2+} channels are present at high levels in the Schwann cells of the neuromuscular junction (Robitaille et al. 1996). For a more detailed information on ion channels in Schwann cells, see Chiu (1991) and Verkhratsky and Steinhäuser (2000). Mature Schwann cells express a number of receptors, including erbB2, erbB3 (Meyer and Birchmeyer 1995), P2X, P2Y (Kobayashi et al. 2006), and laminin receptors (Einheber et al. 1993; Matsumura et al. 1993).

4d. Structure and Chemical Composition of the Peripheral Myelin

Polarized light studies have suggested that myelin has a highly ordered molecular organization (Schmidt 1924, 1936, 1937; Schmitt and Bear 1937). X-ray diffraction studies have shown that fresh myelin is composed of radially arranged units with a repeat period of 17–18 nm (Schmitt et al. 1935). Under the electron microscope, myelin appears to consist of concentrically arranged lamellae (Figs. VII.9 and VII.10) which form a radially repeating, highly ordered structure (Fernández-Morán 1950, 1952, 1954; Sjöstrand 1950, 1953c). In electron micrographs of fixed and embedded peripheral nerve fibers, the repeat period of myelin is about 12 nm (Sjöstrand 1953c; Karlsson 1966a) and hence smaller than that measured by X-ray diffraction in fresh myelin; the difference between these values results from shrinkage mainly due to fixation and dehydration (Fernández-Morán and Finean 1957). In electron micrographs of freeze-etched material, the repeat period of myelin is, in fact, about 18 nm (Bischoff and Moor 1967).

Electron microscope studies have shown that in the myelin sheath regular and dense lines (major dense lines, about 3.5 nm in thickness) alternate (Fig. VII.10) with less regular and less dense lines (intraperiod lines, about 5.5 nm in thickness). Closer examination at high resolution reveals that the intraperiod line is often split into two discrete lines (minor dense lines) separated by a gap, which measures 2–3 nm in sectioned material and 1.5 nm in fresh material analyzed by X-ray diffraction (Mizuhira and Ozawa 1967; Napolitano and Scallen 1969; Revel and Hamilton 1969).

Two mesaxons (see Sect. VII.B.3a) are found at the inner and outer extremities of the spiral formed by the Schwann cell plasma membrane (Figs. VII.9 and VII.10). At the mesaxons, two areas of the plasma membrane of the same Schwann cell come into contact, and from the apposition of their outer faces, the intraperiod line of the myelin is formed; the gap between the two minor dense lines can definitely be recognized as a narrow, spiralled extracellular channel extending from the extracellular space under the basal lamina through the myelin spiral to the periaxonal space. The close apposition of the inner (=cytoplasmic) faces of the spiral turns of the Schwann cell plasma membrane gives rise to the major dense line. The alternation of lines of different appearance (major dense lines and intraperiod lines or minor dense lines) is clearly related to the asymmetry in structure of the Schwann cell plasma membrane.

Reflexive (see Sect. VII.A.6) adherent, tight, and gap junctions have been observed in inner and outer mesaxons (for a review, see Scherer and Arroyo 2002). Moreover, reflexive tight junctions have also been observed in the compact myelin of nerves from teleosts, amphibians, and mammals (Bertaud 1978; Shinowara et al. 1980). Freeze-fractured preparations have shown that the tight junctions occurring at the inner and outer mesaxons and in the compact myelin, as well as those observed at the Schmidt-Lanterman incisures (see Sect. VII.B.4e) and at the paranodal region (see Sect. VII.B.4f), differ in some respects from the typical tight junctions found in epithelia. None of these myelin tight junctions display the network of continuous strands and grooves of typical tight junctions but rather are characterized by linear rows of particles, discontinuous strands and corresponding pits and grooves (Bertaud 1978; Shinowara et al. 1980).

Since myelin consists principally of the plasma membranes of cells enveloping axons (Schwann cells in the PNS, and oligodendrocytes in the CNS), its chemical composition is basically similar to that of other plasma membranes. Compared with such membranes, however, myelin has a higher lipid/protein ratio, lipids accounting for 70–85 % and proteins 15–30 % of the dry weight of myelin. Furthermore, myelin has considerably more cholesterol, phosphoglycerides, and galactocerebroside than other plasma membranes. Of the phospholipids, phosphatidylethanolamine is more abundant than phosphatidylcholine: the proportion of these two phospholipids is, therefore, inverted relative to that found in other membranes. The lipid composition of myelin is remarkably constant from species to species.

The major proteins that have been identified in the peripheral myelin are protein zero, P_2 protein, peripheral myelin protein 22, myelin basic protein, myelin-associated glycoprotein, and myocilin (for a review, see Snipes and Suter 1995). Protein zero (P_0, molecular weight about 28 kD) is a transmembrane glycoprotein which accounts for 50–60 % of the total peripheral myelin protein; this protein links adjacent lamellae and is essential for myelin compaction. P_2 protein (molecular weight about 15 kD) is the antigen responsible for inducing an autoimmune disease (experimental allergic neuritis). Peripheral myelin protein 22 (PMP 22, molecular weight about 22 kD) is a glycoprotein accounting for less than 5 % of the total peripheral myelin protein. Myelin basic protein (MBP, molecular weight about 18 kD) comprises 15–20 % of the total peripheral myelin protein; this protein participates in myelin compaction. Myelin-associated glycoprotein (MAG, molecular weight about 100 kD) is present in small amounts in the peripheral myelin and is located in the periaxonal Schwann cell membrane, in Schmidt-Lanterman incisures, and in paranodal bulbs; this protein modulates axon diameter through phosphorylation of axonal neurofilament proteins. Myocilin is a glycoprotein of molecular weight about 55 kD.

Originally, myelin was considered to be metabolically stable, but subsequent autoradiographic evidence has modified this view by demonstrating the occurrence of a continuous turnover and recycling of its constituents. For example, myelin lipids and proteins become labeled when radioactive precursors are parenterally administered to adult animals and when peripheral nerves from adult animals are incubated with radioactive precursors in vitro (for a review, see Benjamins and Smith 1984). Moreover, continuous exchanges of cholesterol have been detected between Schwann cell cytoplasm, myelin, and axoplasm. However, the constituent turnover in myelin is slower than that in other membranes. Myelin lacks synthetic capacity, and there is evidence that the proteins and glycoproteins of peripheral myelin are synthesized within the Schwann cell and subsequently transferred to and incorporated into myelin (see, among others, Gould 1977). Also the neuron probably contributes to myelin maintenance (see Sect. V.D.4) by providing some

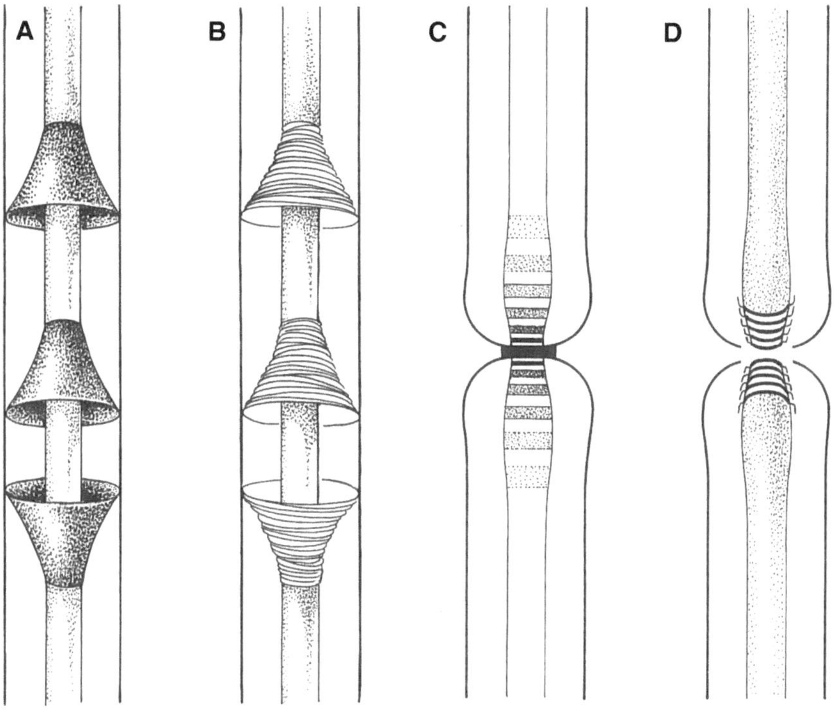

Fig. VII.12 Various features of myelinated nerve fibers which can be seen by using different technical procedures. **A** and **B** internodal segments, **C** and **D** nodal and paranodal regions. **A** funnel-shaped membranes, corresponding to Schmidt-Lanterman incisures, in a fiber impregnated with reduced silver nitrate; **B** spiralled filaments (Golgi-Rezzonico spiral apparatus) corresponding to Schmidt-Lanterman incisures in a fiber impregnated with the Golgi chrome-silver technique; **C** Frommann's lines in a fiber immersed in a dilute solution of silver nitrate; **D** spiny bracelet of Nageotte in a fiber fixed in potassium dichromate and stained with acid fuchsin

metabolites (e.g., phospholipid molecules and bases such as choline or its derivatives) for the renewal of specific myelin components (Droz et al. 1978, 1981; Haley and Ledeen 1979; Gould et al. 1982).

4e. Schmidt-Lanterman Incisures

In the fibers of peripheral nerves observed under the light microscope, these incisures appear as funnel-shaped clefts (Fig. VII.7) occurring at irregular intervals along the myelin sheath (Schmidt 1874; Lanterman 1877). These incisures were also observed in living nerves (Lanterman 1877; Ranvier 1878), in which they appear as dark lines; notwithstanding their detection in vivo, they were for a long time interpreted as mere artifacts (see, among others, Koelliker 1896). In material impregnated with reduced silver nitrate, each incisure appear as a funnel image (Fig. VII.12) which was interpreted as a special infundibulum membrane, as a cement substance, or even as a thickened portion of the neurokeratin net (see Ramón y Cajal 1928). In material impregnated with the Golgi chrome-silver technique, a tenuous filament was observed at each incisure which, starting near the outer surface of the myelin sheath, winds itself in an ever-tightening spiral to end near the axon (Golgi 1881); this spiralled filament (Fig. VII.12) was named the Golgi-Rezzonico spiral apparatus.

Under the electron microscope, it can be seen at the incisures that each major dense line of the myelin sheath (see Sect. VII.B.4d) is split to accommodate a cytoplasmic layer of the Schwann cell (Robertson 1958), which forms a continuous spiral and connects the inner and outer cytoplasmic compartments of the Schwann cell. This spiralled cytoplasmic layer of the Schwann cell, which may contain some microtubules following a spiral course around the axon, probably corresponds to the Golgi-Rezzonico spiral apparatus. At the incisures, the intraperiod line of the myelin sheath is also split with a 10-nm gap intervening between the two apposed outer leaflets of the Schwann cell plasma membrane. This gap provides a potential channel extending from the extracellular space under the basal lamina to the periaxonal space. F-actin, ezrin, radixin, and moesin have all been localized in incisures (Trapp et al. 1989). In addition to the complete circumferential incisures described above, incomplete circumferential and longitudinally oriented incisures have also been reported (Robertson 1958; Mugnaini et al. 1977). These circumferential and longitudinal incisures may be continuous with each other at certain points (Mugnaini et al. 1977).

Series of reflexive (see Sect. VII.A.6) adherent, tight, and gap junctions have been observed at the Schmidt-Lanterman incisures (for a review, see Scherer and Arroyo 2002). The reflexive adherent junctions contain E-cadherin, α-catenin, and β-catenin (Fannon et al. 1995). The reflexive tight junctions found here, which differ in some respects from the typical tight junctions found in epithelia, contain, among other constituents, claudin-19 (Miyamoto et al. 2005). These junctions are permeable to ionic lanthanum but impermeable to the following tracers (Revel and Hamilton 1969; Hall and Williams 1971; Mackenzie et al. 1984): microperoxidase (molecular weight about 1.9 kD, diameter 2 nm), horseradish peroxidase (molecular weight about 40 kD, diameter 5–6 nm), and ferritin (molecular weight about 900 kD, diameter 10–12 nm). Reflexive junctions exist also across the inner and outer mesaxons (see Sect. VII.B.4d) and at the myelin terminal loops of the paranodal region (see Sect. VII.B.4f). The reflexive gap junctions that have been found at Schmidt-Lanterman incisures contain connexin32. These gap junctions could provide a short radial pathway for the rapid transfer of ions and small molecules between outermost and innermost periaxonal cytoplasmic domains of the Schwann cell, short circuiting the much longer circumferential route (Bruzzone et al. 1996; Balice-Gordon et al. 1998).

4f. Nodes of Ranvier

As noted previously, on both sides of the node, the myelin sheaths terminate as dilated bulbs (paranodal bulbs). The proximal bulb (i.e., that closer to the neuronal body from which the axon takes its origin) is nearly always larger than the distal one. In nerve fibers of mammals, the length of the nodal axon (i.e., of the axon segment not covered by myelin) increases with increasing fiber size (Berthold 1968); for example, in felines this length is 0.9 μm in small fibers and up to 1.4 μm in large fibers. In the electric organ of a gymnotid fish (*Sternarchus*), the nodes of Ranvier can reach the remarkable length of 50 μm and can show unusual structural features (Waxman et al. 1972); unlike typical nodes, these do not propagate action potentials. At a node, the axon diameter is usually reduced, reaching the maximum reduction at the paranodes; here, the axon diameter may be one-third to one-sixth of the internodal diameter, while midway along the node it may be a half to a third of the internodal diameter. Hence, the nodal axon often appears barrel shaped. The reduction in the axon diameter at the node of Ranvier is greater in large than in small fibers.

If unfixed nerves are immersed in a dilute solution (e.g., 0.3 %) of silver nitrate and subsequently exposed to light, a small black or brown Latin cross, the so-called cross of Ranvier (Fig. VII.13), can be observed at each node (Ranvier 1871a). Ranvier stated that the transverse component of the cross was due to the occurrence of a cementing ring around the

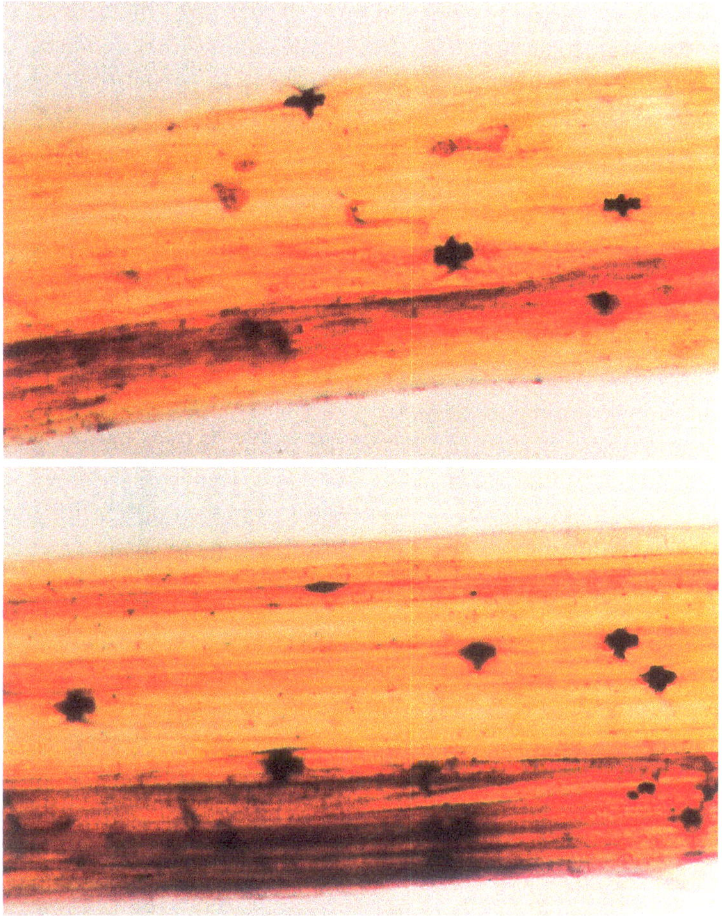

Fig. VII.13 Crosses of Ranvier. Light micrographs of an unfixed nerve immersed in a solution of silver nitrate and subsequently exposed to light. Rat thoracic nerve, ×500 (Preparation by the late Professor A.C. Bruni)

nodal axon and that the longitudinal component of the cross was simply a stained segment of the axon; the length of the longitudinal component of the cross increases with the duration of fiber immersion in the solution of silver nitrate. After immersion of myelinated nerve fibers in a dilute solution of silver nitrate for a time longer than that required to obtain crosses of Ranvier, transverse striations across the nodal and paranodal axon can be observed (Frommann 1864); these striations, whose thickness varies from about 1 to 10 μm, are known as Frommann's lines (Fig. VII.12). The thickness and separation of these striations increase with increasing distance from the node. In myelinated fibers fixed in potassium dichromate and stained with acid fuchsin, Nageotte (1910b) observed a structure consisting of five or six rings encircling the axon where it passes through the end of the myelin sheath. Each of these rings bore a number of curved spines which pointed into the paranodal bulb following its curve. This structure, called the spiny bracelet of Nageotte (Fig. VII.12), seems to reflect the arrangement of the cytoplasmic pockets of the Schwann cell in the paranodal region as seen under the electron microscope (see later, this section).

At paranodes, the outer surface of the myelin displays a crenated appearance due to longitudinal grooves. These grooves, which are evident not only in chemically fixed but also in freeze-substituted nerve fibers (Okamura and Tsukita 1986), are deeper in large fibers and shallower in small fibers and filled with columns of Schwann cell cytoplasm, so that the overall outline of the nerve fiber is circular (Fig. VII.14). These longitudinal columns of Schwann cell cytoplasm contain large numbers of small, rod-shaped mitochondria (Hess and Lansing 1953), whose concentration may be 10–20 times greater than that found in the remainder of the Schwann cell cytoplasm. They also contain small cisternae of the granular endoplasmic reticulum, free polysomes, glycogen particles, dense granules having a mean diameter of 100 nm, microtubules, and filaments. Toward the node, the longitudinal myelin grooves become shallower until, in the immediate vicinity of the node, they disappear; here, the Schwann cell longitudinal cytoplasmic columns become confluent, forming a collar which contains a few mitochondria and many small osmiophilic granules.

Numerous microvilli (Robertson 1957, 1959b; Elfvin 1961), 70–80 nm in diameter, issue from this collar and project into the nodal gap substance (see Sect. VII.B.4g) surrounding the nodal axon (Figs. VII.14 and VII.18). These microvilli, which interdigitate with those from the adjacent Schwann cell and may contact the nodal axolemma, contain actin, ezrin, radixin, and moesin (Melendez-Vasquez et al. 2001; Scherer et al. 2001). The protein gliomedin, which is released from Schwann cells, is tethered to these microvilli (Feinberg et al. 2010). In myelinated fibers of spinal nerve roots from normal, adult specimens, microvilli derived from the paranodal (cat, Berthold 1968; rat, Steer 1971) and the internodal (mouse, Ishii et al. 1985; lizard, Pannese et al. 1988a, 1989) Schwann cell cytoplasm have occasionally been found in addition to nodal microvilli. The dimensions and structure of these microvilli are very similar to those of nodal microvilli of the same cell.

The lamellae of the myelin sheath end in the paranodal region in a characteristic manner. The innermost lamellae terminate farthest from the node, while the outermost ones terminate nearest to the node, thereby overhanging the inner ones (Fig. VII.14). Thus, the sheath as a whole gradually becomes thinner as the distance from the node decreases. Near the end of these lamellae, each major dense line splits to accommodate a portion of the Schwann cell cytoplasm. In longitudinal sections, these cytoplasmic portions appear as two series of pockets lined up along both sides of the axon (Figs. VII.14 and VII.15). They indent the surface of the axon, whereby the axolemma exhibits a scalloped appearance (Fig. VII.16). These pockets actually represent sections of a continuous cytoplasmic spiral connecting the inner to the outer cytoplasmic compartment of the

Fig. VII.14 Nodal and paranodal regions of a myelinated nerve fiber of the PNS. In this slightly oblique section, several fingerlike processes of the Schwann cell are evident in the nodal region (**nr**), and the ending of the myelin sheath lamellae can be seen in the paranodal region. The innermost lamellae terminate farthest from the node, the outermost lamellae terminate nearest the node, the myelin sheath becoming thinner as the distance from the node decreases. Note the regular arrangement of the cytoplasmic pockets of the Schwann cell (some of which are marked *). Between the two *arrows*, these pockets appear linked to one another by a stack of reflexive adherent junctions. The pockets actually represent sections of a continuous cytoplasmic spiral connecting the inner (**ic**) to the outer (**oc**) cytoplasmic compartment of the Schwann cell. Tubules of the agranular endoplasmic reticulum (some indicated by *crossed arrows*), neurofilaments, and microtubules (some indicated by *arrowheads*) are arranged parallel to the long axis of the axon (**A**). **ct** connective tissue, **my** myelin sheath. A transverse section through the paranodal region of a myelinated fiber is shown in the *inset*. The myelin sheath displays a crenated contour due to longitudinal grooves. These grooves are filled with columns of Schwann cell cytoplasm, so that the overall outline of the nerve fiber is circular. **A** axon. Rabbit dorsal root, ×56,000; *inset*, ×8,900

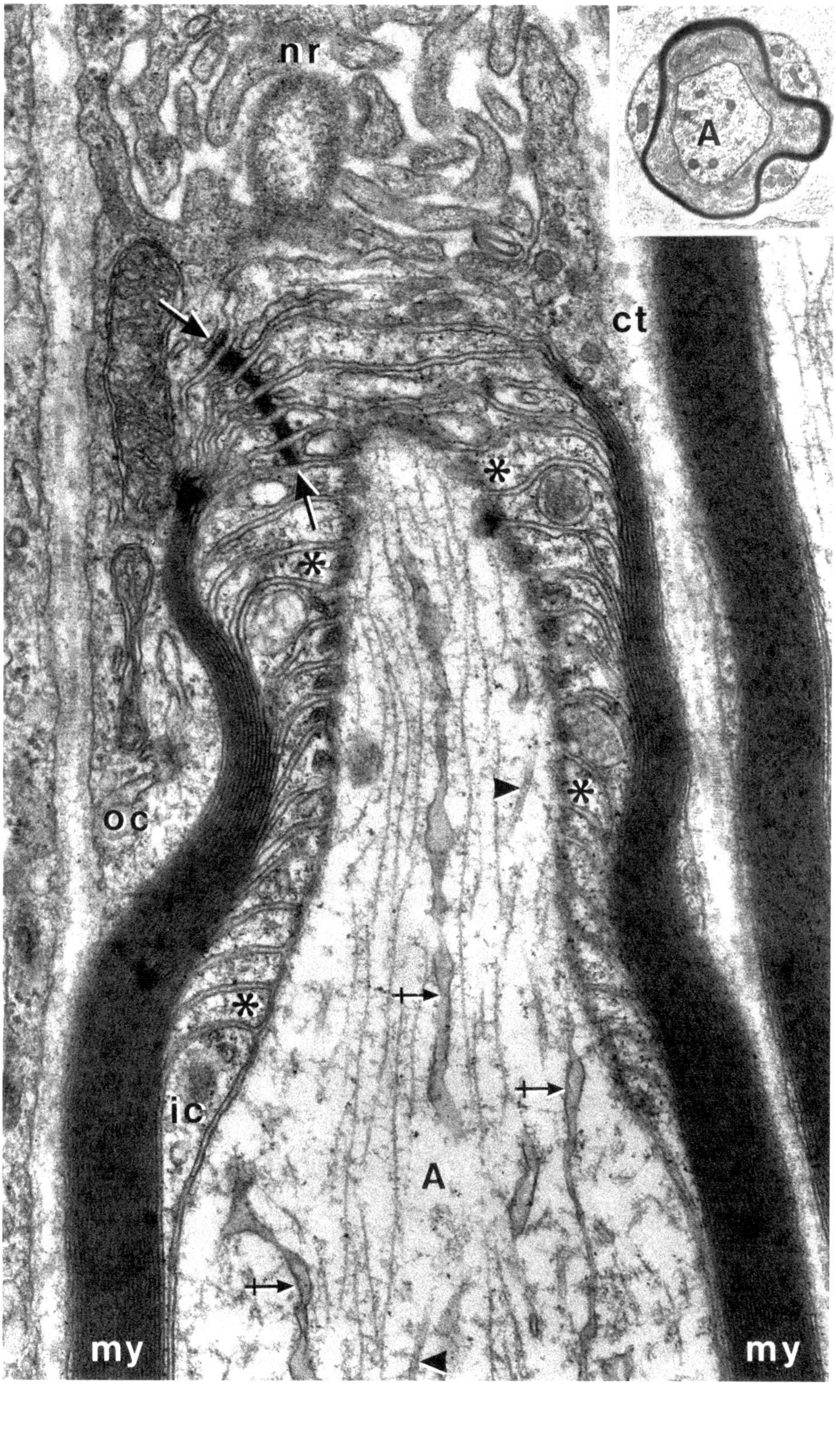
nr
A
ct
*
*
*
oc
*
ic
A
my
my

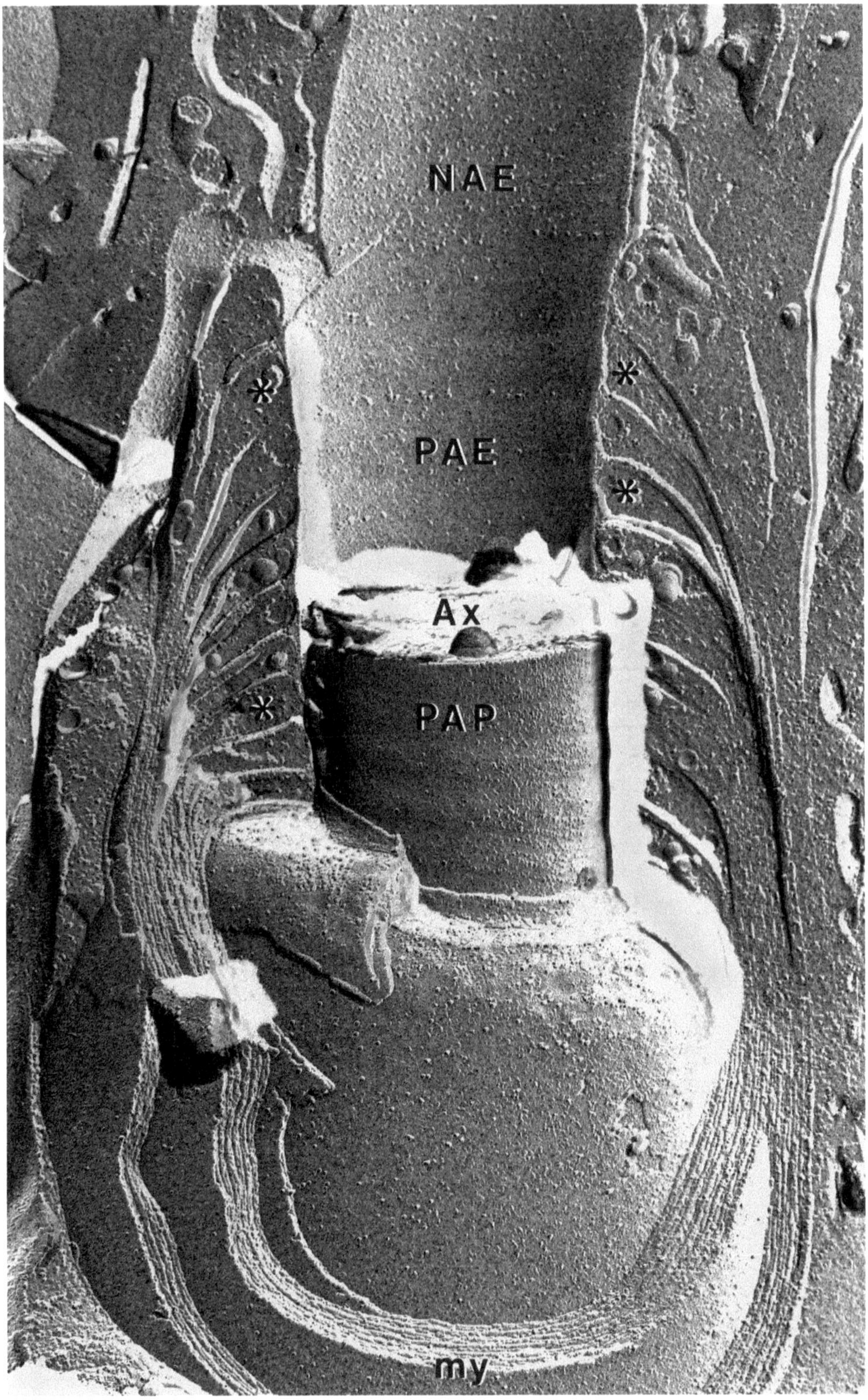
NAE
*
*
PAE
*
Ax
*
PAP
my

Schwann cell. This cytoplasmic spiral contains one or more microtubules which follow a spiral course around the axon like those in the Schmidt-Lanterman incisures. Whereas in small fibers the cytoplasmic pockets of the Schwann cell are usually regularly arranged, in large fibers often some of these pockets do not reach the axolemma and remain piled up between the necks of the pockets apposed to the axolemma. According to Berthold (1978), the double rows of these terminal pockets unapposed to the axolemma would be the ultrastructural correlate of the spiny bracelet of Nageotte (see above, this subsection). The cytoplasmic pockets of the Schwann cell in the paranodal region are linked to one another by reflexive (see Sect. VII.A.6) tight junctions (Schnapp and Mugnaini 1975; Sandri et al. 1977; Tetzlaff 1978; Shinowara et al. 1980) and sometimes also by stacks of reflexive adherent (Fig. VII.14) junctions (Harkin 1964). These reflexive adherent junctions contain E-cadherin, α-catenin, and β-catenin (Fannon et al. 1995). The reflexive tight junctions are constituted by claudin-19 (Miyamoto et al. 2005). These tight junctions, like those occurring at the inner and outer mesaxons (see Sect. VII.B.4d), in compact myelin (see Sect. VII.B.4d), and Schmidt-Lanterman incisures (see Sect. VII.B.4e), are homologous to but not identical with the typical tight junctions found in epithelia. All these junctions are involved in the mechanical role of stabilizing the myelin sheath (Tabira et al. 1978). The reflexive tight junctions of the Schwann cell in the paranodal region, like those at the Schmidt-Lanterman incisures, are permeable to ionic lanthanum but impermeable to the following tracers: microperoxidase, horseradish peroxidase, and ferritin (Revel and Hamilton 1969; Hall and Williams 1971; Mackenzie et al. 1984). The molecular weight and diameter of these tracers are reported in Sect. VII.B.4e.

The gap between the axolemma and the Schwann cell plasma membrane, usually named the periaxonal space, while at least 12 nm in width through most of the internode, is reduced to about 3 nm in the paranodal region. In longitudinal sections through the nerve fiber, a row of densities regularly spaced at 25–30-nm intervals can be seen in this intercellular space (Bargmann and Lindner 1964). These densities are sections of a helically arranged junction which links the axolemma with the Schwann cell plasma membrane. Freeze-fractured preparations have revealed that this axoglial junction (Fig.VII.17) is similar to the septate junctions found in invertebrates (Schnapp and Mugnaini 1975). Axonal (Caspr, also called paranodin, and contactin) and glial (neurofascin) adhesion molecules have been identified at this junction (Bellen et al. 1998). The helically arranged axoglial junction is permeable to ionic lanthanum, microperoxidase, horseradish peroxidase, and ferritin. The amount and velocity of diffusion of these tracers are inversely proportional to their molecular weight. This junction may allow the diffusion of water-soluble nutrients and metabolites to and from the internodal periaxonal space (for a review, see Rosenbluth 2009).

Fig. VII.15 Myelinated axon in a freeze-fractured preparation. In the *lower part* of the picture, the cleavage plane runs through the myelin (**my**) and proceeding upward exposes the P face of the paranodal axolemma (**PAP**); then, the cleavage plane passes through the axoplasm (**Ax**) and exposes the E face of the paranodal (**PAE**) and nodal (**NAE**) axolemma. Two series of cytoplasmic pockets of the myelin-forming cell (some indicated by *) can be seen along both sides of the axon at the paranodal region. Note that intramembranous large particles are more highly concentrated in the nodal than the paranodal axolemma. Adult cat, ×69,000 (Courtesy of P.T. Massa and E. Mugnaini)

Fig. VII.16 Myelinated axon in a freeze-fractured preparation. The picture shows the E face of the nodal (**NAE**) and paranodal (**PAE**) axolemma. The latter exhibits a scalloped appearance produced by the cytoplasmic pockets of the myelin-forming cells. Some of these pockets, which can be seen along both sides of the axon, are indicated by *. Note that intramembranous large particles are more highly concentrated in the nodal than the paranodal axolemma. Adult cat, ×67,500 (Courtesy of P.T. Massa and E. Mugnaini)

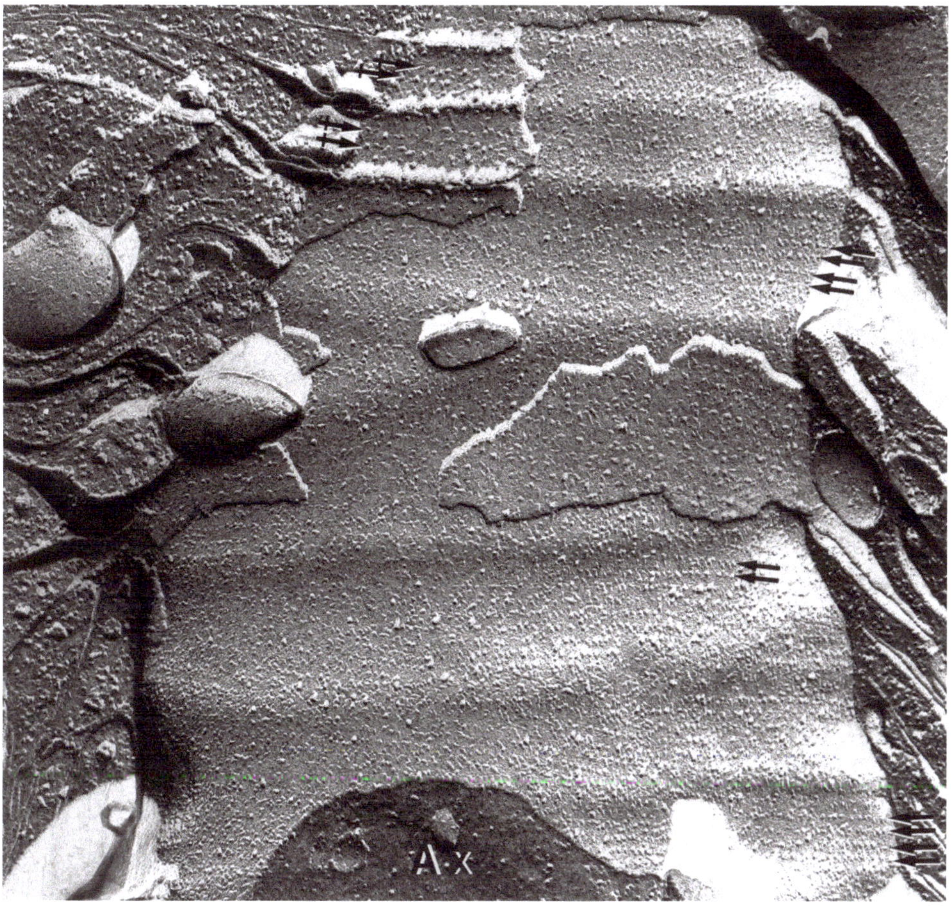

Fig. VII.17 Myelinated axon in a freeze-fractured preparation. Some of the particles of the junction which links the axolemma to the plasma membrane of the myelin-forming cell can be seen on the P face of the paranodal axolemma (*arrows*). Rows of particles (*crossed arrows*) can be seen on the E face of the plasma membrane of the myelin-forming cell. **Ax** axoplasm. Adult cat, ×69,000 (Courtesy of P.T. Massa and E. Mugnaini)

4g. Nodal Axon

A precise knowledge of the surface area of the nodal axolemma is important for the evaluation of the changes in charge density and ionic flux associated with the propagation of the action potential. In feline ventral roots, the surface area of the nodal axolemma measures 3–5 μm^2 in small fibers and 15–25 μm^2 in large fibers (Berthold 1968).

Intramembranous large particles, which with the freeze-fracture technique appear in the E fracture face of the plasma membrane (Figs. VII.15 and VII.16), are highly concentrated in the nodal axolemma (Rosenbluth 1976). The density of these particles at the node correlates well with electrophysiological estimates of the concentration of Na^+ channels. Therefore, it has been suggested that each of these particles corresponds to one Na^+ channel (see Sect. V.E). As known, nodal Na^+ channels enable saltatory conduction (see Sect. VII.B.4h). Much evidence indicates that interaction between Schwann cells and axolemma is required for the initial assembly of both these particles and Na^+ channels at the nodal axolemma (for reviews, see Rosenbluth 1988; Ching et al. 1999). K^+ channels, which are thought to be important for repolarizing the nodal membrane after action potential conduction, are clustered at the juxtaparanodal regions (Wang et al. 1993). The spatially separate clustering of

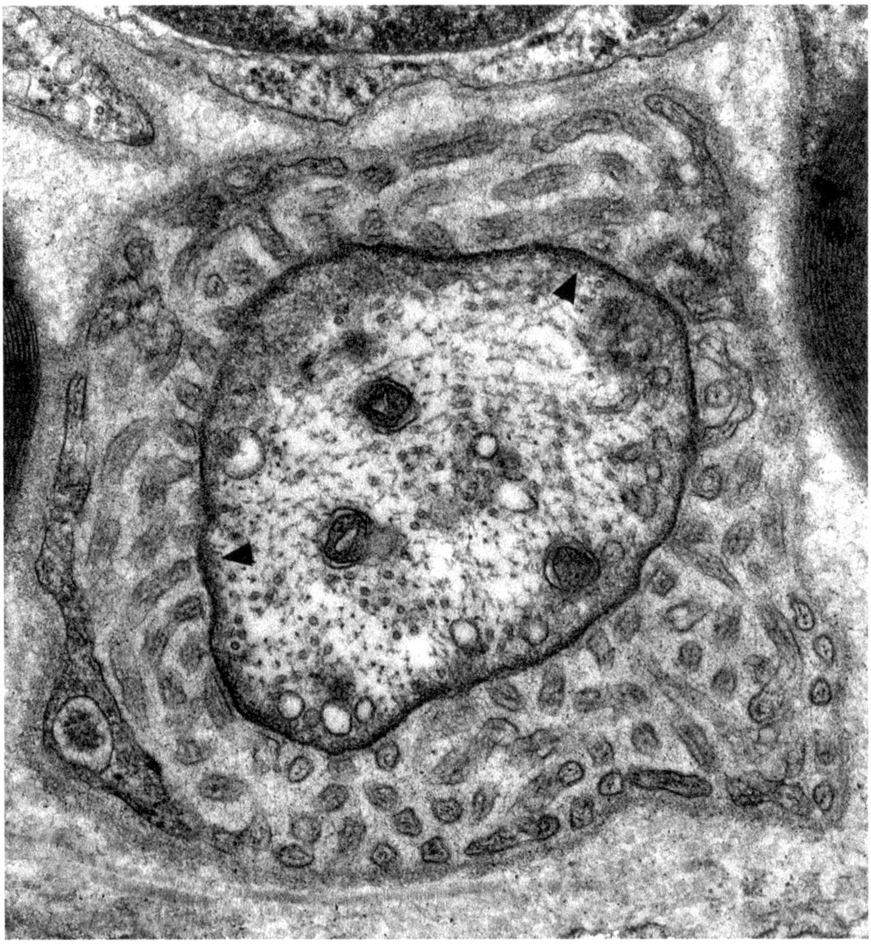

Fig. VII.18 Node of Ranvier. Transverse section through the nodal region of a myelinated nerve fiber of the PNS. Microtubules, neurofilaments, and a dense subaxolemmal layer (▲) can be seen within the nodal axon, which is surrounded by fingerlike processes from the two adjacent Schwann cells. These processes are embedded in an amorphous substance filling the nodal gap. Lizard dorsal root, ×42,000

Na^+ and K^+ channels is maintained by the axoglial junctions located in the paranodes and by axonal ankyrins and spectrins present at the nodes and juxtaparanodes. The junctions function as barriers to the lateral movement of ion channels, whereas ankyrins and spectrins anchor Na^+ and K^+ channels to the axonal cytoskeleton.

Beneath the nodal axolemma is a layer of dense material (Elfvin 1961), 15–25 nm thick, which mainly consists of actin microfilaments and is separated from the axolemmal inner surface by a distance of 6–10 nm (see Sect. V.E). This layer (Fig. VII.18) is similar to that situated beneath the axolemma of the initial segment of the axon (see Sect. V.D.1). The following molecules are abundant at the nodes: the neuron cell adhesion molecules (CAMs) neurofascin-186 (NF-186) and neuron glia-related CAM (NrCAM), ankyrin$_G$, and βIV spectrin. Na^+ channels are associated with NrCAM and NF-186 and these multiprotein complexes are linked to the axonal cytoskeleton through ankyrin$_G$ and βIV spectrin. Na^+ channels, NF-186, NrCAM, ankyrins, and spectrins are also abundant in the initial segment of the axon (see Sect. V.D.1). Neither the subaxolemmal layer of dense material nor the high density of E face particles is present at the inactive nodes of the electric organ of the fish *Sternarchus* (Waxman et al. 1972; Kristol et al. 1977; see also Sect. VII.B.4f).

In marked contrast to the initial segment of the axon, the nodal axon does not contain ribosomes. At the node, the number of neurofilaments is much less than in the internodal segment (Tsukita and Ishikawa 1981); given that the transverse diameter of the nodal axon is, however, generally less than that of the internodal segment (see Sect. VII.B.4f), the concentration of neurofilaments in the axoplasm of the nodal axon does not differ greatly from that observed in the internodal segment. Unlike neurofilaments, microtubules are more concentrated in the nodal than in the internodal segment. In the nodal axon, microtubules are usually present as single units; only in the larger nerve fibers do they sometimes appear assembled in fascicles.

The nodal gap is filled with an amorphous substance of medium electron density, which probably corresponds to the cementing ring observed by Ranvier (1871a, 1878) after immersion of the nerve in a dilute solution of silver nitrate (see Sect. VII.B.4f). The microvilli projecting from the adjacent Schwann cells are embedded in this nodal gap substance (Fig. VII.18). This extracellular substance consists of several proteoglycans (including brevican, versican, and syndecans), and mucopolysaccharides. The basal lamina which surrounds the Schwann cells is continuous across the nodal gap (Robertson 1957), thus forming a tubular investment around the fiber. It contains specific laminin isoforms.

4h. Functional Aspects of the Myelin Sheath

As is known, the conduction velocity along the axon is proportional to its transverse diameter (Gasser and Erlanger 1927). In the course of evolution, the increase in conduction velocity necessary to improve the functional efficiency of the organism and therefore assure its survival and domination of the environment has been attained by two major mechanisms: (1) the increase in the diameter of the axon and (2) the ensheathment of the axon with envelopes restricting the interchange of material between the axon and the space outside the nerve fiber (see also Sect. VII.B.2). Giant axons are found in circuits that mediate escape responses (e.g., in squid, lobster, crayfish, and insects). The axons of these invertebrates may be up to 1 mm in diameter. Myelin ensheathment results in a great economy of space, as illustrated by the fact that a 15-µm-diameter frog axon enveloped by a myelin sheath conducts at about the same rate as a squid giant axon of 500 µm diameter. Furthermore, since the axon diameter correlates directly with the nerve cell body volume (see Chap. III), any increase in the axon diameter implies a proportionate increase in the size of the nerve cell body.

In order to explain how it is that, for axons of the same diameter, conduction is faster in myelinated than in unmyelinated fibers, Lillie (1925) advanced the saltatory conduction theory, which has been amply supported by experimental work (see, among others, Tasaki and Takeuchi 1941, 1942; Huxley and Stämpfli 1949) and is now generally accepted. This theory assumes that, unlike the situation in unmyelinated fibers where the depolarization propagates along the axon affecting the whole of the axolemma, in myelinated fibers the ionic fluxes between the axon and the space outside the nerve fiber are confined to the region of the nodal axolemma, so that the current flows longitudinally from node to node without depolarizing the internodal axolemma. According to this theory, therefore, the nodes of Ranvier act as current generators. Since ion current flows only at the nodes in myelinated fibers, much less energy is expended to restore the ion concentration gradients after an impulse passes in these fibers than in unmyelinated fibers of equal diameter.

For saltatory conduction to take place, the internodal periaxonal space must be insulated from the space outside the myelin sheath. In myelinated nerve fibers, the internodal periaxonal space communicates with the space outside the myelin sheath via elongated and tortuous channels filled with glycoproteins. At each internode, two of these channels are located at the level of the axoglial junctions existing in the paranodal regions; other channels, bounded by the Schwann cell plasma membrane and not completely obliterated by many atypical tight junctions, are located in the compact myelin, at the paranodal regions, and at the Schmidt-Lanterman incisures. These junctional systems do not completely insulate the internodal periaxonal space from the space outside the myelin sheath but, together with the substance filling the extracellular channels, serve to retard substantially the diffusion of ions and molecules from one compartment to the other. The diffusion of certain small molecules through the myelin sheath, which probably has a metabolic role, has only a minimal effect on such a rapid event as the action potential propagation. The junctional systems of the myelin sheath and the substance filling the extracellular channels thus seem to have reached a compromise between the need to assure rapid propagation of the action potential and that of allowing the interchange necessary for the maintenance of the myelin sheath.

For the origin of some of the materials used in the maintenance of the myelin sheath, see Sect. V.D.4.

4i. Mitotic Activity of Schwann Cells

Schwann cell division is a common occurrence during growth (amphibians, Koelliker 1905, Speidel 1933; mammals, Peters and Muir 1959, Diner 1965, Asbury 1967, Martin and Webster 1973, Terry et al. 1974). Conversely, during adult life under normal conditions mitotic Schwann cells (reptiles, Pannese et al. 1987a) and Schwann cells labeled with tritiated thymidine (mammals, Friede and Johnstone 1967, Bradley and Asbury 1970) are a rare occurrence. Mitotic Schwann cells retain their relationships with axons (reptiles, Pannese et al. 1987a; mammals, Diner 1965, Martin and Webster 1973). The rate of Schwann cell proliferation in adult life increases noticeably under various experimental conditions (e.g., after axonal section and following the administration of certain chemical substances) and under some pathological conditions (see Pannese et al. 1987a for references).

In vitro studies have shown that axons stimulate Schwann cells to divide, provided that the latter are in direct contact with the axon (Wood and Bunge 1975). The mitogenic factor seems to be an axonal surface protein (Salzer et al. 1980). Note also that a number of other factors (e.g., factors present in pituitary extracts – Dubois-Dalcq et al. 1981; Lemke and Brockes 1984) can stimulate Schwann cells to divide.

4j. Phagocytic Activity of Schwann Cells

Schwann cells are capable of phagocytosis, as the following examples illustrate. In vitro, Schwann cells may turn into macrophages (Weiss 1944; Weiss and Wang 1945). Schwann cells engulf fragments of degenerating axons and myelin remnants after nerve crush or transection (Blümcke 1963; Nathaniel and Pease 1963; Berner et al. 1973) and during chronic lead intoxication (Lampert and Schochet 1968). In embryos, Schwann cells may sequester fragments of axons undergoing spontaneous degeneration and necrotic Schwann cells (Chu-Wang and Oppenheim 1978). Schwann cells phagocytose nerve terminals in the postnatal elimination of neuromuscular junctions (Smith et al. 2013). Injury of mouse peripheral nerves induces Schwann cells to express the galactose-specific lectin MAC-2 (Reichert et al. 1994), a molecule characteristic of macrophages.

4k. Age-Related Changes

The following unusual features have been described in Schwann cells of old rats: tenuous adaxonal processes invaginating the axon and sometimes sequestering portions of axoplasm (Grover-Johnson and Spencer 1981), the presence of residual bodies and myelin debris (van den Bosch de Aguilar et al. 1992), and a significant decrease in the mean percentage of cytoplasmic volume occupied by mitochondria (Choo et al. 1990). Finally, in Schwann cells of fowl, intranuclear inclusions were found to be exceedingly rare in young animals but more frequent in old animals (Fiori 1987).

Myelin alterations reported in the PNS of aged subjects are similar to those observed in the CNS (see Sect. VIII.F.3). They include segmental demyelination followed by remyelination (e.g., Griffiths and Duncan 1975; Sharma et al. 1980; Knox et al. 1989; Adinolfi et al. 1991; Robertson et al. 1993; Ceballos et al. 1999), loss of myelin tightness (decompaction) due to the occurrence of splits between lamellae, presence of wide incisures, and myelin loops (Knox et al. 1989; Ceballos et al. 1999). These alterations may lead to decreased conduction velocity along nerve fibers, which has been reported to occur in the PNS of aged subjects (for a review, see Verdú et al. 2000).

C. Other Neuroglial Cells of the PNS

As shown by many authors (e.g., Richardson 1958; Taxi 1965; Gabella 1972b; Cook and Burnstock 1976; Wilson et al. 1981; Gershon and Rothman 1991), both connective tissue and blood vessels are excluded from the small ganglia and the internodal strands of intramural plexuses. Small ganglia and internodal strands are compact structures, within which nerve cell bodies, dendrites, axons, and glial cells are closely packed (Fig. VII.19), and the intercellular gaps are in the range of 20–30 nm. This organization is more similar to that typical of the CNS than that of sensory and autonomic ganglia. Unlike the satellite cells of sensory and autonomic ganglia and Schwann cells, the glial cells of autonomic intramural plexuses do not produce a basal lamina and bear numerous long branching processes. These processes do not ensheath individual axons but ensheath bundles of tightly packed axons (Richardson 1960; Gabella 1972a). The glial cells of these plexuses, which are coupled by gap junctions, are not closely associated with neurons. Of the whole surface of the nerve cell body, only a portion is in contact with glial cells, while the remainder is covered by axons and synaptic endings, or may lie immediately beneath the basal lamina surrounding the ganglion (Fig. VII.19). Like astrocytes but unlike the satellite cells of sensory and autonomic ganglia and Schwann cells, the glial cells of autonomic intramural plexuses express the central type of glial fibrillary acidic protein. These glial cells also express vimentin, L-arginine, and a surface molecule called the rat neural antigen-2 (Ran-2). Thus, the glial cells of the autonomic intramural plexuses differ significantly from the satellite cells of sensory and autonomic ganglia and also from Schwann cells; they present a number of morphological and molecular similarities to astrocytes.

The ensheathing cells of the olfactory nerve, which originate from the olfactory placode, share the following characteristics with the glial cells of the autonomic intramural plexuses. Both types of glia bear long, branching processes enfolding bundles of tightly packed axons (Gasser 1956; de Lorenzo 1957), fail to form a basal lamina, have cellular glycogen, and express the central type of glial fibrillary acidic protein (for a review, see Ramón-Cueto and Valverde 1995). Unlike the satellite cells of sensory and autonomic ganglia and Schwann cells, the olfactory ensheathing cells have nuclei with a karyoplasm of fairly even, low density, except for a thin condensed rim adjacent to the nuclear envelope. These cells express fibronectin, laminin, aquaporin-1,

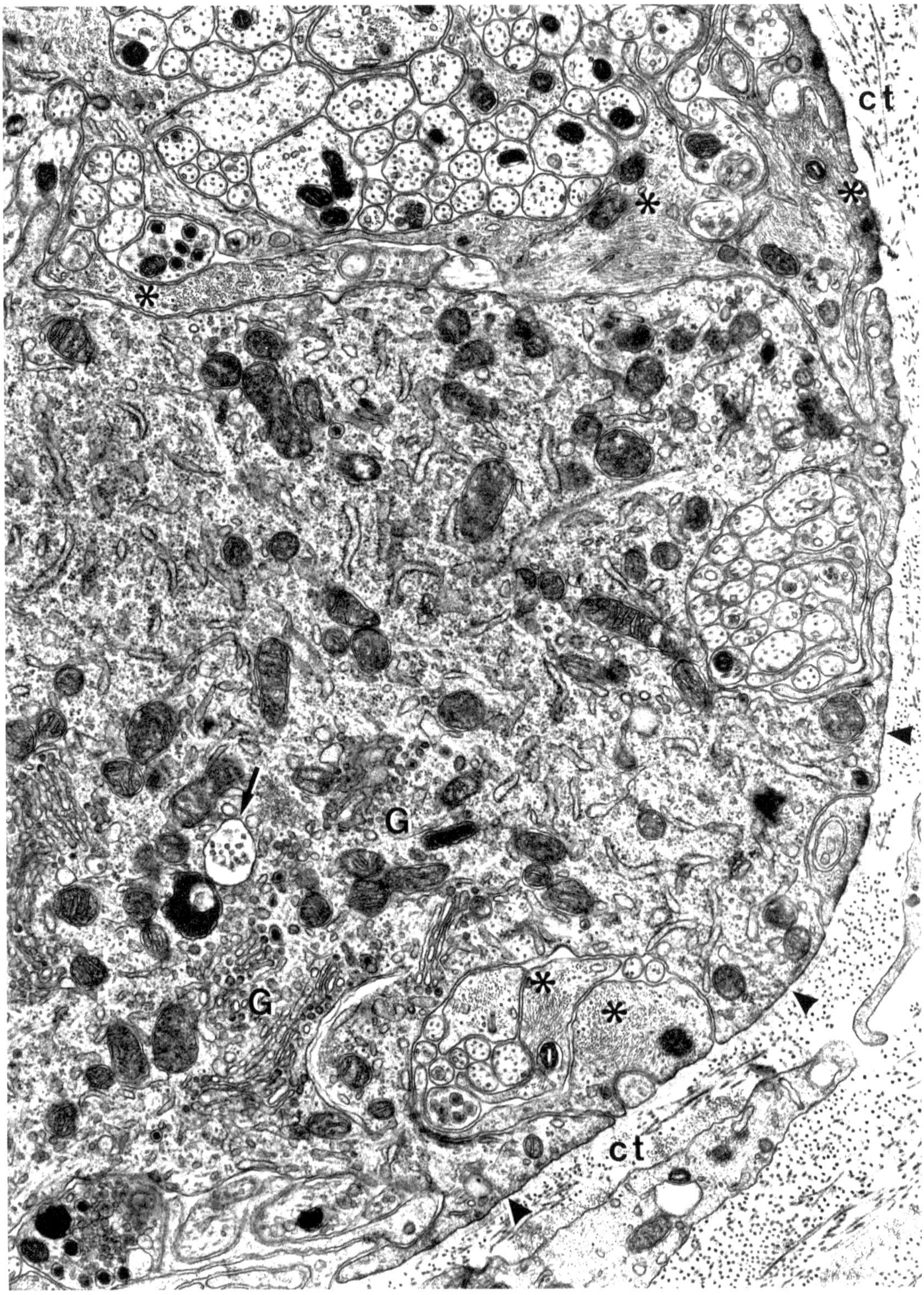
ct
*
*
*
G
G
*
*
ct

04 antigen, and cell adhesion molecules (L1 and N-CAM). Thus, the olfactory ensheathing cells, like the glial cells of the autonomic intramural plexuses, differ significantly from the satellite cells of sensory and autonomic ganglia and from Schwann cells. For these reasons, the glia of the olfactory nerve and autonomic intramural plexuses should be regarded as a category of peripheral glial cells distinct from satellite and Schwann cells.

D. Functions of the PNS Neuroglia

1. Control of Traffic to Neurons

In sensory and autonomic ganglia, each neuron is usually enveloped by an individual satellite cell sheath (see Sect. VII.A.2) whose outer contour faces the interstitial connective tissue containing capillaries. Thus, all substances from the blood must pass through the satellite cell sheath to reach the neuron. Since neurons in sensory and autonomic ganglia lack the protection of a vascular barrier such as is present in the CNS (see Sect. XI.B), only the satellite cell sheath controls the traffic of substances to the ganglionic neuron.

Electron microscope studies using electron-opaque tracers have provided useful information on this function. The concentration of tracers in the connective tissue space is much higher than immediately outside the neuronal plasma membrane (Rosenbluth and Wissig 1964; Arvidson 1979). Thus, the perineuronal sheath acts as a barrier, although partially, to the movement of substances toward the neuron. Obviously, both the basal lamina and the satellite cells are involved in this function. The basal lamina acts at least as a mechanical filter excluding large particles and retarding the passage of other particles. After crossing the basal lamina, the tracers reach the neuronal surface by diffusing along the extracellular channels of the sheath (Rosenbluth and Wissig 1964; Villegas and Villegas 1964; Baker 1965; Holtzman and Peterson 1969; Jacobs et al. 1976; Shinder and Devor 1994). Lanthanum penetrates the satellite cell sheath more readily at the axon hillock-initial segment than elsewhere (Shinder and Devor 1994). The chemical composition and physical state of the substance filling the intercellular channels of the sheath, as well as fixed charges on the plasma membrane of satellite cells, may influence the diffusion of the materials along these channels. Satellite cells might control the traffic of materials along the intercellular channels by influencing the composition and density of the substance filling these channels and/or by the action of enzymes located on the plasma membranes bounding the said channels. These enzymes, whose occurrence in these sites was demonstrated by Novikoff et al. (1966), could degrade materials that might adversely affect neuronal activities. The conspicuous length and tortuosity of the pathway between the basal lamina and the neuronal surface (see Sect. VII.A.6) may slow down the diffusion of materials through the perineuronal sheath. As a consequence, an increase in the efficiency of the hypothetical control exerted by satellite cells over the traffic of materials to the neuron would take place. The barrier function, at least partial, of the perineuronal satellite cell sheath may be important when substances moving toward the neuron are toxic.

Fig. VII.19 Cellular organization in an autonomic intramural plexus. Most of the field is occupied by a portion of the neuronal perikaryon in which various organelles can be seen: Golgi complexes (**G**), profiles of the granular endoplasmic reticulum, mitochondria, a multivesicular body (*arrow*), dense bodies, etc. Only a portion of the surface of the neuronal perikaryon is covered by axons and nonneuronal cells (*), the remainder (*arrowheads*) lies immediately beneath the basal lamina. The neuronal perikaryon, axons, and nonneuronal cells appear closely packed. Intermediate filaments are prominent in the nonneuronal cells. **ct** connective tissue. Ganglion of the myenteric plexus of the guinea pig ileum, ×16,300 (Courtesy of G. Gabella)

2. Homeostasis of the Perineuronal Environment

It is known that neuronal activity causes an increase in the extracellular K^+ concentration, which in turn results in an increased neuronal excitability. Satellite cells have a major role in K^+ buffering within the perineuronal microenvironment. These cells have a highly negative resting membrane potential (Hösli et al. 1978) and express the K^+ channel Kir4.1 (see Sect. VII.A.5), which plays a central role in the uptake of extracellular K^+. Once K^+ has entered satellite cells, it is rapidly redistributed between adjacent cells through gap junctions (Huang et al. 2005).

As regards Schwann cells, it has been suggested that the paranodal apparatus, comprising the Schwann cell cytoplasmic columns with their mitochondrial content, Schwann cell microvilli, and nodal gap substance, is concerned with the control and maintenance of the ionic environment of the node (Landon and Langley 1971).

Glutamate is the principal neurotransmitter used by sensory ganglion neurons, but high levels of glutamate are neurotoxic. Perineuronal satellite cells contain the proteins necessary for the uptake and recycling of glutamate (see Sect. VII.A.5), and it is thought that these cells take up excess glutamate. In this way, they would maintain glutamate in the perineuronal microenvironment below neurotoxic levels. Glutamate taken up by satellite cells is converted to glutamine via glutamine synthetase; glutamine is then returned to the neuron for conversion to glutamate.

γ-Aminobutyric acid can depolarize the neurons of both sensory and autonomic ganglia. Perineuronal satellite cells, which express transporters for γ-aminobutyric acid (see Sect. VII.A.5), may control the concentration of this amino acid in the perineuronal microenvironment (Young et al. 1973; Brown and Galvan 1977). Thus, these cells may protect the neuronal plasma membrane from the effects of excessive concentrations of γ-aminobutyric acid. The gap junctional coupling of satellite cells (see Sect. VII.A.6) would improve the efficiency of these functions.

3. Neuroprotection

Some of the neuroprotective functions of satellite cells were described in Sects. VII.D.1 and VII.D.2. Furthermore, in vitro studies have shown that the presence of satellite cells in culture prolongs the survival of ganglionic neurons (Shimizu 1965; Varon and Raiborn 1972; Varon et al. 1973; Mudge 1984). To explain these results, Varon (1976) suggested that satellite cells produce a trophic factor and release it to the neurons. The idea that satellite cells produce neurotrophic factors and release them to the neuron with which they are associated has recently been invoked to explain the effects that satellite cells have on injured neurons (e.g., see Thippeswamy et al. 2005). Nitric oxide produced by injured ganglionic neurons seems to stimulate their associated satellite cells to generate cyclic guanosine 3′,5′-monophosphate. The latter seems to activate processes leading to the synthesis of neurotrophic factors in satellite cells. The factors released by satellite cells would facilitate the survival of the injured neurons.

4. Metabolic Cooperation with the Neuron

Schwann cells support axonal survival and function. It has been suggested that these cells cooperate metabolically with the axon through the synthesis of materials which are subsequently transferred to the axon (for the "channel transfer hypothesis," see Sect. VII.B.4c). Evidence from invertebrates indicates that the neuroglial cells surrounding the axon of a peripheral nerve can transfer proteins and RNA to the axon (for a more detailed information, see Sotelo-Silveira et al. 2006). This mechanism may prolong the survival time of axonal segments severed from their cell bodies. Note that isolated axonal segments of invertebrates and lower vertebrates may survive for months, provided they are surrounded by their neuroglial sheaths. There is evidence that in mammals Schwann cells can also transfer RNA and ribosomes to axons (Court et al. 2008). Furthermore, Schwann cells may promote axonal elongation by delivering neurotrophic factors to the neuron.

Some of the roles that PNS neuroglial cells play toward the neuron suggest therapeutic use of these cells after CNS injury. Results of recent studies suggest that transplants of Schwann cells and particularly olfactory ensheathing cells may create a permissive environment for axonal outgrowth in the injured spinal cord.

5. Influence on Neuronal Morphology

Like glial cells of the CNS (see Chap. III), satellite cells and Schwann cells influence neuronal shape. For example, Schwann cells can stimulate neurons of the rat superior cervical ganglion to extend dendrites (Tropea et al. 1988). Conversely, satellite cells can prevent dendrite formation on sensory neurons: when co-cultured with satellite cells, neurons of the rat nodose ganglion do not grow dendrites, but in the absence of satellite cells, they do extend dendrites (De Koninck et al. 1993). Satellite cells and Schwann cells also influence neuronal maturation. Embryonic chick sensory neurons maintain a bipolar form for periods of up to one month if grown in vitro in the absence of nonneuronal cells; however, when purified Schwann cells are added to the culture, the neurons change dramatically from the bipolar to the pseudounipolar form (Mudge 1984).

6. Influence on Axon Diameter

As described in Chap. III, the diameter of the axon depends on a variety of factors, both intrinsic and extrinsic to the neuron. Among the latter, interactions between an axon and its enveloping cells (e.g., Schwann cells) play an important role.

7. Modulation of Synaptic Transmission

There is evidence that perisynaptic Schwann cells (also known as terminal Schwann cells, see Sect. VI.D) are involved in the modulation of synaptic transmission at the neuromuscular junction (Robitaille 1998). This Schwann cell role is similar to that played by astrocytes in the CNS (see Sect. VIII.E.6). It has been hypothesized that satellite cells also influence synaptic transmission in autonomic ganglia.

E. Neuron-Glia Communication

It is well established that neurons of the PNS are in functional communication with their associated glial cells, but the mechanisms of this communication are still poorly understood. Since peripheral glial cells and related neurons are not linked by gap junctions, the signals involved could be (a) diffusible molecules released by neurons (or glial cells) which interact with receptors on glial cells (or neurons), or (b) molecules acting at close range such as adhesion molecules. Some findings show that neuregulins expressed by spinal ganglion neurons, and their receptors erbB2 and erbB3 expressed by Schwann cells, play an essential role in axon-Schwann cell interactions; for example, they determine axon myelination and control myelin sheath thickness (for a review, see Corfas et al. 2004). Recent findings suggest that, during regeneration, communication from Schwann cells to axons is also mediated by small (50–100 nm) vesicles (exosomes) released by Schwann cells and selectively internalized by axons (Lopez-Verrilli et al. 2013). Exosomes contain, among other molecules, RNA and neurotrophic factors.

Adenosine triphosphate seems to play a crucial role in neuron-satellite cell communication. This molecule is released from the body of spinal ganglion neurons and may activate ensheathing satellite cells through purinergic receptors (Zhang et al. 2007; Gu et al. 2010). Nitric oxide also seems to be involved in neuron-satellite cell communication (for further details, see Hanani 2005, 2010). Less is known about the mechanism of satellite cell-neuron communication. It has been shown that satellite cells synthesize and release prostaglandins; the latter might interact with prostaglandin receptors on adjacent neurons (Vesin et al. 1995).

VIII. The Neuroglia of the CNS

A. Historical Note

The presence of a ciliated epithelium lining the walls of the cerebral ventricles was first mentioned by Purkinje (1836), who was also able to detect the beating of the cilia.

The earliest mention of an interstitial substance in the CNS was made by Virchow (1846, R. Virchow [1821–1902]) in a paper concerned with the walls of cerebral ventricles. Beneath the ependyma, he found a diffuse, nonneuronal substance in which fibrils and oval nuclei were embedded. Later, Virchow (1854a) stated that this substance filled the spaces between nerve cells throughout the CNS and was made up of an interstitial material of soft consistency in which very delicate cells were suspended. He called this substance *nervenkitt*, meaning nerve glue (neuroglia).

Deiters was able to isolate the neuroglial cells, which later were called astrocytes, and called them cell equivalents or connective tissue cells. In the book published posthumously in 1865 and edited by Max Schultze, Deiters wrote that many fibers radiate in every direction from the zone immediately surrounding the nucleus of the abovementioned cells. These fibers are highly refringent, smooth surfaced, and branched. Deiters was also concerned with the definition of the criteria useful to distinguish these cells from nerve cells. In consideration of the contribution made by Deiters to the knowledge of these neuroglial cells, they were for many years referred to as Deiters cells. The term astrocyte was introduced by Lenhossék (1893).

Henle (1871, J. Henle [1809–1885]) described the neuroglia as an interstitial, granular matter of a nervous type in which a large number of connective fibrils are embedded. Max Schultze and a number of investigators after him maintained that the neuroglia is in some respects comparable to the connective tissue of other organs. This view found support in the opinion that the neuroglial fibrils are formed within the neuroglial cells and later become extracellular (Ranvier 1873, 1883; Weigert 1890, 1895, C. Weigert [1845–1904]; Huber 1901, G. C. Huber [1865–1934]). However, Boll (1873, F. Boll [1849–1879]) pointed out the existence of some differences between the neuroglia and the connective tissue, although likening them to each other to some extent. Golgi (1885) and later Held (1909b) stated that neuroglial fibrils are extensions of neuroglial cell bodies. Ramón y Cajal (1909) unequivocally showed that neuroglial fibrils and connective tissue fibrils differ in their histogenetic characters and display different staining reactions. In spite of the contributions by Golgi, Held, Ramón y Cajal, and other researchers, the view that the neuroglia was in some way comparable to the connective tissue was very

E. Pannese, *Neurocytology: Fine Structure of Neurons, Nerve Processes, and Neuroglial Cells*, 2nd Edition,
DOI 10.1007/978-3-319-06856-5_8,

popular and was dismissed altogether only after it was shown, by examination of well-preserved tissues from the CNS under the electron microscope, that the neuroglial fibrils are not extracellular but lie entirely within the neuroglial cells (Palay 1958b).

The procedures used initially to study the neuroglia (disintegration of specimens by acetic acid, isolation techniques by manual tearing and crushing of fresh or previously hardened specimens, staining with carmine) allowed only the gross features and main characteristics of neuroglial cells to be identified (Pannese 1996). Metallic impregnation procedures employed later (Golgi's chrome-silver technique or black reaction, Ramón y Cajal's technique of reduced silver nitrate, Ramón y Cajal's gold chloride-sublimate technique, and so forth) provided a more detailed information on both the morphology and arrangement of these cells, while the methyl violet dyeing technique introduced by Weigert allowed a systematic examination of the distribution of neuroglial fibrils.

Thanks to the use of his chrome-silver impregnation technique, Golgi (1885) was able to describe with great accuracy the shape of the somata and processes of the neuroglial cells and to illustrate in great detail the intimate relationships of these cells to blood vessels in many regions of the CNS. According to Golgi (1885), the neuroglia consists only of cells richly endowed with branching processes; in other words, in the neuroglia neither interstitial substance in which the cells are embedded nor fibers independent of cells occur. The following example clearly illustrates the impact of the black reaction on our knowledge of the neuroglia. Using the procedures that were available prior to the invention of the black reaction, Bergmann (1857, C.G.L.C. Bergmann [1814–1865]) described long, straight fibers running across the molecular layer of the cerebellar cortex and terminating at the pial surface. These fibers are known as Bergmann fibers. Using his reaction, which stains the whole neuroglial cell, Golgi (1885) recognized that Bergmann fibers were actually the processes of neuroglial cells whose bodies were located near those of the Purkinje cells. These neuroglial cells were called Golgi epithelial cells.

Held (1909b), who like Hardesty (1904, I. Hardesty [1866–1944]) maintained that the neuroglia was syncytial in nature, described the neuroglial limiting membrane, which for a long time was designated as the "marginal glia of Held." He also studied the perivascular neuroglia. In addition to the ependyma and choroidal epithelium, Ramón y Cajal (1913) distinguished three main types of neuroglial cells: star-shaped cells with long processes found mainly in the white matter (fibrous astrocytes), star-shaped cells with short processes found mainly in the gray matter (protoplasmic astrocytes), and cells devoid of processes ("third element").

By means of his ammoniacal silver carbonate impregnation technique, Río Hortega (1919, 1921, P. del Río Hortega [1882–1945]) later identified, within the "third element" described by Ramón y Cajal, a neuroglial cell type that clearly differed from astrocytes, possessing comparatively few, sparsely branching processes. Río Hortega therefore called these cells oligodendrocytes. The oligodendrocyte probably corresponds to one of the cells previously described by Robertson (1899) by means of his platinum impregnation technique. Subsequently, Río Hortega (1928) described four types of oligodendrocyte according to the number, length, and arrangement of their processes.

For over a century, it was thought that neuroglia played simply the passive role of filling the gaps between neurons. In recent decades, it has been demonstrated that astrocytes receive information from neurons and in their turn send signals to neurons. These findings have radically changed our views on the roles of the neuroglia and the functioning of the nervous system.

B. Some Evolutionary Aspects

Of the neuroglia in invertebrates, little is known. Some invertebrates (e.g., lower Platyhelminthes) seem to have a nervous system lacking neuroglial cells. However, these cells occur in most invertebrates (for a review on the evolutionary origins of neuroglia, see Hartline 2011). Often in neuroglial cells of invertebrates, certain features of astrocytes (many filaments, relatively few ribosomes, and a large number of glycogen particles) may be found together with some of those of oligodendrocytes (abundant ribosomes, numerous microtubules, and large numbers of dense bodies) in a great variety of combinations (see, among others, Nicaise 1973; Roots 1986). In molluscs and annelids, there is some separation of cell types. In these groups, one may find cells which are rich in filaments and poor in dense bodies, and other cells which contain only scattered filaments and many dense bodies. It is interesting that in the arthropods studied so far, neuroglial cells seem to lack intermediate filaments but are rich in microtubules.

Ependymal cells are believed to be the main component of the neuroglia in the lancelet (Müller 1900; Achúcarro 1915); these cells have long processes reaching the surface of the nervous system to form end-feet. In some lower vertebrates, too, such as cyclostomes and elasmobranchs, ependymal cells and tanycytes constitute the predominant neuroglial cells; tanycytes are particularly numerous in some cyclostomes, for example, lampreys. In the vertebrates, the main evolutionary development appears to be a reduction in the proportion of ependymal cells and an increase in the proportion of nonependymal neuroglial cells. This phylogenetic shift seems to be associated with an increase in the vascularization of the CNS (see, among others, Sarnat et al. 1975). Amphibian tanycytes have the features of both ependymal cells and astrocytes (Korte and Rosenbluth 1981). In reptiles, although nonependymal neuroglial cells are very abundant, ependymal cells are still prominent (Kruger and Maxwell 1967). Nonependymal neuroglial cells become quantitatively predominant in birds (Achúcarro 1915). Astrocytes are generally of the fibrous type in fish and amphibians; protoplasmic astrocytes are numerous in reptiles and become very abundant in birds (Achúcarro 1915).

Glial fibrillary acidic protein, the major component of neuroglial intermediate filaments in adult animals (see Sect. VIII.E.1), is similar in both lower vertebrates and mammals. However, there are exceptions. The intermediate filaments of neuroglial cells in cyclostomes mainly consist of keratin-like molecules (Merrick et al. 1995). In fish and amphibians, the neuroglial cells of the optic nerve, but not those of the brain and spinal cord, express keratins as their predominant intermediate filament proteins (Giordano et al. 1989; Rungger-Brändle et al. 1989).

As regards the quantitative relationships between neuroglial cells and nerve cells, some textbooks state that neuroglial cells outnumber neurons by a factor of ten in the human CNS and that the neuroglia to neuron ratio has increased with the phylogenetic development. However, studies using a variety of techniques have led to the conclusion that the number of neuroglial cells does not exceed the number of neurons across different species.

C. The Ependyma

The ependyma is a single-layered epithelium which lines the walls of the cavities derived from the lumen of the neural tube (central canal of the spinal cord, fourth ventricle, cerebral aqueduct, third ventricle, and lateral ventricles).

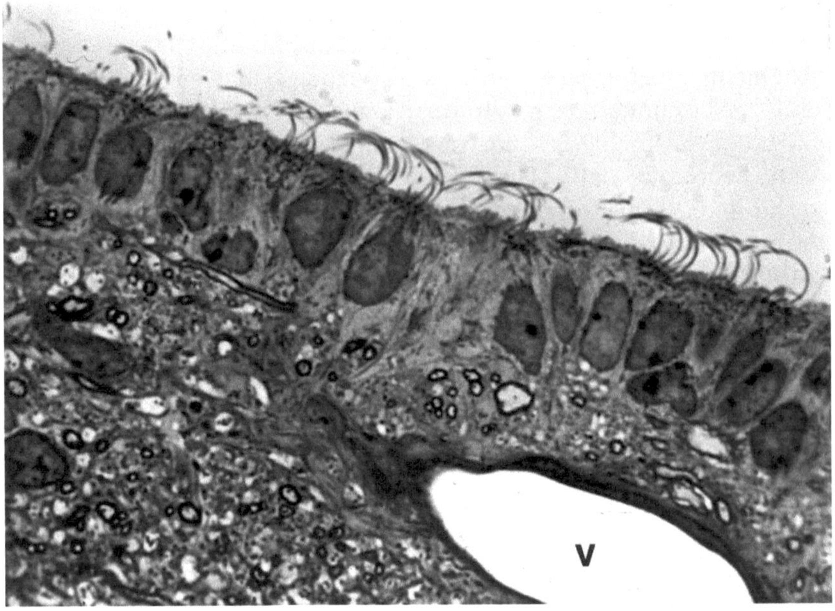

Fig. VIII.1 **Ependymal cells**. The ependymal cells in this light micrograph are columnar with oval nuclei and contain cilia emanating from basal bodies. Above the nuclei, there are numerous mitochondria. **v** blood vessel. Central canal of the rabbit spinal cord, 1-μm section stained with toluidine blue, ×1,400

1. Cell Shape and Intercellular Relationships

Ependymal cells may be flattened, cuboidal, or columnar (Fig. VIII.1); in mammals their height varies from 0.2 to 15 μm (Westergaard 1970). The luminal surface, which is in contact with the cerebrospinal fluid, is coated by a glycocalyx. Glycocalyx components include sialic acid, poly-N-lactosamine, and D-galactose. This surface almost always exhibits microvilli and frequently also cilia (Purkinje 1836). Microvilli (Figs. VIII.2 and VIII.3), which greatly increase the luminal free surface of the cell, vary considerably in length and diameter and are not arranged uniformly; while some arise singly from the cell surface, others originate from a common pedicle. Cilia (Figs. VIII.1, VIII.2, and VIII.3) are numerous in some regions, while in others they may be infrequent or almost absent; for example, in the third ventricle, the upper region is lined by heavily ciliated ependymal cells, whereas the lower region has an ependymal surface almost devoid of cilia (mouse, rat, rabbit, and man, Bruni et al. 1972; sheep, Kozlowski et al. 1973; mink, Scott et al. 1973). In certain regions, each cell is provided with several dozen cilia (up to 60 in the rat, Brightman and Palay 1963), often clustered into a tuft (Fig. VIII.3), but elsewhere only one cilium per cell is present (e.g., in the central canal of the spinal cord of the goldfish, frog, turtle, and pigeon, Nakayama and Kohno 1974; in the infundibular recess of the rat and mouse, Bruni et al. 1972). The flattened cells have only a few microvilli and cilia, the latter arising from the perinuclear area; the free surface of these cells is, therefore, mainly smooth. The ciliary shaft in the rat may vary in length from 5–8 μm (Peters 1974) to 15–20 μm (Brightman and Palay 1963) and in diameter from 0.2 to 0.4 μm. In some areas, the free luminal surface of the ependymal cell may also show protrusions varying in shape and size. Sometimes, the

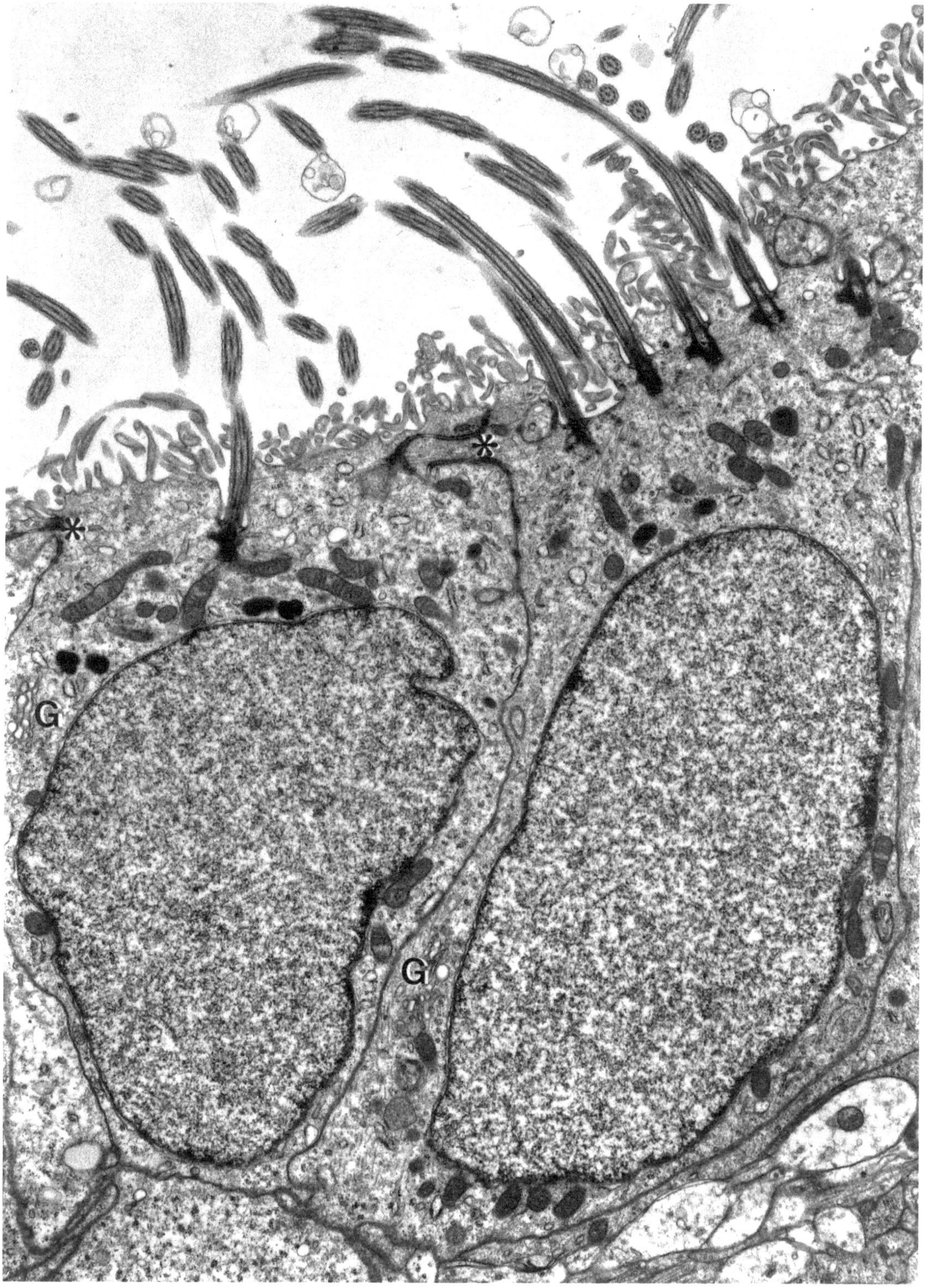

Fig. VIII.2 Ependymal cells. At their luminal free surface, the ependymal cells in this picture show many microvilli and several cilia, each cilium arising from a basal body just under the luminal surface. Mitochondria, dense bodies, short cisternae of the granular endoplasmic reticulum, and Golgi complexes (**G**) can be seen in the cytoplasm. Near the luminal free surface, the lateral membranes of contiguous cells form complex intercellular junctions (*). Central canal of the rabbit spinal cord, ×12,600

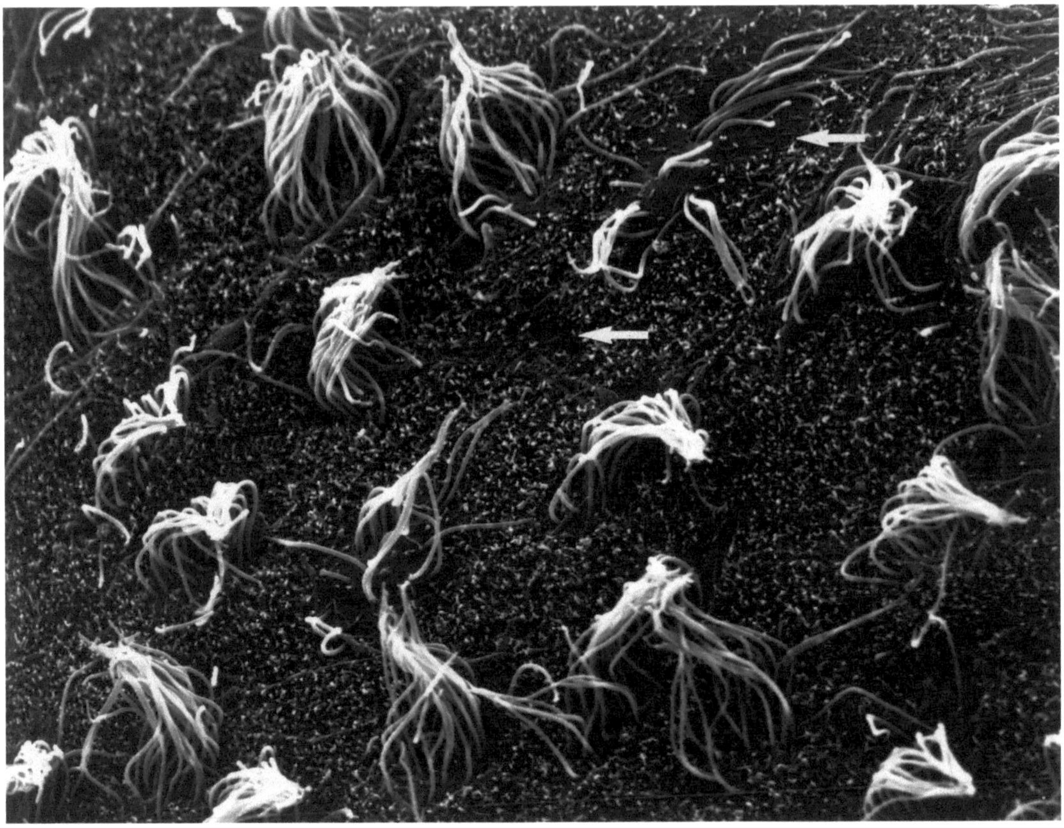

Fig. VIII.3 Ependymal cells. The luminal free surface of the ependyma is shown in this scanning electron micrograph. A tuft of cilia projects from the surface of each ependymal cell, which is otherwise covered with microvilli. Note, however, that circumscribed areas (*arrows*) are denuded of microvilli. Lateral recess of the rat fourth cerebral ventricle, ×5,200 (Courtesy of J.E. Bruni)

luminal portion of the cell forms a laminar expansion which overlaps part of the surface of an adjacent cell.

The lateral cell surfaces may show short, digit-like or lamellar projections which interdigitate with those of adjacent cells (Tennyson and Pappas 1962). Near the apical surfaces of contiguous cells, the confronted lateral membranes form complex intercellular junctions (Fig. VIII.2) which were named terminal bars by optical microscopists. Electron microscope studies have shown that these terminal bars consist of both adherent and gap junctions (Blinzinger 1962; Brightman and Palay 1963; Klinkerfuss 1964; Brightman and Reese 1969). Adherent junctions, which often occur in groups of two or more, usually completely encircle the cells and are, therefore, evident in all sections perpendicular to the luminal surface of the ependyma. Conversely, gap junctions (often situated between the adherent ones) being macular cannot always be identified in these sections. The presence of these gap junctions explains how the injection of a colored substance into a single cell is followed by spreading of the dye to other ependymal cells. The following junctional proteins have been identified in ependymal cells: the adherent junction cadherins and the gap junction connexins 26, 30, 43, and 45.

The basal surface of the ependymal cell may be closely related either to the cells of the subependymal layer (see Sect. VIII.C.8), to myelinated or unmyelinated axons, or to the basal lamina of a blood capillary. This surface is rather irregular. It may show some indentations; occasionally it extends laterally as a sheetlike cytoplasmic projection which courses beneath adjacent cells, parallel to the luminal surface of the ependymal lining. The capillary basal laminae often appear ramified into an extensive labyrinth (Leonhardt 1970; Westergaard 1970). Sometimes, digit-like extensions of these basal laminae indent the basal surfaces of the ependymal cells.

The ependymal cells which cover the circumventricular organs (median eminence, subcommissural organ, vascular organ of the lamina terminalis, subfornical organ, area postrema, neurohypophysis, and pineal gland) are provided with many microvilli but are devoid of cilia on their luminal surface. In addition, some cells show cytoplasmic protrusions which extend from this surface into the lumen of the ventricle. The nonciliated ependymal cells of the circumventricular organs are linked to each other not only by gap junctions but also by tight junctions (Brightman et al. 1975) or by junctions of a character transitional between the adherent and occluding types (Gotow and Hashimoto 1979). Both tight junctions and those described by Gotow and Hashimoto completely encircle these cells. While the former constitute an impermeable barrier, the latter only slow down the movement of many substances between the cerebrospinal fluid and the brain tissue.

2. Cell Structure

Assemblies of orthogonally arranged particles are present in the freeze-fractured plasma membrane of ependymal cells. These assemblies are composed of aquaporin-4 (Verbavatz et al. 1997; Rash et al. 1998), the most abundant aquaporin in the CNS. Aquaporins are channel-forming membrane proteins that are involved in water transport across the plasma membrane. Unlike aquaporin-4, other aquaporins do not usually form assemblies of orthogonally arranged particles. Such assemblies are found in the ependymal cells of higher vertebrates, but are often absent in the ependymal cells of lower vertebrates (e.g., elasmobranchs, Gotow and Hashimoto 1984; frog, Korte and Rosenbluth 1981; lizard, Neuhaus and Wolburg 1985). These specializations also occur in astrocytes (see Sect. VIII.E.1) and in the satellite cells of PNS ganglia (see Sect. VII.A.4).

The ciliary shaft of ependymal cells usually contains nine pairs of peripheral microtubules surrounding a central pair (Blinzinger 1962; Brightman and Palay 1963; Klinkerfuss 1964). However, there is at least one exception: motile ependymal cilia in the central canal of the zebrafish spinal cord lack the central pair of microtubules (Kramer-Zucker et al. 2005). Each cilium arises just under the luminal surface from a basal body (Figs. VIII.1 and VIII.2), which often bears a coarse appendage extending radially from its wall and a cross-striated rootlet (Brightman and Palay 1963; Klinkerfuss 1964). A network of actin microfilaments is found directly inside the luminal free surface of the cell.

The nucleus, which usually lies in a central position, may be round, oval with its long axis perpendicular to the luminal surface of the cell (Fig. VIII.1), or flattened. It contains a rather light, granular karyoplasm (Fig. VIII.2), in which a bundle of filaments (6 nm in thickness) has occasionally been found (Hirano and Zimmerman 1967; Warchol 1978). The nucleolus is usually eccentric.

As a rule, in columnar and cuboidal ependymal cells, the following cytoplasmic organelles are situated above the nucleus: numerous mitochondria (Figs. VIII.1 and VIII.2), a Golgi complex, a few small cisternae of the granular endoplasmic reticulum, dense bodies, and multivesicular bodies. Free polysomes, microtubules, 8–10-nm filaments frequently arranged in bundles, vesicles, and tubules of the agranular endoplasmic reticulum can be found throughout the cytoplasm. In the flattened ependymal cells, the majority of cytoplasmic organelles are situated lateral to the nucleus. Ependymal cells of adult vertebrates do not express glial fibrillary acidic protein (Ludwin et al. 1976; Roessmann et al. 1980) but express vimentin (Schnitzer et al. 1981).

Ependymal cells derive from radial neuroglia during development (Spassky et al. 2005) and do not divide in the adult CNS of mammals (Doetsch et al. 1999; Spassky et al. 2005). There is evidence that in adult mammals some cells originating by mitosis in the subependymal layer replace lost ependymal cells (Luo et al. 2008). However, this replacement does not occur during aging.

In old age, the cilia of ependymal cells are reduced in density. Furthermore, in restricted regions, ependymal cells may be entirely lost, and at these sites the subependymal astrocytes may become hypertrophic. Biondi bodies (see Sect. VIII.D.2) have been described in the ependymal cells of aged humans.

3. Functions of the Ependyma

Ependymal cilia have been shown to beat in a coordinated manner. It has been claimed that this beating is crucial for the normal flow of the cerebrospinal fluid (Kramer-Zucker et al. 2005). Furthermore, ependymal cells form a barrier thought to protect the brain from potentially harmful substances in the cerebrospinal fluid. To establish whether the substances which enter the brain tissue from the cerebrospinal fluid move through the ependymal cells or along the intercellular clefts between these cells, electron-opaque tracers (ferritin, horseradish peroxidase, and lanthanum) have been injected into the cerebral ventricles. Electron microscope studies showed that these tracers move along the clefts between the ependymal cells in greater amount than those pinocytosed by these cells (Brightman 1965a, b; Brightman and Reese 1969). Although the tracers used until now are unlike the substances normally diffusing through the ependyma, the above studies have provided useful information as to some functional aspects of this cellular layer. According to the results obtained, it seems likely that the ependyma may slightly reduce the amounts of the substances reaching the brain tissue and may exert some influence over their diffusion. Ependymal cells take up particles of above-quoted tracers by pinocytosis and segregate them into dense and multivesicular bodies, thus contributing to the restriction of the amount of substances reaching the brain tissue. The presence of lysosomal enzymes (e.g., acid phosphatase and cathepsins) in ependymal cells suggests a role for these cells in the degradation of some substances moving from the cerebrospinal fluid to the brain tissue. The rather complex configuration of the clefts between ependymal cells increases the area of contact between the plasma membrane of these cells and tracer particles, thereby increasing the opportunity for ependymal cells to incorporate particles diffusing through the intercellular clefts. Note that pinocytotic vesicles, both coated and uncoated, are usually more numerous along the lateral surfaces than along the free luminal surfaces of ependymal cells. It is conceivable that also the chemical composition and physical state of the material filling the intercellular clefts between ependymal cells, as well as fixed charges on the plasma membrane of these cells, may influence the movement of substances along these clefts.

Finally, the release of secretory products into the cerebrospinal fluid has been claimed for ependymal cells (Wittkowski 1969; Kozlowski et al. 1973; Booz 1975). However, little is known about the chemical nature of these substances.

4. Tanycytes

Some ependymal cells have a long basal process; these cells are called tanycytes (Horstmann 1954). During development, the basal process of tanycytes extends to the subpial surface of the CNS. In some lower vertebrates this situation may persist throughout life in those regions in which the distance between the wall of the cavities derived from the lumen of the neural tube and the outer surface of the CNS remains small. In other regions, however, with increasing thickness of the CNS the basal process of tanycytes becomes disconnected from the subpial surface and comes into contact with blood capillaries. The latter condition usually takes place in higher vertebrates.

Tanycytes are commonly furnished with a particularly large number of microvilli and very few cilia (Brawer 1972; Millhouse 1972; Rafols and Goshgarian 1985). From their lateral surfaces, many lamellar processes are seen to arise and interdigitate with those of contiguous tanycytes. Near the apical surfaces of contiguous tanycytes, the confronted lateral membranes are linked by tight junctions (Brightman and Reese 1969). The nucleus of tanycytes is often less regular in shape than that of the ordinary ependymal cells (Leonhardt 1966; Brawer 1972; Millhouse 1972). Free ribosomes, cisternae of the agranular endoplasmic reticulum, and intermediate filaments are present within the cytoplasm of tanycytes. These cells coexpress glial fibrillary acidic protein and vimentin (Yen and Fields 1981). The Golgi apparatus is highly developed and also extends into the basal process (Leonhardt 1966). The latter, that often contains many microtubules, may be unbranched or give off two or three branches; laminar, spinous, and knob-like appendages may arise from the basal process and from its branches (Millhouse 1972; Korte and Rosenbluth 1981). The basal process or its branches form end-feet facing the basal lamina covering the capillary wall or the subpial surface of the CNS; these end-feet may be reciprocally connected by gap junctions (Korte and Rosenbluth 1981). In the frog cerebellum, the laminar appendages which arise from the basal process or from its branches ensheath the synapses between parallel fiber boutons and Purkinje cell dendritic spines (Korte and Rosenbluth 1981). Tanycytes express the low-affinity neurotrophin receptor p75, 04 antigen, insulin-like growth factor-1, and α-type estrogen receptor.

Age-related changes in tanycytes include the accumulation in their cytoplasm of lipid droplets and of bodies consisting of a dense material and myelin figures.

Tanycytes are probably a heterogeneous cell population. There is evidence that some are involved in the transport of particular substances from the cerebrospinal fluid to the blood or to neurons (e.g., see Peruzzo et al. 2004). Other tanycytes are probably modified radial neuroglial cells that persist in the adult CNS (for a review, see Rakic 2003a). Radial neuroglial cells are present in the developing CNS of all vertebrates. It has long been known that during development these cells, which rarely express glial fibrillary acidic protein, have the dual functions of providing support for the migration of young neurons from their birthplace to their final location and of giving rise to other neuroglial cells. More recently it has been found that in the developing CNS of vertebrates, radial neuroglial cells may give rise to neurons (e.g., Malatesta et al. 2000; Hartfuss et al. 2001; Alvarez-Buylla and Garcia-Verdugo 2002). In mammals, at the end of development, most radial neuroglial cells transform into astrocytes, whereas in many nonmammalian vertebrate species, these cells persist into adult life (e.g., see Stevenson and Yoon 1982; Alvarez- Buylla et al. 1987). There is evidence that in restricted regions of the CNS, radial neuroglial cells continue to generate new neurons in the adult (for reviews, see Rakic 2003b; Tramontin et al. 2003).

5. Golgi Epithelial Cells and Müller Cells

Golgi epithelial cells (also known as Bergmann cells) are cerebellar neuroglial cells whose bodies lie near those of Purkinje cells (see also Sect. VIII.A. They are provided with long, straight processes that bifurcate once or twice and run across the molecular layer terminating at the pial surface. These processes bear laminar, spinous, and knob-like appendages which insinuate between parallel and climbing fibers (Fig. VIII.4).

Müller cells of the retina are slender cells, which extend radially from the outer to the inner limiting membrane, and give off many lamellar and filiform branches. The surface of these cells is provided with numerous socket-like excavations which contain bodies of neuronal cells. For a review of the functions of Müller cells, see Reichenbach and Bringmann (2013).

Golgi epithelial cells and Müller cells are usually regarded as specialized astrocytic cells. However, these cells retain some of the morphological and biochemical features of radial neuroglial cells (e.g., see Das et al. 2006; Sottile et al. 2006). Thus, they may be considered descendents of the embryonic radial neuroglial cells that persist in the adult mammalian brain (Rakic 2003a). Golgi epithelial cells and Müller cells have also been shown to possess some neural stem cell characteristics (e.g., see Das et al. 2006; Sottile et al. 2006).

6. Axons and Neurons Associated with the Ependyma

Axons, usually unmyelinated but sometimes also myelinated, can be found in the intercellular space between the lateral surfaces of adjacent ependymal cells as well as on the luminal surface of the ependyma (see, among others, Kolmer 1930; Pesonen 1940; Brightman and Palay 1963; Rinne 1966; Takeichi 1966; Leonhardt and Lindner 1967; Leonhardt 1968; Vigh-Teichmann et al. 1970; Westergaard 1970; Leonhardt and Lindemann 1973; Lorez and Richards 1973). Supraependymal axons often terminate in bulb-like endings containing vesicles in which 5-hydroxytryptamine and γ-aminobutyric acid have been demonstrated. Structures resembling adherent junctions can be found between these axonal endings and the luminal surface of ependymal cells (see, among others, Takeichi 1966; Leonhardt and Prien 1968; Westergaard 1972; Leonhardt and Lindemann 1973). Some authors (e.g., Leonhardt and Prien 1968; Leonhardt and Lindemann 1973) have interpreted these junctions as synapses.

Neurons have been found scattered among the ependymal cells lining the walls of the cerebral ventricles and of the central canal of the spinal cord of all zoological classes, from cyclostomes to mammals (Tretjakoff 1913; Agduhr 1922; Kolmer 1931; Pesonen 1940). Each of these intraependymal neurons is in contact with the cerebrospinal fluid usually via a short thick apical process which terminates in a bulb-like end bearing stereocilia and/or a single kinocilium (see, among others, Arnold 1970; Vigh et al. 1971). There are also neurons similar to those just described, whose cell body is situated immediately beneath the ependyma; a thick apical process emerging from the body of these neurons passes between the ependymal cells and at its end contacts the cerebrospinal fluid (see, among others, McKenna and Rosenbluth 1974). Axons extending from the cell body of these intraependymal and immediately subependymal neurons may terminate in synaptic areas of various brain regions or may form endings of neurohormonal type at the external cerebrospinal fluid space. Sensory functions have been attributed to axons and nerve cells associated with the ependyma (Westergaard 1972; McKenna and Rosenbluth 1974; Vigh and Vigh-Teichmann 1998). Some of them might detect changes in the chemical composition of the cerebrospinal fluid (e.g., changes in hormone levels) and relay the relative information to the neurons with which they are in synaptic contact (McKenna and Rosenbluth 1974). Other cerebrospinal fluid-contacting neurons, whose structure resembles that of

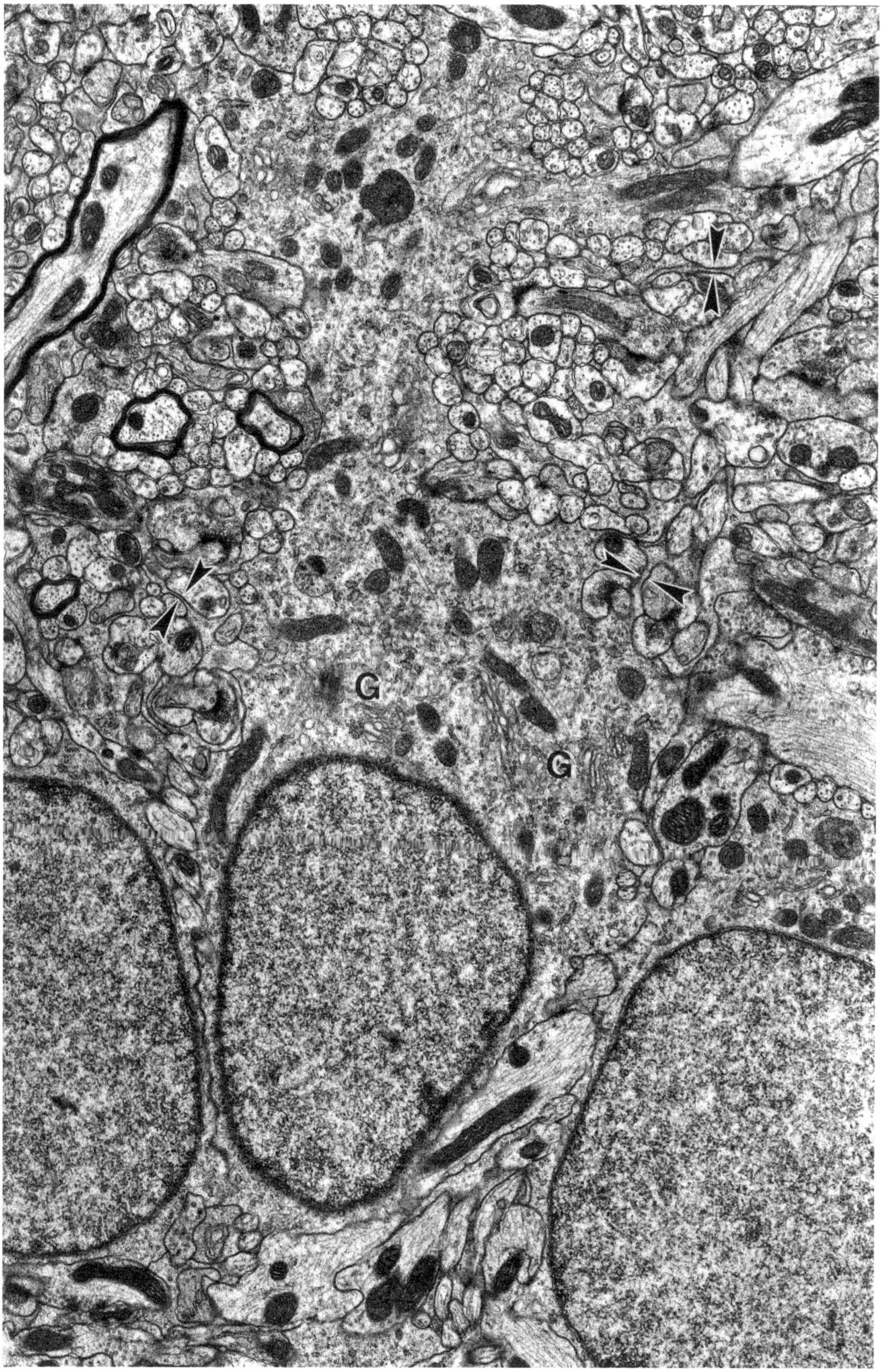

Fig. VIII.4 Golgi epithelial cells. The karyoplasm is fairly homogeneous in density; the cytoplasm contains mitochondria, Golgi complexes (**G**), short cisternae of the granular endoplasmic reticulum, and very few intermediate filaments. These cells have irregular outlines with thin lamellar processes (some indicated by *arrowheads*) which insinuate between the neighboring elements of the neuropil. Rat cerebellar cortex, ×10,700 (Courtesy of S. Matsuda)

mechanoreceptor cells, may detect changes in cerebrospinal fluid pressure.

Small neurons containing a large amount of catecholamines have been found scattered among ependymal cells (see, among others, Braak 1967; Baumgarten et al. 1970); these neurons are thought to release catecholamines into the cerebrospinal fluid. For a review of the nerve cells associated with the ependyma which are assumed to play a sensory or secretory role, see Fleischhauer (1972).

7. Supraependymal Cells

Cells lying on the luminal surface of the ependyma were first detected by light microscopy. This finding prompted intensive study by scanning electron microscopy in various species and in various regions of the cerebral ventricles (Noack et al. 1972; Allen and Low 1973; Coates 1973; Hosoya and Fujita 1973; Leonhardt and Lindemann 1973; Scott et al. 1973). Supraependymal cells, which occur singly or occasionally in small clusters, include neurons, neuroglial cells, and macrophages.

8. The Subependymal Layer

A cell layer is often present beneath the ependyma. This layer, also referred to as the subventricular zone, is prominent during embryonic life but gradually decreases in thickness after birth. This decrease does not take place to the same extent in all species and in all periventricular regions. In many adult rodents, the subependymal layer is thin, whereas it is usually thicker in carnivores and primates; moreover, in a given species, many regional differences in the thickness, organization, and cellular composition of this layer are normally present (Fleischhauer 1972).

The following cell types can be distinguished in the subependymal layer (e.g., see Blakemore and Jolly 1972; Stensaas and Gilson 1972; Doetsch et al. 1997; Alvarez-Buylla and García-Verdugo 2002): (a) cells with astrocytic characteristics that have large, pale nuclei and fine processes; (b) ectopic ependymal cells, occurring singly or in small groups, that bear microvilli and cilia, the latter intervene between adjacent cells or project into small, closed cavities; (c) microglial cells that have small, dark nuclei and an electron-dense cytoplasm in which long, narrow cisternae of the granular endoplasmic reticulum and numerous dense bodies are present; (d) immature cells often arranged in clusters; these cells have a cytoplasmic rim containing a Golgi apparatus, free polysomes, some microtubules but few other organelles; (e) and young neurons that form chains.

Under normal conditions, a large number of proliferating cells is concentrated in the subependymal layer even in adult life, as shown by the presence of mitotic figures (Allen 1912; Bryans 1959; Privat and Leblond 1972) and by the incorporation of specific labeled DNA precursors (e.g., see Smart 1961; Noetzel and Rox 1964; Altman 1966; Lewis 1968; Korr 1980). The cells most frequently labeled are immature cells but also young neurons and cells with astrocytic characteristics divide. Some cells originating by mitosis in the subependymal layer migrate into neighboring regions where they may differentiate into astrocytes or oligodendrocytes (Lewis 1968; Privat and Leblond 1972; Paterson et al. 1973; Korr 1980). It was long thought that the proliferation of cells with astrocytic characteristics gave rise only to new neuroglial cells. However, more recent work (Doetsch et al. 1999) has shown that cells with astrocytic characteristics divide to generate immature cells that in turn may give rise to young neurons. The latter successively migrate into the olfactory bulb, where they mature into granule neurons and periglomerular interneurons. The cells with astrocytic characteristics present in the subependymal layer therefore have properties of neural stem cells, and this layer is a source of new neurons even in adult life.

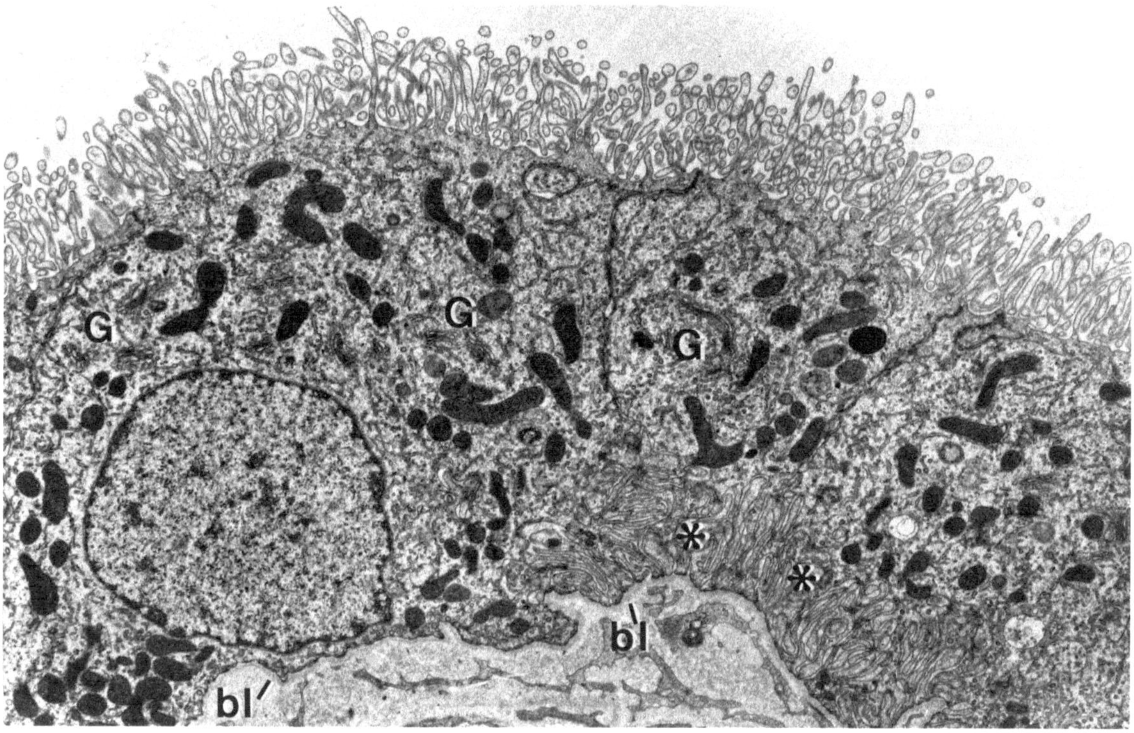

Fig. VIII.5 Choroid epithelium. At their luminal free surface, the choroid cells display many tightly packed microvilli. Numerous mitochondria, Golgi complexes (**G**), and short endoplasmic reticulum cisternae can be seen in the cytoplasm. The lateral surfaces of adjacent cells form complex interdigitations (*) near the base of the cells. The basal surface of these cells rests on a basal lamina (**bl**); below this are thin laminar cell processes. Choroid plexus of the rat fourth ventricle, ×9,900 (Courtesy of S. Matsuda and M. Ledda)

In some mammalian species, postnatal neurogenesis also occurs in the subgranular zone of the hippocampal dentate gyrus (Altman and Das 1965) from where the new cells migrate into the granular cell layer of the hippocampus and mature into granule neurons. In the CNS of both mammals and lower vertebrates, new neurons continue to be generated throughout life, but in adult mammals neurogenesis progressively declines with age. In mammals, the number of neurons generated during adult life is low, relative to the total number of neurons present in the CNS. By contrast, in lower vertebrates new neurons are produced throughout life in numerous regions of the CNS, and the number of neurons born in adult life is much higher than in mammals.

The discovery that new neurons are produced in discrete regions of the mammalian CNS throughout life has profoundly changed our thinking on neurogenesis in the mammalian CNS. It was long believed that in mammals all neurons formed during embryonic or early postnatal development, and that no new neurons were added to the CNS subsequently.

D. The Choroid Epithelium

1. Cell Shape and Intercellular Relationships

In the adult, the choroid epithelium, which lines the ventricular surface of the choroid plexuses, consists of a single layer of cuboidal cells (Fig. VIII.5).

When examined with a scanning electron microscope, the luminal surface of this

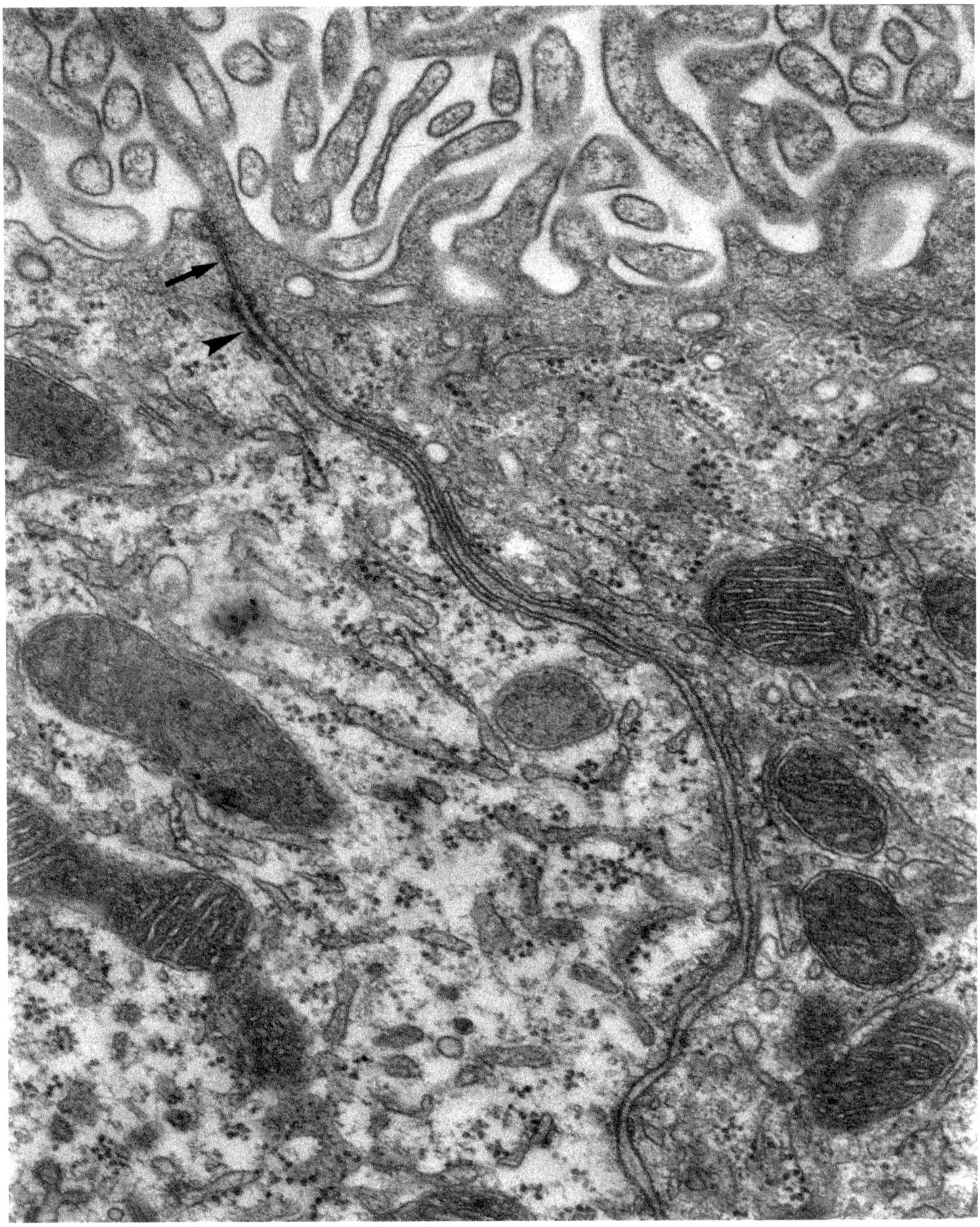

Fig. VIII.6 Choroid epithelium. The relations between the lateral surfaces of adjacent choroid cells are shown. Near their apical ends, the apposed lateral membranes form a tight junction (*arrow*) and, immediately beneath this, an adherent junction (*arrowhead*). Deeper, the intercellular space varies in width, and some endoplasmic reticulum cisternae lie close to and parallel with the lateral plasma membranes. Mitochondria, short endoplasmic reticulum cisternae, and vesicles can be seen in the cytoplasm. Choroid plexus of the rat fourth ventricle, ×51,000 (Courtesy of S. Matsuda and M. Ledda)

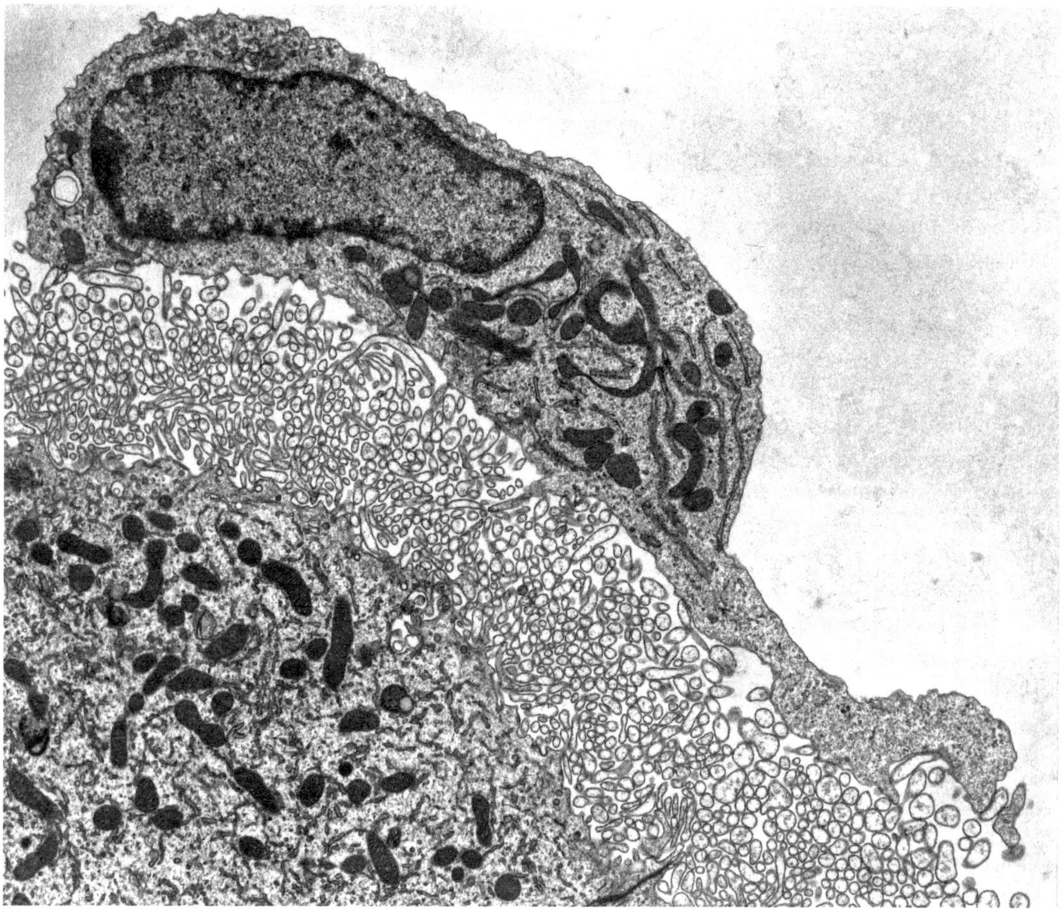

Fig. VIII.7 Epiplexus cell. This cell, which rests on microvilli of the choroid epithelium, has an oval nucleus with the chromatin mainly condensed in a peripheral band. Cisternae of the granular endoplasmic reticulum, mitochondria, and a vacuole can be seen in the cytoplasm. Choroid plexus of the rat fourth ventricle, ×9,000 (Courtesy of S. Matsuda and M. Ledda)

epithelium appears composed of domed protrusions separated by deep furrows (Chamberlain 1973; Peters 1974; Scott et al. 1974). The free surface of each cell bears many tightly packed polymorphic microvilli (Figs. VIII.5, VIII.6, and VIII.7), and either a single cilium or a tuft of cilia. It has been suggested that the luminal surface of the choroid epithelium cells is usually in a highly dynamic state and the microvilli are transient formations (Santolaya and Rodriguez Echandia 1968).

Intercellular junctions and elaborate interdigitations are present along the lateral cell surfaces. Near the apical surfaces of contiguous cells, the confronted lateral membranes are linked by tight junctions (Fig. VIII.6), which probably completely encircle the cells (Brightman 1968; Brightman and Reese 1969; Bohr and Møllgård 1974). The tight junction proteins occludin and claudins have been identified in choroid epithelium cells. Immediately beneath the tight junctions, there are adherent junctions (Fig. VIII.6). The lateral surfaces of adjacent cells form complex interdigitations (Fig. VIII.5) near the base of the cells (Tennyson and Pappas 1961).

The basal surface of the cell rests on a basal lamina (Fig. VIII.5) which separates the choroid epithelium from the underlying connective tissue. The latter is rich in blood vessels and has capillaries with a fenestrated endothelium.

2. Cell Structure

With regard to cilia, contrasting data have been reported. Cilia containing nine pairs of peripheral microtubules surrounding a central pair were described in the rabbit (Tennyson and Pappas 1961), frog (Paul 1968), and pig (Davis et al. 1973), whereas cilia lacking the central pair of microtubules were described in the cat (Santolaya and Rodriguez Echandia 1968) and rat (Peters et al. 1991).

The nucleus, usually round, lies in a central position. It contains a rather light, homogeneous karyoplasm in which bundles of filaments may be present (Bargmann and Katritsis 1966).

The cytoplasm, which in many but not all cells has a light appearance, is rich in organelles (Fig. VIII.5). Mitochondria are mainly concentrated in the apical region. Here, small Golgi complexes are likewise situated, while others lie on both sides of the nucleus. Short cisternae of the granular endoplasmic reticulum, free polysomes, vesicles and tubules of the agranular endoplasmic reticulum, dense bodies, lipofuscin granules, bundles of filaments, and a large number of clear vesicles (30–40 nm in diameter), both coated and uncoated, can be found throughout the cytoplasm. Choroid epithelium cells seem to lack the assemblies of orthogonally arranged particles present in the plasma membranes of ependymal cells (see Sect. VIII.C.2) and astrocytes (see Sect. VIII.E.1). However, choroid epithelium cells express both aquaporin-1 (Nielsen et al. 1993) and K-Cl cotransporters (Kanaka et al. 2001) involved in water transport.

Mitotic cells and cells labeled with tritiated thymidine have been observed in the choroid epithelium of adult animals (see, among others, Kaplan 1980). These findings suggest that, under normal conditions, choroid epithelium cells in the adult can be renewed.

In aged humans, the choroid epithelium cells may become flattened and may contain iron-positive inclusions and Biondi bodies (Biondi 1933). Electron microscope studies (Bargmann and Katritsis 1966; Oksche and Kirschstein 1972; Wen et al. 1999) have shown that Biondi bodies consist of a bundle of filaments (8–10 nm in thickness) among which small lipid droplets and dense granules are interspersed. Lipofuscin granules may also be associated with the above structures.

3. Functions of the Choroid Epithelium

The cerebrospinal fluid is replaced continuously and at a rapid rate. Most of this fluid (up to 90 % in humans) is produced by the choroid plexuses, while a limited amount is probably produced by the ependyma. The formation of cerebrospinal fluid is believed to occur in two stages: passive filtration of fluid across the fenestrated endothelium of the choroid capillaries, followed by active secretion by the choroid epithelium. Note that choroid epithelium cells have some structural characteristics of secretory cells (basal infoldings, many microvilli, and numerous mitochondria).

Many substances, even when present in high concentrations in the blood, enter the cerebrospinal fluid slowly and in small amounts since the choroid plexus acts as a selective barrier between the blood and the cerebrospinal fluid. As shown by studies carried out with electron-dense tracers (Brightman 1968; Brightman and Reese 1969; Davis and Milhorat 1975), these substances cross the fenestrated endothelium of the choroid capillaries, penetrate both the basal lamina apposed to the endothelium and that apposed to the choroid epithelium, and subsequently permeate the intercellular spaces between the choroid epithelium cells, reaching the level of the tight junctions that connect these cells to each another. The tight junctions block the diffusion of the above substances along this extracellular route and force them to pass through the cytoplasm of the choroid epithelium cells, thus enabling selective mechanisms of uptake or exclusion to come into play.

4. Epiplexus Cells

These cells lie on the ventricular surface of the choroid epithelium (Fig. VIII.7). They were first described in lower vertebrates by Kolmer (1921) and were thus called Kolmer cells. In scanning electron micrographs, these cells appear highly polymorphic (star-shaped, spherical, bipolar, or pyramid shaped) and may display processes varying in length and thickness (Hosoya and Fujita 1973; Peters 1974; Chamberlain 1974; Allen 1975). Kolmer (1921) regarded these cells as macrophages, a hypothesis later confirmed experimentally (Kappers 1953; Carpenter et al. 1970). Kolmer cells, later named epiplexus cells by Kappers (1953), have been shown to play a scavenger role and may be involved in immune responses in the ventricular system (Ling et al. 1998).

E. Astrocytes

As their name indicates, astrocytes are star-shaped cells (Fig. VIII.8). Many processes radiate from the perikaryon and can often be traced as far as the walls of blood vessels (Fig. VIII.8a) or to the subpial surface of the CNS. Other features of help in identifying these cells in light microscope preparations are a light nucleus, larger than those belonging to oligodendrocytes and microglial cells, and the presence of fibrils within their cytoplasm. The quantity of these glial fibrils, which are birefringent (Fig. VIII.8c), is greater within fibrous astrocytes than in protoplasmic astrocytes. In electron micrographs, astrocytes are characterized mainly by the low density of their cytoplasmic matrix and by the presence of bundles of filaments within their cytoplasm (see, among others, Mugnaini and Walberg 1964; Schultz 1964; Wendell-Smith et al. 1966; Kruger and Maxwell 1967; Mori and Leblond 1969b).

1. Fibrous Astrocytes

These cells occur mainly in the white matter but may also be found in the gray matter (e.g., in the human thalamus and in the inferior olive of the cat). In light microscope preparations (Figs. VIII.8 and VIII.9), these cells show 10–30 long, thin, smooth surfaced, and poorly branched processes; both the body and processes contain numerous fibrils. The nucleus may be round, oval, or irregular in shape. Unlike the processes of protoplasmic astrocytes (see Sect. VIII.E.2), those of fibrous astrocytes interdigitate extensively with those of adjacent fibrous astrocytes.

Under the electron microscope, the karyoplasm usually shows a fairly even, low density, except for a thin, condensed rim adjacent to the nuclear envelope (Fig. VIII.10). Bundles of filaments have occasionally been found in the karyoplasm (Mugnaini 1967). In the cytoplasm, which has a light appearance, intermediate filaments are the most prominent component. They are about 8 nm in thickness and occur in closely packed bundles (Fig. VIII.10), which correspond to the glial fibrils observed in light microscope preparations. Within the perikaryon, these bundles seem to be randomly arranged, but within the processes, they are disposed lengthwise. Vimentin is the major filament component in immature astrocytes (Dahl and Bignami 1985). Vimentin progressively decreases, and glial fibrillary acidic protein progressively increases in astrocytes during development (Dahl and Bignami 1985). Thus, glial fibrillary acidic protein usually

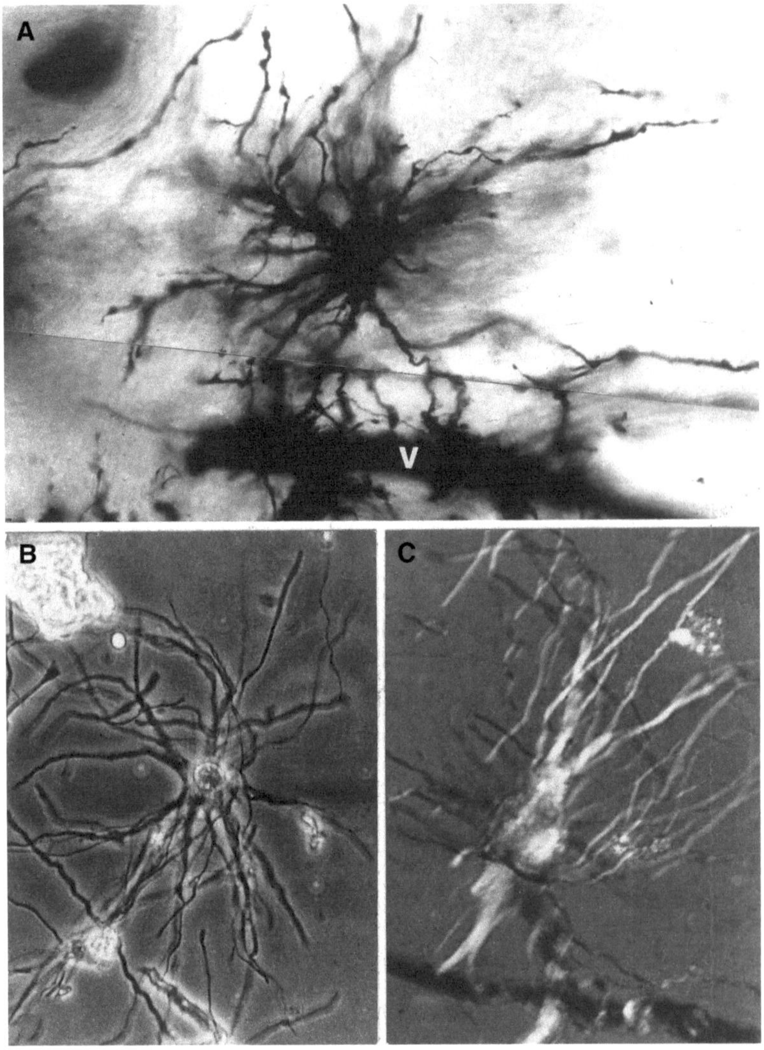

Fig. VIII.8 (A-C) **Fibrous astrocytes**. (**A**) This astrocyte sends a number of processes to the wall of a blood vessel (**v**). Cat spinal cord, light micrograph of a Golgi preparation, ×450. (**B**) Fibrous astrocyte isolated from the human brain. Phase contrast photomicrograph, ×450. (**C**) Both the body and processes of this fibrous astrocyte, isolated from the human brain, appear birefringent. Polarized light photomicrograph, × 600

becomes the major constituent of intermediate filaments in mature astrocytes. However, certain subpopulations of mature fibrous astrocytes, for example, astrocytes of the rat optic nerve (Calvo et al. 1990), coexpress glial fibrillary acidic protein and vimentin.

Intermediate filaments are very numerous, but microtubules are scarce in mature fibrous astrocytes, while the reverse is true in immature astrocytes, in that microtubules are numerous but filaments are extremely rare (Vaughn and Peters 1967). Hence, during the maturation of these cells, the number of microtubules decreases, and the number of intermediate filaments increases. Finally, there are findings suggesting that the intermediate filaments of astrocytes are in dynamic equilibrium with a pool of unassembled subunits (Nakamura et al. 1991).

Glial fibrillary acidic protein is frequently used as a molecular marker to identify astrocytes, but not all astrocytes express detectable levels of this protein. Other markers that

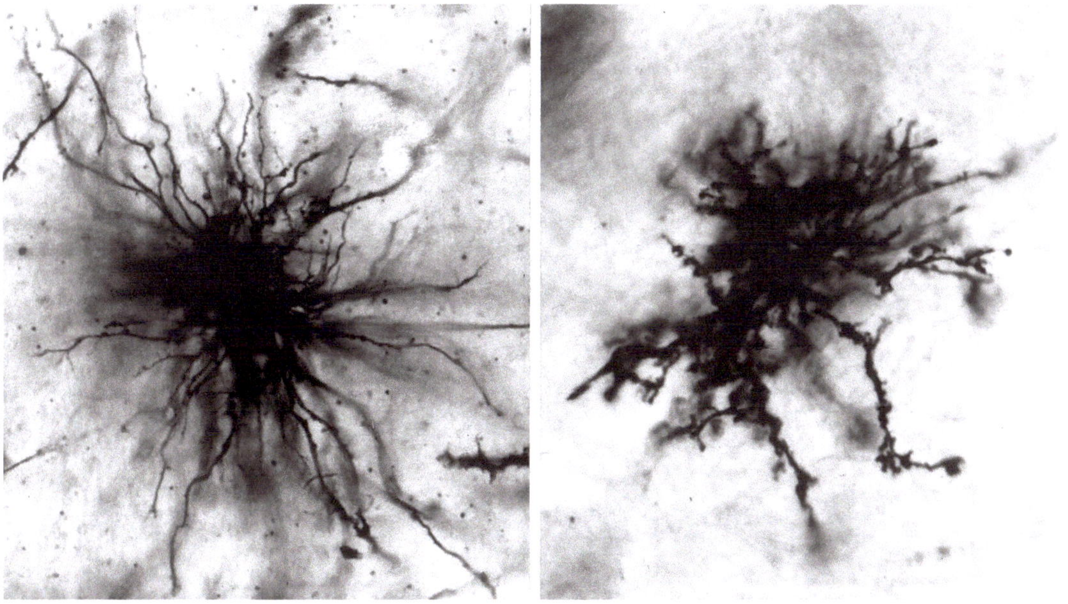

Fig. VIII.9 Astrocytes. A fibrous (*left*) and a protoplasmic (*right*) astrocytes are shown in these light micrographs. Cat spinal cord, Golgi method, ×450

are used to identify astrocytes are glutamine synthetase and S-100 proteins, but these molecules have been found in other cell types of the nervous system.

In comparison with other types of glial cells, the mitochondria, endoplasmic reticulum, and Golgi complexes are sparse within the cytoplasm of fibrous astrocytes (Fig. VIII.10). As to mitochondria, see also Sect. VIII.E.2. The granular endoplasmic reticulum is represented by a small number of short cisternae; only a few polysomes occur free in the cytoplasmic matrix; the Golgi complexes are poorly developed. Centrioles and one cilium (Fig. VIII.10) containing nine peripheral pairs but lacking the central pair of microtubules have also been seen in astrocytes (see, among others, Dahl 1963).

Heavily electron-dense glycogen granules, 20–40 nm in diameter, are very common within the cytoplasm of both fibrous and protoplasmic astrocytes. At times, however, these granules cannot be found, possibly because glycogen is difficult to preserve in electron microscopy. Peroxisomes and pigment granules (Fig. VIII.10) have also been found in both fibrous and protoplasmic astrocytes.

In transverse sections, the processes of fibrous astrocytes display regular contours and often have roundish or oval profiles; inside these processes, a bundle of filaments is always evident (Figs. VIII.13, VIII.14, and VIII.15). As mentioned above, the processes of both fibrous and protoplasmic astrocytes often end on blood vessels (Figs. VIII.8 and VIII.10). In the specimens stained by metallic impregnation procedures, these processes can be seen to end on the vascular wall, forming expansions known as end-feet or sucker feet by light microscopists. However, electron microscope studies have shown that these processes form a continuous layer around each blood vessel (Wolff 1963; Mugnaini and Walberg 1964) and that the thickness of this layer may vary considerably (Fig. XI.2) even from one segment to another in the same vessel. The images of end-feet seen under the light microscope are probably the result of artifacts caused by technical procedures. A gap of 40–100 nm, occupied by a basal lamina, usually intervenes between the plasma membrane of astrocytes and that of endothelial cells (Figs. XI.3 and XI.4).

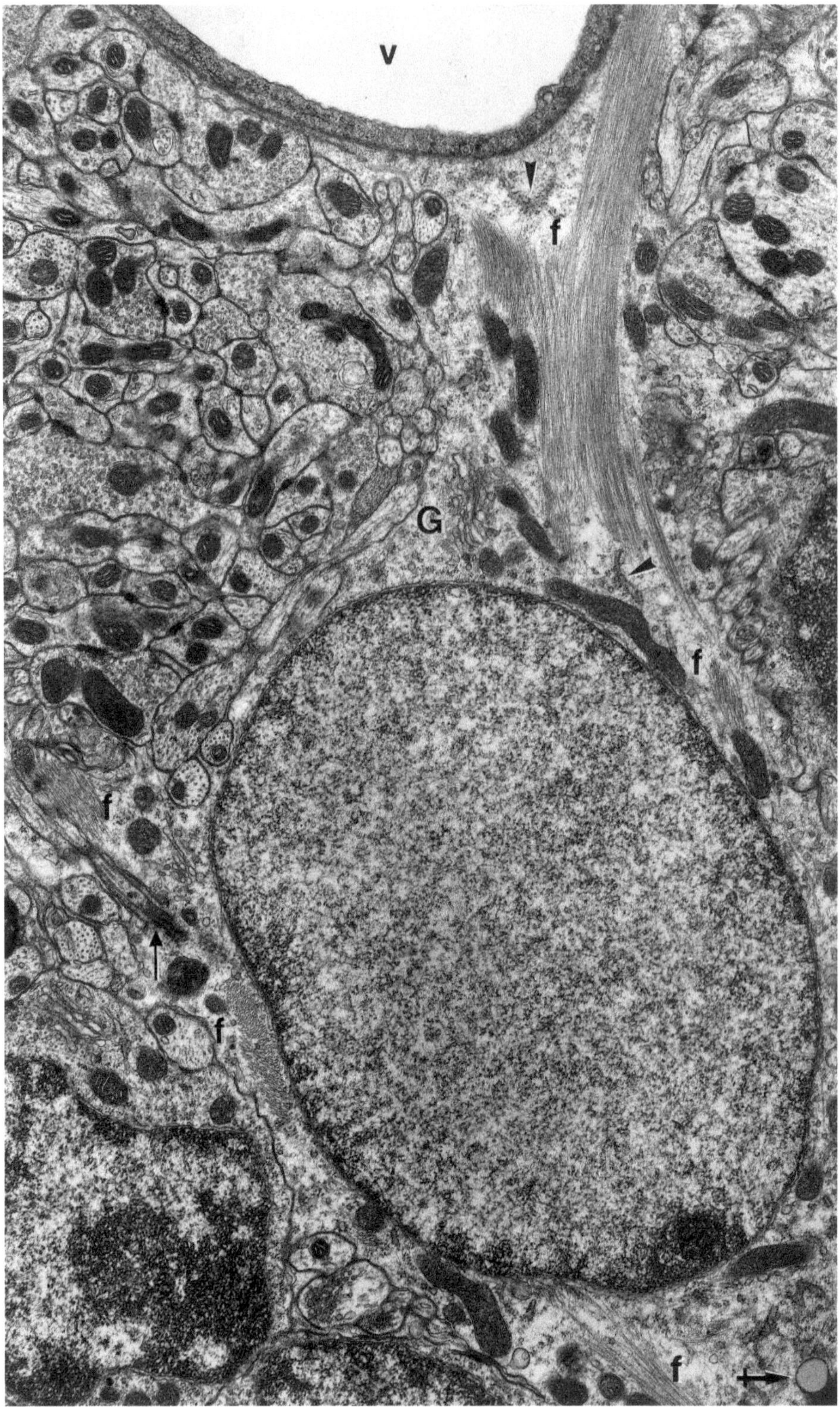
V
f
G
f
f
f
f

Where this gap is wider, two basal laminae can be seen, one applied to the endothelial cells and the other to the astrocytes; in the space between these two basal laminae, collagen fibers and sometimes also pericytes (see Sect. XI.D.1) can be found.

At the subpial surface of the CNS, the processes of both fibrous and protoplasmic astrocytes again form a continuous sheet known as the neuroglial limiting membrane or glia limitans (Held 1909b). The thickness of this neuroglial sheet varies greatly in different species and in different regions of the CNS (Figs. V.20 and VIII.11) of the same animal (Bondareff and McLone 1973). The surface of the neuroglial limiting membrane facing the pia mater, like that of the perivascular neuroglia, is covered by a basal lamina (Figs. V.20 and VIII.11). As regards the nature of the cells which produce the basal laminae associated with astrocytes, note that both in vitro (Liesi et al.1983) and in vivo (Bernstein et al. 1985) astrocytes secrete laminin, a major basal lamina component.

Assemblies of orthogonally arranged particles (Fig. VIII.12) are present in the freeze-fractured plasma membrane of astrocytes (Dermietzel 1973, 1974c; Landis and Reese 1974b; Nabeshima et al. 1975; Hanna et al. 1976; Sandri et al. 1977; Anders and Brightman 1979; Brightman et al. 1980; Massa and Mugnaini 1982). The center-to-center spacing of these particles measures about 7 nm. These specializations also occur in ependymal cells (see Sect. VIII.C.2) and the satellite cells of PNS ganglia (see Sect. VII.A.4), whereas they have never been recognized within the plasma membranes of oligodendrocytes or neurons. The concentration of assemblies is highest in the regions of the astrocytic membrane apposed to the blood vessels or facing the pia mater (Fig. VIII.11), while it is much lower in the regions apposed to other astrocytes or neurons (Landis and Reese 1982). As noted previously (see Sect. VIII.C.2), the orthogonally arranged particles are composed of aquaporin-4 (Verbavatz et al. 1997; Rash et al. 1998). Other aquaporins do not usually form assemblies of orthogonally arranged particles, so that astrocytes in teleosts express aquaporins but are devoid of orthogonally arranged particles (Wolburg et al. 1983; Grupp et al. 2010). Astrocytes of elasmobranchs (Gotow and Hashimoto 1984) and amphibians (Korte and Rosenbluth 1981) also often lack orthogonally arranged particles.

Glycoproteins which mediate intercellular adhesion (cell adhesion molecules, see Sect. V.E) are present on the plasma membrane of astrocytes. Moreover, astrocytes express a large variety of ion channels and neurotransmitter receptors, the most abundant of which are glutamate receptors and purinergic receptors (for reviews, see Verkhatsky and Steinhäuser 2000; Shelton and McCarthy 2001). Contiguous astrocytes are mutually linked by gap junctions (Fig. VIII.12) and by small, button-like, adherent junctions (Brightman and Reese 1969; Tani et al. 1973; Landis and Reese 1974b; Hanna et al. 1976; Massa and Mugnaini 1982). Astrocytic gap junctions,

Fig. VIII.10 Fibrous astrocyte. The karyoplasm exhibits a fairly even, low density except for the thin, condensed rim adjacent to the nuclear envelope. In the cytoplasm, which has a light appearance, mitochondria, short cisternae of the granular endoplasmic reticulum (*arrowheads*), and Golgi complexes (**G**) are sparse, whereas closely packed bundles of intermediate filaments (**f**) are the most prominent component. *Arrow* points to a cilium, *crossed arrow* points to a pigment granule. Note that a portion of this astrocyte surrounds a blood vessel (**v**). Rat cerebellar cortex, ×9,500 (Courtesy of S. Matsuda)

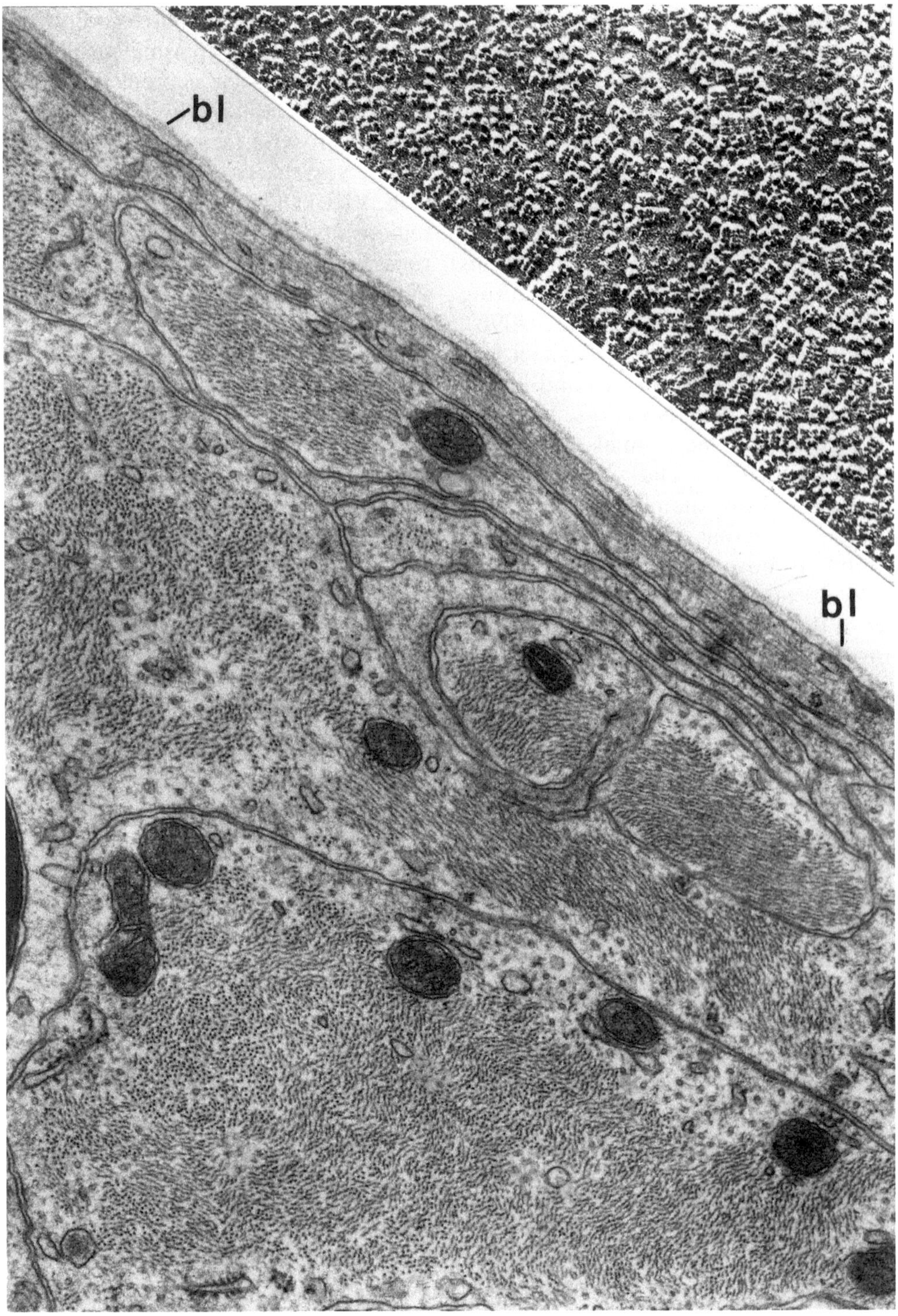

Fig. VIII.11 Neuroglial limiting membrane. At the subpial surface of the CNS, astrocytes form a continuous layer whose surface facing the pia mater is covered by a basal lamina (**bl**). This layer may contain prominent bundles of intermediate filaments, as shown in this picture. A freeze-fractured preparation of an astrocytic plasma membrane facing the pia mater is shown in the *inset*. The cleavage plane exposes the P face of this plasma membrane and reveals many assemblies of orthogonally arranged particles. Rat spinal cord, ×39,500; *inset*, ×122,000 (*Inset* courtesy of E. Reale and L. Luciano)

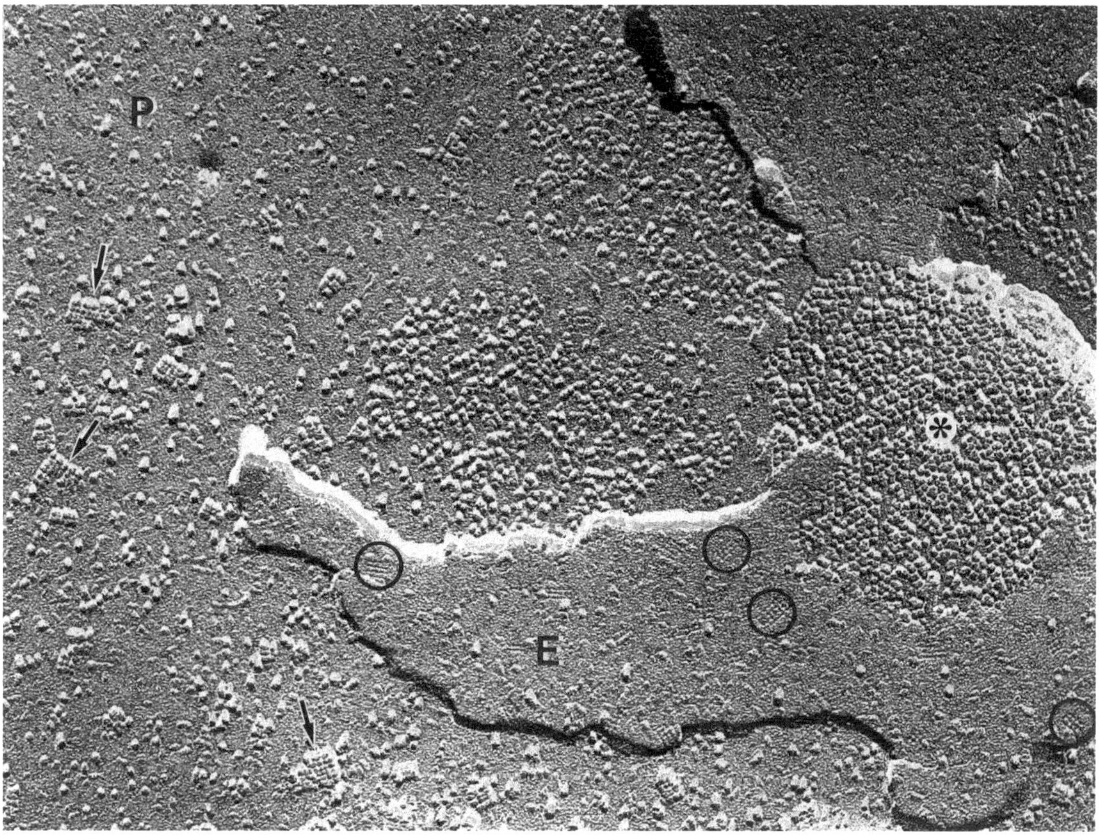

Fig. VIII.12 Astrocytic plasma membranes in a freeze-fractured preparation. The cleavage plane shifts from the P face (**P**) to the E face (**E**) of apposed astrocytic plasma membranes and reveals assemblies of orthogonally arranged particles (some *arrowed*) on the P face, and complementary assemblies of pits (some *encircled*) on the E face. * gap junction. Adult cat, ×247,000 (Courtesy of P.T. Massa and E. Mugnaini)

which are composed mainly by connexins 43 and 30 (for a review, see Nagy et al. 2004), are permeable not only to ions but also to second messengers, amino acids, glucose, and lactate. Dye transfer and electrophysiology experiments have revealed that astrocytes in vivo are strongly coupled to each other via gap junctions (Nagy and Rash 2000). The role of these gap junctions is considered in Sects. VIII.E.5b and VIII.E.6. Gap junctions have also been detected between astrocytes and oligodendrocytes (see Sect. VIII.F.1) in mammals (Massa and Mugnaini 1982), and tight junctions have been detected between astrocytes in fish (Wolburg et al. 1983; Gotow and Hashimoto 1984).

2. Protoplasmic Astrocytes

These cells occur mainly in the gray matter. In light microscope preparations, they exhibit many processes; with respect to those of fibrous astrocytes, the processes of these cells are more numerous, thicker, shorter, and more extensively branched and have an irregular surface (Fig. VIII.9). Comparative studies have shown that protoplasmic astrocytes in humans are significantly larger and extend many more primary processes than those in rodents (Oberheim et al. 2006). Protoplasmic astrocytes are not distributed randomly in the CNS gray matter. Each astrocyte occupies a distinct domain (Bushong et al. 2002), and the processes of two adjacent astrocytes only slightly overlap at the borders between the two cells. Therefore, the gray matter appears to be parcellated into discrete astrocytic microdomains.

Glial fibrils, fewer in number than those in fibrous astrocytes, occur in both the cell body and proximal portions of the processes. It has long been known that protoplasmic astrocytes are more readily damaged than fibrous ones (Golgi 1885), a characteristic possibly related to the smaller number of fibrils they contain. Granules which can be impregnated with Achúcarro's ammoniacal silver technique or with the formol-uranium technique have been observed by light microscopists within the cytoplasm of astrocytes, most frequently in protoplasmic ones. These granules, called *gliosomen* (gliosomes) by Fieandt (1910), have been interpreted by different authors as a particular type of mitochondria, as products of a secretory activity, or as organelles which participate in the formation of glial fibrils (for further details, see Ramón y Cajal 1913).

Under the electron microscope, the karyoplasm shows a fairly even, low density similar to that of fibrous astrocytes, and most of the cytoplasmic organelles (endoplasmic reticulum, Golgi complexes, centrioles, microtubules, etc.) and inclusions (glycogen and pigment granules) show the same characteristics as their counterparts in fibrous astrocytes (Fig. VIII.13). The quantity of filaments in protoplasmic astrocytes is less than in fibrous astrocytes but may increase under pathological conditions (see Sect. VIII.E.7). Astrocytic mitochondria, like neuronal mitochondria, are mobile; however, mitochondria in astrocytes seem to move slower than those in neurons. Mitochondria with particular characteristics have been found in astrocytes in addition to those with the usual features (Gray 1960b; Mugnaini and Walberg 1964; Blinzinger et al. 1965; Srebro 1965; Hashimoto 1969; Morales and Duncan 1971). Some of these mitochondria have dilated intracristal spaces containing helical filaments; others show very few cristae and an abundant, dense matrix consisting of closely packed filaments; and still others exhibit ordered arrays of prism-shaped cristae. These mitochondria with special features could correspond to gliosomes.

The processes of protoplasmic astrocytes usually have an irregular outline; the distal portions of these processes often exhibit a sheetlike shape (Fig. VIII.13) and may be as thin as about 20 nm. The lamellar portions of these processes may be in close contact with interneuronal synapses (Špaček 1971, 1985b). Not all synapses are in contact with astrocytes. The proportions of synapses contacted by astroglia vary according to the region of the CNS. About 75 % of synapses on Purkinje cell spines of the mouse cerebellum (Špaček 1985b), about 57 % of synapses in the stratum radiatum of the rat hippocampal area CA1 (Ventura and Harris 1999), and about 29 % of those on pyramidal cell spines of the mouse visual cortex (Špaček 1985b) are contacted by astroglia. Furthermore, astroglia ensheathment of the synapse may vary in extent: astroglial process may completely envelope the synapse, may cover only the pre- or the postsynaptic element, or may seal the synaptic cleft laterally (e.g., see Špaček 1985b). It has been estimated that a single astrocyte can contact about 90,000 synapses in rodents, and much more than a million in humans (Oberheim et al. 2006).

The sheetlike distal portions of astrocytic processes are able to change their shape. Actin-associated proteins ezrin and radixin have been localized in these thin lamellae (Derouiche and Frotscher 2001) and could be involved in their motility. Dynamic changes of the perisynaptic astroglial processes have been described, for example, in the rat hippocampus (Wenzel et al. 1991; Lushnikova et al. 2009). The use of confocal two-photon

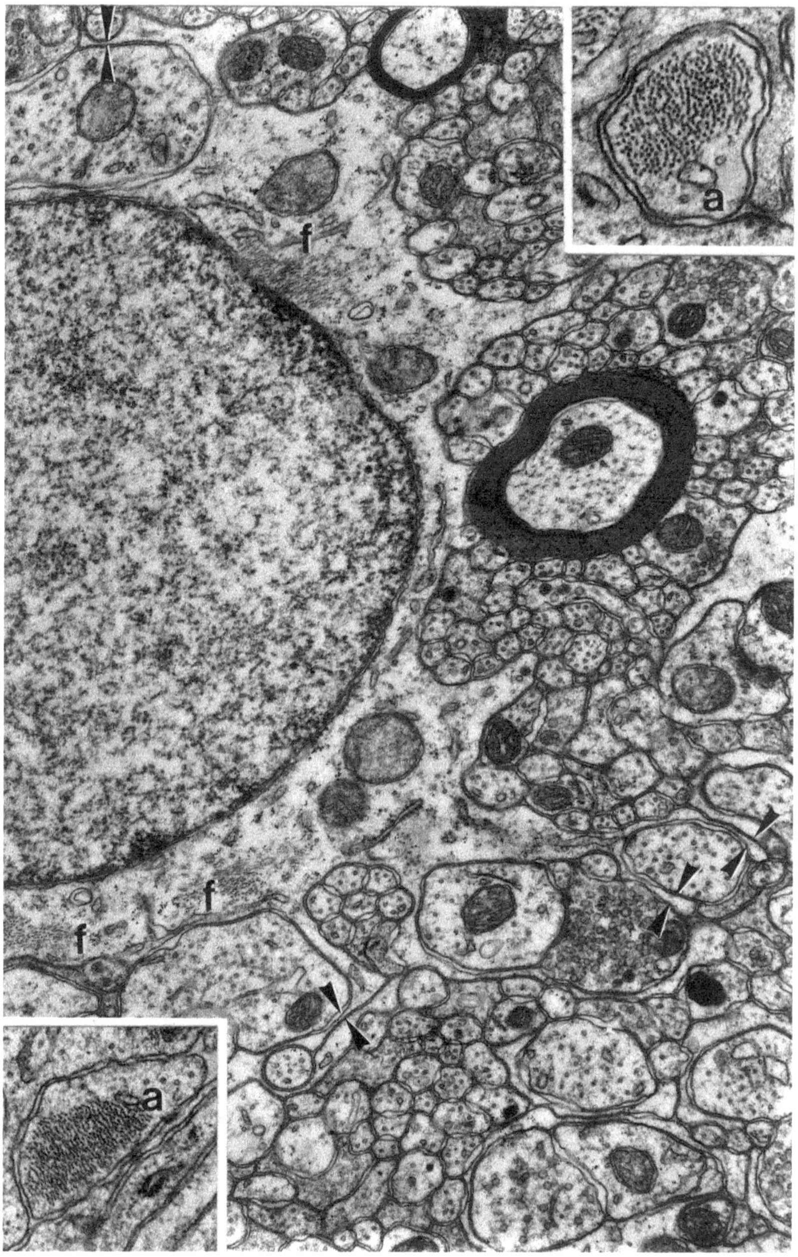

Fig. VIII.13 Protoplasmic astrocyte. The karyoplasm exhibits a rather homogeneous density. Sparse mitochondria and cisternae of the granular endoplasmic reticulum can be seen in the cytoplasm, which has a light appearance. The intermediate filaments (**f**) are fewer than in fibrous astrocytes. The processes of this cell appear as thin lamellae (*arrowheads*) which intervene between the surrounding neuropil elements. Processes of fibrous astrocytes (**a**) are shown in the *insets* to compare them with those of the protoplasmic astrocyte. Fibrous astrocyte processes usually have roundish or oval profiles and contain a bundle of filaments and few microtubules. Rat spinal cord, ×24,500; *upper inset*, ×55,000; *lower inset*, ×32,700

microscopy and specific fluorescent dyes, which selectively label astrocytes in vivo, has revealed the gliding of thin lamellar astroglial processes along the neuronal surface, and the transient extension of filopodia-like processes from astrocytes (Hirrlinger et al. 2004). The junctions linking adjacent astrocytes to each another are considered in Sect. VIII.E.1, and the communication between astrocytes and synapses are discussed in Sect. VIII.E.6.

Cells whose structure is intermediate between that of protoplasmic and that of fibrous astrocytes have been described (see, among others, Ramón y Cajal 1909; Mugnaini and Walberg 1964; Privat and Rataboul 1986). In this context, note that protoplasmic astrocytes may transform into fibrous astrocytes both in vitro and in vivo (e.g., in vitro after changing the culture medium, and in vivo following injury).

3. Astrocyte Heterogeneity

The division of astrocytes into fibrous and protoplasmic types, mainly based on their morphology and anatomical location, dates back to the end of the nineteenth century. Over the last 50 years, results obtained with immunocytochemical techniques have suggested that astrocytes are considerably more heterogeneous than originally thought. A number of examples of astroglial heterogeneity are listed below.

Two subpopulations of astrocytes have been distinguished in cultures of developing rat optic nerve (Raff and Miller 1984). Type 1 astrocytes express glial fibrillary acidic protein and do not bind tetanus toxin or the monoclonal antibody A2B5, which recognizes plasma membrane gangliosides. Type 1 astrocytes bear processes specialized to form perivascular envelopes and the subpial glia limitans. Type 2 astrocytes express both glial fibrillary acidic protein and the A2B5 antibody and also bear processes that end exclusively at nodes of Ranvier. However, these two subpopulations have not been identified in vivo. In the mature optic nerve of the mouse, for example, only a single type of astrocyte is seen, the processes of which have branches that end on the pial surface, blood vessels, and also at nodes of Ranvier (Butt et al. 1994).

Two subpopulations of astrocytes have also been distinguished on the basis of glial fibrillary acidic protein expression, one with high expression and the other with undetectable expression (Walz and Lang 1998). However, this might be a reflection of the ability of astrocytes to dynamically regulate levels of fibrillary acidic protein under the influence of environmental factors rather than the consequence of there being two distinct subpopulations of astrocytes (Walz 2000).

Astrocytes can differ in the expression of transmitter receptors, transporters, and ion channels (for a review, see Matyash and Kettenmann 2010). Research in this field is at an early stage, and the currently available data do not provide a satisfyingly unified picture.

4. Age-Related Changes

The mean number of astrocytes does not change significantly with age (Vaughan and Peters 1974; Lindsey et al. 1979; Bhatnagar et al. 1997; Long et al. 1998; Peters and Sethares 2002, 2004; Sandell and Peters 2002; Pelvig et al. 2008; Peters et al. 2008b).

Astrocytes undergo hypertrophy with age, more prominently in the white matter than gray matter. In aged subjects, these cells have larger bodies, thicker processes, and contain increased amounts of intermediate filaments than those in young adults (e.g., Landfield et al. 1977; Geinisman et al. 1978b; Sloane et al. 2000; Sandell and Peters 2002). The age-related increase in intermediate filaments is reflected in enhanced expression of glial fibrillary acidic protein (O'Callaghan and Miller 1991; Jalenques et al. 1995; Kohama et al. 1995), and also its messenger RNA (Nichols et al. 1993; Kane et al. 1997). A consequence of the astrocytic hypertrophy is that the neuroglial limiting membrane is much thicker in aged than young subjects.

With age, astrocytes accumulate dense inclusions within their cytoplasm presumably through phagocytosis (Vaughan and Peters 1974). Since some components of these inclusions are labeled with antibodies to myelin basic protein, it has been suggested that in old age astrocytes participate in the phagocytosis of the degenerating myelin (Peters and Sethares 2003). With increasing age, astrocytes accumulate lipofuscin (Peters and Sethares 2002). Lipofuscin accumulation seems to have little or no effect on either the metabolism or functional activity of neurons (for a review, see Ledda et al. 1999). This is probably also the case for astrocytes.

Round basophilic bodies, which are periodic-acid-Schiff-positive and 5-20 μm or more in diameter, may be present in the astrocytic cytoplasm and as extracellular deposits in normal aged humans. These bodies, called *corpora amilacea*, accumulate to a greater extent in Alzheimer disease and other degenerative disorders. Electron microscope studies have shown that they are devoid of a limiting membrane and consist of randomly arranged filaments, among which a dense, flocculent material is interspersed. They are made up of a glucose polymer and small amounts of proteins. The origin of *corpora amylacea* is unclear, and it is unknown whether they are harmful to astrocytes or neurons (Wiśniewski and Terry 1973).

5. Functions of Astrocytes

Astrocytes were originally thought to provide only structural support to neurons. Later it was shown that astrocytes have further complex and dynamic functions in the CNS. Recent evidence suggests that these cells actively interact with neurons.

5a. Structural Support

That astrocytes play this role was stated, among others, by Robertson (1897), Müller (1900), and Niessing (1936). Evidence for this function is that astrocytes usually contain a medium to large quantity of filaments arranged in bundles, are held together by intercellular junctions, and are firmly attached to the vascular tree, which in turn forms a robust skeleton-like structure within the CNS. This view has been strengthened by comparative studies of the framework formed by astrocytes in the white matter of the spinal cord from many vertebrates and by studying changes undergone by single astrocytes under experimental conditions. For example, the abundance of fibrillar structures within the neuroglial framework of the spinal cord white matter correlates with spinal cord size and nerve fiber thickness (Pannese 1957, 1958a). Moreover, following intense and prolonged mechanical stress, astrocytes develop larger bodies and contain increased amounts of fibrils (Pannese 1956).

5b. Homeostasis of the Extracellular Environment

Astrocytes play a critical role in regulating water movements and the composition of the fluid in the extracellular space of the CNS. The astrocytic plasma membrane contains a number of proteins (e.g., aquaporins, ion channels, and ion cotransporters) that are involved in transmembrane water movements (Simard and Nedergaard 2004). The concentration of aquaporins is particularly high in the regions of the astrocytic membrane apposed to blood vessels (see Sect. VIII.E.1).

The K^+ concentration in the intercellular clefts around neurons increases as a result of neuronal activity (for a review, see Somjen 1975). If unregulated, this phenomenon could result in continued depolarization of neuronal membranes and could affect transmitter release at synapses. There is experimental evidence that astrocytes, which express voltage-gated K^+ channels, remove excess extracellular K^+ (see, among others, Orkand 1977; Hertz 1978; Walz and Hertz 1983; Simard and Nedergaard 2004). K^+ taken up by astrocytes could be redistributed from high to low concentration regions (Orkand et al. 1966; Gardner-Medwin et al. 1981) through gap junctions. Alternatively, K^+ could be stored as KCl within the astrocytes immediately adjacent to the active neuron and returned to the latter subsequently (Gray and Ritchie 1985).

High levels of transmitters in the extracellular space affect synaptic transmission and may be toxic for neurons, and uptake mechanisms exist to prevent the accumulation in the extracellular space of transmitters released by active neurons. Astrocytes express receptors for a variety of neurotransmitters (e.g., glutamate, γ-aminobutyric acid, and glycine) and actively take up these neurotransmitters (for reviews, see Verkhratsky and Steinhäuser 2000; Shelton and McCarthy 2001). After uptake, the transmitters may be converted into precursors and returned to neurons. For example, the glutamate taken up by astrocytes is converted into glutamine by glutamine synthetase, and glutamine is then transferred back to neurons (for a review, see Edwards 2007). Lugaro (1907) was probably the first to suggest that neuroglial cells might remove waste products potentially toxic for neurons from the extracellular space of the CNS.

5c. Local Regulation of Blood Flow and Contribution to the Energy Metabolism of the Neuron

Increase in local neuronal activity leads to an increase in blood flow, which is restricted to the activated region. This response assures that active neurons receive adequate supply of oxygen and energy substrates. There is evidence that astrocytes are primary mediators of changes in local CNS blood flow in response to changes in neuronal activity (e.g., see Koehler et al. 2009). Astrocytes are well suited to play this role because they are in close contact with both neurons and blood vessels and produce and release vasoactive agents (Gordon et al. 2007; Iadecola and Nedergaard 2007). These agents include prostaglandins, nitric oxide, arachidonic acid, and adenosine triphosphate. The release of vasoactive agents by astrocytes results in arteriole dilation, and in an increase blood flow.

Furthermore, some experimental results indicate that astrocytes take up glucose from circulating blood and deliver energy substrate (e.g., lactate) to neurons (for reviews, see Brown and Ransom 2007; Pellerin et al. 2007). It is noteworthy that astrocytes are the principal storage site for glycogen in the CNS and they contain the enzymatic machinery for glycogen metabolism. Because of the strategic localization of astrocytes between neurons and blood capillaries, as early as 1885 Golgi attributed the function of transferring nutrients from blood vessels to neurons.

5d. Neuroprotection

As noted in Sect. VIII.E.5b, astrocytes remove excess extracellular ions and are involved in inactivating neurotransmitters present in the extracellular space. Since high concentrations of ions and transmitters in the extracellular space may have negative effects for neuronal activities, these astrocyte functions are neuroprotective.

Astrocytes also produce several trophic factors involved in neuroprotection. For example, tumor necrosis factor-α, transforming growth factor-β, nerve growth factor, brain-derived neurotrophic factor, glial cell line-derived neurotrophic factor, ciliary neurotrophic factor, basic fibroblast growth factor, and erythropoietin (for a review, see Trendelenburg and Dirnagl 2005). In vitro studies have shown that trophic factors released by astrocytes support the survival of central neurons. Neurons cultured at low density can survive for many weeks in the presence of astrocytes (Banker 1980). Astrocyte removal is followed by the neuronal death within 2–4 days, which is however prevented by the addition of neurotrophic factors (Ohgoh et al. 1998). Results of in vitro studies also suggest that astrocytes protect neurons against oxidative stress (Desagher et al. 1996).

Finally, astrocytes also play neuroprotective roles in pathological conditions. In brain inflammation, astrocytes regulate immune responses by releasing cytokines and components of the complement system (for a review, see Trendelenburg and Dirnagl 2005).

Further functions of astrocytes are discussed in other sections. For astrocyte activity at nodes of Ranvier, see Sect. VIII.F.2; for astrocyte involvement in synaptic function, see Sect. VIII.E.6; and for astrocyte function in relation to the blood-brain barrier, see Sect. XI.B.

6. Neuron-Astrocyte Communication

Most of the astrocyte functions described above imply the existence of bidirectional communication between neurons and astrocytes.

Neurons communicate with astrocytes by releasing neurotransmitters such as glutamate (Dani et al. 1992; Porter and McCarthy 1996). Astrocytes express a large variety of neurotransmitter receptors, including metabotropic glutamate receptors, and respond to stimulation with transient rises in intracellular Ca^{2+}. These Ca^{2+} oscillations, also known as calcium waves, propagate to contiguous astrocytes by passage of second messengers through gap junctions (Cornell-Bell et al. 1990; Charles et al. 1991; Dani et al. 1992). Calcium oscillations can also propagate between astrocytes that are not in direct physical contact (Hassinger et al. 1996) via signals that diffuse in extracellular spaces (probably through release of adenosine triphosphate, which activates purinergic receptors on neighboring but not contiguous astrocytes). In both cases, the rise in Ca^{2+} concentration is due to the release of Ca^{2+} from intracellular stores (e.g., from a specialized portion of the endoplasmic reticulum). In turn astrocytes are able to release a number of neuroactive agents (e.g., glutamate, glycine, D-serine, adenosine triphosphate, and peptides) that may trigger a Ca^{2+} response in neurons (e.g., see Parpura et al. 1994). Furthermore, there is evidence that astrocytes are able to modulate synaptic activity by releasing neuroactive agents which may influence both the pre- and postsynaptic element (for a review, see Perea and Araque 2010).

7. Reactive Astrocytes

It has long been known that astrocytes react to CNS injury (see, among others, Río Hortega and Penfield 1927). In the area damaged by an experimentally produced injury or by certain diseases (e.g., demyelinating diseases), astrocytes (or a subpopulation of these cells, Miller et al. 1986) undergo a sequence of changes and are usually designated as reactive astrocytes. These changes begin with transient cellular swelling (Norenberg 1994) and continue with glycogen accumulation, formation of various cytoplasmic inclusions, increase in cell number, and cell hypertrophy. A marked increase in glycogen content is often observed in the cytoplasm of astrocytes near the injury site (Shimizu and Hamuro 1958; Klatzo et al. 1961; Maxwell and Kruger 1965; Nathaniel and Nathaniel 1981). These cells often contain fat inclusions, myelin fragments, dense bodies, and other material. Accordingly, some authors (McMahan 1967; Mugnaini et al. 1967a; Vaughn and Pease 1970; Ghetti et al. 1972; Nathaniel and Nathaniel 1981) have suggested that reactive astrocytes, although not primary phagocytes (see Sect. IX.C), phagocytose degenerated structures. The increase in astrocyte number is believed to occur by division of differentiated astrocytes and also by proliferation of precursor cells. Mitotic division of differentiated astrocytes has often been reported (e.g., see Bignami and Ralston 1969; Cavanagh 1970; Skoff and Vaughn 1971). Many astrocytes in the injured area also undergo hypertrophy. Cell bodies and processes increase in size, the cytoplasm comes to contain a highly elaborated Golgi apparatus, a well-developed endoplasmic reticulum, and an increased amount of intermediate filaments (Nathaniel and Nathaniel 1981; Norenberg 1994). Reactive astrocytes express high levels of glial fibrillary acidic protein. Furthermore, they express vimentin and sometimes also nestin and synemin, intermediate filament proteins that are not detectable in mature astrocytes under normal conditions. Many different types of intercellular signaling molecules are thought to be involved in triggering the astrocytic reaction or regulating specific aspects of this reaction (for a list of these molecules, see Sofroniew and Vinters 2010).

Reactive astrocytes may exert both detrimental and beneficial effects following CNS injury (e.g., see Sofroniew and Vinters 2010). The best-known detrimental effect is the inhibition of axon regeneration. Extracellular matrix molecules (chondroitin sulfate proteoglycans, heparan sulfate proteoglycans, dermatan sulfate proteoglycans, and keratan sulfate proteoglycans) produced by reactive astrocytes mediate the inhibition of axon regeneration (McKeon et al. 1991). In addition, the dense mat of intertwined astroglial processes, usually referred to as glial scar, may form a physical barrier to axon regrowth. NG2-expressing neuroglial cells (see Sect. VIII.G) are also involved in the inhibition of axon regeneration. Furthermore, the damaged axons are exposed to myelin-associated inhibitory proteins (see Sect. VIII.F.2) which also have detrimental effects on regeneration. Beneficial effects exerted by reactive astrocytes include the production of factors trophic to neurons (see Sect. VIII.E.5d); furthermore, the glial scar circumscribes the injury site and restricts the spread of cellular damage. Whether harmful or beneficial effects prevail depends on species, age, and environmental conditions.

Note, finally, that reactive astrocytes may be found far from the damaged area (Ludwin 1985; Schiffer et al. 1986), possibly as a consequence of the diffusion of signals from the injury site.

F. Oligodendrocytes

Oligodendrocytes occur in both white and gray matter. In the white matter, they may be arranged in rows among the nerve fibers and are called interfascicular oligodendrocytes; in the gray matter, they may be associated with groups of nerve fibers or closely applied to the surface of nerve cell bodies. In the latter case, they are referred to as satellite oligodendrocytes. However, oligodendrocytes are not the only cells to take up a satellite position since this may also be shown, although less frequently, by astrocytes and microglial cells.

1. Cell Shape and Structure

Compared with astrocytes, oligodendrocytes are smaller and possess smaller nuclei, which often lie in an eccentric position and stain more deeply with basic dyes. These cells have a spherical or polyhedral body, from which a few slender processes radiate. Unlike those of astrocytes, the processes of oligodendrocytes give off few branches, do not contain evident fibrils, and do not form end-feet on the vascular wall.

In electron microscope preparations, the chromatin of oligodendrocytes appears clumped (Fig. VIII.14); the cytoplasmic matrix appears denser than in astrocytes (Farquhar and Hartmann 1957; Mugnaini and Walberg 1964; Schultz 1964; Kruger and Maxwell 1966; Wendell-Smith et al. 1966). The granular endoplasmic reticulum is represented by numerous, rather short cisternae, some of which may be arranged in stacks (Figs. VIII.14 and VIII.15); free polysomes are more abundant than in astrocytes, whereas glycogen granules are uncommon. There is

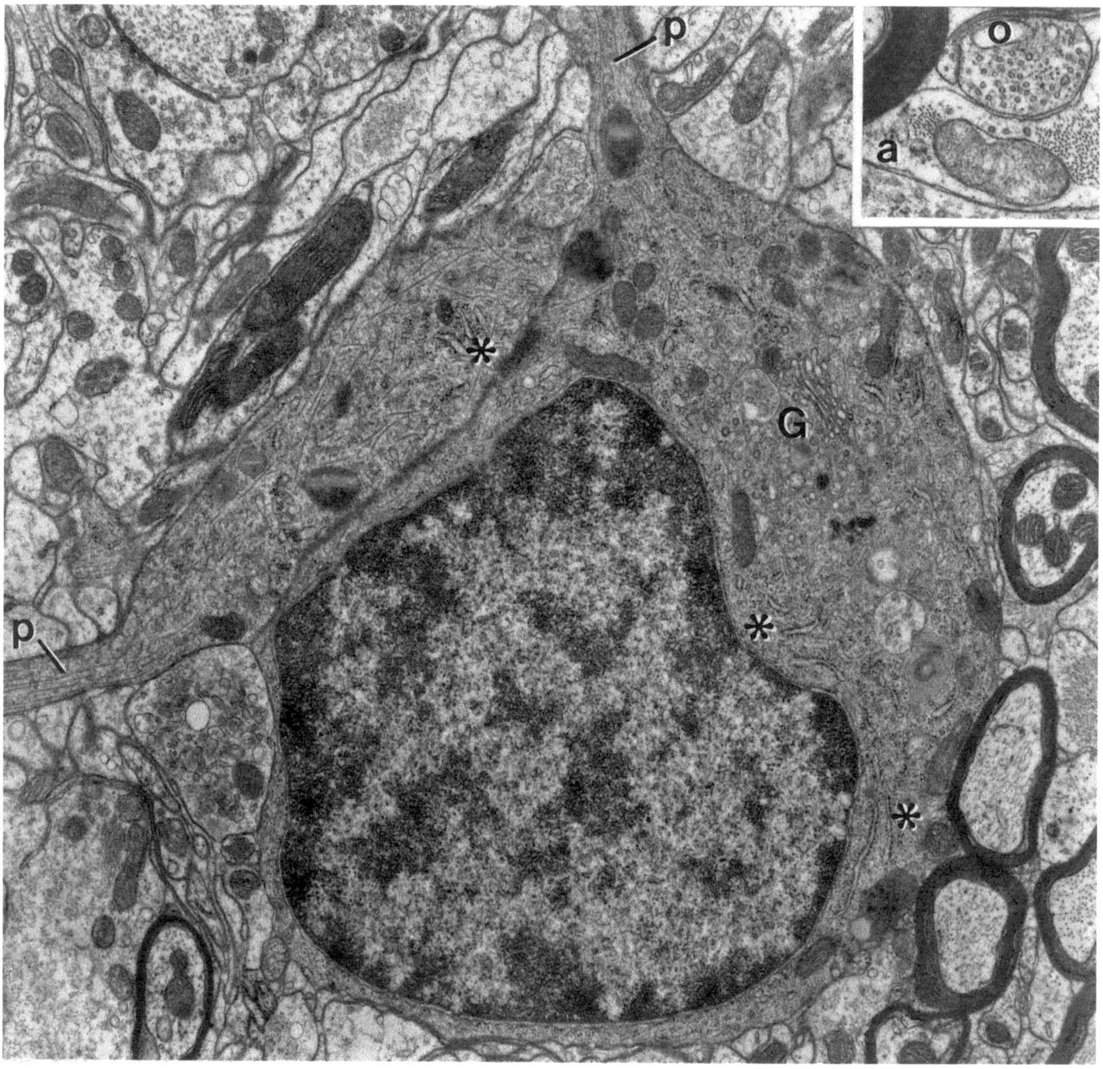

Fig. VIII.14 Oligodendrocytes. Dense clumps of chromatin are prominent in the nucleus. Mitochondria, dense bodies, Golgi complexes (**G**), and numerous cisternae of the granular endoplasmic reticulum, some arranged in stacks (*), can be seen in the cytoplasm, which has a dense appearance. Microtubules are numerous, both in the perikaryon and processes (**p**). Processes of an oligodendrocyte (**o**) and of a fibrous astrocyte (**a**), both in transverse section, are shown in the *inset*; the former contains many microtubules but few intermediate filaments, whereas the latter contains few microtubules but many filaments arranged in bundles. Rat spinal cord, ×16,500; *inset*, ×32,000

some evidence that the membrane-attached and free polysomes in oligodendrocytes synthesize different myelin proteins (see Sect. VIII.F.2). The Golgi apparatus is well developed, and mitochondria exhibit the usual features. Dense bodies with a homogeneous or heterogeneous content can also be found. Filaments are scarce, and microtubules are numerous; the quantitative ratio between filaments and microtubules is, therefore, the reverse of that found in mature astrocytes.

The processes of oligodendrocytes contain many microtubules (Figs. VIII.14 and VIII.15)

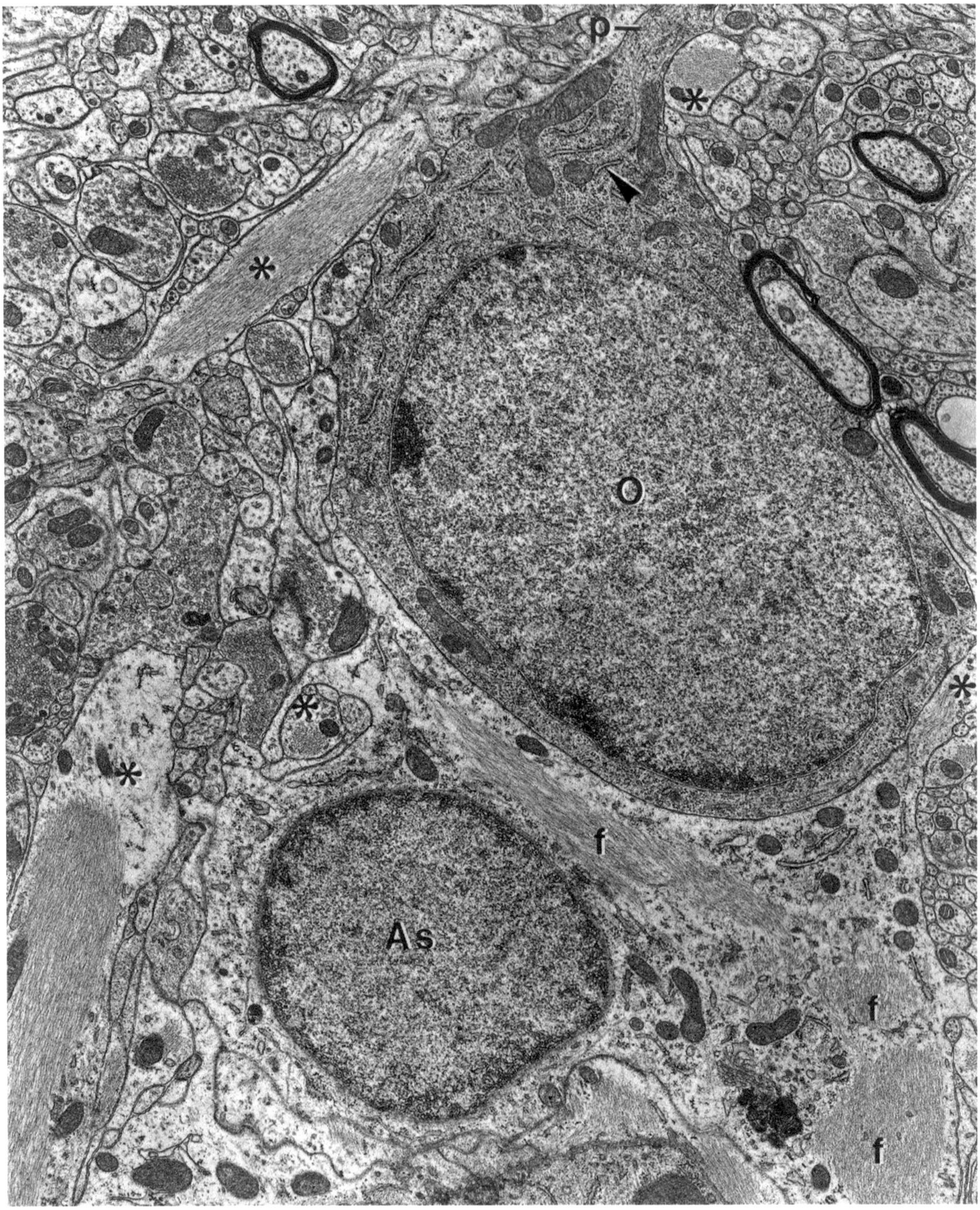

Fig. VIII.15 Oligodendrocyte and astrocyte. In this field, an oligodendrocyte (**O**) lies next to an astrocyte (**As**). Note that the cytoplasm of the oligodendrocyte is denser than that of the astrocyte. A stack of granular endoplasmic reticulum cisternae is marked by an *arrowhead* in the perikaryon of the oligodendrocyte; **f** bundles of intermediate filaments in the perikaryon of the astrocyte, **p** process of the oligodendrocyte, * processes of astrocytes. Rat spinal cord, ×13,600

which are disposed lengthwise. The abundance of microtubules makes it easy to distinguish the processes of oligodendrocytes from those of astrocytes (Figs. VIII.14 and VIII.15), although it sometimes hinders their discrimination from dendrites. Dendrites, however, can be distinguished from oligodendrocyte processes in that the latter have a denser cytoplasm and lack synapses on their surface.

Oligodendrocytes display a wide range of nuclear and cytoplasmic densities, varying from usually larger, relatively pale cells to generally smaller, dark cells (e.g., see Kruger and Maxwell 1966; Mori and Leblond 1970). These observations have been made after resorting to various methods of fixation; hence, the above differences between oligodendrocytes may confidently be regarded as real and not due to artifacts. The results of autoradiographic studies (Mori and Leblond 1970; Paterson et al. 1973; Imamoto et al. 1978) indicate that the various forms of oligodendrocytes which can be seen in electron micrographs represent different maturational stages within this cell population, dark oligodendrocytes representing fully mature cells.

With the freeze-fracture technique, it has been shown that the internal structure of the oligodendrocyte plasma membrane differs from that of astrocytes (Massa and Mugnaini 1982). The plasma membrane of oligodendrocytes does not contain the assemblies of orthogonally arranged particles which have been demonstrated on the split plasma membrane of most astrocytes (Fig. VIII.12) but shows large, tall globular particles, small globular particles, small ellipsoidal particles, and thin, short, rectilinear strands (Massa and Mugnaini 1982). Galactocerebroside is thought to be a specific constituent of the oligodendrocyte plasma membrane and for this reason has been proposed as a marker to identify this type of cell in culture (Raff et al. 1978). Other markers used to identify oligodendrocytes are the 2′,3′-cyclic nucleotide 3′-phosphodiesterase (CNPase) and the myelin-associated glycoprotein (MAG). The role of this protein is considered in Sect. VIII.F.2. Receptors expressed by oligodendrocytes include those for glutamate, N-methyl-D-aspartate, γ-aminobutyric acid, and glycine (Káradóttir and Attwell 2007).

Oligodendrocytes form gap junctions predominantly with adjacent astrocytes (Massa and Mugnaini 1982; Rash et al. 2001) and with each other in certain areas of the CNS (Maglione et al. 2010). While astrocytes are highly interconnected, coupling between astrocytes and oligodendrocytes is more limited. Oligodendrocytes express connexins 29, 32, and 47, whereas astrocytes mainly express connexins 30 and 43 (see Sect. VIII.E.1). Therefore, the gap junctions between astrocytes and oligodendrocytes are heterotypic, i.e., contain two or more different connexins. In fact, these gap junctions have often been found labeled for connexins 30 or 43 on the astrocytic side, and for connexin32 on the oligodendrocyte side (for a review, see Nagy et al. 2004). Finally, tight junctions may occur between adjacent oligodendrocytes (Tani et al. 1977; Massa and Mugnaini 1982).

2. Functions of Oligodendrocytes

While in the PNS myelin is formed by Schwann cells (see Sect. VII.B.4), in the CNS myelin is provided by oligodendrocytes. That oligodendrocytes are myelinating cells in the CNS was suggested by Río Hortega (1921, 1928) and was demonstrated with the electron microscope (Luse 1956; Peters 1964; Sternberger et al. 1978). A Schwann cell forms only one internode, while an oligodendrocyte can produce many internodes around more than one axon, each internode being formed by an oligodendrocyte process. It has been estimated that a single oligodendrocyte may form as many as 60 internodes (Peters and Proskauer 1969; Matthews and Duncan 1971). In normal adult subjects, the processes connecting the body of an oligodendrocyte and the myelin sheaths are very thin, long, and tortuous. For these reasons, the demonstration of direct continuity between the body of an oligodendrocyte and a myelin sheath is very difficult; indeed, this continuity has only rarely been shown (Kruger and Maxwell 1966; Hirano 1968).

Many structural and chemical similarities but also some differences exist between central and peripheral myelin. Both central and peripheral myelin sheaths are subdivided into segments separated from each other at nodes. The segments are usually called internodes. Interestingly, Ranvier, who was the first to describe the nodes in peripheral nerve fibers (1871a), maintained that there were none in central nerve fibers (1873). Later, however, Tourneux and Le Goff (1875), Schiefferdecker (1887), and Porter (1890) described myelin interruptions in the CNS. While the Schwann cell remains closely associated with the single internode to which it has given rise, the body of the oligodendrocyte usually lies at some distance from its internodes. As a result, in the PNS, the compartment of the Schwann cell cytoplasm situated on the outside of the myelin sheath is expanded at the midpoint of each internode to contain the Schwann cell nucleus, while this does not occur in the CNS.

Central and peripheral myelin sheaths differ also in the amounts of neuroglial cytoplasm found on the outside and the inside of the myelin sheath (Maturana 1960; Peters 1960). In fact, in central myelin, the neuroglial cytoplasm occurring within both these compartments is usually confined to a small part of the circumference, thus forming two narrow, longitudinal ridges which in cross section appear as two small, tonguelike processes. The neuroglial cytoplasm of central myelin sheaths is much more prominent during development. In the CNS of the adult, due to the fact that the neuroglial cytoplasm within the outer compartment is limited, the adjacent myelin sheaths are often in close contact with each other, and here all the more so because of the absence of both basal lamina and connective structures as found between adjacent myelin sheaths in the PNS.

The minimum axon diameter for myelination is of 0.2 μm in the CNS, whereas it is of about 1 μm in the PNS (see Sect. VII.B.4). A positive correlation between the diameter of the axon and the thickness of the related myelin sheath has been found in the CNS (Bernstein 1966; Samorajski and Friede 1968) as in peripheral nerves (see Sect. VII.B.4a). However, given equal axonal diameters, the myelin sheath is generally thinner in the CNS than in the PNS (e.g., see Fraher 1978). In the CNS, Schmidt-Lanterman incisures are present only in the sheaths of thick fibers.

The nodes also show some differences between central and peripheral myelin sheaths. There is no basal lamina around the nodal axon in the CNS (Fig. VIII.16). Given the same transverse diameter of the nerve fiber, the nodes are usually longer in the CNS than in the peripheral nerves; moreover, in the PNS, the nodes are surrounded by finger-like processes projecting from the Schwann cells (see Sect. VII.B.4F), while in the CNS processes from astrocytes and NG2-expressing cells (see Sect. VIII.G) may come into close contact with the nodal axon membrane (Peters 1966; Hildebrand 1971; Raine 1984; Waxman and Black 1984). The nodal axon is surrounded by an extracellular matrix, some components of which have been identified. They include a glycoprotein (tenascin) and some chondroitin sulfate proteoglycans (brevican, versican, and phosphacan).

Little is known regarding the relationship between the internodal length and transverse diameter of the axon (or nerve fiber) in the CNS; the information available (Hess and Young 1949; Bodian 1951) seems to indicate that in the CNS, as in peripheral nerves (see Sect. VII.B.4a), there is a roughly linear relationship between the axon (or nerve fiber) transverse diameter and the internodal length. At a given axon diameter, internodes are shorter in the CNS than in the PNS.

The lipid composition of central and peripheral myelin differs only quantitatively (for a review, see Morell 1984); for example, central myelin has a higher concentration of

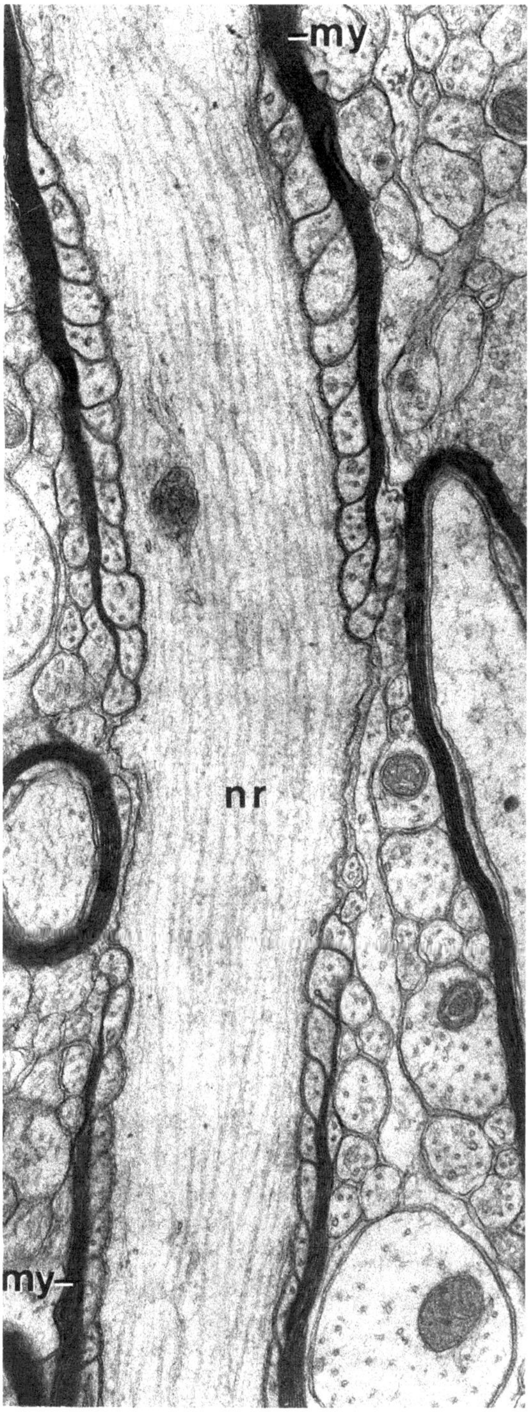

Fig. VIII.16 Nodal and paranodal regions of a myelinated nerve fiber of the CNS. In the nodal region (**nr**), the axon is bare, i.e., there are neither fingerlike processes nor basal lamina around it. In the paranodal regions, the lamellae of the myelin sheath terminate regularly beginning with the innermost lamellae, which end farthest from the node. Thus, the myelin sheath becomes progressively thinner toward the node. The regularly arranged cytoplasmic pockets of the enveloping cell indent the surface of the axon, giving the axolemma a scalloped appearance. Note that the myelin sheath (**my**) enveloping this longitudinally sectioned axon is in close contact with the neighboring elements of neuropil. Rat spinal cord, ×36,000

galactocerebroside and lower concentrations of phosphatidylcholine and sphingomyelin than peripheral myelin. The proteins of these two myelin types, however, differ markedly. The major proteins which have been characterized in central myelin are myelin basic protein (MBP, molecular weight about 18 kD), proteolipid protein (PLP, molecular weight about 25 kD), and intermediate protein ($DM_{20,}$ molecular weight about 20 kD). MBP and PLP (which account, respectively, for about 30–35 and 50 % of central myelin proteins) are adhesion molecules involved in myelin compaction. MBP is localized at the major dense line, and PLP is believed to occur at the intraperiod line. The MBP protein is the antigen responsible for inducing an autoimmune disease (experimental allergic encephalomyelitis). Some evidence suggests that PLP is synthesized on membrane-attached polysomes largely confined to the oligodendrocyte perikaryon, whereas MBP is synthesized on free polysomes concentrated in the oligodendrocyte processes (e.g., see Colman et al. 1982). Some of the differences in the size and distribution of the intramembrane particles observed between central and peripheral myelin (Schnapp and Mugnaini 1978) may be related to the different myelin protein compositions. Some molecules associated with CNS myelin exert a strong inhibitory effect on growing axons (e.g., see Schwab and Caroni 1988; Xie and Zheng 2008). These molecules play an important role in the restriction of the axonal regeneration after injury in the adult mammalian CNS (see Sect. VIII.E.7). Myelin-associated glycoprotein (MAG) is considered later, in this section.

Both the central and peripheral myelin sheaths exhibit a regular, lamellar structure, but the period of central myelin is approximately 10 % smaller. The period has been found to be about 15 nm in the central myelin and about 18 nm in the peripheral myelin in fresh material examined with X-ray diffraction (for a review, see Gregson 1976), 16 and about 18 nm, respectively, in freeze-fractured material (Bischoff and Moor 1967), and 10.7 and 11.9 nm, respectively, in fixed and embedded material (Karlsson 1966a). In both central and peripheral myelin, the external leaflets of the plasma membrane which constitutes the myelin sheath are often not fused but separated by a gap. In fixed and embedded material, this gap is 2–3 nm wide in peripheral myelin, but it is only 1–1.5 nm wide in central myelin, and this may account for at least part of the difference in the periodicity of central and peripheral myelin.

Tight junctions (Fig. VIII.17) have been observed in the central myelin sheaths (Dermietzel 1974a; Reale et al. 1975). They consist of several strands which follow a straight or gently undulating course along the axis of the nerve fiber and appear as ridges on the P face and as grooves on the E face of each split myelin lamella. These junctions are localized such that they give rise to a series of radial lines visible in transverse sections of central myelin sheaths after potassium permanganate staining. These radial lines, which extend for varying distances across the thickness of the sheath, are known as the radial component of the myelin sheath (Peters 1961). It has been suggested that these tight junctions, which are formed by proteins of the claudin family, potentiate the insulating function of the myelin mainly in small diameter fibers (Devaux and Gow 2008). Reflexive (see Sect. VII.A.6) gap junctions containing connexin32 have been found at paranodal loops, at Schmidt-Lanterman incisures, and within compact myelin (for a review, see Nagy et al. 2004).

While myelin generally envelops only the axonal process of a neuron, it sometimes also surrounds dendrites (see Sect. V.C.1) and/or nerve cell bodies, both in the PNS (see Sect. VII.A.7) and CNS. Partially or completely myelinated nerve cell bodies have been observed in various regions of the CNS of normal subjects in a wide range of species (cerebellar cortex, Rosenbluth 1966, Suyeoka 1968, Mugnaini 1972, Palay and

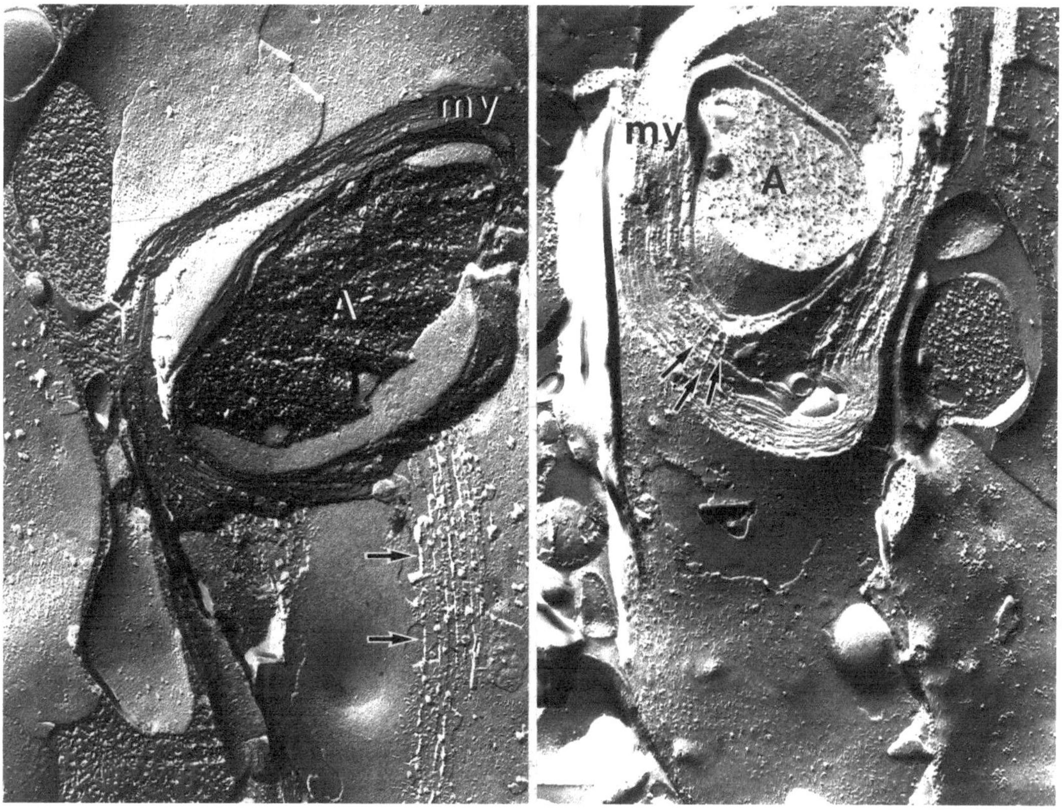

Fig. VIII.17 Tight junctions of central myelin sheaths in a freeze-fractured preparation. *Left*: Ridges (*arrows*) are exposed on the P face of one of the outermost myelin lamellae. *Right*: Ridges (*arrows*) are exposed on some consecutive deep lamellae of the steplike fractured myelin. **A** axon, **my** myelin sheath. Optic nerve of rabbit, ×40,000 (Courtesy of E. Reale and L. Luciano)

Chan-Palay 1974, Cooper and Beal 1977; spinal cord, Blinzinger et al. 1972, Beal and Cooper 1976; olfactory bulb, Willey 1973, Braak et al. 1977, Burd 1980, Tigges and Tigges 1980, Remahl and Hildebrand 1985, Valverde and Lopez-Mascaraque 1991; habenular nuclei, Kemali and Sada 1973). In some instances, the myelin sheath covering a nerve cell body seems to represent an extension of an axonal myelin sheath; in other cases, it seems to be formed by oligodendrocytes apposed to the neuronal perikaryon. It is possible that these perikaryal sheaths develop as a result of an overgrowth of myelin, in response to stimuli of still unknown nature.

Oligodendrocytes influence both the composition and properties of the axonal cytoskeleton thereby taking part in the control of axon diameter (Sánchez et al. 1996; see also Chap. III and Sect. VII.D.6). The myelin-associated glycoprotein (MAG) is thought to be involved in this process. This protein, which has both adhesive and signal properties, is located in the periaxonal plasma membrane of myelinating oligodendrocytes and Schwann cells. It locally influences the phosphorylation of axonal neurofilaments (Dashiell et al. 2002).

Furthermore, oligodendrocytes induce Na^+ channel clustering at the nodal axolemma (Kaplan et al. 1997), thus allowing the saltatory conduction characteristic of myelinated axons. These cells also play a crucial role in the localization of K^+ channels in juxtaparanodal regions (Rasband et al. 1999). As in the PNS (see Sect. VII.B.4F), in the CNS a helically arranged junction links the axolemma with the glial plasma membrane. The following adhesion molecules have been identified at this axoglial junction: neurofascin, Caspr, and contactin. It has been suggested that the paranodal axoglial junction acts as a barrier separating the nodal and juxtaparanodal domains, enriched in Na^+ channels and K^+ channels, respectively.

Finally, oligodendrocytes may complement astrocytes in their ability to provide trophic support for neurons (see Sect. VIII.E.5d). It has been suggested that oligodendrocytes transfer metabolites, glycolytic enzymes, messengerRNA, and microRNA to axons via exosomes (for a review, see Du and Dreyfus 2002).

3. Vulnerability of Oligodendrocytes to Injury and Age-Related Changes in the Oligodendrocyte-Myelin Complex

It has been estimated that the plasma membrane of a myelinating oligodendrocyte may reach a surface area of 20×10^5 μm^2. The production and support of such a large amount of plasma membrane require an enormous metabolic load with consequent accumulation of abundant toxic byproducts. Furthermore, since iron is a cofactor of many myelin synthetic enzymes, oligodendrocytes have large stores of iron which can elicit the formation of free radicals and lipid peroxidation. For these reasons, oligodendrocytes are much more vulnerable to oxidative damage than other cells of the nervous system (for a review, see McTigue and Tripathi 2008), and are strongly affected by the aging process.

In old age, aggregates of dense inclusions are often present in the perikaryon (Fig. VIII.18) and in bulbous enlargements of the primary processes of oligodendrocytes (Peters 1996). These inclusions have irregular shapes and can be as large as 1 μm in diameter. Some are composed of a homogeneously dense material, whereas others contain patches of a component of lower density embedded in a dense material. Sometimes the pale component appears to consist of stacks of thin lamellae. Oligodendrocytes are not generally considered to have phagocytic activity, so those containing such inclusions are likely to be breaking down and resorbing their own myelin sheaths (Peters 1996).

During aging, some oligodendrocytes degenerate, and new ones are produced. If these two processes are balanced, the number of these cells does not change significantly with age, as verified in the auditory (Vaughan and Peters 1974) and prefrontal (Peters and Sethares 2002) cortices of the rat, and the anterior commissure (Sandell and Peters 2003) of the monkey. The imbalance between these two processes may lead to a significant increase or decrease in the number of these cells. For example, the number of oligodendrocytes has been reported to increase with age in the optic nerve (Sandell and Peters 2002), visual cortex (Peters and Sethares 2004; Peters et al. 2008b), and fornix (Peters et al. 2010) of the monkey but to decrease in the human neocortex (Pelvig et al. 2008), and in the rat subcortical white matter (Chen et al. 2011). As regards the origin of new oligodendrocytes, the prevailing view is that mature oligodendrocytes do not divide (Keirstead and Blakemore 1997) and that new oligodendrocytes originate from progenitor cells present in white and gray matter. In old age, oligodendrocytes are often arranged in clusters (Peters 1996; Peters and Sethares 2004). This arrangement suggests that the cluster is the result of recent cell division.

Myelin sheaths themselves may undergo degeneration in old age (e.g., see Goemaere-Vanneste and van den Bosch de Aguilar 1987; Tigges et al. 1992). A common type of age-related degenerative change in myelin sheaths is a local splitting of the major dense line with enclosure of a pocket of cytoplasm

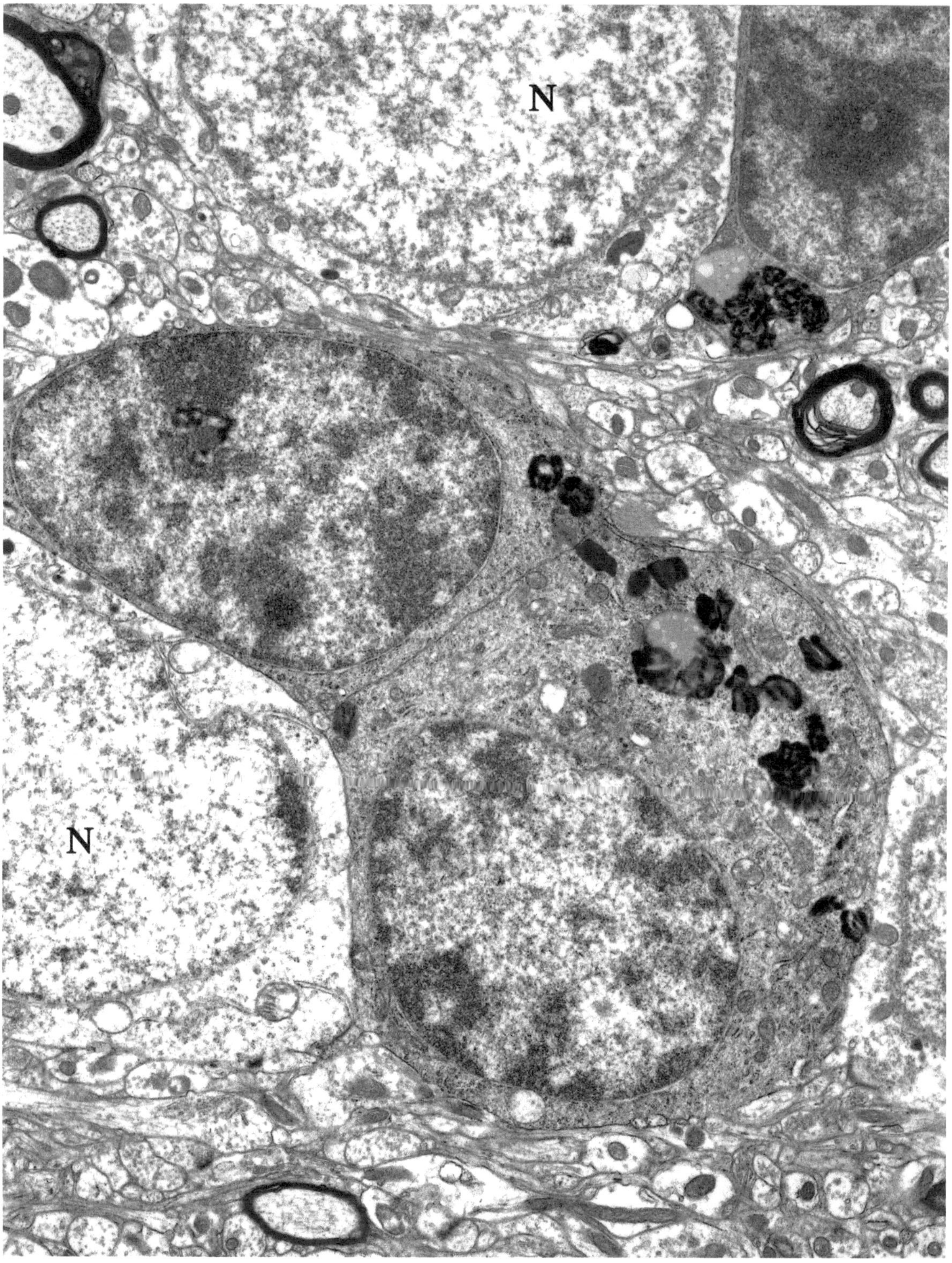

Fig. VIII.18 Oligodendrocytes from an old monkey. Their cytoplasm contains dense inclusions that are typical of aging oligodendrocytes. **N** nerve cell nucleus. Primary visual cortex of a 29-year-old rhesus monkey, ×13,000 (Courtesy of A. Peters)

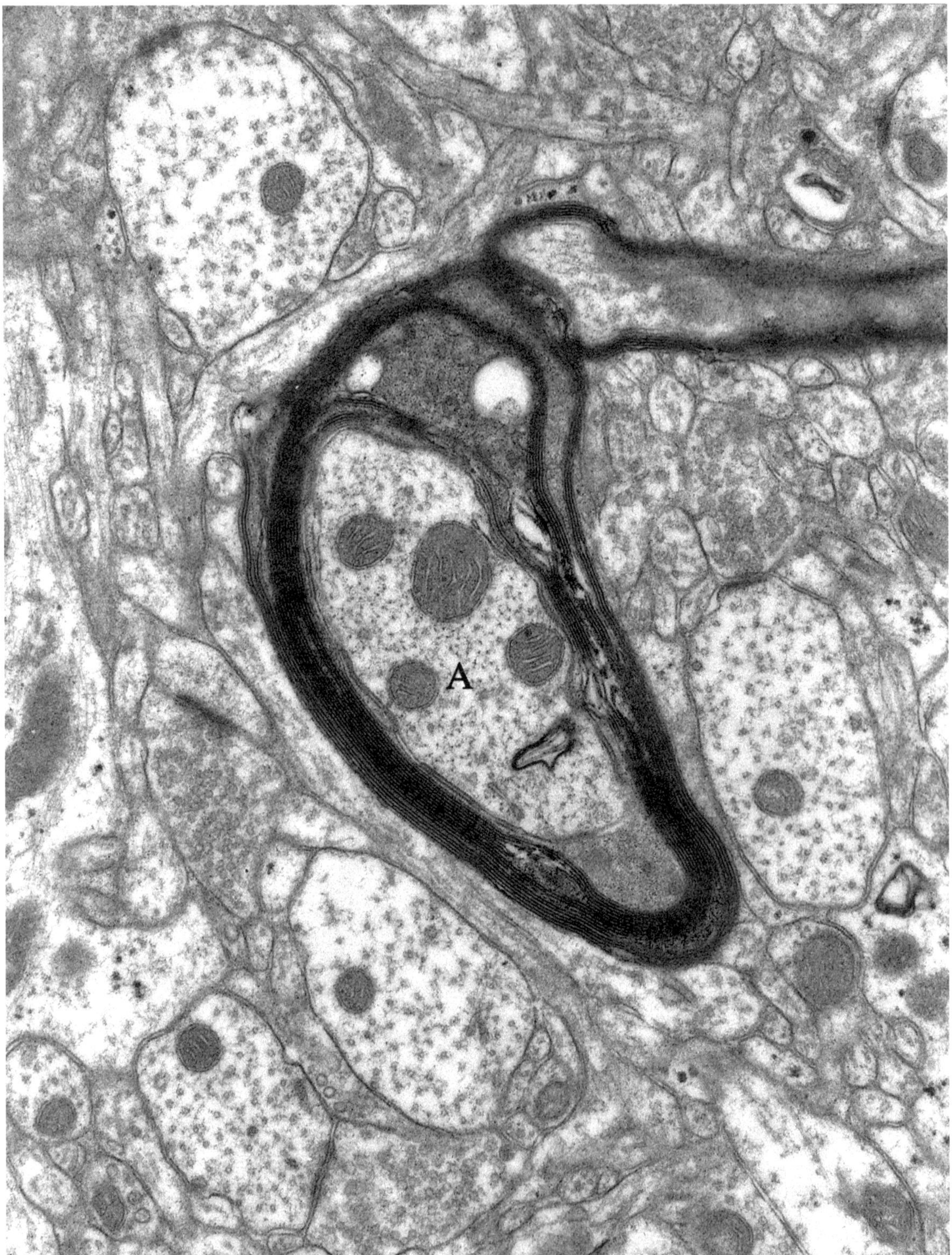

Fig. VIII.19 Myelinated nerve fiber from an old monkey. The axon (**A**) is intact, but the sheath has begun to degenerate, as shown by the presence of dense cytoplasm between some of the turns of the lamellae of myelin. Prefrontal cortex of a 27-year-old rhesus monkey, ×35,000 (Courtesy of A. Peters)

within the split (Peters et al. 2000). The cytoplasm in the pocket, probably belonging to the oligodendrocyte that formed the sheath, appears to be degenerating, because it is dense (Fig. VIII.19) and often contains amorphous dense bodies (Peters et al. 2000).

Another type of age-related degenerative change in myelin sheath is the formation of blebs. These, which have round profiles and may be up to 10 μm in diameter (Feldman and Peters 1998), are produced by a splitting of the intraperiod line of the compact myelin. All these changes can lead to myelin degeneration.

Other age-related myelin changes are the occurrence of sheaths of redundant myelin (Sturrock 1976), and the appearance of circumferential splits in thick sheaths (Peters et al. 2000). These changes suggest that myelin continues to form in old age. In fact, the degeneration of myelin internodes is followed by remyelination (Peters and Sethares 2003). The degenerated myelin internodes are replaced by internodes which are shorter than preexisting ones. The alterations in myelin sheaths plus the formation of shorter internodes are expected to decrease conduction velocity along axons (e.g., see Gutiérrez et al. 1995). Such a decrease has been recorded in the CNS of old humans and animals (Rogers et al. 1981; Aston-Jones et al. 1985; Xi et al. 1999). It is likely that the reduced conduction velocity along nerve fibers contributes to the cognitive decline that often occurs in old age (Peters 2002).

G. NG2-Expressing Cells

The chondroitin sulfate proteoglycan NG2 (Nerve Glia antigen 2) is a transmembrane glycoprotein (see Sect. V.E) expressed by various cells within and outside the CNS. In the CNS, NG2-expressing cells are present in both white and gray matter. In the rat, they comprise 8–9 % of the total cell population in white matter, and 2–3 % of the total cell population in gray matter (Dawson et al. 2003). These cells have been referred to as β-astrocytes (Reyners et al. 1982), polydendrocytes (Nishiyama et al. 2002), synantocytes (Butt et al. 2002), β-neuroglial cells (Peters 2004), and NG2-glia (Wigley et al. 2007).

In light microscope preparations, NG2-expressing cells have multiple (8–30), thin, branched processes extending from a small cell body. The cells in gray matter have processes that extend in all directions and have therefore a stellate morphology; by contrast, the cells in white matter have processes that extend parallel to axons.

The currently prevailing view is that NG2-expressing cells represent a glial cell population distinct from astrocytes, oligodendrocytes, and microglia. These cells differ in shape from mature oligodendrocytes and are distinct from resting microglia since they do not label with microglial antigens (Nishiyama et al. 1997). Although these cells appear similar in shape to protoplasmic astrocytes, they are distinct from them since they do not express glial fibrillary acidic protein, vimentin, S-100 protein, or glutamine synthetase. These cells express several functional receptors (e.g., glutamate, GABA kainate, nicotinic, and purinergic receptors).

Under the electron microscope, the nucleus exhibits a rather irregular outline. The karyoplasm shows a low density except for a condensed rim beneath the nuclear envelope, and a few clumps dispersed throughout the nucleus. The cytoplasm, which has a light appearance, contains a few cisternae of the granular endoplasmic reticulum, free polysomes, Golgi complexes, mitochondria, and microtubules but not intermediate filaments (Peters 2004).

It is generally agreed that NG2-expressing cells give rise to oligodendrocytes both during development and in adult life. A remarkable feature of NG2-expressing cells is that they receive synapses. Synaptic contacts formed by axons with processes of NG2-expressing cells have been described in some regions of the rat CNS (Bergles et al. 2000; Ziskin et al.

2007; see also Sect. VI.J). It has been suggested that these synapses allow neurons to control oligodendrogenesis and thus myelination. However, the exact roles played by these cells in the adult CNS are poorly understood. Since processes of these cells, in addition to those of astrocytes, contact nodes of Ranvier (Butt et al. 1999) and synapses (Ong and Levine 1999), it has been suggested that these cells carry out some of the physiological roles which have been attributed to astrocytes at nodes (see Sect. VIII.F.2) and synapses (see Sect. VIII.E.6).

Following CNS injury, NG2-expressing cells become reactive, undergo proliferation, and contribute to the glial scar (see Sect. VIII.E.7). These cells produce a number of extracellular matrix molecules which inhibit axon regrowth (Tan et al. 2005).

H. Renewal of the Neuroglial Cell Population

Throughout life, some astrocytes and oligodendrocytes undergo degeneration and are lost from the mammalian CNS, as shown by the fact that pyknotic neuroglial cells are found in the adult CNS under normal conditions (Niessing 1936; Pannese 1958b; Smart and Leblond 1961; Pannese and Ferrannini 1967; Korr et al. 1975). Since the size of the astroglial population remains roughly constant with time (e.g., Sandell and Peters 2002; Peters et al. 2008b), new astrocytes must be generated to maintain their steady-state number. As regards oligodendrocytes, the number of these cells has been reported to increase significantly with age in several regions of the CNS (e.g., Sandell and Peters 2002; Peters et al. 2008b, 2010). Thus, new oligodendrocytes continue to be generated throughout life.

In the adult, new neuroglial cells probably arise within the subependymal layer and also from immature cells persisting throughout the CNS. Cells with proliferative capacity are present within the subependymal layer (see Sect. VIII.C.8) under normal conditions both during development and in adult life (Allen 1912; Bryans 1959; Smart 1961; Noetzel and Rox 1964; Altman 1966; Lewis 1968; Privat and Leblond 1972; Korr 1980). The number of these cells varies considerably according to the zoological species and area of the subependymal layer. Note that this layer is also a source of new neurons (see Sect. VIII.C.8). Some of the cells originating by mitosis in this location probably replace lost ependymal cells, while others migrate into neighboring regions (Altman 1966; Lewis 1968). In these regions the cells differentiate into astrocytes or oligodendrocytes (Lewis 1968; Privat and Leblond 1972; Paterson et al. 1973; Korr 1980) although many remain immature. In adult life, subependymal layer cells continue to divide, but migration from this layer slows considerably and continues only toward spatially restricted regions of the CNS. The immature cells (glial progenitors), which are widely distributed throughout the CNS, retain their proliferative capacity and continue to generate new neuroglial cells. Therefore, it may be concluded that a turnover of astrocytes and oligodendrocytes occurs throughout life in the mammalian CNS, as suggested by Robertson as early as 1897. Oligodendrocyte turnover is slower than that of astrocytes.

IX. Microglial Cells

A. Historical Note

Using his ammoniacal silver carbonate impregnation technique, Río Hortega (1919, 1921) was able to clearly distinguish within Ramón y Cajal's "third element" (see Sect. VIII.A) two distinctly different types of cell, which he named oligodendrocytes (see Sect. VIII.F) and microglial cells. He described these last as small cells with branched processes bearing spine-like projections, showing migratory and phagocytic activities in pathological and experimental conditions. In recognition of the fundamental contribution made by Río Hortega in this field, these cells were called Hortega cells by a number of authors. According to Río Hortega (1921, 1930) microglial cells, unlike macroglia, are of mesodermal origin and first appear shortly before birth or in the newborn animal, i.e., considerably later than macroglia. Later research provided strong evidence in favor of the mesodermal origin of microglial cells (e.g., see Gozzano 1931; Kershman 1939; Field 1955; Sturrock 1974; Perry et al. 1985; Streit et al. 1988) but also revealed that microglia are present in early development (e.g., see Kershman 1939; Field 1955; Sturrock 1984), much earlier than the perinatal period indicated by Río Hortega (1930).

Such cells had previously been observed by other authors who referred to them as *Stäbchenzellen* (rod cells) and mesoglia, or *Körnchenzellen*, *Gitterzellen* and compound granular corpuscles when observed in phagocytic activity. In 1899 Nissl had described cells with a rod-shaped nucleus which in pathological material displayed phagocytic abilities and marked proliferative potential. However, this author was unable to establish whether these cells were also present in normal material, where he found them difficult to distinguish from capillary wall cells. Using his platinum impregnation technique, Robertson (1900) had shown that in the CNS in addition to neuroglial cells, there are also elements which are entirely different in shape from neuroglial cells, have processes which are not attached to the vascular wall, and under pathological conditions seem to be endowed with phagocytic activity.

Comprehensive accounts of microglia are to be found in the reviews by Glees (1955), Cammermeyer (1970), Ling (1981), Dickson and Lee (1997), and Graeber and Streit (2010).

B. Resting Microglia

This term refers to microglial cells found in the healthy CNS. According to Río Hortega's original description (1919), in normal material a microglial cell appears as a small element whose polymorphic nucleus (elongated, oval or roughly triangular) stains deeply with basic dyes and is surrounded by little cytoplasm which extends into a few branched processes covered with many small spines.

E. Pannese, *Neurocytology: Fine Structure of Neurons, Nerve Processes, and Neuroglial Cells*, 2nd Edition,
DOI 10.1007/978-3-319-06856-5_9,

In normal material studied under the electron microscope, identification of cells corresponding to the classical microglial cells described in normal tissue by light microscopists proved difficult. Initially there were contradictory results. It was particularly difficult to differentiate between microglial cells and oligodendrocytes. As a consequence, the occurrence of resting microglia in normal material, which had been widely accepted by light microscopists, was questioned or even frankly denied by several electron microscopists (e.g., Kruger and Maxwell 1966; Wendell-Smith et al. 1966; King 1968).

Clear evidence of the existence of microglia in the healthy CNS was provided at the electron microscope level by Mori and Leblond (1969a) and Blakemore (1969). Mori and Leblond (1969a) examined in the electron microscope normal material stained by a modification of the ammoniacal silver carbonate impregnation technique of Río Hortega (1919) and were able to identify microglial cells in consequence of the electron opaque reaction product located over them; subsequently they compared these specifically stained cells with those occurring in routine electron microscope preparations. Within the normal CNS, Blakemore (1969) searched for microglial cells in areas where oligodendrocytes are usually absent, such as the periependymal zones. These studies, along with those of other authors (e.g., Blinzinger and Hager 1964; Stensaas and Stensaas 1968; Vaughn and Peters 1971; Ling et al. 1973; Vaughan and Peters 1974; Stensaas 1977; Ling 1981), strengthened the view that microglial cells are a distinct cell type commonly occurring within the healthy CNS and shed light on the features typifying resting microglia at the electron microscope level.

In electron microscope preparations, the contours of microglial cells appear very irregular and follow those of the surrounding elements in the neuropil (Fig. IX.1). Their processes display roundish profiles in transverse section and bear short and angular projections, which insinuate themselves between neighboring elements; these projections are the counterparts of the spines observed by Río Hortega (1919) in light microscope preparations. Dense clumps of chromatin are prominent in the nucleus (Figs. IX.1 and IX.2). As in oligodendrocytes, in microglial cells the cytoplasm appears dense (Figs. IX.1 and IX.2), and the Golgi apparatus is well developed. Unlike oligodendrocytes (see Sect. VIII.F.1), however, microglial cells possess few microtubules and have a granular endoplasmic reticulum consisting of long, narrow cisternae (Fig. IX.2), which follow a winding course in the cytoplasm. Unlike astrocytes (see Sects. VIII.E.1 and VIII.E.2), microglial cells contain neither glycogen granules nor bundles of filaments. Dense bodies (Fig. IX.2) and lipofuscin granules are much more commonly encountered in resting microglial than macroglial cells. Possibly some of these inclusions are counterparts of the granules stained with the periodic-acid-Schiff reaction, which were observed by Cammermeyer (1970) in light microscope preparations. Various markers have been employed to identify resting microglia but all have shown some limitations.

Concerning the distribution of resting microglia, Río Hortega (1919, 1930) maintained that these cells are scattered throughout the CNS, though more numerous in gray than white matter. This was confirmed by Ling and Leblond (1973) and Ling (1976). These cells may be in a satellite position with respect to neurons or closely associated with blood vessels but, unlike astrocytes (see Sect. VIII.E.1), in the specimens stained by metallic impregnation procedures their processes do not form end feet on the vascular wall (Fig. IX.2).

In the optic nerve, microglial cells constitute 2–6 and 4–5 % of the neuroglial cell population in urodele amphibians (Stensaas 1977) and the rat (Vaughn and Peters 1971), respectively. In the adult rat, the proportion of microglia in the total glial population averages 5 % in the corpus callosum (Mori and Leblond 1969a; Ling and Leblond 1973), 8 % in the white matter of the spinal

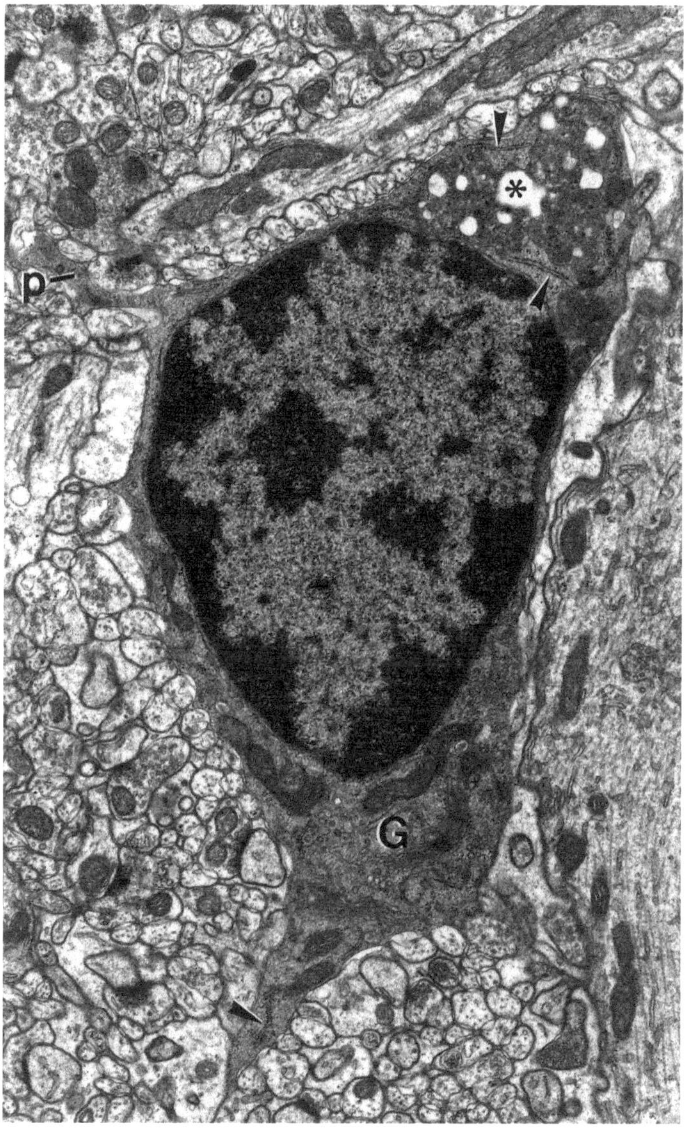

Fig. IX.1 Microglial cell. The chromatin is clumped; mitochondria, cisternae of the granular endoplasmic reticulum (*arrowheads*), and a Golgi complex (**G**) can be seen in the cytoplasm, which has a dense appearance. A mass of granular debris (*) enclosed in a vacuole is evident at one end of the cell. The cell outline is irregular. **p** cell process. Rat cerebellar cortex, ×13,800

cord (Ling 1976), 12 % in the gray matter of the spinal cord (Ling 1976), and 18–20 % in the cerebral cortex (Ling and Leblond 1973; Vaughan and Peters 1974).

Emerging data indicate that microglial cells derive from progenitors that arise in the yolk sac and migrate into the CNS parenchyma during early development (Ginhoux et al. 2010). Resting microglial cells retain their proliferative capacity under normal conditions in vivo even in adult life (Lawson et al. 1992). This cell population is thought to be renewed throughout life, although at a slow rate. The proliferative activity of these cells increases greatly in a variety of diseases.

Resting microglial cells express receptors for a variety of signaling molecules (e.g., acetylcholine, noradrenaline, glutamate,

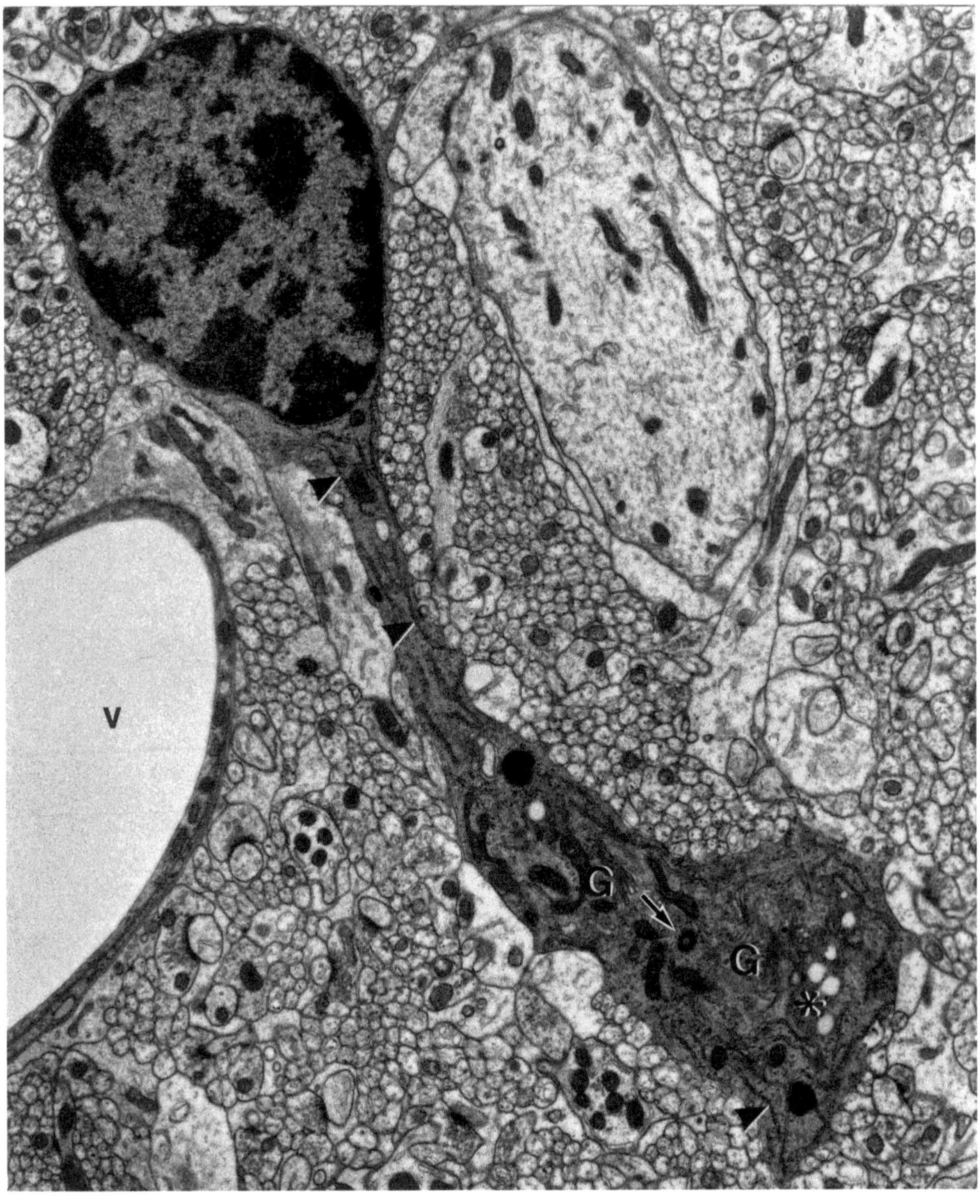

Fig. IX.2 Microglial cell. Dense clumps of chromatin are prominent in the nucleus, and the cytoplasm is dense. Long cisternae of the granular endoplasmic reticulum (some indicated by *arrowheads*), Golgi complexes (**G**), mitochondria, dense bodies, and a centriole (*arrow*) can be seen in the cytoplasm. A mass of granular debris (*) enclosed in a vacuole is visible at one end of the cell. **v** blood vessel. Rat cerebellar cortex, ×12,000

γ-aminobutyric acid, purines, dopamine, cannabinoids, and neuropeptides; for a review, see Pocock and Kettenmann 2007). Furthermore, resting microglial cells express ion channels and adhesion molecules. While the functions of microglia in acute pathological events and after experimental lesions have been extensively investigated (see Sect. IX.C), the roles of microglia in the healthy CNS long remained unknown. Microglial cells are involved in the immune surveillance in the CNS parenchyma.

Furthermore, recent studies using two-photon laser-scanning microscopy have shown that under normal conditions microglial cells remain in a stable position, but continuously move their processes with the apparent purpose of surveying their microenvironment (Nimmerjahn et al. 2005). Microglial process movement, in which actin is involved (Sogn et al. 2013), seems preferentially directed toward synapses. Hence, it was suggested that these cells are concerned with monitoring and potentially modulating synaptic activity (Wake et al. 2009). In this regard, note that microglial cells produce many signaling molecules known to regulate synaptic function, including cytokines and neurotransmitters. It is thought that under normal conditions, microglial cells have a moderate phagocytic activity (Río Hortega 1921, 1930; Stensaas and Stensaas 1968; Vaughan and Peters 1974) eliminating debris resulting from neuronal or neuroglial degeneration. Degenerative changes commonly occur within the nervous system of apparently normal adult animals and are considered to be part of a continuing structural remodeling of the nervous system (see Sects. V.C.3 and VI.F). Finally, several studies have shown that microglia may secrete neurotrophic factors.

C. Neural Macrophages

In acute pathological events and after certain experimentally induced lesions, macrophages appear within the CNS and actively engulf neuronal debris, myelin fragments, etc. The origin of these cells long remained controversial. It now seems clear that these cells derive from resting microglia and also from circulating monocytes under certain conditions. In acute pathological events and after experimental lesioning, resting microglial cells are activated by signaling molecules released from neurons, macroglial cells, and vascular cells. Activated microglial cells withdraw their processes and become roundish, with larger nuclei, more polysomes, and more highly elaborated Golgi complexes than in the resting state. These activated cells proliferate (Cammermeyer 1965; Sjöstrand 1965; Kreutzberg 1966; Price 1972; Skoff 1975; Rieske-Shows et al. 1987) and migrate to the site of injury, where they may act as antigen-presenting cells (for a review, see Aloisi 2001), and transform into macrophages (Río Hortega 1921, 1930; Penfield 1928; Blinzinger and Hager 1964; Bignami and Ralston 1969; Torvik and Skjörten 1971; Blakemore 1972; Stenwig 1972; Wong-Riley 1972; Barron et al. 1974; Skoff 1975; Rieske-Shows et al. 1987). Activated microglial cells release cytokines and prostaglandins and express prostaglandin receptors. Under pathological and experimental conditions involving vascular damage, circulating monocytes enter the lesion site (Adrian and Walker 1962; Rieske-Shows et al. 1987) and transform into macrophages (Konigsmark and Sidman 1963; Huntington and Terry 1966; Roessmann and Friede 1968; Kitamura 1973; Oehmichen and Huber 1976; Imamoto and Leblond 1977; Ling 1979). The relative proportions of microglia-derived macrophages and monocyte-derived macrophages seem to depend on the nature and extent of the damage. Macrophages derived from these two sources show the same morphological features and cannot be distinguished by known markers, hence the confusion in the literature.

Once the process provoking the appearance of neural macrophages has run its course, macrophages derived from resident microglia can return to quiescence while monocyte-derived macrophages do not transform into resident microglia and are completely removed from the CNS (Ajami et al. 2011).

D. Age-Related Changes

The mean number of microglial cells does not change significantly with age either in the cortical area 2 of Krieg of the rat (Brizzee et al. 1968) or in the visual cortex (Peters et al. 2008b) and fornix (Peters et al. 2010) of the monkey. By contrast, microglial cell number increases in the auditory cortex of the rat (Vaughan and Peters 1974) and in the optic nerve (Sandell and Peters 2002) of the monkey. These findings illustrate that age-related changes in the number of microglial cells have a high degree of regional and species selectivity. Gender differences may also contribute. Thus, in two studies that used the same tissue processing and the same stereological method, it was found that the number of microglial cells did not change with age in the hippocampus of male mice (Long et al. 1998) but increased significantly in the hippocampus of female mice (Mouton et al. 2002). Different environmental conditions under which the animals were raised may also have contributed to the variation in findings.

Microglial cells also undergo morphological changes with aging. Such changes include hypertrophy of the perinuclear cytoplasm and shortening, twisting, and fragmentation of processes (Miller and Streit 2007). Some microglial cells have inclusions even in young animals. However, frequency, size, and heterogeneity of inclusions increase markedly with age (e.g., Vaughan and Peters 1974; Peters and Sethares 2002) suggesting that endocytic activities of microglial cells increase with age. Age-related changes in microglial cells are accompanied by increased expression of a number of antigens (e.g., Perry et al. 1993; Sheffield and Berman 1998; for a review, see Conde and Streit 2006).

In the aging CNS, microglia and astrocytes are involved in the formation of neuritic plaques, also known as senile plaques (for a review, see Wiśniewski and Terry 1973). These plaques have been observed in the cortex and basal ganglia of normal-aged individuals of several mammalian species (e.g., humans, dogs, cats, and monkeys) but have not been found in laboratory rodents. In humans, neuritic plaques are particularly numerous in subjects with senile dementia and other diseases. The plaques consist of a core of extracellular amyloid fibrils surrounded by degenerating neuronal processes (axons and dendrites) and by reactive astrocytes and microglial cells.

X. The Cellular Organization of the CNS

Histological techniques employed to study the CNS at the light microscope level generally reveal either neurons or neuroglial cells, and the selective techniques for revealing neurons often stain only a small proportion of the total number present. This selectivity undoubtedly had the merit of facilitating the study of the cells of the nervous system in the thick sections usually employed in light microscopy, since if all the cells present were stained, the preparation would be so complicated as to render the identification of individual cells impossible. However, for a long time the selectivity of these techniques hindered the understanding of the precise organization of the cells within the CNS and kept alive the idea that the extracellular space of the CNS was very extensive.

The advent of the electron microscope, permitting the simultaneous examination of all parts of cells present in a thin section and having a much greater resolving power than the light microscope, made it possible to demonstrate that in the CNS nerve and neuroglial cells are in close apposition to one another with narrow clefts intervening between the apposed plasma membranes of adjacent cells (Figs. V.20 and X.1). The extracellular space is more extensive at specific sites, for example, at the nodes of Ranvier. Determinations, using various methods, indicate that the extracellular space in the vertebrate CNS forms about 20 % of the total volume (for reviews, see Van Harreveld 1972; Syková 1997). The volume of the extracellular space may change as result of cellular shrinkage or swelling. For example, a sustained neuronal activity results in a reduction of the extracellular space, due mainly to astrocytic swelling. A significant decrease in the volume of the extracellular space has also been detected in old age (Bondareff and Narotzky 1972). This decrease is probably a result of the changes in cellular morphology that occur during aging (e.g., see Sect. VIII.E.4). Because of the complex shapes of nerve and neuroglial cells and their staggered arrangement, the extracellular space of the CNS is geometrically highly irregular. This irregularity and the presence of the extracellular matrix (see next paragraph) may influence the movement of substances through this communication channel.

The pioneer work by Bairati (1953) and Hess (1953) showed that the extracellular space of the CNS is filled with an amorphous matrix that is positive to the periodic-acid-Schiff reaction. This matrix consists of a mixture of laminins, proteoglycans (e.g., chondroitin sulfate proteoglycans and heparan sulfate proteoglycans), hyaluronan, fibronectins, tenascins, and reelin (for a review, see Reichardt and Prokop 2011). The components of the extracellular matrix are produced by both nerve and neuroglial cells. It was once thought that the extracellular matrix had the passive role of simply filling the

E. Pannese, *Neurocytology: Fine Structure of Neurons, Nerve Processes, and Neuroglial Cells*, 2nd Edition,
DOI 10.1007/978-3-319-06856-5_10,

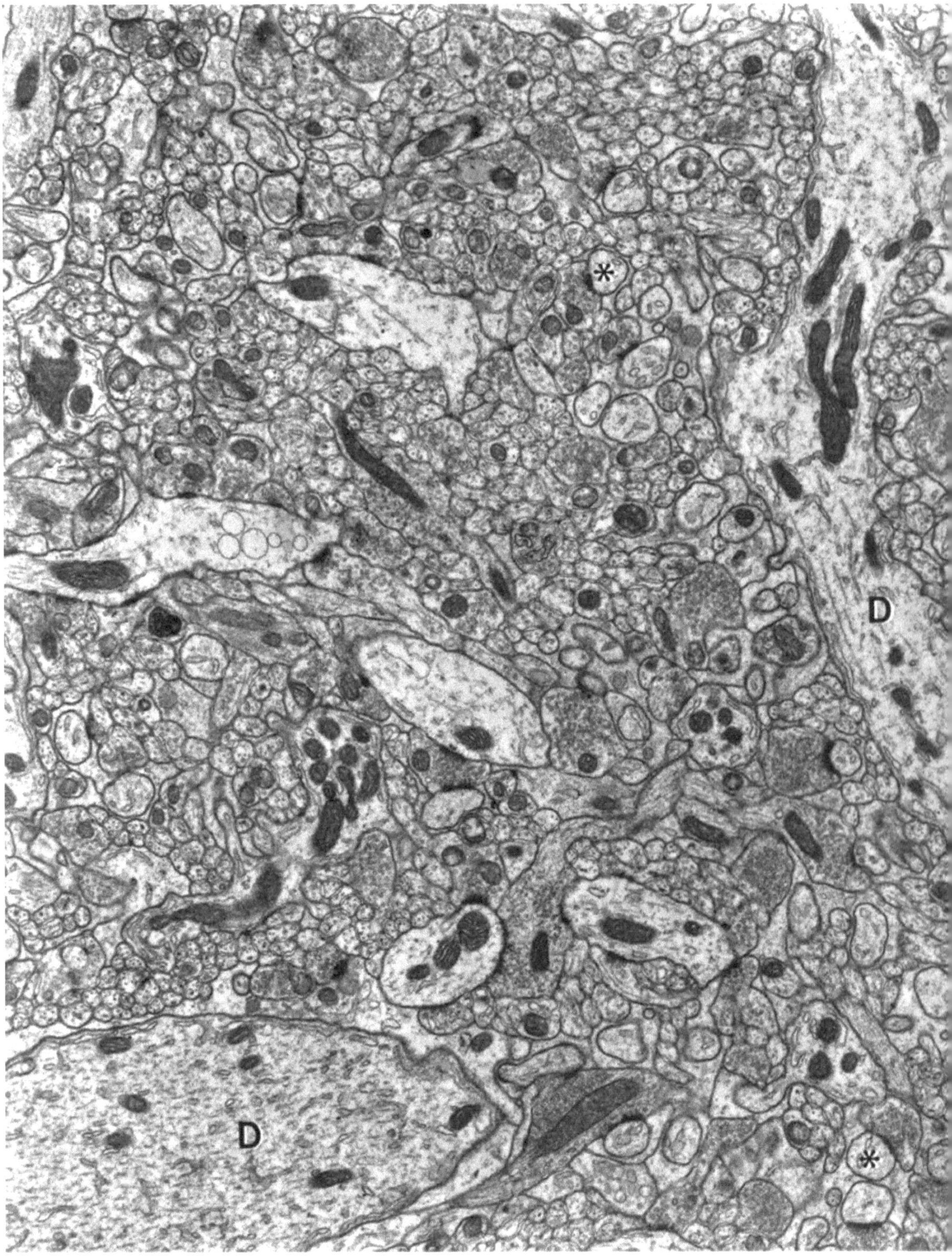

Fig. X.1 Neuropil of the cerebellar cortex. Some dendrites (**D**) of Purkinje cells appear in longitudinal or oblique section, and many axons of granule cells (parallel fibers) appear in transverse section. Numerous synapses (some of which are marked *) between these axons and the spines of Purkinje cell dendrites are

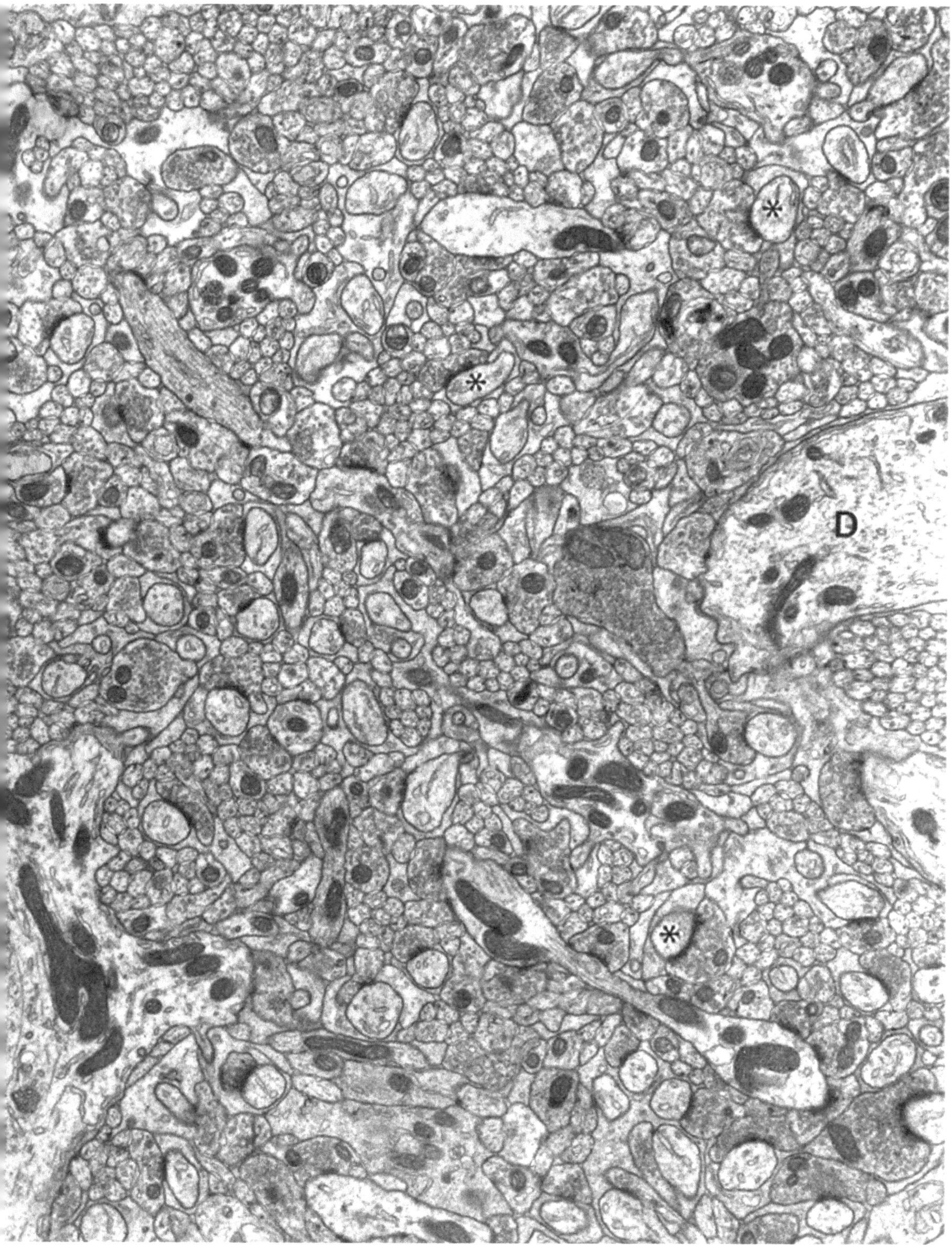

evident. Note that all the cells are in close apposition to each other. Molecular layer of the rat cerebellar cortex, ×15,000

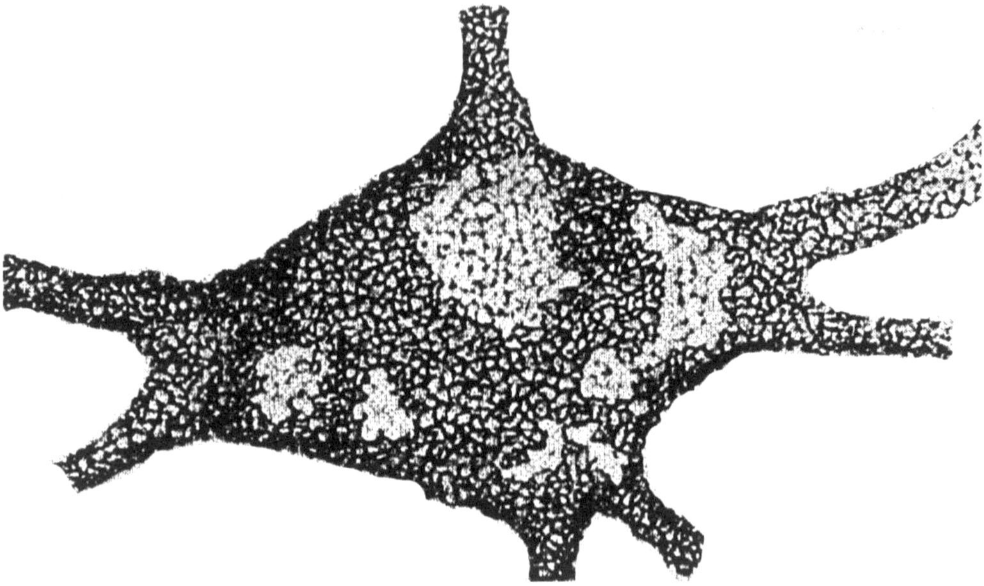

Fig. X.2 Drawing showing a nerve cell with reticular covering (From Golgi 1898a)

clefts between adjacent nerve and glial cells. Subsequently, however, transmembrane receptors that link molecules of the extracellular matrix to the intracellular cytoskeleton were identified. Thus, molecules of the extracellular matrix also perform signaling and regulatory functions in various cellular processes. These transmembrane receptors include integrins, dystroglycan, and syndecans (for a review, see Reichardt and Popkop 2011).

The extracellular matrix forms networks surrounding the cell body and proximal dendritic segments of certain neurons in the CNS of adult amphibians, reptiles, birds, and mammals (Fig. X.2). These networks, first described by Golgi (1893; see also 1898a), are known as perineuronal nets. They consist of accumulations of condensed, lattice-like extracellular matrix components interposed between astrocytic processes and the neuronal surface. Major components of perineuronal nets are hyaluronan, tenascins, and a link protein that interacts with the extracellular matrix molecules and maintains their connections stable. Although various hypotheses have been proposed (for a list, see Celio et al. 1998), the role of perineuronal nets remains enigmatic.

In the CNS, basal laminae are limited to well-defined sites. In general, they are found at the interface between cells of epithelial origin and cells of mesenchymal origin, e.g., at the subpial surface (Figs. V.20 and VIII.11) and surrounding blood vessels (Figs. XI.2, XI.3, and XI.4).

The general organization of the CNS, as revealed by electron microscope studies, is similar to that of the epithelium in that it consists predominantly of cells, with the extracellular space occupying only about 20 % of the total volume. The CNS retains in several aspects, therefore, a similar organization to that of the tissue from which nerve and neuroglial cells take their origin. Upon closer comparison, however, significant differences can be noted between the cellular organization of the CNS and that of the ordinary epithelium. In a given epithelium, uniformity of cell shape is generally the rule, and a morphological and functional polarity exists along an axis perpendicular to the basal lamina upon which the tissue rests. Epithelial cells are mainly concerned with exchange of materials from the free surface toward the basal lamina or vice versa, while exchanges in the lateral direction occur over short distances and to only a modest extent. In the CNS, however, primitive polarization persists in a few places only, for example, in the ependyma and choroidal epithelium. Neurons, which often show extremely complex and varied shapes (see Chaps. III and IV), no longer retain the polarization of the epithelium from which they are derived. Neurons are polarized with respect to one another in as much as they are almost always interconnected by means of specialized junctions (synapses), which allow communication from one cell to the next in a single direction only. They are usually arranged in chains with a highly ordered spatial organization. The cells forming a given chain are strictly interdependent functionally, and it is by means of these cellular chains that each neuron can exert its own functional influence, possibly at a great distance. Furthermore, while most cells in ordinary epithelium are more or less equivalent and interchangeable, each neuron has its own peculiar connections; as a result, the organization of which each neuron partakes is highly specific.

XI. The Blood Vessels of the CNS

A. Arteries

In mammals, each artery which penetrates the CNS (Fig. XI.1) is surrounded by a perivascular space, also known as the Virchow-Robin space. This space is lined by two basal laminae: the inner one attached to the tunica adventitia of the vessel and the outer one upon which astrocytic processes are applied. An extension of the pia mater accompanies the vessel into the CNS for a short distance, occupying the perivascular space. Deeper into the CNS, the vessel diminishes in caliber, its muscle layer becomes thinner, the pial sheath becomes discontinuous and finally disappears, and the perivascular space becomes narrower. The perivascular space persists until the precapillary arterioles. At the level of the capillaries, both basal laminae (that of the blood vessel and that of the marginal neuroglia) fuse so that the perivascular space is obliterated (Fig. XI.1). Note that, contrary to what has been thought for many years, the perivascular space does not communicate with the subarachnoid space (Krahn 1982; Krisch et al. 1984).

B. Capillaries and the Blood-Brain Barrier

Capillaries are formed essentially of an endothelial tube which rests on a basal lamina (Figs. XI.2, XI.3, and XI.4). As a rule, in most vertebrates, they are continuous capillaries because they have an uninterrupted (i.e., nonfenestrated) endothelium (Maynard et al. 1957), consisting of thin cells which appear thickened only in the nuclear regions (Fig. XI.2). The cytoplasmic matrix of these cells is rather dense; a small Golgi apparatus (Fig. XI.2) and most of the mitochondria are located near the nucleus, but other mitochondria may be occasionally found in the thin region of the cell. Short cisternae of the granular endoplasmic reticulum, free polysomes, microtubules, and filaments are also present. These endothelial cells, unlike those found elsewhere in the vascular system, show few pinocytotic vesicles (Fig. XI.3).

Adjacent endothelial cells in CNS capillaries are joined by extensive tight junctions which seal the intercellular clefts (Muir and Peters 1962). These junctions form a continuous belt (zonula occludens) around each endothelial cell. In freeze-fractured preparations of capillaries of the CNS (Fig. XI.3), the junction between endothelial cells appears as a network of anastomosing thin ridges on the P face and a corresponding pattern of grooves on the E face (Connell and Mercer 1974). Many of the proteins comprising these junctions, such as zonula occludens 1, 2, 3, claudins, and occludin, have been identified.

The endothelium is surrounded by a basal lamina which contains laminin, fibronectin, heparan sulfate proteoglycan, and collagen

E. Pannese, *Neurocytology: Fine Structure of Neurons, Nerve Processes, and Neuroglial Cells*, 2nd Edition, DOI 10.1007/978-3-319-06856-5_11,

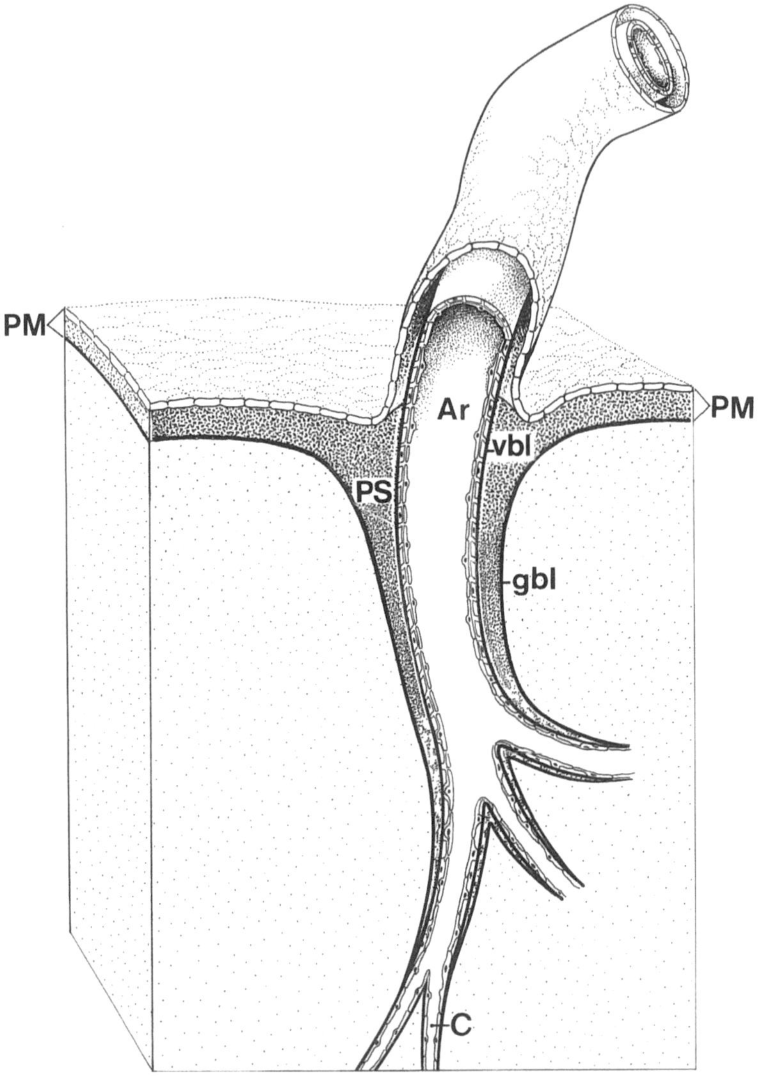

Fig. XI.1 Blood vessels of the CNS. An artery (**Ar**) penetrates the CNS and gives rise to a number of capillaries (**C**). External to the nervous system, the artery is coated with a sheath of cells derived from the pia mater (**PM**). A perivascular space (**PS**) surrounds the artery. Within the CNS, this space is lined by two basal laminae (the inner one, **vbl**, attached to the tunica adventitia of the vessel and the outer one, **gbl**, to which astrocytic processes are applied) and is occupied by an extension of the pia mater. Deeper in the CNS, the perivascular space becomes very narrow, and the pial sheath becomes discontinuous. Deeper still, the inner and outer basal laminae fuse so that the perivascular space is obliterated and the vessel becomes a capillary

type IV. Wherever pericytes (see Sect. XI.D.1) occur, the basal lamina splits to enclose them. CNS capillaries are completely surrounded by an astrocytic sheath external to the basal lamina (Figs. XI.2, XI.3, and XI.4). Astrocytes are attached to basal lamina components through dystroglycan, which in turn is connected to the intracellular cytoskeleton. The astrocytic plasma membrane apposed to the pericapillary basal lamina is highly enriched in orthogonally arranged particles composed of aquaporin-4 (see Sect. VIII.E.1).

Toward the end of the nineteenth century, it was demonstrated that a dye injected into the

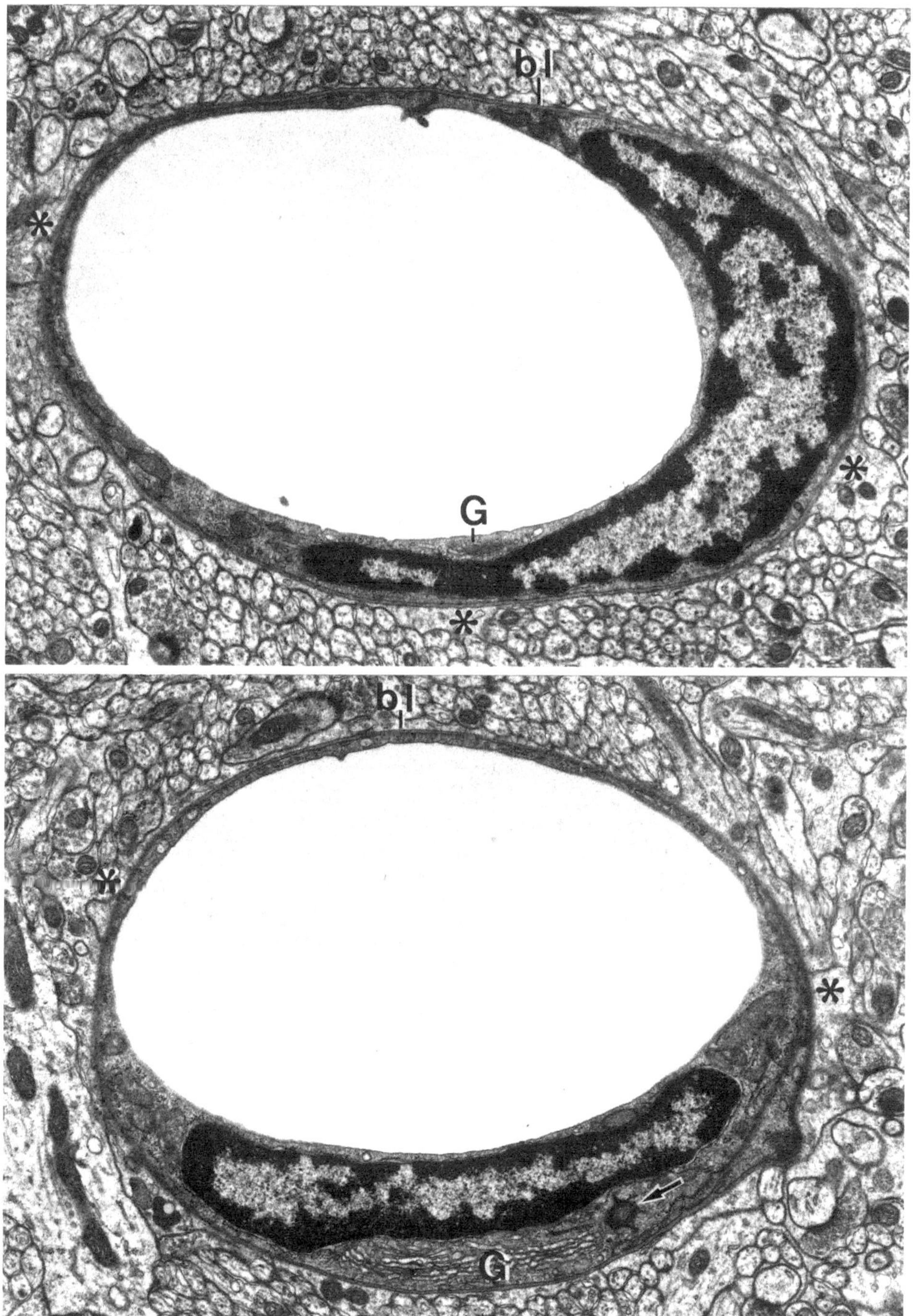

Fig. XI.2 Capillaries of the CNS. These microvessels, which in this picture appear in transverse section, consist of an endothelial tube which rests on a basal lamina (**bl**). The latter is completely surrounded by an astrocytic sheath (*). The endothelial cells are highly attenuated but are thickened in the nuclear regions. Mitochondria, short cisternae of the granular endoplasmic reticulum, Golgi complexes (**G**), and a centriole (*arrow*) can be seen in the endothelial cell cytoplasm, which has a dense appearance. Rat cerebellar cortex, ×13,500

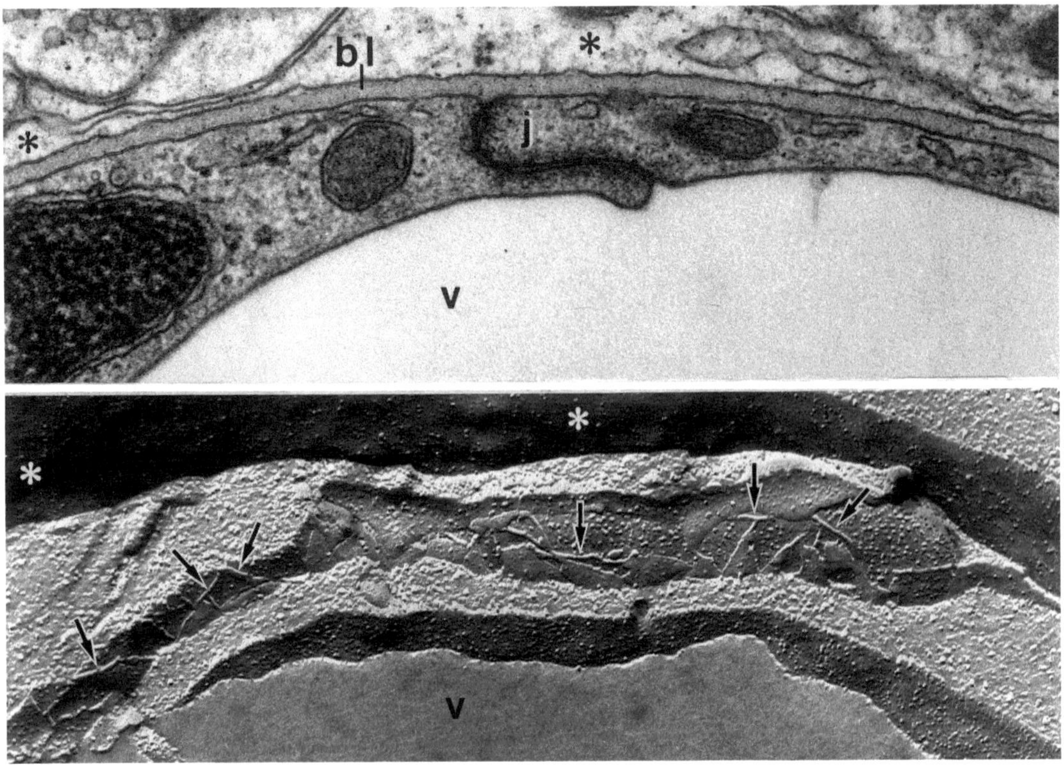

Fig. XI.3 Tight junctions between endothelial cells in the wall of CNS capillaries. In the *upper picture* (thin section), the junction is marked **j**. The endothelial cells, whose cytoplasm has a dense appearance, rest on a basal lamina (**bl**), external to which there is an astrocytic sheath (*). Note that these endothelial cells do not show pinocytotic vesicles. **v** vessel lumen. Rat cerebellar cortex, ×60,000. In the *lower picture* (freeze-fractured preparation), one can see that the junction is composed of interconnected strands (some *arrowed*) marking the transition between the E and P faces of the adjoining endothelial cell plasma membranes. **v** vessel lumen, * P face of an astrocyte plasma membrane. Chick optic tectum, ×60,000 (*Lower picture* courtesy of B. Nico, D. Cantino, and L. Roncali)

vascular system stained most organs but not the CNS (Ehrlich 1885), with the exception of certain restricted areas (for reviews, see Bradbury 1979; Van Deurs 1980). It was later observed that many substances other than dyes enter the CNS very slowly and in small amounts. These observations led to the concept of a barrier between blood and CNS. This barrier is believed to protect the neurons of the CNS from blood-borne agents and to contribute to maintaining homeostasis in the CNS microenvironment. Light microscope investigations carried out by numerous authors failed to establish whether the barrier lies at the level of the pericapillary astrocytic sheath or at the level of the capillary endothelium (for a review, see Bradbury 1979). Experiments with intravascularly injected dense tracers that can be localized with the electron microscope have solved this problem.

The CNS of elasmobranchs has capillaries with a leaky fenestrated endothelium which is readily crossed by macromolecular tracers (Brightman et al. 1971; Hashimoto 1972). In these animals, the barrier is located at the pericapillary neuroglial sheath, which is formed of astrocytic processes linked by tight junctions (Bundgaard and Cserr 1981). These junctions appear to constitute a continuous belt around each

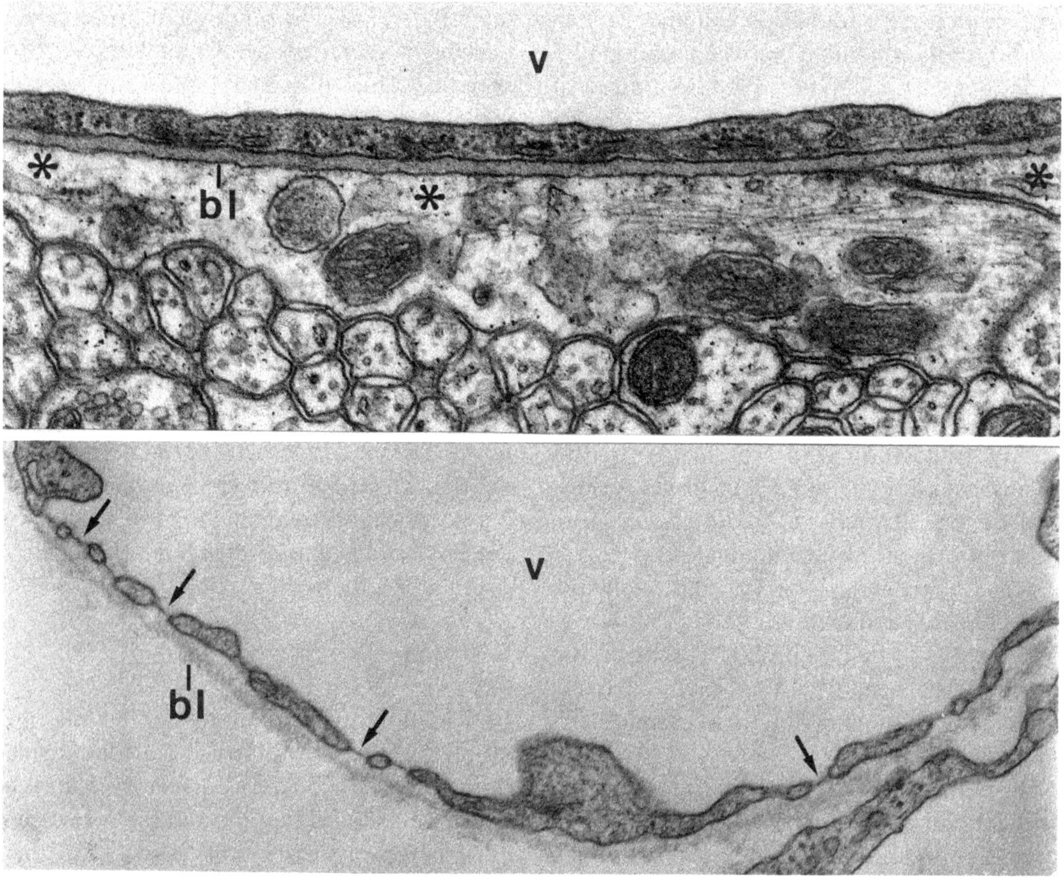

Fig. XI.4 Capillaries of the CNS. The wall of a continuous capillary (*upper picture*) and the wall of a fenestrated capillary (*lower picture*) are shown at the same magnification for comparison. Some fenestrations closed by very thin diaphragms are *arrowed*. **bl** basal lamina, **v** vessel lumen, * astrocytic sheath. *Upper picture*, rat cerebellar cortex; *lower picture*, choroid plexus of the rat fourth ventricle; ×50,000 (*Lower picture* courtesy of S. Matsuda and M. Ledda)

astrocytic process (Gotow and Hashimoto 1984).

The situation is quite different in most other vertebrates. In species representing cyclostomes, teleosts, amphibians, reptiles, birds, and mammals, dense tracers such as horseradish peroxidase and lanthanum, when injected into the blood, remain confined within the capillary lumen (Reese and Karnovsky 1967; Brightman and Reese 1969; Bundgaard 1982). The barrier for these substances is therefore located within the capillary endothelium and not in pericapillary structures. The tight junctions which seal the spaces between adjacent endothelial cells (Fig. XI.3) inhibit the free diffusion of molecules along the intercellular clefts, so that substances may reach the brain tissue only by moving through the endothelial cells. These cells have an extremely low pinocytotic activity as well as a number of active transport systems which ensure the passage of certain substances and prevent the passage of others between blood and brain tissue. Therefore, endothelial cells control what substances enter the CNS, ensuring neuronal protection. In non-elasmobranch vertebrates, the astrocytes ensheathing capillaries do not play a direct role in the blood-brain barrier but have an indirect role in that they influence the formation and maintenance of the tight junctions between endothelial cells (Stewart and Wiley 1981; Arthur et al. 1987; Tao-Cheng et al. 1987).

It is generally accepted that the blood-brain barrier is more permeable in fetal than in adult brains. The developmental tightening of the barrier seems to correlate with changes in the interendothelial tight junctions (Stewart and Hayakawa 1987). An increased permeability of CNS capillaries with leakage of plasma proteins has been recorded in certain experimental situations and in association with a number of pathological conditions (e.g., injection or local application of hypertonic solutions, inhalation of carbon dioxide, mechanical head injury, repeated seizures, hypoxia and ischemia, hypertension, acute inflammatory conditions, and metabolic disturbances). This increased capillary permeability has been attributed mainly to the opening of tight interendothelial junctions.

In certain discrete regions of the CNS, the capillary endothelium is permeable to dyes and proteins. These regions lacking the blood-brain barrier are the following: the area postrema, pineal gland, neurohypophysis, median eminence, subfornical organ, preoptic recess, and choroid plexuses (for a review, see Bradbury 1979). The capillaries of these specialized regions, unlike those found elsewhere in the CNS, are of the fenestrated type (Hashimoto 1966; Rivera-Pomar 1966) and also have endothelial cells showing numerous pinocytotic vesicles, so that diffusion of molecules between the blood and these regions occurs readily. The fenestrations are approximately circular in shape, about 80 nm in diameter, and appear closed by a very thin diaphragm (Fig. XI.4). These capillaries are surrounded by a large perivascular space containing collagen fibrils and sometimes also fibroblasts. This space is lined by two basal laminae: the inner one covers the capillary endothelium, while the outer is contiguous with the astrocytic sheath. In some regions (e.g., the neurohypophysis and median eminence), the lack of a barrier facilitates the release of neurosecretory products into the bloodstream, while in others (e.g., the area postrema) this would allow certain substances to reach the sensitive cells.

Dyes and macromolecular tracers injected into the vascular system remain confined within the lumen of endoneurial capillaries in peripheral nerves but leak out of epineurial and perineurial capillaries. These observations demonstrate the existence of a blood-nerve barrier which lies at the level of the endothelium of the endoneurial capillaries (Waksman 1961; Olsson 1966; Olsson and Reese 1971). Differences in the permeability of the blood-nerve barrier seem to exist between species and in different regions of the PNS of a single species (Olsson 1984).

C. Veins

In the CNS, it is difficult to identify the transition between a capillary and a venule since both appear to consist of an endothelial tube resting on a basal lamina, and the sole difference between them seems to be the lumen caliber (Frederickson and Low 1969). In progressing toward the surface of the CNS, the vein, which is usually devoid of typical smooth muscle cells, becomes surrounded by a gradually wider perivascular space. Collagen fibers, occasional isolated cells, but no complete pial sheath can be seen within this space.

D. Cells Associated with Microvessels

1. Pericytes

These cells were first described by Rouget (1873) and for many years were referred to as Rouget cells. Later Zimmermann (1923) renamed them pericytes. They are associated with arterioles, capillaries, and venules of almost all organs but are particularly numerous on microvessels of the CNS. Pericytes are thought to be of mesodermal origin and migrate into the CNS during vascularization.

Pericytes, which are ensheathed by the basal lamina of the microvessel, have small oval or elongated cell bodies and several branching processes. They have a rather dense cytoplasmic matrix and a nucleus displaying some clumps of chromatin beneath the nuclear envelope. A Golgi apparatus, mitochondria, some cisternae of the granular endoplasmic reticulum, free polysomes, and dense bodies are found within the cytoplasm. Microfilaments have been detected both in the cell body and within processes. Pericytes contain contractile proteins (muscle and nonmuscle actins, and nonmuscle myosin; e.g., see Herman and D'Amore 1985; Bandopadhyay et al. 2001). The contraction of capillary pericytes was directly observed in vivo by two-photon laser scanning microscopy (Fernández-Klett et al. 2010). Pericytes are usually identified by their location and by the expression of α-smooth muscle actin.

Several hypotheses have been formulated regarding the function of pericytes. Since they may modulate the capillary diameter, it has been suggested that they are involved in the process linking neuronal activity to local regulation of blood flow. However, under normal conditions, it seems that astrocytes exert this function of neurovascular coupling (see Sect. VIII.E.5c). Furthermore, the regulation of the local blood flow seems to occur at the level of precapillary arterioles rather than capillaries (Fernández-Klett et al. 2010). Other functions attributed to pericytes are phagocytic activity and maintenance of the blood-brain barrier (e.g., see Thomas 1999). Finally, pericytes have been suggested to be pluripotential and serve as precursors for several cell types (neurons and neuroglia in the CNS).

2. Mast Cells

Mast cells, first described by Ehrlich (1879), occur in the healthy CNS of many species of amphibians, birds, and mammals where they are often associated with intraparenchymal arterioles and venules (for a review, see Dropp 1972). These cells are also located in the meninges and choroid plexuses. In the CNS parenchyma, they are surrounded by normal elements of the neuropil. Mast cells derive from pluripotential stem cells of the bone marrow and penetrate the CNS from the circulation. Their number varies markedly in the CNS of different species and individuals of the same species.

In light microscope preparations, mast cells are usually spherical or oval with a small centrally located nucleus. They are easily identified by the abundance of metachromatic granules within their cytoplasm. In electron microscope preparations, dense clumps of chromatin occur mainly beneath the nuclear envelope. Membrane-bounded spherical granules with a finely granular or heterogeneous content fill the cytoplasm. Following activation, granules can be discharged whole or release their contents out of the cell. Mast cells store a wide variety of biologically active molecules, including histamine, heparin, serotonin, proteases, vasoactive intestinal polypeptide, trophic factors, and nitric oxide (for a review, see Silver et al. 1996).

The precise functional role of mast cells in the healthy CNS is unclear. However, because of their close association with blood vessels and the powerful effects of some of the molecules they store (e.g., histamine and serotonin) on muscle cells of blood vessels, mast cells of the CNS are considered to be implicated in the control of the local blood flow and the vascular permeability (e.g., see Gross et al. 1981; Zhuang et al. 1996). Furthermore, mast cells participate in inflammatory processes in various CNS pathologies.

E. Age-Related Changes

Data on the alterations of microvessels that occur during the normal CNS aging are controversial. Generally accepted alterations are here listed (for a review, see Rascher and

Wolburg 2001). During the normal aging, the density of capillaries decreases, the subendothelial basal lamina is thickened, collagen fibrils are deposited in the vascular wall, lipofuscin-like and vesicular inclusions become more common in pericytes, and lipofuscin-like inclusions accumulate in perivascular neuroglial processes. Blood flow is significantly reduced in several regions of the CNS. The permeability of the blood-brain barrier does not appear to be altered in the intact aged CNS, whereas following injury the permeability increases much more in the old CNS than in the young CNS.

References

Achúcarro N. De l'évolution de la névroglie, et spécialement de ses relations avec l'appareil vasculaire. Trabajos del Laboratorio de Investigaciones biológicas de la Universidad de Madrid **13**, 169–212 (1915)

Adams I., Jones D.G. Quantitative ultrastructural changes in rat cortical synapses during early-, mid- and late-adulthood. Brain Research **239**, 349–363 (1982)

Adinolfi A.M., Yamuy J., Morales F.R., Chase M.H. Segmental demyelination in peripheral nerves of old cats. Neurobiology of Aging **12**, 175–179 (1991)

Adrian E.K.Jr., Walker B.E. Incorporation of thymidine-^{3}H by cells in normal and injured mouse spinal cord. Journal of Neuropathology and Experimental Neurology **21**, 597–609 (1962)

Agduhr E. Über ein zentrales Sinnesorgan (?) bei den Vertebraten. Zeitschrift für Anatomie und Entwicklungsgeschichte **66**, 223–360 (1922)

Agnati L.F., Zoli M., Strömberg I., Fuxe K. Intercellular communication in the brain: wiring versus volume transmission. Neuroscience **69**, 711–726 (1995)

Aguayo A.J., Bray G.M., Perkins C.S., Duncan I.D. Axon-sheath cell interactions in peripheral and central nervous system transplants. In: Aspects of Developmental Neurobiology (J.A. Ferrendelli Ed.) pp. 361–383. Society for Neuroscience, Bethesda 1979

Aguayo A.J., Charron L., Bray G.M. Potential of Schwann cells from unmyelinated nerves to produce myelin: a quantitative ultrastructural and radiographic study. Journal of Neurocytology **5**, 565–573 (1976)

Aguayo A.J., Martin J.B., Bray G.M. Effects of nerve growth factor antiserum on peripheral unmyelinated nerve fibers. Acta Neuropathologica **20**, 288–298 (1972)

Ahmad A., Spear P.D. Effects of aging on the size, density, and number of rhesus monkey lateral geniculate neurons. Journal of Comparative Neurology **334**, 631–643 (1993)

Ajami B., Bennett J.L., Krieger C., McNagny K.M., Rossi F.M.V. Infiltrating monocytes trigger EAE progression, but do not contribute to the resident microglia pool. Nature Neuroscience **14**, 1142–1149 (2011)

Ajima H., Kawano Y., Takagi R., Aita M., Gomi H., Byers M.R., Maeda T. The exact expression of glial fibrillary acidic protein (GFAP) in trigeminal ganglion and dental pulp. Archives of Histology and Cytology **64**, 503–511 (2001)

Akert K., Pfenninger K. Synaptic fine structure and neural dynamics. In: Cellular Dynamics of the Neuron (S.H. Barondes Ed.) pp. 245–260. Academic Press, New York 1969

Akert K., Pfenninger K., Sandri C., Moor H. Freeze etching and cytochemistry of vesicles and membrane complexes in synapses of the central nervous system. In: Structure and Function of Synapses (G.D. Pappas, D.P. Purpura Eds.) pp. 67–86. Raven Press, New York 1972

Albe-Fessard D., Martins-Ferreira H. Role de la commande nerveuse dans la synchronisation du fonctionnement des éléments de l'organe électrique du Gymnote, *Electrophorus electricus* L. Journal de Physiologie **45**, 533–546 (1953)

Aldskogius H. Fine structural changes in nerve cell bodies of the adult rabbit dorsal motor vagal nucleus during axon reaction. Neuropathology and Applied Neurobiology **4**, 323–341 (1978)

Aldskogius H., Risling M. Number of dorsal root ganglion neurons and axons in cats of different ages. Experimental Neurology **106**, 70–73 (1989)

Allen D.J. Scanning electron microscopy of epiplexus macrophages (Kolmer cells) in the dog. Journal of Comparative Neurology **161**, 197–214 (1975)

Allen D.J., Low F.N. The ependymal surface of the lateral ventricle of the dog as revealed by scanning electron microscopy. American Journal of Anatomy **137**, 483–489 (1973)

Allen D.M., Hamilton K.A. Ultrastructural identification of synapses between mitral/tufted cell dendrites. Brain Research **860**, 170–173 (2000)

Allen E. The cessation of mitosis in the central nervous system of the albino rat. Journal of Comparative Neurology **22**, 547–568 (1912)

Allen R.A. Isolated cilia in inner retinal neurons and in retinal pigment epithelium. Journal of Ultrastructure Research **12**,730–747 (1965)

Alley K.E. Morphogenesis of the trigeminal mesencephalic nucleus in the hamster: cytogenesis and neurone

DOI 10.1007/978-3-319-06856-5,

death. Journal of Embryology and Experimental Morphology **31**, 99–121 (1974)

Aloisi F. Immune function of microglia. Glia **36**, 165–179 (2001)

Altman J. Proliferation and migration of undifferentiated precursors cells in the rat during postnatal gliogenesis. Experimental Neurology **16**, 263–278 (1966)

Altman J., Anderson W.J. Experimental reorganization of the cerebellar cortex. I. Morphological effects of elimination of all microneurons with prolonged x-irradiation started at birth. Journal of Comparative Neurology **146**, 355–406 (1972)

Altman J., Das G.D. Autoradiographic and histological evidence of postnatal hippocampal neurogenesis in rats. Journal of Comparative Neurology **124**, 319–335 (1965)

Alvarez J., Giuditta A., Koenig E. Protein synthesis in axons and terminals: significance for maintenance, plasticity and regulation of phenotype. With a critique of slow transport theory. Progress in Neurobiology **62**, 1–62 (2000)

Alvarez M.P., Solas M.T., Suarez I., Fernandez B. Glial fibrillary acidic protein-like immunoreactivity in cat satellite cells of sympathetic ganglia. Acta Anatomica **136**, 9–11 (1989)

Alvarez V.A., Sabatini B.L. Anatomical and physiological plasticity of dendritic spines. Annual Review of Neuroscience **30**, 79–97 (2007)

Alvarez-Buylla A., Buskirk D.R., Nottebohm F. Monoclonal antibody reveals radial glia in adult avian brain. Journal of Comparative Neurology **264**, 159–170 (1987)

Alvarez-Buylla A., García-Verdugo J.M. Neurogenesis in adult subventricular zone. Journal of Neuroscience **22**, 629–634 (2002)

Anders J.J., Brightman M.W. Assemblies of particles in the cell membranes of developing, mature and reactive astrocytes. Journal of Neurocytology **8**, 777–795 (1979)

Anderson L.E., McClure W.O. Differential transport of protein in axons: comparison between the sciatic nerve and dorsal columns of cats. Proceedings of the National Academy of Sciences of the U.S.A. **70**, 1521–1525 (1973)

Andersson E., Edström A., Jarlstedt J. Properties of RNA from giant axons of the crayfish. Acta Physiologica Scandinavica **78**, 491–502 (1970)

Andersson-Cedergren E. Ultrastructure of motor end plate and sarcoplasmic components of mouse skeletal muscle fiber as revealed by three-dimensional reconstructions from serial sections. Journal of Ultrastructure Research Suppl. 1, 1–191 (1959)

Andres K.H. Der Feinbau des Bulbus olfactorius der Ratte unter besonderer Berücksichtigung der synaptischen Verbindungen. Zeitschrift für Zellforschung und mikroskopische Anatomie **65**, 530–561 (1965a)

Andres K.H. Über die Feinstruktur besonderer Einrichtungen in markhaltigen Nervenfasern des Kleinhirns der Ratte. Zeitschrift für Zellforschung und mikroskopische Anatomie **65**, 701–712 (1965b)

Andres K.H. Anatomy and ultrastructure of the olfactory bulb in fish, amphibia, reptiles, birds and mammals. In: Taste and Smell in Vertebrates (G.E.W. Wolstenholme, J. Knight Eds.) pp. 177–196. J.& A. Churchill, London 1970

Andrews T.J., Li D., Halliwell J., Cowen T. The effect of age on dendrites in the rat superior cervical ganglion. Journal of Anatomy **184**, 111–117 (1994)

Angelides K.J., Smith K.E., Takeda M. Assembly and exchange of intermediate filament proteins of neurons: neurofilaments are dynamic structures. Journal of Cell Biology **108**, 1495–1506 (1989)

Apáthy S. Das leitende Element des Nervensystems und seine topographischen Beziehungen zu den Zellen. Mittheilungen aus der Zoologischen Station zu Neapel **12**, 495–748 (1897)

Apolant H. Ueber die sympathischen Ganglienzellen der Nager. Archiv für mikroskopische Anatomie und Entwicklungsgeschichte **47**, 461–471 (1896)

Appelt D.M., Balin B.J. Analysis of paired helical filaments (PHFs) found in Alzheimer's disease using freeze-drying/rotary shadowing. Journal of Structural Biology **111**, 85–95 (1993)

Appenzeller O. Electron microscopic study of the innervation of the auricular artery in the rat. Journal of Anatomy **98**, 87–91 (1964)

Armstrong J., Young J.Z. End-feet in the cerebral cortex. Journal of Physiology **137**, 10P–11P (1957)

Arndt R. Untersuchungen über die Ganglienkörper der Spinalganglien. Archiv für mikroskopische Anatomie **11**, 140–168 (1875)

Arnold A.P. Gonadal steroid-induced organization and reorganization of neural circuits involved in bird song. In: Synaptic Plasticity (C.W. Cotman Ed.) pp. 263–285. Guilford Press, New York 1985

Arnold W. Über eigentümliche neuronale Zellelemente im Ependym des Zentralkanals von *Salamandra maculosa*. Zeitschrift für Zellforschung und mikroskopische Anatomie **105**, 176–187 (1970)

Arthur F.E., Shivers R.R., Bowman P.D. Astrocyte-mediated induction of tight junctions in brain capillary endothelium: an efficient in vitro model. Developmental Brain Research **36**, 155–159 (1987)

Arvanitaki A. Effects evoked in an axon by the activity of a contiguous one. Journal of Neurophysiology **5**, 89–108 (1942)

Arvidson B. Distribution of intravenously injected protein tracers in peripheral ganglia of adult mice. Experimental Neurology **63**, 388–410 (1979)

Asbury A.K. Schwann cell proliferation in developing mouse sciatic nerve. A radioautographic study. Journal of Cell Biology **34**, 735–743 (1967)

Aston-Jones G., Rogers J., Shaver R.D., Dinan T.G., Moss D.E. Age-impaired impulse flow from nucleus basalis to cortex. Nature **318**, 462–464 (1985)

Atwood H.L., Kwan I. Reciprocal axo-axonal synapses between excitatory and inhibitory neurons in crustaceans. Brain Research **174**, 324–328 (1979)

Atwood H.L., Morin W.A. Neuromuscular and axoaxonal synapses of the crayfish opener muscle. Journal of Ultrastructure Research **32**, 351–369 (1970)

Autilio-Gambetti L., Crane R., Gambetti P. Binding of Bodian's silver and monoclonal antibodies to defined regions of human neurofilament subunits: Bodian's silver reacts with a highly charged unique domain of neurofilaments. Journal of Neurochemistry **46**, 366–370 (1986)

Autilio-Gambetti L., Sipple J., Sudilovsky O., Gambetti P. Intermediate filaments of Schwann cells. Journal of Neurochemistry **38**, 774–780 (1982)

Baas P.W., Deitch J.S., Black M.M., Banker G.A. Polarity orientation of microtubules in hippocampal neurons: uniformity in the axon and nonuniformity in the dendrite. Proceedings of the National Academy of Sciences of the U.S.A. **85**, 8335–8339 (1988)

Bacci A., Huguenard J.R., Prince D.A. Functional autaptic neurotransmission in fast-spiking interneurons: A novel form of feed-back inhibition in the neocortex. Journal of Neuroscience **23**, 859–866 (2003)

Bairati A. Spreading factor and mucopolysaccharides in the central nervous system of vertebrates. Experientia **9**, 461 (1953)

Baird D.H., Hatten M.E., Mason C.A. Cerebellar target neurons provide a stop signal for afferent neurite extension *in vitro*. Journal of Neuroscience **12**, 619–634 (1992)

Baker P.F. A method for the location of extracellular space in crab nerve. Journal of Physiology **180**, 439–447 (1965)

Baker P.F., Hodgkin A.L., Shaw T.I. Replacement of the axoplasm of giant nerve fibres with artificial solutions. Journal of Physiology **164**, 330–354 (1962)

Baker R., Llinás R. Electrotonic coupling between neurones in the rat mesencephalic nucleus. Journal of Physiology **212**, 45–63 (1971)

Balice-Gordon R.J., Bone L.J., Scherer S.S. Functional gap junctions in the Schwann cell myelin sheath. Journal of Cell Biology **142**, 1095–1104 (1998)

Bandopadhyay R., Orte C., Lawrenson J.G., Reid A.R., De Silva S., Allt G. Contractile proteins in pericytes at the blood–brain and blood-retinal barriers. Journal of Neurocytology **30**, 35–44 (2001)

Banker G.A. Trophic interactions between astroglial cells and hippocampal neurons in culture. Science **209**, 809–810 (1980)

Banker G.A., Cowan W.M. Further observations on hippocampal neurons in dispersed cell culture. Journal of Comparative Neurology **187**, 469–494 (1979)

Bannister N.J., Larkman A.U. Dendritic morphology of CA1 pyramidal neurones from the rat hippocampus: II. Spine distributions. Journal of Comparative Neurology **360**, 161–171 (1995)

Barañano D.E., Ferris C.D., Snyder S.H. Atypical neural messengers. Trends in Neurosciences **24**, 99–106 (2001)

Barasa A. Forma, grandezza e densità dei neuroni della corteccia cerebrale in mammiferi di grandezza corporea differente. Zeitschrift für Zellforschung und mikroskopische Anatomie **53**, 69–89 (1960)

Barbe M.-P., Taxi J. The innervation of the myometrium of the cynomolgus monkey (*Macaca fascicularis*). A quantitative electron-microscopic study. Cell and Tissue Research **246**, 575–581 (1986)

Barber P.C., Lindsay R.M. Schwann cells of the olfactory nerves contain glial fibrillary acidic protein and resemble astrocytes. Neuroscience **7**, 3077–3090 (1982)

Bargmann W., Katritsis E. Über die sog. Filamente und das Pigment im Plexus chorioideus des Menschen. Zeitschrift für Zellforschung und mikroskopische Anatomie **75**, 366–370 (1966)

Bargmann W., Lindner E. Über den Feinbau des Nebennerenmarkes des Igels (*Erinaceus europaeus* L.). Zeitschrift für Zellforschung und mikroskopische Anatomie **64**, 868–912 (1964)

Barr M.L. Some observations on the morphology of the synapse in the cat's spinal cord. Journal of Anatomy **74**, 1–11 (1939)

Barr M.L., Bertram L.F., Lindsay H.A. The morphology of the nerve cell nucleus, according to sex. Anatomical Record **107**, 283–297(1950)

Barron K.D., Doolin P.F., Oldershaw J.B. Ultrastructural observations on retrograde atrophy of lateral geniculate body. Journal of Neuropathology and Experimental Neurology **26**, 300–326 (1967)

Barron K.D., Means E.D., Feng T., Harris H. Ultrastructure of retrograde degeneration in thalamus of rat. 2. Changes in vascular elements and transvascular migration of leukocytes. Experimental and Molecular Pathology **20**, 344–362 (1974)

Barron K.D., Sklar S. Response of lysosomes of bulbospinal motoneurons to axon section. Neurology **11**, 866–875 (1961)

Baumgarten H.G., Falck B., Wartenberg H. Adrenergic neurons in the spinal cord of the pike (*Esox lucius*) and their relation to the caudal neurosecretory system. Zeitschrift für Zellforschung und mikroskopische Anatomie **107**, 479–498 (1970)

Baumgarten H.G., Holstein A.F., Rosengren E. Arrangement, ultrastructure, and adrenergic innervation of smooth musculature of the ductuli efferentes, ductus epididymidis and ductus deferens of man. Zeitschrift für Zellforschung und mikroskopische Anatomie **120**, 37–79 (1971)

Baumgold J., Terakawa S., Iwasa K., Gainer H. Membrane-associated cytoskeletal proteins in squid giant axons. Journal of Neurochemistry **36**, 759–764 (1981)

Beal J.A., Cooper M.H. Myelinated nerve cell bodies in the dorsal horn of the monkey (*Saimiri sciureus*). American Journal of Anatomy **147**, 33–48 (1976)

Beal M.F., Walker L.C., Storey E., Segar L., Price D.L., Cork L.C. Neurotransmitters in neocortex of aged rhesus monkeys. Neurobiology of Aging **12**, 407–412 (1991)

Becker N.H., Goldfischer S., Shin W.-Y., Novikoff A.B. The localization of enzyme activities in the rat brain. Journal of Biophysical and Biochemical Cytology **8**, 649–663 (1960)

Bellen H.J., Lu Y., Beckstead R., Bhat M.A. Neurexin IV, caspr and paranodin – novel members of the neurexin family: encounters of axons and glia. Trends in Neurosciences **21**, 444–449 (1998)

Benavides-Piccione R., Ballesteros-Yáñez I., DeFelipe J., Yuste R. Cortical area and species differences in dendritic spine morphology. Journal of Neurocytology **31**, 337–346 (2002)

Benjamins J.A., Smith M.E. Metabolism of myelin. In: Myelin (P. Morell Ed.) 2nd ed., pp. 225–258. Plenum Press, New York 1984

Bennett M.R., Rogers D.C. A study of the innervation of the taenia coli. Journal of Cell Biology **33**, 573–596 (1967)

Bennett M.V.L., Nakajima Y., Pappas G.D. Physiology and ultrastructure of electrotonic junctions. I. Supramedullary neurons. Journal of Neurophysiology **30**, 161–179 (1967a)

Bennett M.V.L., Pappas G.D., Aljure E., Nakajima Y. Physiology and ultrastructure of electrotonic junctions. II. Spinal and medullary electromotor nuclei in Mormyrid fish. Journal of Neurophysiology **30**, 180–208 (1967b)

Bennett M.V.L., Nakajima Y., Pappas G.D. Physiology and ultrastructure of electrotonic junctions. III. Giant electromotor neurons of *Malapterurus electricus*. Journal of Neurophysiology **30**, 209–235 (1967c)

Bennett M.V.L., Pappas G.D., Giménez M., Nakajima Y. Physiology and ultrastructure of electrotonic junctions. IV. Medullary electromotor nuclei in Gymnotid fish. Journal of Neurophysiology **30**, 236–300 (1967d)

Bennett M.V.L., Sandri C., Akert K. Neuronal gap junctions and morphologically mixed synapses in the spinal cord of a teleost, *Sternarchus albifrons* (Gymnotoidei). Brain Research **143**, 43–60 (1978)

Bennett M.V.L., Zukin R.S. Electrical coupling and neuronal synchronization in the mammalian brain. Neuron **41**, 495–511 (2004)

Bennett T., Cobb J.L.S. Studies on the avian gizzard: morphology and innervation of the smooth muscle. Zeitschrift für Zellforschung und mikroskopische Anatomie **96**, 173–185 (1969)

Berger U.V., Hediger M.A. Distribution of the glutamate transporters GLAST and GLT-1 in rat circumventricular organs, meninges, and dorsal root ganglia. Journal of Comparative Neurology **421**, 385–399 (2000)

Bergeron J.J.M., Ehrenreich J.H., Siekevitz P., Palade G.E. Golgi fractions prepared from rat liver homogenates. II. Biochemical characterization. Journal of Cell Biology **59**, 73–88 (1973)

Berghs S., Aggujaro D., Dirkx R.Jr., Maksimova E., Stabach P., Hermel J.-M., Zhang J.-P., Philbrick W., Slepnev V., Ort T., Solimena M. βIV Spectrin, a new spectrin localized at axon initial segments and nodes of Ranvier in the central and peripheral nervous system. Journal of Cell Biology **151**, 985–1001 (2000)

Bergles D.E., Roberts J.D.B., Somogyi P., Jahr C.E. Glutamatergic synapses on oligodendrocyte precursor cells in the hippocampus. Nature **405**, 187–191 (2000)

Bergmann C. Notiz über einige Structurverhältnisse des Cerebellum und Rückenmarks. Zeitschrift für rationelle Medicin **8**, 360–363 (1857)

Berkley H.J. Studies on the lesions produced by the action of certain poisons on the nerve-cell. Part II. Experimental lesions produced by the action of dog's serum on the cortical nerve-cell of the rabbit's brain. Medical News **67**, 225–231 (1895)

Bernardini N., Levey A.I., Augusti-Tocco G. Rat dorsal root ganglia express m1-m4 muscarinic receptor proteins. Journal of the Peripheral Nervous System **4**, 222–232 (1999)

Berner A., Torvik A., Stenwig A.E. Origin of macrophages in traumatic lesions and Wallerian degeneration in peripheral nerves. Acta Neuropathologica **25**, 228–236 (1973)

Bernstein J.J. Relationship of cortico-spinal tract growth to age and body weight in the rat. Journal of Comparative Neurology **127**, 207–218 (1966)

Bernstein J.J., Getz R., Jefferson M., Kelemen M. Astrocytes secrete basal lamina after hemisection of rat spinal cord. Brain Research **327**, 135–141 (1985)

Bertaud W.S. Membrane junctions in the myelin sheath of goldfish lateral nerve. Journal of Cell Science **30**, 77–85 (1978)

Berthold C.-H. Ultrastructure of the node-paranode region of mature feline ventral lumbar spinal-root fibres. Acta Societatis Medicorum Upsaliensis **73**, Suppl. 9, 37–70 (1968)

Berthold C.-H. Morphology of normal peripheral axons. In: Physiology and Pathobiology of axons (S.G. Waxman Ed.) pp. 3–63. Raven Press, New York 1978

Bethe A. Ueber die Neurofibrillen in den Ganglienzellen von Wirbelthieren und ihre Beziehungen zu den Golginetzen. Archiv für mikroskopische Anatomie und Entwicklungsgeschichte **55**, 513–558 (1900)

Bhatnagar M., Cintra A., Chadi G., Lindberg J., Oitzl M., De Kloet E.R., Möller A., Agnati L.F., Fuxe K. Neurochemical changes in the hippocampus of the brown Norway rat during aging. Neurobiology of Aging **18**, 319–327 (1997)

Bhatt D.H., Zhang S., Gan W.-B. Dendritic spine dynamics. Annual Review of Physiology **71**, 261–282 (2009)

Bignami A., Raju T., Dahl D. Localization of vimentin, the nonspecific intermediate filament protein, in embryonal glia and in early differentiating neurons. *In vivo* and *in vitro* immunofluorescence study of the rat embryo with vimentin and neurofilament antisera. Developmental Biology **91**, 286–295 (1982)

Bignami A., Ralston H.J.III The cellular reaction to Wallerian degeneration in the central nervous system of the cat. Brain Research **13**, 444–461 (1969)

Billings-Gagliardi S., Webster H.deF., O'Connell M.F. In vivo and electron microscopic observations on Schwann cells in developing tadpole nerve fibers. American Journal of Anatomy **141**, 375–392 (1974)

Biondi G. Ein neuer histologischer Befund am Epithel des Plexus chorioideus. Zeitschrift für die gesamte Neurologie und Psychiatrie **144**, 161–165 (1933)

Birks R.I. The relationship of transmitter release and storage to fine structure in a sympathetic ganglion. Journal of Neurocytology **3**, 133–160 (1974)

Birks R., Huxley H.E., Katz B. The fine structure of the neuromuscular junction of the frog. Journal of Physiology **150**, 134–144 (1960)

Bischoff A., Moor H. Ultrastructural differences between the myelin sheaths of peripheral nerve fibres and CNS white matter. Zeitschrift für Zellforschung und mikroskopische Anatomie **81**, 303–310 (1967)

Bishop G.A., Berbari N.F., Lewis J., Mykytyn K. Type III adenylyl cyclase localizes to primary cilia throughout the adult mouse brain. Journal of Comparative Neurology **505**, 562–571 (2007)

Bixby J.L., Harris W.A. Molecular mechanisms of axon growth and guidance. Annual Review of Cell Biology **7**, 117–159 (1991)

Bixby J.L., Maunsell J.H.R., Van Essen D.C. Effects of motor unit size on innervation patterns in neonatal mammals. Experimental Neurology **70**, 516–524 (1980)

Björkman A., Wohlfart G. Faseranalyse der Nn. oculomotorius, trochlearis und abducens des Menschen und des N. abducens verschiedener Tiere. Zeitschrift für mikroskopisch-anatomische Forschung **39**, 631–647 (1936)

Blackstad T., Kjaerheim A. Special axo-dendritic synapses in the hippocampal cortex: electron and light microscopic studies on the layer of mossy fibers. Journal of Comparative Neurology **117**, 133–160 (1961)

Blakemore W.F. The ultrastructure of the subependymal plate in the rat. Journal of Anatomy **104**, 423–433 (1969)

Blakemore W.F. Microglial reactions following thermal necrosis of the rat cortex: an electron microscope study. Acta Neuropathologica **21**, 11–22 (1972)

Blakemore W.F., Jolly R.D. The subependymal plate and associated ependyma in the dog. An ultrastructural study. Journal of Neurocytology **1**, 69–84 (1972)

Blaustein M.P., Golovina V.A. Structural complexity and functional diversity of endoplasmic reticulum Ca^{2+} stores. Trends in Neurosciences **24**, 602–608 (2001)

Blinzinger K. Elektronenmikroskopische Untersuchungen am Ependym der Hirnventrikel des Goldhamsters (Mesocricetus aureatus). Acta Neuropathologica **1**, 527–532 (1962)

Blinzinger K., Anzil A.P., Müller W. Myelinated nerve cell perikaryon in mouse spinal cord. Zeitschrift für Zellforschung und mikroskopische Anatomie **128**, 135–138 (1972)

Blinzinger K., Hager H. Elektronenmikroskopische Untersuchungen zur Feinstruktur ruhender und progressiver Mikrogliazellen im ZNS des Goldhamsters. Progress in Brain Research **6**, 99–112 (1964)

Blinzinger K., Rewcastle N.B., Hager H. Observations on prismatic-type mitochondria within astrocytes of the Syrian hamster brain. Journal of Cell Biology **25**, 293–303 (1965)

Bloom G.D., Carlsöö B., Gustafsson H. Intra-acinar nerve terminals in four rodent parotid glands. Acta Anatomica **97**, 291–299 (1977)

Blümcke S. Elektronenoptische Untersuchungen an Schwannschen Zellen während der initialen Degeneration und frühen Regeneration. Beiträge zur pathologischen Anatomie und zur allgemeinen Pathologie **128**, 238–258 (1963)

Bodian D. A note on nodes of Ranvier in the central nervous system. Journal of Comparative Neurology **94**, 475–483 (1951)

Bodian D. Electron microscopy: two major synaptic types on spinal motoneurons. Science **151**, 1093–1094 (1966)

Bohr V., Møllgård K. Tight junctions in human fetal choroid plexus visualized by freeze-etching. Brain Research **81**, 314–318 (1974)

Boll F. Die Histiologie und Histiogenese der nervösen Centralorgane. Archiv für Psychiatrie und Nervenkrankheiten **4**, 1–138 (1873)

Bombardi C., Grandis A., Nenzi A., Giurisato M., Cozzi B. Immunohistochemical localization of substance P and cholecystokinin in the dorsal root ganglia and spinal cord of the bottlenose dolphin (*Tursiops truncatus*). Anatomical Record **293**, 477–484 (2010)

Bondareff W., Geinisman Y. Loss of synapses in the dentate gyrus of the senescent rat. American Journal of Anatomy **145**, 129–136 (1976)

Bondareff W., McLone D.G. The external glial limiting membrane in Macaca: ultrastructure of a laminated glioepithelium. American Journal of Anatomy **136**, 277–296 (1973)

Bondareff W., Narotzky R. Age changes in the neuronal microenvironment. Science **176**, 1135–1136 (1972)

Booz K.H. Secretory phenomena at the ependyma of the IIIrd ventricle of the embryonic rat. Anatomy and Embryology **147**, 143–159 (1975)

Bortolami R., Manni E. Aspetti morfo-funzionali del nucleo mesencefalico del trigemino e del ganglio di Gasser. Archivio Italiano di Anatomia e di Embriologia **86** Suppl., 63–121 (1981)

Bowley M.P., Cabral H., Rosene D.L., Peters A. Age changes in myelinated nerve fibers of the cingulate bundle and corpus callosum in the rhesus monkey. Journal of Comparative Neurology **518**, 3046–3064 (2010)

Bowman M.H., King J.S. The conformation, cytology and synaptology of the opossum inferior olivary nucleus. Journal of Comparative Neurology **148**, 491–524 (1973)

Boycott B.B., Gray E.G., Guillery R.W. Synaptic structure and its alteration with environmental temperature: a study by light and electron microscopy of the central nervous system of lizards. Proceedings of the Royal Society of London. Series B **154**, 151–172 (1961)

Boyd H.D., McLachlan E.M., Keast J.R., Inokuchi H. Three electro-physiological classes of guinea pig sympathetic postganglionic neurone have distinct morphologies. Journal of Comparative Neurology **369**, 372–387 (1996)

Braak E., Braak H., Strenge H. The fine structure of myelinated nerve cell bodies in the bulbus olfactorius of man. Cell and Tissue Research **182**, 221–233 (1977)

Braak H. Elektronenmikroskopische Untersuchungen an Catecholaminkernen im Hypothalamus vom Goldfisch

(*Carassius auratus*). Zeitschrift für Zellforschung und mikroskopische Anatomie **83**, 398–415 (1967)

Brachet J. La détection histochimique des acides pentosenucléiques. Comptes Rendus des Séances de la Société de Biologie **133**, 88–90 (1940)

Bradbury M. The Concept of a Blood-brain Barrier. John Wiley & Sons, Chichester 1979

Bradley W.G., Asbury A.K. Duration of synthesis phase in neurilemma cells in mouse sciatic nerve during degeneration. Experimental Neurology **26**, 275–282 (1970)

Braun N., Sévigny J., Robson S.C., Hammer K., Hanani M., Zimmermann H. Association of the ecto-ATPase NTPDase2 with glial cells of the peripheral nervous system. Glia **45**, 124–132 (2004)

Brawer J.R. The fine structure of the ependymal tanycytes at the level of the arcuate nucleus. Journal of Comparative Neurology **145**, 25–42 (1972)

Bray D., Bunge M.B. Serial analysis of microtubules in cultured rat sensory axons. Journal of Neurocytology **10**, 589–605 (1981)

Breedlove S.M., Arnold A.P. Sexually dimorphic motor nucleus in the rat lumbar spinal cord: response to adult hormone manipulation, absence in androgen-insensitive rats. Brain Research **225**, 297–307 (1981)

Bridgman P.C. Myosin-dependent transport in neurons. Journal of Neurobiology **58**, 164–174 (2004)

Bridgman P.C. Myosin motor proteins in the cell biology of axons and other neuronal compartments. Results and Problems in Cell Differentiation **48**, 91–105 (2009)

Brightman M.W. The distribution within the brain of ferritin injected into cerebrospinal fluid compartments. I. Ependymal distribution. Journal of Cell Biology **26**, 99–123 (1965a)

Brightman M.W. The distribution within the brain of ferritin injected into cerebrospinal fluid compartments. II. Parenchymal distribution. American Journal of Anatomy **117**, 193–220 (1965b)

Brightman M.W. The intracerebral movement of proteins injected into blood and cerebrospinal fluid of mice. Progress in Brain Research **29**, 19–37 (1968)

Brightman M.W., Anders J.J., Rosenstein J.M. Specializations of nonneuronal cell membranes in the vertebrate nervous system. Advances in Cellular Neurobiology **1**, 3–29 (1980)

Brightman M.W., Palay S.L. The fine structure of ependyma in the brain of the rat. Journal of Cell Biology **19**, 415–439 (1963)

Brightman M.W., Prescott L., Reese T.S. Intercellular junctions of special ependyma. In: Brain-Endocrine Interaction (K.M. Knigge, D.E. Scott, M. Kobayashi, S. Ishii Eds.) **2**, 146–165. Karger, Basel 1975

Brightman M.W., Reese T.S. Junctions between intimately apposed cell membranes in the vertebrate brain. Journal of Cell Biology **40**, 648–677 (1969)

Brightman M.W., Reese T.S., Olsson Y., Klatzo I. Morphologic aspects of the blood–brain barrier to peroxidase in elasmobranchs. Progress in Neuropathology **1**, 146–161 (1971)

Brimijoin S., Olsen J., Rosenson R. Comparison of the temperature-dependence of rapid axonal transport and microtubules in nerves of the rabbit and bullfrog. Journal of Physiology **287**, 303–314 (1979)

Brizzee K.R. Quantitative histological studies on aging changes in cerebral cortex of rhesus monkey and albino rat with notes on effects of prolonged low-dose ionizing irradiation in the rat. Progress in Brain Research **40**, 141–160 (1973)

Brizzee K.R., Cancilla P.A., Sherwood N., Timiras P.S. The amount and distribution of pigments in neurons and glia of the cerebral cortex. Journal of Gerontology **24**, 127–135 (1969)

Brizzee K.R., Johnson F.A. Depth distribution of lipofuscin pigment in cerebral cortex of albino rat. Acta Neuropathologica **16**, 205–219 (1970)

Brizzee K.R., Ordy J.M., Hansche J., Kaack B. Quantitative assessment of changes in neuron and glia cell packing density and lipofuscin accumulation with age in the cerebral cortex of a nonhuman primate (*Macaca mulatta*). Neurobiology of Aging **3**, 229–244 (1976)

Brizzee K.R., Ordy J.M., Kaack B. Early appearance and regional differences in intraneuronal and extraneuronal lipofuscin accumulation with age in the brain of a nonhuman primate (*Macaca mulatta*). Journal of Gerontology **29**, 366–381 (1974)

Brizzee K.R., Sherwood N., Timiras P.S. A comparison of cell populations at various depth levels in cerebral cortex of young adult and aged Long-Evans rats. Journal of Gerontology **23**, 289–297 (1968)

Brody H. Organization of the cerebral cortex. III. A study of aging in the human cerebral cortex. Journal of Comparative Neurology **102**, 511–556 (1955)

Brody H. The deposition of aging pigment in the human cerebral cortex. Journal of Gerontology **15**, 258–261 (1960)

Brown A. Axonal transport of membranous and nonmembranous cargoes: a unified perspective. Journal of Cell Biology **160**, 817–821 (2003)

Brown A.M., Ransom B.R. Astrocyte glycogen and brain energy metabolism. Glia **55**, 1263–1271 (2007)

Brown D.A., Galvan M. Influence of neuroglial transport on the action of γ-aminobutyric acid on mammalian ganglion cells. British Journal of Pharmacology **59**, 373–378 (1977)

Brown M.C., Jansen J.K.S., Van Essen D.C. Polyneuronal innervation of skeletal muscle in newborn rats and its elimination during maturation. Journal of Physiology (London) **261**, 387–422 (1976)

Browning K.N., Lees G.M. Myenteric neurons of the rat descending colon: electrophysiological and correlated morphological properties. Neuroscience **73**, 1029–1047 (1996)

Bruckenstein D.A., Higgins D. Morphological differentiation of embryonic rat sympathetic neurons in tissue culture. I. Conditions under which neurons form axons but not dendrites. Developmental Biology **128**, 324–336 (1988)

Bruni J.E., Montemurro D.G., Clattenburg R.E., Singh R.P. A scanning electron microscopic study of the ependymal surface of the third ventricle of the rabbit, rat, mouse and human brain. Anatomical Record **174**, 407–420 (1972)

Bruzzone R., White T.W., Paul D.L. Connections with connexins: The molecular basis of direct intercellular signaling. European Journal of Biochemistry **238**, 1–27 (1996)

Bryans W.A. Mitotic activity in the brain of the adult rat. Anatomical Record **133**, 65–74 (1959)

Buard I., Steinmetz C.C., Claudepierre T., Pfrieger F.W. Glial cells promote dendrite formation and the reception of synaptic input in Purkinje cells from postnatal mice. Glia **58**, 538–545 (2010)

Buell S.J., Coleman P.D. Quantitative evidence for selective dendritic growth in normal human aging but not in senile dementia. Brain Research **214**, 23–41 (1981)

Bundgaard M. Brain barrier systems in the lamprey. I. Ultrastructure and permeability of cerebral blood vessels. Brain Research **240**, 55–64 (1982)

Bundgaard M., Cserr H. A glial blood–brain barrier in elasmobranchs. Brain Research **226**, 61–73 (1981)

Bunge M.B., Bunge R.P., Peterson E.R., Murray M.R. A light and electron microscope study of long-term organized cultures of rat dorsal root ganglia. Journal of Cell Biology **32**, 439–466 (1967)

Bunge R.P., Bunge M.B., Eldridge C.F. Linkage between axonal ensheathment and basal lamina production by Schwann cells. Annual Review of Neuroscience **9**, 305–328 (1986)

Burd G.D. Myelinated dendrites and neuronal perikarya in the olfactory bulb of the mouse. Brain Research **181**, 450–454 (1980)

Burke R.E., Levine D.N., Salcman M., Tsairis P. Motor units in cat soleus muscle: physiological, histochemical and morphological characteristics. Journal of Physiology **238**, 503–514 (1974)

Burke R.E., Tsairis P. Anatomy and innervation ratios in motor units of cat gastrocnemius. Journal of Physiology **234**, 749–765 (1973)

Burnstock G., Costa M. Adrenergic Neurons. Their Organization, Function and Development in the Peripheral Nervous System. Chapman and Hall, London 1975

Burton P.R. Microtubules of frog olfactory axons: their length and number/axon. Brain Research **409**, 71–78 (1987)

Burton P.R. Dendrites of mitral cell neurons contain microtubules of opposite polarity. Brain Research **473**, 107–115 (1988)

Burton P.R., Laveri L.A. The distribution, relationships to other organelles, and calcium-sequestering ability of smooth endoplasmic reticulum in frog olfactory axons. Journal of Neuroscience **5**, 3047–3060 (1985)

Burton P.R., Paige J.L. Polarity of axoplasmic microtubules in the olfactory nerve of the frog. Proceedings of the National Academy of Sciences of the U.S.A. **78**, 3269–3273 (1981)

Bushong E.A., Martone M.E., Jones Y.Z., Ellisman M.H. Protoplasmic astrocytes in CA1 stratum radiatum occupy separate anatomical domains. Journal of Neuroscience **22**, 183–192 (2002)

Büssow H. Schwann cell myelin ensheathing C.N.S. axons in the nerve fibre layer of the cat retina. Journal of Neurocytology **7**, 207–214 (1978)

Butler M.H., David C., Ochoa G.-C., Freyberg Z., Daniell L., Grabs D., Cremona O., De Camilli P. Amphiphysin II (SH3P9; BIN1), a member of the amphiphysin/Rvs family, is concentrated in the cortical cytomatrix of axon initial segments and nodes of Ranvier in brain and around T tubules in skeletal muscle. Journal of Cell Biology **137**, 1355–1367 (1997)

Butt A.M., Colquhoun K., Tutton M., Berry M. Three-dimensional morphology of astrocytes and oligodendrocytes in the intact mouse optic nerve. Journal of Neurocytology **23**, 469–485 (1994)

Butt A.M., Duncan A., Fraser Hornby M., Kirvell S.L., Hunter A., Levine J.M., Berry M. Cells expressing the NG2 antigen contact nodes of Ranvier in adult CNS white matter. Glia **26**, 84–91 (1999)

Butt A.M., Kiff J., Hubbard P., Berry M. Synantocytes: New functions for novel NG2 expressing glia. Journal of Neurocytology **31**, 551–565 (2002)

Cajal S.R. see Ramón y Cajal S.

Calvo J.L., Carbonell A.L., Boya J. Coexpression of vimentin and glial fibrillary acidic protein in astrocytes of the adult rat optic nerve. Brain Research **532**, 355–357 (1990)

Cammermeyer J. Histiocytes, juxtavascular mitotic cells and microglia cells during retrograde changes in the facial nucleus of rabbits of varying age. Ergebnisse der Anatomie und Entwicklungsgeschichte **38**, 195–229 (1965)

Cammermeyer J. The life history of the microglial cell: a light microscopic study. Neurosciences Research **3**, 43–129 (1970)

Canger A.K., Passini M.A., Asch W.S., Leake D., Zafonte B.T., Glasgow E., Schechter N. Restricted expression of the neuronal intermediate filament protein plasticin during zebrafish development. Journal of Comparative Neurology **399**, 561–572 (1998)

Cantino D., Mugnaini E. The structural basis for electrotonic coupling in the avian ciliary ganglion. A study with thin sectioning and freeze-fracturing. Journal of Neurocytology **4**, 505–536 (1975)

Carasso N. Ultra-structure des cellules visuelles de larves d'Amphibiens. Comptes Rendus des Séances de l'Académie des Sciences **247**, 527–531 (1958)

Card J.P. Exploring brain circuitry with neurotropic viruses: New horizons in Neuroanatomy. Anatomical Record (New Anatomy) **253**, 176–185 (1998)

Carden M.J., Schlaepfer W.W., Lee V.M.-Y. The structure, biochemical properties, and immuno-genicity of neurofilament peripheral regions are determined by phosphorylation state. Journal of Biological Chemistry **260**, 9805–9817 (1985)

Carlsen F., Behse F. Three dimensional analysis of Schwann cells associated with unmyelinated nerve fibres in human sural nerve. Journal of Anatomy **130**, 545–557 (1980)

Carpenter F.W., Conel J.L. A study of ganglion cells in the sympathetic nervous system, with special reference to intrinsic sensory neurones. Journal of Comparative Neurology **24**, 269–281 (1914)

Carpenter S.J., McCarthy L.E., Borison H.L. Electron microscopic study on the epiplexus (Kolmer) cells of the cat choroid plexus. Zeitschrift für Zellforschung und mikroskopische Anatomie **110**, 471–486 (1970)

Casey M.A., Feldman M.L. Aging in the rat medial nucleus of the trapezoid body. II. Electron microscopy. Journal of Comparative Neurology **232**, 401–413 (1985)

Castillo C., Norcini M., Martin Hernandez L.A., Correa G., Blanck T.J.J., Recio-Pinto E. Satellite glia cells in dorsal root ganglia express functional NMDA receptors. Neuroscience **240**, 135–146 (2013)

Cavanagh J.B. The proliferation of astrocytes around a needle wound in the rat brain. Journal of Anatomy **106**, 471–487 (1970)

Ceballos D., Cuadras J., Verdú E., Navarro X. Morphometric and ultrastructural changes with ageing in mouse peripheral nerve. Journal of Anatomy **195**, 563–576 (1999)

Ceccarelli B., Hurlbut W.P., Mauro A. Depletion of vesicles from frog neuromuscular junctions by prolonged tetanic stimulation. Journal of Cell Biology **54**, 30–38 (1972)

Ceccarelli B., Hurlbut W.P., Mauro A. Turnover of transmitter and synaptic vesicles at the frog neuromuscular junction. Journal of Cell Biology **57**, 499–524 (1973)

Celio M.R., Spreafico R., De Biasi S., Vitellaro-Zuccarello L. Perineuronal nets: past and present. Trends in Neurosciences **21**, 510–515 (1998)

Chalfie M., Thomson J.N. Organization of neuronal microtubules in the nematode *Caenorhabditis elegans*. Journal of Cell Biology **82**, 278–289 (1979)

Chamak B., Fellous A., Glowinski J., Prochiantz A. MAP2 expression and neuritic outgrowth and branching are coregulated through region-specific neuro-astroglial interactions. Journal of Neuroscience **7**, 3163–3170 (1987)

Chamberlain J.G. Analysis of developing ependymal and choroidal surfaces in rat brains using scanning electron microscopy. Developmental Biology **31**, 22–30 (1973)

Chamberlain J.G. Scanning electron microscopy of epiplexus cells (macrophages) in the fetal rat brain. American Journal of Anatomy **139**, 443–447 (1974)

Chandler R.L., Willis R. An intranuclear fibrillar lattice in neurons. Journal of Cell Science **1**, 283–286 (1966)

Chan-Palay V., Palay S.L. High voltage electron microscopy of rapid Golgi preparations. Neurons and their processes in the cerebellar cortex of monkey and rat. Zeitschrift für Anatomie und Entwicklungsgeschichte **137**, 125–152 (1972)

Chan-Palay V., Palay S.L., Billings-Gagliardi S.M. Meynert cells in the primate visual cortex. Journal of Neurocytology **3**, 631–658 (1974)

Charles A.C., Merrill J.E., Dirksen E.R., Sanderson M.J. Intercellular signaling in glial cells: calcium waves and oscillations in response to mechanical stimulation and glutamate. Neuron **6**, 983–992 (1991)

Charlton B.T., Gray E.G. Comparative electron microscopy of synapses in the vertebrate spinal cord. Journal of Cell Science **1**, 67–80 (1966)

Chen L., Lu W., Yang Z., Yang S., Li C., Shi X., Tang Y. Age-related changes of the oligodendrocytes in rat subcortical white matter. Anatomical Record **294**, 487–493 (2011)

Chen S., Hillman D.E. Dying-back of Purkinje cell dendrites with synapse loss in aging rats. Journal of Neurocytology **28**, 187–196 (1999)

Chevaleyre V., Moos F.C., Desarménien M.G. Correlation between electrophysiological and morphological characteristics during maturation of rat supraoptic neurons. European Journal of Neuroscience **13**, 1136–1146 (2001)

Chicurel M.E., Harris K.M. Three-dimensional analysis of the structure and composition of CA3 branched dendritic spines and their synaptic relationships with mossy fiber boutons in the rat hippocampus. Journal of Comparative Neurology **325**, 169–182 (1992)

Chien C.-L., Mason C.A., Liem R.K.H. α-Internexin is the only neuronal intermediate filament expressed in developing cerebellar granule neurons. Journal of Neurobiology **29**, 304–318 (1996)

Ching W., Zanazzi G., Levinson S.R., Salzer J.L. Clustering of neuronal sodium channels requires contact with myelinating Schwann cells. Journal of Neurocytology **28**, 295–301 (1999)

Chiu S.Y. Functions and distribution of voltage-gated sodium and potassium channels in mammalian Schwann cells. Glia **4**, 541–558 (1991)

Chiu S.Y., Schrager P., Ritchie J.M. Neuronal-type Na^+ and K^+ channels in rabbit cultured Schwann cells. Nature **311**, 156–157 (1984)

Choo D., Malmgren L.T., Rosenberg S.I. Age-related changes in Schwann cells of the internal branch of the rat superior laryngeal nerve. Otolaryngology: Head and Neck Surgery **103**, 628–636 (1990)

Chu-Wang I.-W., Oppenheim R.W. Cell death of motoneurons in the chick embryo spinal cord. II. A quantitative and qualitative analysis of degeneration in the ventral root, including evidence for axon outgrowth and limb innervation prior to cell death. Journal of Comparative Neurology **177**, 59–86 (1978)

Ciaroni S., Cecchini T., Cuppini R., Ferri P., Ambrogini P., Bruno C., Del Grande P. Are there proliferating neuronal precursors in adult rat dorsal root ganglia? Neuroscience Letters **281**, 69–71 (2000)

Citkowitz E., Holtzman E. Peroxisomes in dorsal root ganglia. Journal of Histochemistry and Cytochemistry **21**, 34–41 (1973)

Clark D.A. Muscle counts of motor units: a study in innervation ratios. American Journal of Physiology **96**, 296–304 (1931)

Clattenburg R.E., Singh R.P., Montemurro D.G. Intranuclear filamentous inclusions in neurons of the rabbit hypothalamus. Journal of Ultrastructure Research **39**, 549–555 (1972)

Clendening B., Hume R.I. Cell interactions regulate dendritic morphology and responses to neurotransmitters in embryonic chick sympathetic preganglionic neurons *in vitro*. Journal of Neuroscience **10**, 3992–4005 (1990)

Coates P.W. Supraependymal cells: light and transmission electron microscopy extends scanning electron microscopic demonstration. Brain Research **57**, 502–507 (1973)

Cobb J.L.S., Bennett T. A study of nexuses in visceral smooth muscle. Journal of Cell Biology **41**, 287–297 (1969)

Cochard P., Paulin D. Initial expression of neurofilaments and vimentin in the central and peripheral nervous system of the mouse embryo *in vivo*. Journal of Neuroscience **4**, 2080–2094 (1984)

Coggeshall R.E. A light and electron microscope study of the abdominal ganglion of *Aplysia californica*. Journal of Neurophysiology **30**, 1263–1287 (1967)

Coggeshall R.E., Coulter J.D., Willis W.D.Jr. Unmyelinated axons in the ventral roots of the cat lumbosacral enlargement. Journal of Comparative Neurology **153**, 39–58 (1974)

Coggeshall R.E., Yaksta B.A., Swartz F.J. A cytophotometric analysis of the DNA in the nucleus of the giant cell, R-2, in *Aplysia*. Chromosoma **32**, 205–212 (1970)

Cohen C., DeRosier D., Harrison S.C., Stephens R.E., Thomas J. X-ray patterns from microtubules. Annals of the New York Academy of Sciences **253**, 53–59 (1975)

Cohen J., Mareš V., Lodin Z. DNA content of purified preparations of mouse Purkinje neurons isolated by a velocity sedimentation technique. Journal of Neurochemistry **20**, 651–657 (1973)

Cohen-Cory S., Fraser S.E. Effects of brain-derived neuro-trophic factor on optic axon branching and remodelling *in vivo*. Nature **378**, 192–196 (1995)

Cole J.S., Messing A., Trojanowski J.Q., Lee V.M.-Y. Modulation of axon diameter and neurofilaments by hypomyelinating Schwann cells in transgenic mice. Journal of Neuroscience **14**, 6956–6966 (1994)

Cole K.S. Surface forces of the Arbacia egg. Journal of Cellular and Comparative Physiology **1**, 1–9 (1932)

Cole K.S., Michaelis E.M. Surface forces of fertilized Arbacia eggs. Journal of Cellular and Comparative Physiology **2**, 121–126 (1932)

Colello R.J., Pott U., Schwab M.E. The role of oligodendrocytes and myelin on axon maturation in the developing rat retinofugal pathway. Journal of Neuroscience **14**, 2594–2605 (1994)

Coleman P.D., Flood D.G. Dendritic proliferation in the aging brain as a compensatory repair mechanism. Progress in Brain Research **70**, 227–237 (1986)

Coleman P.D., Riesen A.H. Environmental effects on cortical dendritic fields. I. Rearing in the dark. Journal of Anatomy **102**, 363–374 (1968)

Colman D.R., Kreibich G., Frey A.B., Sabatini D.D. Synthesis and incorporation of myelin polypeptides into CNS myelin. Journal of Cell Biology **95**, 598–608 (1982)

Colonnier M. Synaptic patterns on different cell types in the different laminae of the cat visual cortex. An electron microscope study. Brain Research **9**, 268–287 (1968)

Colonnier M., Guillery R.W. Synaptic organization in the lateral geniculate nucleus of the monkey. Zeitschrift für Zellforschung und mikroskopische Anatomie **62**, 333–355 (1964)

Colonnier M., Tremblay J.P., McLennan H. Synaptic contacts on glial cells in the abdominal ganglion of *Aplysia californica*. Journal of Comparative Neurology **188**, 391–400 (1979)

Conde J.R., Streit W.J. Microglia in the aging brain. Journal of Neuropathology and Experimental Neurology **65**, 199–203 (2006)

Connell C.J., Mercer K.L. Freeze-fracture appearance of the capillary endothelium in the cerebral cortex of mouse brain. American Journal of Anatomy **140**, 595–600 (1974)

Conradi S. Ultrastructural specialization of the initial axon segment of cat lumbar motoneurons. Acta Societatis Medicorum Upsaliensis **71**, 281–284 (1966)

Conradi S. Ultrastructure and distribution of neuronal and glial elements on the surface of the proximal part of a motoneuron dendrite, as analyzed by serial sections. Acta Physiologica Scandinavica Suppl. **332**, 49–64 (1969a)

Conradi S. Observations on the ultrastructure of the axon hillock and initial axon segment of lumbosacral motoneurons in the cat. Acta Physiologica Scandinavica Suppl. **332**, 65–84 (1969b)

Conti F., Hille B., Neumcke B., Nonner W., Stämpfli R. Conductance of the sodium channel in myelinated nerve fibres with modified sodium inactivation. Journal of Physiology **262**, 729–742 (1976)

Cook R.D., Burnstock G. The ultrastructure of Auerbach's plexus in the guinea-pig. I. Neuronal elements. II. Non-neuronal elements. Journal of Neurocytology **5**, 171–194, 195–206 (1976)

Cooke B.M., Woolley C.S. Gonadal hormone modulation of dendrites in the mammalian CNS. Journal of Neurobiology **64**, 34–46 (2005)

Cooper M.H., Beal J.A. Myelinated granule cell bodies in the cerebellum of the monkey (*Saimiri sciureus*). Anatomical Record **187**, 249–256 (1977)

Corfas G., Velardez M.O., Ko C.-P., Ratner N., Peles E. Mechanisms and roles of axon-Schwann cell interactions. Journal of Neuroscience **24**, 9250–9260 (2004)

Cornell-Bell A.H., Finkbeiner S.M., Cooper M.S., Smith S.J. Glutamate induces calcium waves in cultured astrocytes: Long-range glial signaling. Science **247**, 470–473 (1990)

Costa C.B., Paula-Barbosa M.M. Nuclear inclusions in the nerve cells of the sphenopalatine ganglion of the dog. Cell and Tissue Research **201**, 289–294 (1979)

Cotard-Bartley M.P., Secchi J., Glomot R., Cavanagh J.B. Spontaneous degenerative lesions of peripheral nerves in aging rats. Veterinary Pathology **18**, 110–113 (1981)

Cotman C.W., Nieto-Sampedro M., Harris E.W. Synapse replacement in the nervous system of adult vertebrates. Physiological Reviews **61**, 684–784 (1981)

Coupland R.E. Electron microscopic observations on the structure of the rat adrenal medulla. II. Normal innervation. Journal of Anatomy **99**, 255–272 (1965)

Court F.A., Hendriks W.T.J., MacGillavry H.D., Alvarez J., van Minnen J. Schwann cell to axon transfer of ribosomes: Toward a novel understanding of the role of glia in the nervous system. Journal of Neuroscience **28**, 11024–11029 (2008)

Courtney J., Steinbach J.H. Age changes in neuromuscular junction morphology and acetylcholine receptor distribution on rat skeletal muscle fibres. Journal of Physiology **320**, 435–447 (1981)

Couteaux R. Nouvelles observations sur la structure de la plaque motrice et interprétation des rapports myoneuraux. Comptes Rendus des Séances de la Société de Biologie **138**, 976–978 (1944)

Couteaux R. Principaux critères morphologiques et cytochimiques utilisables aujourd'hui pour définir les divers types de synapses. Actualités neurophysiologiques, 3e série, 145–173 (1961)

Cragg B.G. The density of synapses and neurones in the motor and visual areas of the cerebral cortex. Journal of Anatomy **101**, 639–654 (1967)

Cragg B.G. The density of synapses and neurons in normal, mentally defective and ageing human brains. Brain **98**, 81–90 (1975)

Cragg B.G. Ultrastructural features of human cerebral cortex. Journal of Anatomy **121**, 331–362 (1976)

Cross B.A. The neurosecretory impulse. In: Neurosecretion. The Final Neuroendocrine Pathway. (F. Knowles, L. Vollrath Eds.) pp. 115–128. Springer-Verlag, Berlin 1974

Cross D.J., Flexman J.A., Anzai Y., Maravilla K.R., Minoshima S. Age-related decrease in axonal transport measured by MR imaging in vivo. NeuroImage **39**, 915–926 (2008)

Csakvari E., Hoyk Z., Gyenes A., Garcia-Ovejero D., Garcia-Segura L.M., Párducz Á. Fluctuation of synapse density in the arcuate nucleus during the estrous cycle. Neuroscience **144**, 1288–1292 (2007)

Cuénod M., Sandri C., Akert K. Enlarged synaptic vesicles in optic nerve terminals induced by intraocular injection of colchicine. Brain Research **39**, 285–296 (1972)

Cumming R., Burgoyne R.D. Compartmentalization of neuronal cytoskeletal proteins. Bioscience Reports **3**, 997–1006 (1983)

Curcio C.A., Coleman P.D. Stability of neuron number in cortical barrels of aging mice. Journal of Comparative Neurology **212**, 158–172 (1982)

Custer N.V. Structurally specialized contacts between the photoreceptors of the retina of the axolotl. Journal of Comparative Neurology **151**, 35–56 (1973)

Czosnek H., Soifer D., Wisniewski H.M. Studies on the biosynthesis of neurofilament proteins. Journal of Cell Biology **85**, 726–734 (1980)

Dahl D., Bignami A. Intermediate filaments in nervous tissue. Cell and Muscle Motility **6**, 75–96 (1985)

Dahl D., Chi N.H., Miles L.E., Nguyen B.T., Bignami A. Glial fibrillary acidic (GFA) protein in Schwann cells: fact or artifact? Journal of Histochemistry and Cytochemistry **30**, 912–918 (1982)

Dahl E. The fine structure of nuclear inclusions. Journal of Anatomy **106**, 255–262 (1970)

Dahl H.A. Fine structure of cilia in rat cerebral cortex. Zeitschrift für Zellforschung und mikroskopische Anatomie **60**, 369–386 (1963)

Dahlström A., Häggendal J. Some quantitative studies on the noradrenaline content in the cell bodies and terminals of a sympathetic adrenergic neuron system. Acta Physiologica Scandinavica **67**, 271–277 (1966)

Dale H.H., Feldberg W., Vogt M. Release of acetylcholine at voluntary motor nerve endings. Journal of Physiology **86**, 353–380 (1936)

Dani J.W., Chernjavsky A., Smith S.J. Neuronal activity triggers calcium waves in hippocampal astrocyte networks. Neuron **8**, 429–440 (1992)

Danielli J.F., Davson H. A contribution to the theory of permeability of thin films. Journal of Cellular and Comparative Physiology **5**, 495–508 (1935)

Danzer S.C., McMullen N.T., Rance N.E. Testosterone modulates the dendritic architecture of arcuate neuroendocrine neurons in adult male rats. Brain Research **890**, 78–85 (2001)

Das A.V., Mallya K.B., Zhao X., Ahmad F., Bhattacharya S., Thoreson W.B., Hegde G.V., Ahmad I. Neural stem cell properties of Müller glia in the mammalian retina: Regulation by Notch and Wnt signaling. Developmental Biology **299**, 283–302 (2006)

Dashiell S.M., Tanner S.L., Pant H.C., Quarles R.H. Myelin-associated glycoprotein modulates expression and phosphorylation of neuronal cytoskeletal elements and their associated kinases. Journal of Neurochemistry **81**, 1263–1272 (2002)

David G.B. On the structure of the synapse. In: Metabolism of the Nervous System (D. Richter Ed.) pp. 58–66. Pergamon Press, London 1957

David S., Nathaniel E.J.H. Intranuclear inclusions in the developing neurons of the rat cuneate nuclei. Cell and Tissue Research **193**, 525–532 (1978)

Davis D.A., Lloyd B.J.Jr., Milhorat T.H. A comparative ultrastructural study of the choroid plexuses of the immature pig. Anatomical Record **176**, 443–454 (1973)

Davis D.A., Milhorat T.H. The blood-brain barrier of the rat choroid plexus. Anatomical Record **181**, 779–790 (1975)

Davis J.Q., Lambert S., Bennett V. Molecular composition of the node of Ranvier: Identification of ankyrin-binding cell adhesion molecules neurofascin (mucin+/third FNIII domain-) and NrCAM at nodal axon segments. Journal of Cell Biology **135**, 1355–1367 (1996)

Dawson M.R.L., Polito A., Levine J.M., Reynolds R. NG2-expressing glial progenitor cells: an abundant and widespread population of cycling cells in the adult rat CNS. Molecular and Cellular Neuroscience **24**, 476–488 (2003)

de Brabander J.M., Kramers R.J.K., Uylings H.B.M. Layer-specific dendritic regression of pyramidal cells with ageing in the human prefrontal cortex. European Journal of Neuroscience **10**, 1261–1269 (1998)

De Camilli P., Cameron R., Greengard P. Synapsin I (protein I), a nerve terminal-specific phosphoprotein. I. Its general distribution in synapses of the central and peripheral nervous system demonstrated by immunofluorescence in frozen and plastic sections. Journal of Cell Biology **96**, 1337–1354 (1983)

De Camilli P., Miller P.E., Navone F., Theurkauf W.E., Vallee R.B. Distribution of microtubule-associated protein 2 in the nervous system of the rat studied by immunofluorescence. Neuroscience **11**, 819–846 (1984)

de Duve C., Pressman B.C., Gianetto R., Wattiaux R., Appelmans F. Tissue fractionation studies. 6. Intracellular distribution patterns of enzymes in rat-liver tissue. Biochemical Journal **60**, 604–617 (1955)

Deitch A.D., Murray M.R. The Nissl substance of living and fixed spinal ganglion cells. I. A phase contrast study. Journal of Biophysical and Biochemical Cytology **2**, 433–444 (1956)

Deiters O. Untersuchungen über Gehirn und Rückenmark des Menschen und der Säugethiere (M. Schultze Hrsg.). F. Vieweg und Sohn, Braunschweig 1865

De Koninck P., Carbonetto S., Cooper E. NGF induces neonatal rat sensory neurons to extend dendrites in culture after removal of satellite cells. Journal of Neuroscience **13**, 577–585 (1993)

del Castillo J., Katz B. Local activity at a depolarized nerve-muscle junction. Journal of Physiology **128**, 396–411 (1955)

Deller T., Merten T., Roth S.U., Mundel P., Frotscher M. Actin-associated protein synaptopodin in the rat hippocampal formation: Localization in the spine neck and close association with the spine apparatus of principal neurons. Journal of Comparative Neurology **418**, 164–181 (2000)

Delorenzi E. Costanza numerica delle cellule del Purkinje in individui di varia età. Bollettino della Società Italiana di Biologia Sperimentale **6**, 80–82 (1931)

de Lorenzo A.J. Electron microscopic observations of the olfactory mucosa and olfactory nerve. Journal of Biophysical and Biochemical Cytology **3**, 839–850 (1957)

de Lorenzo A.J.D. Electron microscopy: tight junctions in synapses of the chick ciliary ganglion. Science **152**, 76–78 (1966)

de Lorenzo A.J.D., Brzin M., Dettbarn W.D. Fine structure and organization of nerve fibers and giant axons in *Homarus americanus*. Journal of Ultrastructure Research **24**, 367–384 (1968)

Denis-Donini S., Glowinski J., Prochiantz A. Glial heterogeneity may define the three-dimensional shape of mouse mesencephalic dopaminergic neurones. Nature **307**, 641–643 (1984)

Dentler W.L., Granett S., Rosenbaum J.L. Ultrastructural localization of the high molecular weight proteins associated with in vitro-assembled brain microtubules. Journal of Cell Biology **65**, 237–241 (1975)

Dering M.A., Santer R.M., Watson A.H.D. Age-related changes in the morphology of preganglionic neurons projecting to the rat hypogastric ganglion. Journal of Neurocytology **25**, 555–563 (1996)

Dermietzel R. Visualization by freeze-fracturing of regular structures in glial cell membranes. Naturwissenschaften **60**, 208 (1973)

Dermietzel R. Junctions in the central nervous system of the cat. I. Membrane fusion in central myelin. Cell and Tissue Research **148**, 565–576 (1974a)

Dermietzel R. Junctions in the central nervous system of the cat. II. A contribution to the tertiary structure of the axonal-glial junctions in the paranodal region of the node of Ranvier. Cell and Tissue Research **148**, 577–586 (1974b)

Dermietzel R. Junctions in the central nervous system of the cat. III. Gap junctions and membrane-associated orthogonal particle complexes (MOPC) in astrocytic membranes. Cell and Tissue Research **149**, 121–135 (1974c)

De Robertis E. Electron microscope observations on the submicroscopic organization of the retinal rods. Journal of Biophysical and Biochemical Cytology **2**, 319–330 (1956)

De Robertis E. Fine structure of synapses in the CNS. IV International Congress of Neuropathology, Proceedings **2**, 35–38. G. Thieme Verlag, Stuttgart 1962

De Robertis E., Bennett H.S. Some features of the submicroscopic morphology of synapses in frog and earthworm. Journal of Biophysical and Biochemical Cytology **1**, 47–58 (1955)

De Robertis E., Rodriguez de Lores Arnaiz G., Salganicoff L., Pellegrino de Iraldi A., Zieher L.M. Isolation of synaptic vesicles and structural organization of the acetylcholine system within brain nerve endings. Journal of Neurochemistry **10**, 225–235 (1963)

De Robertis E., Schmitt F.O. An electron microscope analysis of certain nerve axon constituents. Journal of Cellular and Comparative Physiology **31**, 1–24 (1948)

Derouiche A., Frotscher M. Peripheral astrocyte processes: monitoring by selective immunostaining for the actin-binding ERM proteins. Glia **36**, 330–341 (2001)

Desagher S., Glowinski J., Premont J. Astrocytes protect neurons from hydrogen peroxide toxicity. Journal of Neuroscience **16**, 2553–2562 (1996)

Deschênes M., Labelle A., Landry P. Morphological characterization of slow and fast pyramidal tract cells in the cat. Brain Research **178**, 251–274 (1979)

Desmond N.L., Levy W.B. Changes in the numerical density of synaptic contacts with long-term potentiation in the hippocampal dentate gyrus. Journal of Comparative Neurology **253**, 466–475 (1986)

Devaux J., Gow A. Tight junctions potentiate the insulative properties of small CNS myelinated axons. Journal of Cell Biology **183**, 909–921 (2008)

DeVoogd T.J., Nottebohm F. Sex differences in dendritic morphology of a song control nucleus in the canary: a quantitative Golgi study. Journal of Comparative Neurology **196**, 309–316 (1981)

Devor M. Unexplained peculiarities of the dorsal root ganglion. Pain, Suppl. 6, S27-S35 (1999)

Devor M., Govrin-Lippmann R., Frank I., Raber P. Proliferation of primary sensory neurons in adult rat dorsal root ganglion and the kinetics of retrograde cell loss after sciatic nerve section. Somatosensory Research **3**, 139–167 (1985)

de Waegh S.M., Lee V.M.-Y., Brady S.T. Local modulation of neurofilament phosphorylation, axonal caliber, and slow axonal transport by myelinating Schwann cells. Cell **68**, 451–463 (1992)

Dewey M.M., Barr L. A study of the structure and distribution of the nexus. Journal of Cell Biology **23**, 553–586 (1964)

Diamond J., Gray E.G., Yasargil G.M. The function of the dendritic spine: an hypothesis. In: Excitatory Synaptic Mechanisms (P. Andersen, J.K.S. Jansen Eds.) pp. 213–222. Universitetsforlaget, Oslo 1970

Diamond M.C., Johnson R.E., Gold M.W. Changes in neuron number and size and glia number in the young, adult, and aging rat medial occipital cortex. Behavioral Biology **20**, 409–418 (1977)

Dichter M.A. Rat cortical neurons in cell culture: culture methods, cell morphology, electrophysiology, and syn apse formation. Brain Research **149**, 279–293 (1978)

Dickson D.W., Lee S.C. Microglia. In: Textbook of Neuropathology , 3rd ed. (R.L. Davis, D.M. Robertson Eds.) pp. 165–205. Williams & Wilkins, Baltimore 1997

Dieterich H.J. Electron microscopic studies of the innervation of the rat kidney. Zeitschrift für Anatomie und Entwicklungsgeschichte **145**, 169–186 (1974)

Diner O. Les cellules de Schwann en mitose et leurs rapports avec les axones au cours du développement du nerf sciatique chez le Rat. Comptes Rendus des Séances de l'Académie des Sciences **261**, 1731–1734 (1965)

Disse J. Ueber die Spinalganglien der Amphibien. Anatomischer Anzeiger, Ergänzungsheft zu **8**, 201–204 (1893)

Dixon A.D. The ultrastructure of nerve fibers in the trigeminal ganglion of the rat. Journal of Ultrastructure Research **8**, 107–121 (1963)

Dixon J.S. Nuclear bodies in normal and chromatolytic sympathetic neurons. Anatomical Record **168**, 179–186 (1970)

Dixon J.S., Gosling J.A. Histochemical and electron microscopic observations on the innervation of the upper segment of the mammalian ureter. Journal of Anatomy **110**, 57–66 (1971)

Doetsch F., Caillé I., Lim D.A., García-Verdugo J.M., Alvarez-Buylla A. Subventricular zone astrocytes are neural stem cells in the adult mammalian brain. Cell **97**, 703–716 (1999)

Doetsch F., García-Verdugo J.M., Alvarez-Buylla A. Cellular composition and three-dimensional organization of the subventricular germinal zone in the adult mammalian brain. Journal of Neuroscience **17**, 5046–5061 (1997)

Dogiel A.S. Der Bau der Spinalganglien bei den Säugetieren. Anatomischer Anzeiger **12**, 140–152 (1896)

Dogiel A.S. Zur Frage über den feineren Bau der Spinalganglien und deren Zellen bei Säugetieren. Internationale Monatsschrift für Anatomie und Physiologie **14**, 73–116 (1897)

Dolapchieva S., Lierse W. On some aspects of Schwann-sheath development and the possible role of their desmosome-like structures. Anatomy and Embryology **147**, 213–224 (1975)

Donaldson H.H., Nagasaka G. On the increase in the diameters of nerve-cell bodies and of the fibers arising from them during the later phases of growth (albino rat). Journal of Comparative Neurology **29**, 529–552 (1918)

Doolin P.P., Barron K.D., Kwak S. Ultrastructural and histochemical analysis of cytoplasmic laminar bodies in lateral geniculate neurons of adult cat. American Journal of Anatomy **121**, 601–622 (1967)

Dotti C.G., Sullivan C.A., Banker G.A. The establishment of polarity by hippocampal neurons in culture. Journal of Neuroscience **8**, 1454–1468 (1988)

Dowling J.E., Boycott B.B. Organization of the primate retina: electron microscopy. Proceedings of the Royal Society of London. Series B **166**, 80–111 (1967)

Doyère L. Mémoire sur les tardigrades. Annales des Sciences Naturelles **14**, 269–361 (1840)

Dräger U.C. Coexistence of neurofilaments and vimentin in a neurone of adult mouse retina. Nature **303**, 169–172 (1983)

Dreyer F., Peper K., Akert K., Sandri C., Moor H. Ultrastructure of the 'active zone' in the frog neuromuscular junction. Brain Research **62**, 373–380 (1973)

Dropp J.J. Mast cells in the central nervous system of several rodents. Anatomical Record **174**, 227–237 (1972)

Dropp J.J., Sodetz F.J. Autoradiographic study of neurons and neuroglia in autonomic ganglia of behaviorally stressed rats. Brain Research **33**, 419–430 (1971)

Droz B., Di Giamberardino L., Koenig H.L. Contribution of axonal transport to the renewal of myelin phospholipids in peripheral nerves. I. Quantitative radioautographic study. Brain Research **219**, 57–71 (1981)

Droz B., Di Giamberardino L., Koenig H.L., Boyenval J., Hassig R. Axon-myelin transfer of phospholipid components in the course of their axonal transport as visualized by radioautography. Brain Research **155**, 347–353 (1978)

Droz B., Leblond C.P. Migration of proteins along the axons of the sciatic nerve. Science **137**, 1047–1048 (1962)

Droz B., Leblond C.P. Axonal migration of proteins in the central nervous system and peripheral nerves as shown by radioautography. Journal of Comparative Neurology **121**, 325–346 (1963)

Droz B., Rambourg A., Koenig H.L. The smooth endoplasmic reticulum: structure and role in the renewal of axonal membrane and synaptic vesicles by fast axonal transport. Brain Research **93**, 1–13 (1975)

Du Y., Dreyfus C.F. Oligodendrocytes as providers of growth factors. Journal of Neuroscience Research **68**, 647–654 (2002)

Duan H., Wearne S.L., Rocher A.B., Macedo A., Morrison J.H., Hof P.R. Age-related dendritic and spine changes in corticocortically projecting neurons in macaque monkeys. Cerebral Cortex **13**, 950–961 (2003)

Dubois-Dalcq M., Rentier B., Baron A., Van Evercooren N., Burge B.W. Structure and behavior of rat primary and secondary Schwann cells in vitro. Experimental Cell Research **131**, 283–297 (1981)

Duce I.R., Keen P. Can neuronal smooth endoplasmic reticulum function as a calcium reservoir? Neuroscience **3**, 837–848 (1978)

Duch C., Levine R.B. Remodeling of membrane properties and dendritic architecture accompanies the postembryonic conversion of a slow into a fast motoneuron. Journal of Neuroscience **20**, 6950–6961 (2000)

Dumitriu D., Hao J., Hara Y., Kaufmann J., Janssen W.G.M., Lou W., Rapp P.R., Morrison J.H. Selective changes in thin spine density and morphology in monkey prefrontal cortex correlate with aging-related cognitive impairment. Journal of Neuroscience **30**, 7507–7515 (2010)

Duque A., Tepper J.M., Detari L., Ascoli G.A., Zaborszky L. Morphological characterization of electrophysiologically and immunohistochemically identified basal forebrain cholinergic and neuropeptide Y-containing neurons. Brain Structure and Function **212**, 55–73 (2007)

Dutrochet H. Recherches anatomiques et physiologiques sur la structure intime des animaux et des végétaux, et sur leur motilité. Baillière, Paris 1824

Eakin R.M., Kuda A. Ultrastructure of sensory receptors in ascidian tadpoles. Zeitschrift für Zellforschung und mikroskopische Anatomie **112**, 287–312 (1971)

Eames R.A., Gamble H.J. Schwann cell relationships in normal human cutaneous nerves. Journal of Anatomy **106**, 417–435 (1970)

Ebendal T. Effects of nerve growth factor on the synthesis of nucleic acids and proteins in cultured chick embryo trigeminal ganglia. Zoon **3**, 159–167 (1975)

Ebner F.F., Colonnier M. Synaptic patterns in the visual cortex of turtle: an electron microscopic study. Journal of Comparative Neurology **160**, 51–80 (1975)

Edinger L. Vergleichend-entwickelungsgeschichtliche und anatomische Studien im Bereiche der Hirnanatomie. Anatomischer Anzeiger **8**, 305–321 (1893)

Edmunds S.M., Parnavelas J.G. 'Presynaptic astrocytes' in rat visual cortex. Brain Research **259**, 285–287 (1983)

Edström A. The ribonucleic acid in the Mauthner neuron of the goldfish. Journal of Neurochemistry **11**, 309–314 (1964)

Edwards R.H. The neurotransmitter cycle and quantal size. Neuron **55**, 835–858 (2007)

Ehrenberg C.G. Nothwendigkeit einer feineren mechanischen Zerlegung des Gehirns und der Nerven vor der chemischen. Poggendorff's Annalen der Physik und Chemie **28**, 449–465 (1833)

Ehrenberg C.G. Beobachtung einer bisher unbekannten auffallenden Structur des Seelenorgans bei Menschen und Thieren. Dümmler, Berlin 1836

Ehrenberg F. Mittheilungen über Resultate bei Anwendung des chromatisch-polarisierten Lichtes für mikroskopische Verhältnisse. Bericht über die Verhandlungen der Königlichen Preussischen Akademie der Wissenschaften zu Berlin, S. 55–76 (1849)

Ehrlich P. Beiträge zur Kenntnis der granulierten Bindegewebszellen und der eosinophilen Leukocyten. Archiv für Anatomie und Physiologie S. 166–169 (1879)

Ehrlich P. Das Sauerstoff-Bedürfnis des Organismus. Eine farben-analytische Studie. PhD thesis. Hirschwald, Berlin 1885

Ehrlich P. Ueber die Methylenblaureaction der lebenden Nervensubstanz. Deutsche medicinische Wochenschrift **12**, 49–52 (1886)

Einheber S., Milner T.A., Giancotti F., Salzer J.L. Axonal regulation of Schwann cell integrin expression suggests a role for α6β4 in myelination. Journal of Cell Biology **123**, 1223–1236 (1993)

Elekes K., Florey E. New types of synaptic connections in crayfish stretch receptor organs: an electron microscopic study. Journal of Neurocytology **16**, 613–626 (1987)

Elfvin L.-G. The ultrastructure of unmyelinated fibers in the splenic nerve of the cat. Journal of Ultrastructure Research **1**, 428–454 (1958)

Elfvin L.-G. The ultrastructure of the nodes of Ranvier in cat sympathetic nerve fibers. Journal of Ultrastructure Research **5**, 374–387 (1961)

Elfvin L.G., Björklund H., Dahl D., Seiger Å. Neurofilament-like and glial fibrillary acidic protein-like immunoreactivities in rat and guinea-pig sympathetic ganglia in situ and after perturbation. Cell and Tissue Research **250**, 79–86 (1987)

Elfvin L.-G., Forsman C. The ultrastructure of junctions between satellite cells in mammalian sympathetic ganglia as revealed by freeze-etching. Journal of Ultrastructure Research **63**, 261–274 (1978)

Elson K., Ribeiro R.M., Perelson A.S., Simmons A., Speck P. The life span of ganglionic glia in murine sensory ganglia estimated by uptake of bromodeoxyuridine. Experimental Neurology **186**, 99–103 (2004)

Elzholz A. Zur Kenntniss der Veränderungen im centralen Stumpfe lädirter gemischter Nerven. Jahrbücher für Psychiatrie und Neurologie **17**, 323–359 (1898)

Endo S., Sakai H., Matsumoto G. Microtubules in squid giant axon. Cell Structure and Function **4**, 285–293 (1979)

Eng H., Lund K., Campenot R.B. Synthesis of β-tubulin, actin, and other proteins in axons of sympathetic neurons in compartmented cultures. Journal of Neuroscience **19**, 1–9 (1999)

Englisch H.-J., Kunz G., Wenzel J. Zur Spines-Verteilung an Pyramidenneuronen der CA-1-Region des Hippocampus der Ratte. Zeitschrift für mikroskopische-anatomische Forschung **88**, 85–102 (1974)

Enriques P. La forma come funzione della grandezza. 2[a] memoria: Ricerche sui gangli nervosi degli Invertebrati. Archiv für Entwicklungsmechanik der Organismen **25**, 655–714 (1908)

Erickson H.P. Microtubule surface lattice and subunit structure and observations on reassembly. Journal of Cell Biology **60**, 153–167 (1974)

Ermilov L.G., Miller S.M., Schmalz P.F., Hanani M., Szurszewski J.H. The three-dimensional structure of neurons in the guinea pig inferior mesenteric and pelvic hypogastric ganglia. Autonomic Neuroscience: Basic and Clinical **83**, 116–126 (2000)

Erulkar S.D., Soller R.W. Interactions among lumbar motoneurons on opposite sides of the frog spinal cord: morphological and electrophysiological studies. Journal of Comparative Neurology **192**, 473–488 (1980)

Evans D.H.L., Evans E.M. The membrane relationships of smooth muscles: an electron microscope study. Journal of Anatomy **98**, 37–46 (1964)

Fain G.L., Gold G.H., Dowling J.E. Receptor coupling in the toad retina. Cold Spring Harbor Symposia on Quantitative Biology **40**, 547–561 (1976)

Fallon J.R. Preferential outgrowth of central nervous system neurites on astrocytes and Schwann cells as compared with nonglial cells in vitro. Journal of Cell Biology **100**, 198–207 (1985)

Famiglietti E.V.Jr. Dendro-dendritic synapses in the lateral geniculate nucleus of the cat. Brain Research **20**, 181–191 (1970)

Famiglietti E.V.Jr., Kolb H. A bistratified amacrine cell and synaptic circuitry in the inner plexiform layer of the retina. Brain Research **84**, 293–300 (1975)

Famiglietti E.V.Jr., Peters A. Schwann cell architecture The synaptic glomerulus and the intrinsic neuron in the dorsal lateral geniculate nucleus of the cat. Journal of Comparative Neurology **144**, 285–334 (1972)

Fannon A.M., Sherman D.L., Ilyina-Gragerova G., Brophy P.J., Friedrich V.L.Jr., Colman D.R. Novel E-cadherin-mediated adhesion in peripheral nerve: Schwann cell architecture is stabilized by autotypic adherens junctions. Journal of Cell Biology **129**, 189–202 (1995)

Farquhar M.G., Hartmann J.F. Neuroglial structure and relationships as revealed by electron microscopy. Journal of Neuropathology and Experimental Neurology **16**, 18–39 (1957)

Fatt P., Katz B. An analysis of the end-plate potential recorded with an intra-cellular electrode. Journal of Physiology **115**, 320–370 (1951)

Fatt P., Katz B. Spontaneous subthreshold activity at motor nerve endings. Journal of Physiology **117**, 109–128 (1952)

Fawcett D.W. On the occurrence of a fibrous lamina on the inner aspect of the nuclear envelope in certain cells of vertebrates. American Journal of Anatomy **119**, 129–146 (1966)

Feinberg K., Eshed-Eisenbach Y., Frechter S., Amor V., Salomon D., Sabanay H., Dupree J.L., Grumet M., Brophy P.J., Shrager P., Peles E. A glial signal consisting of gliomedin and NrCAM clusters axonal Na^+ channels during the formation of nodes of Ranvier. Neuron **65**, 490–502 (2010)

Feinstein B., Lindegård B., Nyman E., Wohlfart G. Morphologic studies of motor units in normal human muscles. Acta Anatomica **23**, 127–142 (1955)

Feldman M.L., Dowd C. Loss of dendritic spines in aging cerebral cortex. Anatomy and Embryology **148**, 279–301 (1975)

Feldman M.L., Peters A. Intranuclear rods and sheets in rat cochlear nucleus. Journal of Neurocytology **1**, 109–127 (1972)

Feldman M.L., Peters A. A technique for estimating total spine numbers on Golgi-impregnated dendrites. Journal of Comparative Neurology **188**, 527–542 (1979)

Feldman M.L., Peters A. Ballooning of myelin sheaths in normally aged macaques. Journal of Neurocytology **27**, 605–614 (1998)

Fenstermaker V., Chen Y., Ghosh A., Yuste R. Regulation of dendritic length and branching by semaphorin 3A. Journal of Neurobiology **58**, 403–412 (2004)

Fernandez B., Suarez I., Gutierrez C. Intranuclear inclusions in hypothalamic neurons: a study in the ageing hamster. Journal of Submicroscopic Cytology **16**, 495–501 (1984)

Fernández E., Cuenca N., Cerezo J.R., De Juan J. Visual experience during postnatal development determines the size of optic nerve axons. NeuroReport **5**, 365–367 (1993)

Fernandez H.L., Burton P.R., Samson F.E. Axoplasmic transport in the crayfish nerve cord. The role of fibrillar constituents of neurons. Journal of Cell Biology **51**, 176–192 (1971)

Fernandez H.L., Huneeus F.C., Davison P.F. Studies on the mechanism of axoplasmic transport in the crayfish cord. Journal of Neurobiology **1**, 395–409 (1970)

Fernández-Klett F., Offenhauser N., Dirnagl U., Priller J., Lindauer U. Pericytes in capillaries are contractile in vivo, but arterioles mediate functional hyperemia in the mouse brain. Proceedings of the National Academy of Sciences of the U.S.A. **107**, 22290–22295 (2010)

Fernández-Morán H. Sheath and axon structures in the internode portion of vertebrate myelinated nerve fibres. An electron microscope study of rat and frog sciatic nerves. Experimental Cell Research **1**, 309–340 (1950)

Fernández-Morán H. The submicroscopic organization of vertebrate nerve fibres. An electron microscope study of myelinated and unmyelinated nerve fibres. Experimental Cell Research **3**, 282–359 (1952)

Fernández-Morán H. The submicroscopic structure of nerve fibres. Progress in Biophysics and Biophysical Chemistry **4**, 112–147 (1954)

Fernández-Morán H., Finean J.B. Electron microscope and low-angle x-ray diffraction studies of the nerve myelin sheath. Journal of Biophysical and Biochemical Cytology **3**, 725–748 (1957)

Ferrer I., Gullotta F. Down's syndrome and Alzheimer's disease: dendritic spine counts in the hippocampus. Acta Neuropathologica **79**, 680–685 (1990)

Ferrendelli J.A., Sedgwick W.G., Suntzeff V. Regional energy metabolism and lipofuscin accumulation in mouse brain during aging. Journal of Neuropathology and Experimental Neurology **30**, 638–649 (1971)

Fiala J.C., Spacek J., Harris K.M. Dendritic spine pathology: cause or consequence of neurological disorders? Brain Research Reviews **39**, 29–54 (2002)

Fieandt H. von Eine neue Methode zur Darstellung des Gliagewebes, nebst Beiträgen zur Kenntnis des Baues und der Anordnung der Neuroglia des Hundehirns. Archiv für mikroskopische Anatomie und Entwicklungsgeschichte **76**, 125–209 (1910)

Field E.J. Observations on the development of microglia together with a note on the influence of cortisone. Journal of Anatomy **89**, 201–208 (1955)

Field E.J., Peat A. Intranuclear inclusions in neurones and glia: a study in the ageing mouse. Gerontologia **17**, 129–138 (1971)

Fifková E. Changes in the visual cortex of rats after unilateral deprivation. Nature **220**, 379–381 (1968)

Fifková E., Delay R.J. Cytoplasmic actin in neuronal processes as a possible mediator of synaptic plasticity. Journal of Cell Biology **95**, 345–350 (1982)

Fifková E., Tonks M., Cullen-Dockstader K. Changes in the nuclear pore complexes of the dentate granule cells in aged rats. Experimental Neurology **95**, 755–762 (1987)

Filliatreau G., Di Giamberardino L. Microtubule polarity in myelinated axons as studied after decoration with tubulin. Biology of the Cell **42**, 69–72 (1981)

Fiori M.G. Intranuclear inclusions in Schwann cells of aged fowl ciliary ganglia. Journal of Anatomy **154**, 201–214 (1987)

Fisher S.K. A somato-somatic synapse between amacrine and bipolar cells in the cat retina. Brain Research **43**, 587–590 (1972)

Fleischer B., Fleischer S., Ozawa H. Isolation and characterization of Golgi membranes from bovine liver. Journal of Cell Biology **43**, 59–79 (1969)

Fleischhauer K. Ependyma and subependymal layer. In: The Structure and Function of Nervous Tissue (G.H. Bourne Ed.) **6**, 1–46. Academic Press, New York 1972

Flemming W. Vom Bau der Spinalganglienzellen. In: Beiträge zur Anatomie und Embryologie als Festgabe für J. Henle, S. 12–25. Cohen, Bonn 1882

Follenius E. Organisation scalariforme du réticulum endoplasmique dans certains processus nerveux de l'hypothalamus de *Gasterosteus aculeatus* L. Zeitschrift für Zellforschung und mikroskopische Anatomie **106**, 61–68 (1970)

Fontana F. Traité sur le venin de la vipere sur les poisons americains sur le laurier-cerise et sur quelques autres poisons végétaux. On y a joint des observations sur la structure primitive du corps animal. Différentes expériences sur la reproduction des nerfs et la description d'un nouveau canal de l'oeil. 2 vol. Florence 1781

Forel A. Einige hirnanatomische Betrachtungen und Ergebnisse. Archiv für Psychiatrie und Nervenkrankheiten **18**, 162–198 (1887)

Forger N.G., Breedlove S.M. Seasonal variation in mammalian striated muscle mass and motoneuron morphology. Journal of Neurobiology **18**, 155–165 (1987)

Foster M., Sherrington C.S. A Text Book of Physiology. Part III: The Central Nervous System (7th ed.). Macmillan and Co., London 1897

Fraher J.P. Quantitative studies on the maturation of central and peripheral parts of individual ventral motoneuron axons. I. Myelin sheath and axon calibre. Journal of Anatomy **126**, 509–533 (1978)

Franca-Netto A.S. Dimensioni e numero dei neuroni in relazione alla mole somatica. Confronti tra due roditori di mole somatica molto differente. Zeitschrift für Zellforschung und mikroskopische Anatomie **36**, 222–234 (1951)

Franke W.W. Structure, biochemistry, and functions of the nuclear envelope. International Review of Cytology Suppl. 4, 71–236 (1974)

Frederickson R.G., Low F.N. Blood vessels and tissue space associated with the brain of the rat. American Journal of Anatomy **125**, 123–146 (1969)

Friede R.L., Johnstone M.A. Responses of thymidine labeling of nuclei in gray matter and nerve following sciatic transection. Acta Neuropathologica **7**, 218–231 (1967)

Friede R.L., Samorajski T. Relation between the number of myelin lamellae and axon circumference in fibers of vagus and sciatic nerves in mice. Journal of Comparative Neurology **130**, 223–232 (1967)

Friede R.L., Samorajski T. Axon caliber related to neurofilaments and microtubules in sciatic nerve fibers of rats and mice. Anatomical Record **167**, 379–388 (1970)

Friend D.S. Cytochemical staining of multivesicular body and Golgi vesicles. Journal of Cell Biology **41**, 269–279 (1969)

Friend D.S., Farquhar M.G. Functions of coated vesicles during protein absorption in the rat vas deferens. Journal of Cell Biology **35**, 357–376 (1967)

Friend D.S., Murray M.J. Osmium impregnation of the Golgi apparatus. American Journal of Anatomy **117**, 135–150 (1965)

Frolkis V.V., Tanin S.A., Gorban Y.N. Age-related changes in axonal transport. Experimental Gerontology **32**, 441–450 (1997)

Frommann C. Zur Silberfärbung der Axencylinder. Archiv für pathologische Anatomie und Physiologie und für klinische Medicin **31**, 151–153 (1864)

Frotscher M. Dendritic plasticity in response to partial deafferentation. In: Neurobiology of the Hippocampus (W. Seifert Ed.) pp. 65–80. Academic Press, London 1983

Fuchs E., Weber K. Intermediate filaments: structure, dynamics, function, and disease. Annual Review of Biochemistry **63**, 345–382 (1994)

Fujisawa H., Morioka H., Nakamura H., Watanabe K. Gap junctions in the differentiated neural retinae of newly hatched chickens. Journal of Cell Science **22**, 597–606 (1976)

Fujisawa K. Rat spinal ganglion neurone: quantitative aspects of its development and involution. In: Development and Involution of Neurones (K. Fujisawa, Y. Morimatsu Eds.) pp. 151–165. Japan Scientific Societies Press, Tokyo 1992

Fujita S. DNA constancy in neurons of the human cerebellum and spinal cord as revealed by Feulgen cytophotometry and cytofluorometry. Journal of Comparative Neurology **155**, 195–202 (1974)

Fukuda T., Kosaka T. Gap junctions linking the dendritic network of GABAergic interneurons in the hippocampus. Journal of Neuroscience **20**, 1519–1528 (2000)

Fukuda T., Kosaka T. Ultrastructural study of gap junctions between dendrites of parvalbumin-containing GABAergic neurons in various neocortical areas of the adult rat. Neuroscience **120**, 5–20 (2003)

Fukuda T., Kosaka T., Singer W., Galuske R.A.W. Gap junctions among dendrites of cortical GABAergic neurons establish a dense and widespread intercolumnar network. Journal of Neuroscience **26**, 3434–3443 (2006)

Furness J.B., Iwayama T. Terminal axons ensheathed in smooth muscle cells of the vas deferens. Zeitschrift für Zellforschung und mikroskopische Anatomie **113**, 259–270 (1971)

Furshpan E.J. "Electrical transmission" at an excitatory synapse in a vertebrate brain. Science **144**, 878–880 (1964)

Furshpan E.J., Potter D.D. Mechanism of nerve-impulse transmission at a crayfish synapse. Nature **180**, 342–343 (1957)

Furshpan E.J., Potter D.D. Transmission at the giant motor synapses of the crayfish. Journal of Physiology **145**, 289–325 (1959)

Gabella G. Innervation of the intestinal muscular coat. Journal of Neurocytology **1**, 341–362 (1972a)

Gabella G. Fine structure of the myenteric plexus in the guinea-pig ileum. Journal of Anatomy **111**, 69–97 (1972b)

Gabella G. The structural relations between nerve fibres and muscle cells in the urinary bladder of the rat. Journal of Neurocytology **24**, 159–187 (1995)

Gabella G., Berggren T., Uvelius B. Hypertrophy and reversal of hypertrophy in rat pelvic ganglion neurons. Journal of Neurocytology **21**, 649–662 (1992)

Gabella G., Trigg P., McPhail H. Quantitative cytology of ganglion neurons and satellite glial cells in the superior cervical ganglion of the sheep. Relationship with ganglion neuron size. Journal of Neurocytology **17**, 753–769 (1988)

Gabella G., Uvelius B. Urinary bladder of rat: fine structure of normal and hypertrophic musculature. Cell and Tissue Research **262**, 67–79 (1990)

Galabov P., Schiebler T.H. The ultrastructure of the developing neural lobe. Cell and Tissue Research **189**, 313–329 (1978)

Gall J.G. Cajal bodies: The first 100 years. Annual Review of Cellular and Developmental Biology **16**, 273–300 (2000)

Gambetti P., Autilio-Gambetti L., Papasozomenos S.Ch. Bodian's silver method stains neurofilament polypeptides. Science **213**, 1521–1522 (1981)

Gambetti P., Autilio-Gambetti L., Shafer B., Pfaff L. Quantitative autoradiographic study of labeled RNA in rabbit optic nerve after intraocular injection of [^{3}H] uridine. Journal of Cell Biology **59**, 677–684 (1973)

Gambetti P., Gonatas N.K. Fibrils and lattice-like intranuclear structures in nuclei of neurons. Rivista di Patologia Nervosa e Mentale **88**, 188–196 (1967)

Gamble H.J. Comparative electron-microscopic observations on the connective tissues of a peripheral nerve and a spinal nerve root in the rat. Journal of Anatomy **98**, 17–25 (1964)

Gamble H.J., Eames R.A. An electron microscopc study of the connective tissues of human peripheral nerve. Journal of Anatomy **98**, 655–663 (1964)

Gamble H.J., Gosset J.M. Specialization of Schwann cell membranes. Nature **212**, 734–735 (1966)

Gardner E. Decrease in human neurones with age. Anatomical Record **77**, 529–536 (1940)

Gardner-Medwin A.R., Coles J.A., Tsacopoulos M. Clearance of extracellular potassium: evidence for spatial buffering by glial cells in the retina of the drone. Brain Research **209**, 452–457 (1981)

Garrett J.R. The autonomic innervation of rabbit salivary glands studied electron microscopically after 5-hydroxydopamine administration. Cell and Tissue Research **178**, 551–562 (1977)

Gasser H.S. Properties of dorsal root unmedullated fibers on the two sides of the ganglion. Journal of General Physiology **38**,709–728 (1955)

Gasser H.S. Olfactory nerve fibers. Journal of General Physiology **39**, 473–496 (1956)

Gasser H.S., Erlanger J. The rôle played by the sizes of the constituent fibers of a nerve trunk in determining the form of its action potential wave. American Journal of Physiology **80**, 522–547 (1927)

Gath I., Stålberg E. In situ measurement of the innervation ratio of motor units in human muscles. Experimental Brain Research **43**, 377–382 (1981)

Gazzaley A.H., Siegel S.J., Kordower J.H., Mufson E.J., Morrison J.H. Circuit-specific alterations of *N*-methyl-D-aspartate receptor subunit 1 in the dentate gyrus of aged monkeys. Proceedings of the National Academy of Sciences of the U.S.A. **93**, 3121–3125 (1996)

Geinisman Y. Loss of axosomatic synapses in the dentate gyrus of aged rats. Brain Research **168**, 485–492 (1979)

Geinisman Y., Bondareff W., Dodge J.T. Partial deafferentation of neurons in the dentate gyrus of the senescent rat. Brain Research **134**, 541–545 (1977a)

Geinisman Y., Bondareff W., Dodge J.T. Dendritic atrophy in the dentate gyrus of the senescent rat. American Journal of Anatomy **152**, 321–330 (1978a)

Geinisman Y., Bondareff W., Dodge J.T. Hypertrophy of astroglial processes in the dentate gyrus of the senescent rat. American Journal of Anatomy **153**, 537–544 (1978b)

Geinisman Y., Bondareff W., Telser A. Diminished axonal transport of glycoproteins in the senescent rat brain. Mechanisms of Ageing and Development **6**, 363–378 (1977b)

Geinisman Y., Detoledo-Morrell L., Morrell F., Persina I.S., Beatty M.A. Synapse restructuring associated with the maintenance phase of hippocampal long-term potentiation. Journal of Comparative Neurology **368**, 413–423 (1996)

Gelfan S., Rapisarda A.F. Synaptic density on spinal neurons of normal dogs and dogs with experimental hind-limb rigidity. Journal of Comparative Neurology **123**, 73–96 (1964)

Geren B.B. The formation from the Schwann cell surface of myelin in the peripheral nerves of chick embryos. Experimental Cell Research **7**, 558–562 (1954)

Geren B.B., Schmitt F.O. The structure of the Schwann cell and its relation to the axon in certain invertebrate nerve fibers. Proceedings of the National Academy of Sciences of the U.S.A. **40**, 863–870 (1954)

Gersh I., Bodian D. Some chemical mechanisms in chromatolysis. Journal of Cellular and Comparative Physiology **21**, 253–279 (1943)

Gershon M.D., Rothman T.P. Enteric glia. Glia **4**, 195–204 (1991)

Ghetti B., Horoupian D.S., Wiśniewski H.M. Transsynaptic response of the lateral geniculate nucleus and the pattern of degeneration of the nerve terminals in the rhesus monkey after eye enucleation. Brain Research **45**, 31–48 (1972)

Ghijsen W.E.J.M., Leenders A.G.M. Differential signaling in presynaptic neurotransmitter release. Cellular and Molecular Life Sciences **62**, 937–954 (2005)

Giannaris E.L., Rosene D.L. A stereological study of the numbers of neurons and glia in the primary visual cortex across the lifespan of male and female rhesus monkeys. Journal of Comparative Neurology **520**, 3492–3508 (2012)

Gibson P.H. EM study of the numbers of cortical synapses in the brains of ageing people and people with Alzheimer-type dementia. Acta Neuropathologica **62**, 127–133 (1983)

Gilbert D.S., Newby B.J., Anderton B.H. Neurofilament disguise, destruction and discipline. Nature **256**, 586–589 (1975)

Gilmore S.A. Spinal nerve root degeneration in aging laboratory rats: A light microscopic study. Anatomical Record **174**, 251–257 (1972)

Ginhoux F., Greter M., Leboeuf M., Nandi S., See P., Gokhan S., Mehler M.F., Conway S.J., Ng L.G., Stanley E.R., Samokhvalov I.M., Merad M. Fate mapping analysis reveals that adult microglia derive from primitive macrophages. Science **330**, 841–845 (2010)

Giordano S., Glasgow E., Tesser P., Schechter N. A type II keratin is expressed in glial cells of the goldfish visual pathway. Neuron **2**, 1507–1516 (1989)

Giuditta A., Chun J.T., Eyman M., Cefaliello C., Bruno A.P., Crispino M. Local gene expression in axons and nerve endings: The glia-neuron unit. Physiological Reviews **88**, 515–555 (2008)

Glasgow E., Druger R.K., Levine E.M., Fuchs C., Schechter N. Plasticin, a novel type III neurofilament protein from goldfish retina: increased expression during optic nerve regeneration. Neuron **9**, 373–381 (1992)

Glees P. Neuroglia. Morphology and Function. Blackwell, Oxford 1955

Glees P., Gopinath G. Age changes in the centrally and peripherally located sensory neurons in rat. Zeitschrift für Zellforschung und mikroskopische Anatomie **141**, 285–298 (1973)

Globus A., Lux H.D., Schubert P. Somadendritic spread of intracellularly injected tritiated glycine in cat spinal motoneurons. Brain Research **11**, 440–445 (1968)

Globus A., Scheibel A.B. Synaptic loci on visual cortical neurons of the rabbit: the specific afferent radiation. Experimental Neurology **18**, 116–131 (1967a)

Globus A., Scheibel A.B. Synaptic loci on parietal cortical neurons: terminations of corpus callosum fibers. Science **156**, 1127–1129 (1967b)

Globus A., Scheibel A.B. Pattern and field in cortical structure: the rabbit. Journal of Comparative Neurology **131**, 155–172 (1967c)

Gobel S. Axo-axonic septate junctions in the basket formations of the cat cerebellar cortex. Journal of Cell Biology **51**, 328–333 (1971)

Gobel S., Dubner R. Fine structural studies of the main sensory trigeminal nucleus in the cat and rat. Journal of Comparative Neurology **137**, 459–494 (1969)

Gobel S., Falls W.M., Bennett G.J., Abdelmoumene M., Hayashi H., Humphrey E. An EM analysis of the synaptic connections of horseradish peroxidase-filled stalked cells and islet cells in the substantia gelatinosa of adult cat spinal cord. Journal of Comparative Neurology **194**, 781–807 (1980)

Goemaere-Vanneste J., van den Bosch de Aguilar P. Étude des fibres nerveuses périphériques au cours du vieillissement chez le rat. La Cellule **74**, 263–280 (1987)

Goldstein L.S.B., Yang Z. Microtubule-based transport systems in neurons: The roles of kinesins and dyneins. Annual Review of Neuroscience **23**, 39–71 (2000)

Golgi C. Sulla struttura della sostanza grigia del cervello. Gazzetta Medica Italiana - Lombardia **6**, 244–246 (1873)

Golgi C. Sulla struttura delle fibre nervose midollate periferiche e centrali. Archivio per le Scienze Mediche **4**, 221–246 (1881)

Golgi C. Sulla fina anatomia degli organi centrali del sistema nervoso. Rivista Sperimentale di Freniatria e di Medicina Legale **8**, 165–195, 361–391 (1882); **9**, 1–17, 161–192, 385–402 (1883); **11**, 72–123, 193–220 (1885)

Golgi C. Intorno all'origine del quarto nervo cerebrale (patetico o trocleare) e di una questione di Isto-fisiologia generale che a questo argomento si collega. Rendiconti della Reale Accademia dei Lincei Serie 5[a], **2**, 379–389, 443–450 (1893)

Golgi C. Intorno alla struttura delle cellule nervose. Bollettino della Società Medico-Chirurgica di Pavia, pp.1–14 (1898a)

Golgi C. Sulla struttura delle cellule nervose del gangli spinali. Bollettino della Società Medico-Chirurgica di Pavia, pp. 53–63 (1898b)

Gonzales R.B., DeLeon Galvan C.J., Rangel Y.M., Claiborne B.J. Distribution of thorny excrescences on CA3 pyramidal neurons in the rat hippocampus. Journal of Comparative Neurology **430**, 357–368 (2001)

Goodman L.A., Model P.G. Superinnervation enhances the dendritic branching pattern of the Mauthner cell in the developing axolotl. Journal of Neuroscience **8**, 776–791 (1988)

Gordon G.R.J., Mulligan S.J., MacVicar B.A. Astrocyte control of the cerebrovasculature. Glia **55**, 1214–1221 (2007)

Gorgels T.G.M.F. Junctional specializations between growth cones and glia in the developing rat pyramidal tract: synapse-like contacts and invaginations. Journal of Comparative Neurology **306**, 117–128 (1991)

Gorski R.A., Harlan R.E., Jacobson C.D., Shryne J.E., Southam A.M. Evidence for the existence of a sexually dimorphic nucleus in the preoptic area of the rat. Journal of Comparative Neurology **193**, 529–539 (1980)

Gorter E., Grendel F. On bimolecular layers of lipoids on the chromocytes of the blood. Journal of Experimental Medicine **41**, 439–443 (1925)

Gotow T., Hashimoto P.H. Fine structure of the ependyma and intercellular junctions in the area postrema of the rat. Cell and Tissue Research **201**, 207–225 (1979)

Gotow T., Hashimoto P.H. Plasma membrane organization of astrocytes in elasmobranchs with special reference to the brain barrier system. Journal of Neurocytology **13**, 727–742 (1984)

Gotow T., Yoshikawa H., Hashimoto P.H. Distribution patterns of orthogonal arrays and alkaline phosphatase in plasma membranes of satellite cells in rat spinal ganglia. Anatomy and Embryology **171**, 171–179 (1985)

Gottlieb D.I., Cowan W.M. On the distribution of axonal terminals containing spheroidal and flattened synaptic vesicles in the hippocampus and dentate gyrus of the rat and cat. Zeitschrift für Zellforschung und mikroskopische Anatomie **129**, 413–429 (1972)

Gould E., Butcher L.L. Developing cholinergic basal forebrain neurons are sensitive to thyroid hormone. Journal of Neuroscience **9**, 3347–3358 (1989)

Gould E., Woolley C.S., Frankfurt M., McEwen B.S. Gonadal steroids regulate dendritic spine density in hippocampal pyramidal cells in adulthood. Journal of Neuroscience **10**, 1286–1291 (1990)

Gould R.M. Incorporation of glycoproteins into peripheral nerve myelin. Journal of Cell Biology **75**, 326–338 (1977)

Gould R.M., Spivack W.D., Sinatra R.S., Lindquist T.D., Ingoglia N.A. Axonal transport of choline lipids in normal and regenerating rat sciatic nerve. Journal of Neurochemistry **39**, 1569–1578 (1982)

Gozzano M. L'istogenesi della microglia. Rivista di Neurologia **4**, 225–265; 374–412 (1931)

Graeber M.B., Streit W.J. Microglia: biology and pathology. Acta Neuropathologica **119**, 89–105 (2010)

Grafstein B. Transport of protein by goldfish optic nerve fibers. Science **157**, 196–198 (1967)

Grafstein B. Transneuronal transfer of radioactivity in the central nervous system. Science **172**, 177–179 (1971)

Grafstein B., Forman D.S. Intracellular transport in neurons. Physiological Reviews **60**, 1167–1283 (1980)

Gray E.G. Axo-somatic and axo-dendritic synapses of the cerebral cortex: an electron microscope study. Journal of Anatomy **93**, 420–433 (1959)

Gray E.G. The fine structure of the insect ear. Philosophical Transactions of the Royal Society of London B **243**, 75–94 (1960a)

Gray E.G. Regular organisation of material in certain mitochondria of neuroglia of lizard brain. Journal of Biophysical and Biochemical Cytology **8**, 282–285 (1960b)

Gray E.G. The granule cells, mossy synapses and Purkinje spine synapses of the cerebellum: light and electron microscope observations. Journal of Anatomy **95**, 345–356 (1961)

Gray E.G. A morphological basis for pre-synaptic inhibition? Nature **193**, 82–83 (1962)

Gray E.G. Electron microscopy of presynaptic organelles of the spinal cord. Journal of Anatomy **97**, 101–106 (1963)

Gray E.G., Guillery R.W. The basis for silver staining of synapses of the mammalian spinal cord: a light and electron microscope study. Journal of Physiology **157**, 581–588 (1961)

Gray E.G., Guillery R.W. On nuclear structure in the ventral nerve cord of the leech *Hirudo medicinalis*. Zeitschrift für Zellforschung und mikroskopische Anatomie **59**, 738–745 (1963a)

Gray E.G., Guillery R.W. A note on the dendritic spine apparatus. Journal of Anatomy **97**, 389–392 (1963b)

Gray L., Smith Z., Rubel E.W. Developmental and experiential changes in dendritic symmetry in n.

laminaris of the chick. Brain Research **244**, 360–364 (1982)

Gray P.T.A., Ritchie J.M. Ion channels in Schwann and glial cells. Trends in Neurosciences **8**, 411–415, (1985)

Green E.J., Greenough W.T., Schlumpf B.E. Effects of complex or isolated environments on cortical dendrites of middle-aged rats. Brain Research **264**, 233–240 (1983)

Greenough W.T., Larson J.R., Whithers G.S. Effects of unilateral and bilateral training in a reaching task on dendritic branching of neurons in the rat motor-sensory forelimb cortex. Behavioral and Neural Biology **44**, 301–314 (1985)

Gregson N.A. The chemistry and structure of myelin. In: The Peripheral Nerve (D.N. Landon Ed.) pp. 512–604. Chapman and Hall, London 1976

Griffiths G., Simons K. The trans Golgi network: sorting at the exit site of the Golgi complex. Science **234**, 438–443 (1986)

Griffiths I.R., Duncan I.D., McQueen A. Age changes in the dorsal and ventral lumbar nerve roots of dogs. Acta Neuropathologica **32**, 75–85 (1975)

Grill J.D., Riddle D.R. Age-related and laminar-specific dendritic changes in the medial frontal cortex of the rat. Brain Research **937**, 8–21 (2002)

Grillo M.A. Cytoplasmic inclusions resembling nucleoli in sympathetic neurons of adult rats. Journal of Cell Biology **45**, 100–117 (1970)

Grillo M.A., Palay S.L. Ciliated Schwann cells in the autonomic nervous system of the adult rat. Journal of Cell Biology **16**, 430–436 (1963)

Grimmelikhuijzen C.J.P., Carstensen K., Darmer D., Moosler A., Nothacker H.-P., Reinscheid R.K., Schmutzler C., Vollert H., McFarlane I., Rinehart K.L. Coelenterate neuropeptides: structure, action and biosynthesis. American Zoologist **32**, 1–12 (1992)

Grinnell A.D. A study of the interaction between motoneurones in the frog spinal cord. Journal of Physiology **182**, 612–648 (1966)

Groc L., Petanjek Z., Gustafsson B., Ben-Ari Y., Hanse E., Khazipov R. *In vivo* blockade of neural activity alters dendritic development of neonatal CA1 pyramidal cells. European Journal of Neuroscience **16**, 1931–1938 (2002)

Gross P.M., Teasdale G.M., Angerson W.J., Harper A.M. H_2-receptors mediate increases in permeability of the blood-brain barrier during arterial histamine infusion. Brain Research **210**, 396–400 (1981)

Grover-Johnson N., Spencer P.S. Peripheral nerve abnormalities in aging rats. Journal of Neuropathology and Experimental Neurology **40**, 155–165 (1981)

Grundke-Iqbal I., Vorbrodt A.W., Iqbal K., Tung Y.-C., Wang G.P., Wisniewski H.M. Microtubule-associated polypeptides tau are altered in Alzheimer paired helical filaments. Molecular Brain Research **4**, 43–52 (1988)

Grupp L., Wolburg H., Mack A.F. Astroglial structures in the zebrafish brain. Journal of Comparative Neurology **518**, 4277–4287 (2010)

Gu J., Firestein B.L., Zheng J.Q. Microtubules in dendritic spine development. Journal of Neuroscience **28**, 12120–12124 (2008)

Gu Y., Chen Y., Zhang X., Li G.-W., Wang C., Huang L.-Y. M. Neuronal soma-satellite glial cell interactions in sensory ganglia and the participation of purinergic receptors. Neuron Glia Biology **6**, 53–62 (2010)

Güldner F.-H. Synaptology of the rat suprachiasmatic nucleus. Cell and Tissue Research **165**, 509–544 (1976)

Güldner F.-H., Wolff J.R. Neurono-glial synaptoid contacts in the median eminence of the rat: ultrastructure, staining properties and distribution on tanycytes. Brain Research **61**, 217–234 (1973)

Güldner F.-H., Wolff J.R. Dendro-dendritic synapses in the suprachiasmatic nucleus of the rat hypothalamus. Journal of Neurocytology **3**, 245–250 (1974)

Günther J. Impulse conduction in the myelinated giant fibers of the earthworm. Structure and function of the dorsal nodes in the median giant fiber. Journal of Comparative Neurology **168**, 505–532 (1976)

Guillery R.W. The organization of synaptic interconnections in the laminae of the dorsal lateral geniculate nucleus of the cat. Zeitschrift für Zellforschung und mikroskopische Anatomie **96**, 1–38 (1969)

Guillery R.W., Colonnier M. Synaptic patterns in the dorsal lateral geniculate nucleus of the monkey. Zeitschrift für Zellforschung und mikroskopische Anatomie **103**, 90–108 (1970)

Gutiérrez R., Boison D., Heinemann U., Stoffel W. Decompaction of CNS myelin leads to a reduction of the conduction velocity of action potentials in optic nerve. Neuroscience Letters **195**, 93–96 (1995)

Hagbarth K.-E., Wohlfart G. The number of muscle-spindles in certain muscles in cat in relation to the composition of the muscle nerves. Acta Anatomica **15**, 85–104 (1952)

Hagedorn L., Paratore C., Brugnoli G., Baert J.-L., Mercader N., Suter U., Sommer L. The Ets domain transcription factor Erm distinguishes rat satellite glia from Schwann cells and is regulated in satellite cells by neuregulin signaling. Developmental Biology **219**, 44–58 (2000)

Hahn A. Einige Beobachtungen an Riesenlarven von Rana esculenta. Archiv für mikroskopische Anatomie (I) **80**, 1–38 (1912)

Haimann C., Torri-Tarelli F., Fesce R., Ceccarelli B. Measurement of quantal secretion induced by ouabain and its correlation with depletion of synaptic vesicles. Journal of Cell Biology **101**, 1953–1965 (1985)

Hajdu F., Hassler R., Somogyi Gy. Neuronal and synaptic organization of the lateral geniculate nucleus of the tree shrew, *Tupaia glis*. Cell and Tissue Research **224**, 207–223 (1982)

Haley J.E., Ledeen R.W. Incorporation of axonally transported substances into myelin lipids. Journal of Neurochemistry **32**, 735–742 (1979)

Hall S.M., Williams P.L. The distribution of electron-dense tracers in peripheral nerve fibres. Journal of Cell Science **8**, 541–555 (1971)

Halls S.A., Sheller R.A., Bittner G.D., Mandy W.J. Intermediate filaments in crayfish axons. Journal of Cell Biology **105**, 209a (1987)

Halpain S. Actin and the agile spine: how and why do dendritic spines dance? Trends in Neurosciences **23**, 141–146 (2000)

Hama K. Some observations on the fine structure of the giant nerve fibers of the earthworm, *Eisenia foetida*. Journal of Biophysical and Biochemical Cytology **6**, 61–66 (1959)

Hama K. The fine structure of the Schwann cell sheath of the nerve fiber in the shrimp (*Penaeus japonicus*). Journal of Cell Biology **31**, 624–632 (1966)

Hamada I., Sakai M., Kubota K. Morphological differences between fast and slow pyramidal tract neurons in the monkey motor cortex as revealed by intracellular injection of horseradish peroxidase by pressure. Neuroscience Letters **22**, 233–238 (1981)

Hamlyn L.H. The fine structure of the mossy fibre endings in the hippocampus of the rabbit. Journal of Anatomy **96**, 112–120 (1962)

Hammerschlag R., Stone G.C., Bolen F.A., Lindsey J.D., Ellisman M.H. Evidence that all newly synthesized proteins destined for fast axonal transport pass through the Golgi apparatus. Journal of Cell Biology **93**, 568–575 (1982)

Hámori J., Szentágothai J. The Purkinje cell baskets: ultrastructure of an inhibitory synapse. Acta Biologica Academiae Scientiarum Hungaricae **15**, 465–479 (1965)

Hámori J., Szentágothai J. Participation of Golgi neuron processes in the cerebellar glomeruli: an electron microscope study. Experimental Brain Research **2**, 35–48 (1966)

Hámori J., Szentágothai J. Identification of synapses formed in the cerebellar cortex by Purkinje axon collaterals: an electron microscope study. Experimental Brain Research **5**, 118–128 (1968)

Hámori J., Takács J., Petrusz P. Immunogold electron microscopic demonstration of glutamate and GABA in normal and deafferented cerebellar cortex: Correlation between transmitter content and synaptic vesicle size. Journal of Histochemistry and Cytochemistry **38**, 1767–1777 (1990)

Hamzei-Sichani F., Davidson K.G.V., Yasumura T., Janssen W.G.M., Wearne S.L., Hof P.R., Traub R.D., Gutiérrez R., Ottersen O.P., Rash J.E. Mixed electrical-chemical synapses in adult rat hippocampus are primarily glutamatergic and coupled by connexin-36. Frontiers in Neuroanatomy **6**, 13 (2012)

Hanani M. Satellite glial cells in sensory ganglia: from form to function. Brain Research Reviews **48**, 457–476 (2005)

Hanani M. Satellite glial cells in sympathetic and parasympathetic ganglia: In search of function. Brain Research Reviews **64**, 304–327 (2010)

Hanani M., Huang T.Y., Cherkas P.S., Ledda M., Pannese E. Glial cell plasticity in sensory ganglia induced by nerve damage. Neuroscience **114**, 279–283 (2002)

Hanna R.B., Hirano A., Pappas G.D. Membrane specializations of dendritic spines and glia in the weaver mouse cerebellum: a freeze-fracture study. Journal of Cell Biology **68**, 403–410 (1976)

Hannover A. Die Chromsäure, ein vorzügliches Mittel bei mikroskopischen Untersuchungen. Müller's Archiv für Anatomie, Physiologie und wissenschaftliche Medicin, S. 549–558 (1840)

Hardesty I. Observations on the medulla spinalis of the elephant with some comparative studies of the *intumescentia cervicalis* and the neurones of the *columna anterior*. Journal of Comparative Neurology **12**, 125–182 (1902)

Hardesty I. On the development and nature of the neuroglia. American Journal of Anatomy **3**, 229–268 (1904)

Harding B.N. Dendro-dendritic synapses, including reciprocal synapses, in the ventrolateral nucleus of the monkey thalamus. Brain Research **34**, 181–185 (1971)

Harkin J.C. A series of desmosomal attachments in the Schwann sheath of myelinated mammalian nerves. Zeitschrift für Zellforschung und mikroskopische Anatomie **64**, 189–195 (1964)

Hartfuss E., Galli R., Heins N., Götz M. Characterization of CNS precursor subtypes and radial glia. Developmental Biology **229**, 15–30 (2001)

Hartline D.K. The evolutionary origins of glia. Glia **59**, 1215–1236 (2011)

Hartmann J.F. An electron optical study of sections of central nervous system. Journal of Comparative Neurology **99**, 201–249 (1953)

Harvey E.N., Fankhauser G. The tension at the surface of the eggs of the salamander, Triturus (Diemyctylus) viridescens. Journal of Cellular and Comparative Physiology **3**, 463–475 (1933)

Harvey E.N., Marsland D.A. The tension at the surface of Amoeba dubia with direct observations on the movement of cytoplasmic particles at high centrifugal speeds. Journal of Cellular and Comparative Physiology **2**, 75–97 (1932)

Hashimoto P.H. On the fenestrated capillary endothelium in the area postrema of the rat and cat. Acta Anatomica Nipponica **41**, 154–155 (1966)

Hashimoto P.H. Electron microscopic study on gliosome formation in postnatal development of spinal cord in the cat. Journal of Comparative Neurology **137**, 251–266 (1969)

Hashimoto P.H. Intracellular channels as a route for protein passage in the capillary endothelium of the shark brain. American Journal of Anatomy **134**, 41–58 (1972)

Hassinger T.D., Guthrie P.B., Atkinson P.B., Bennett M.V.L., Kater S.B. An extracellular signaling component in propagation of astrocytic calcium waves. Proceedings of the National Academy of Sciences of the U.S.A. **93**, 13268–13273 (1996)

Hatai S. On the presence of the centrosome in certain nerve cells of the white rat. Journal of Comparative Neurology **11**, 25–39 (1901)

Haug H. Die Gestalt des Neurons aus der Sicht quantitativer Untersuchungen (Referat). Verhandlungen der anatomischen Gesellschaft **76**, 73–80 (1982)

Haug H. Are neurons of the human cerebral cortex really lost during aging? A morphometric examination. In: Senile Dementia of the Alzheimer Type (J. Traber, W.H. Gispen Eds.) pp. 150–163. Springer, Berlin 1985

Haug H., Kühl S., Mecke E., Sass N.-L., Wasner K. The significance of morphometric procedures in the investigation of age changes in cytoarchitectonic structures of human brain. Journal für Hirnforschung **25**, 353–374 (1984)

Haydon P.G. Neuroglial networks: Neurons and glia talk to each other. Current Biology **10**, R712-R714 (2000)

Hegstrom C.D., Breedlove S.M. Seasonal plasticity of neuromuscular junctions in adult male Siberian hamster (*Phodopus sungorus*). Brain Research **819**, 83–88 (1999)

Heidemann S.R., Landers J.M., Hamborg M.A. Polarity orientation of axonal microtubules. Journal of Cell Biology **91**, 661–665 (1981)

Heidenhain M. Plasma und Zelle. Eine allgemeine Anatomie der lebendigen Masse. Band 2. In: Handbuch der Anatomie des Menschen (K. von Bardeleben Hrsg.). G. Fischer, Jena 1911

Held H. Beiträge zur Structur der Nervenzellen und ihrer Fortsätze. I, II, III. Archiv für Anatomie und Entwickelungsgeschichte, 396–416 (1895); 204–294 (1897a); 273–312 Suppl.-Band (1897b)

Held H. Die Entwicklung des Nervengewebes bei den Wirbeltieren. J.A. Barth, Leipzig 1909a

Held H. Ueber die Neuroglia marginalis der menschlichen Grosshirnrinde. Monatsschrift für Psychiatrie und Neurologie **26** (Ergänzungsheft) 360–416 (1909b)

Helfert R.H., Sommer T.J., Meeks J., Hofstetter P., Hughes L.F. Age-related synaptic changes in the central nucleus of the inferior colliculus of Fischer-344 rats. Journal of Comparative Neurology **406**, 285–298 (1999)

Helmholtz A. [=H.L.F.] De fabrica systematis nervosi evertebratorum. Inaugural Dissertation. Typis Nietackianis, Berolini 1842

Hemminki K. Relative turnover of tubulin subunits in rat brain. Biochimica et Biophysica Acta **310**, 285–288 (1973)

Henderson G., Tomlinson B.E., Gibson P.H. Cell counts in human cerebral cortex in normal adults throughout life using an image analysing computer. Journal of the Neurological Sciences **46**, 113–136 (1980)

Henle J. Handbuch der Nervenlehre des Menschen. F. Vieweg und Sohn, Braunschweig 1871

Henneman E. Relation between size of neurons and their susceptibility to discharge. Science **126**, 1345–1347 (1957)

Henneman E., Somjen G., Carpenter D.O. Functional significance of cell size in spinal motoneurons. Journal of Neurophysiology **28**, 560–580 (1965)

Henrikson C.K., Vaughn J.E. Fine structural relationships between neurites and radial glial processes in developing mouse spinal cord. Journal of Neurocytology **3**, 659–675 (1974)

Herman C.J., Lapham L.W. Neuronal polyploidy and nuclear volumes in the cat central nervous system. Brain Research **15**, 35–48 (1969)

Herman I.M., D'Amore P.A. Microvascular pericytes contain muscle and nonmuscle actin. Journal of Cell Biology **101**, 43–52 (1985)

Herman M.M., Ralston H.J.III Laminated cytoplasmic bodies and annulate lamellae in the cat ventrobasal and posterior thalamus. Anatomical Record **167**, 183–196 (1970)

Hernández C., Blackburn E., Alvarez J. Calibre and microtubule content of the non-medullated and myelinated domains of optic nerve axons of rats. European Journal of Neuroscience **1**, 654–658 (1989)

Herrera A.A., Banner L.R., Nagaya N. Repeated, *in vivo* observation of frog neuromuscular junctions: remodelling involves concurrent growth and retraction. Journal of Neurocytology **19**, 85–99 (1990)

Hertz L. An intense potassium uptake into astrocytes, its further enhancement by high concentrations of potassium, and its possible involvement in potassium homeostasis at the cellular level. Brain Research **145**, 202–208 (1978)

Hervonen A., Koistinaho J., Alho H., Helen P., Santer R.M., Rapoport S.I. Age related heterogeneity of lipopigments in human sympathetic ganglia. Mechanisms of Ageing and Development **35**, 17–29 (1986)

Hess A. The ground substance of the central nervous system revealed by histochemical staining. Journal of Comparative Neurology **98**, 69–92 (1953)

Hess A. The fine structure of young and old spinal ganglia. Anatomical Record **123**, 399–423 (1955)

Hess A. The fine structure and morphological organization of the peripheral nerve-fibres and trunks of the cockroach (*Periplaneta americana*). Quarterly Journal of Microscopical Science **99**, 333–340 (1958)

Hess A., Lansing A.I. The fine structure of peripheral nerve fibers. Anatomical Record **117**, 175–199 (1953)

Hess A., Young J.Z. Correlation of internodal length and fibre diameter in the central nervous system. Nature **164**, 490–491 (1949)

Heuser J.E., Doggenweiler C.F. The fine structural organization of nerve fibers, sheaths, and glial cells in the prawn, *Palaemonetes vulgaris*. Journal of Cell Biology **30**, 381–403 (1966)

Heuser J.E., Reese T.S. Evidence for recycling of synaptic vesicle membrane during transmitter release at the frog neuromuscular junction. Journal of Cell Biology **57**, 315–344 (1973)

Heuser J.E., Reese T.S. Structural changes after transmitter release at the frog neuromuscular junction. Journal of Cell Biology **88**, 564–580 (1981)

Heuser J.E., Reese T.S., Dennis M.J., Jan Y., Jan L., Evans L. Synaptic vesicle exocytosis captured by quick freezing and correlated with quantal transmitter release. Journal of Cell Biology **81**, 275–300 (1979)

Heuser J.E., Reese T.S., Landis D.M.D. Functional changes in frog neuromuscular junctions studied with freeze-fracture. Journal of Neurocytology **3**, 109–131 (1974)

Hibino H., Horio Y., Fujita A., Inanobe A., Doi K., Gotow T., Uchiyama Y., Kubo T., Kurachi Y. Expression of an inwardly rectifying K^+ channel, Kir4.1, in satellite cells of rat cochlear ganglia. American Journal of Physiology **277**, C638-C644 (1999)

Hidaka S., Akahori Y., Kurosawa Y. Dendrodendritic electrical synapses between mammalian retinal ganglion cells. Journal of Neuroscience **24**, 10553–10567 (2004)

Hild W. Das morphologische, kinetische und endokrinologische Verhalten von hypothalamischem und neurohypophysären Gewebe in vitro. Zeitschrift für Zellforschung und mikroskopische Anatomie **40**, 257–312 (1954)

Hildebrand C. Ultrastructural and light-microscopic studies of the nodal region in large myelinated fibres of the adult feline spinal cord white matter. Acta Physiologica Scandinavica Suppl. 364, 43–71 (1971)

Hinds J.W. Early neuron differentiation in the mouse olfactory bulb. II. Electron microscopy. Journal of Comparative Neurology **146**, 253–276 (1972)

Hinds J.W., McNelly N.A. Aging of the rat olfactory bulb: growth and atrophy of constituent layers and changes in size and number of mitral cells. Journal of Comparative Neurology **171**, 345–368 (1977)

Hinds J.W., McNelly N.A. Aging in the rat olfactory bulb: quantitative changes in mitral cell organelles and somato-dendritic synapses. Journal of Comparative Neurology **184**, 811–819 (1979)

Hinojosa R. Synaptic ultrastructure in the tangential nucleus of the goldfish (*Carassius auratus*). American Journal of Anatomy **137**, 159–186 (1973)

Hinojosa R., Robertson J.D. Ultrastructure of the spoon type synaptic endings in the nucleus vestibularis tangentialis of the chick. Journal of Cell Biology **34**, 421–430 (1967)

Hinrichsen C.F.L. Coupling between cells of the trigeminal mesencephalic nucleus. Dental Research **49**, 1369–1373 (1970)

Hinrichsen C.F.L., Larramendi L.M.H. The trigeminal mesencephalic nucleus. II. Electron microscopy. American Journal of Anatomy **127**, 303–320 (1970)

Hirano A. Pathology of amyotrophic lateral sclerosis. In: Slow, Latent, and Temperate Virus Infections (D.C. Gaidusek, C.J. Gibbs Jr., M. Alpers Eds.). NINDB Monograph No. 2, pp. 23–37. US Department of Health, Education and Welfare, Washington D.C., 1965

Hirano A. A confirmation of the oligodendroglial origin of myelin in the adult rat. Journal of Cell Biology **38**, 637–640 (1968)

Hirano A., Dembitzer H.M. Cerebellar alterations in the weaver mouse. Journal of Cell Biology **56**, 478–486 (1973)

Hirano A., Zimmerman H.M. Some new cytological observations of the normal rat ependymal cell. Anatomical Record **158**, 293–302 (1967)

Hirata Y. Some observations of the fine structure of the synapses in the olfactory bulb of the mouse, with particular reference to the atypical synaptic configurations. Archivum Histologicum Japonicum **24**, 293–302 (1964)

Hirata Y. Occurrence of cylindrical synaptic vesicles in the central nervous system perfused with buffered formalin solution prior to OsO_4-fixation. Archivum Histologicum Japonicum **26**, 269–279 (1966)

Hirokawa N. Cross-linker system between neurofilaments, microtubules, and membranous organelles in frog axons revealed by the quick-freeze, deep-etching method. Journal of Cell Biology **94**, 129–142 (1982)

Hirokawa N., Glicksman M.A., Willard M.B. Organization of mammalian neurofilament polypeptides within the neuronal cytoskeleton. Journal of Cell Biology **98**, 1523–1536 (1984)

Hirrlinger J., Hülsmann S., Kirchhoff F. Astroglial processes show spontaneous motility at active synaptic terminals *in situ*. European Journal of Neuroscience **20**, 2235–2239 (2004)

His W. Zur Geschichte des menschlichen Rückenmarkes und der Nervenwurzeln. Abhandlungen der mathematisch-physischen Classe der Königlich Sächsischen Gesellschaft der Wissenschaften **13**, 147–209, 477–514 (1886)

His W. Die Neuroblasten und deren Entstehung im embryonalen Mark. Abhandlungen der mathematisch-physischen Classe der Königlich Sächsischen Gesellschaft der Wissenschaften **15**, 311–372 (1889)

Hisanaga S.-i., Hirokawa N. Structure of the peripheral domains of neurofilaments revealed by low angle rotary shadowing. Journal of Molecular Biology **202**, 297–305 (1988)

Hobi R., Studer M., Ruch F., Kuenzle C.C. The DNA content of cerebral cortex neurons. Determinations by cytophotometry and high performance liquid chromatography. Brain Research **305**, 209–219 (1984)

Hodge C.F. A microscopical study of changes due to functional activity in nerve cells. Journal of Morphology **7**, 95–164 (1892)

Hodge C.F. Changes in ganglion cells from birth to senile death. Observations on man and honey-bee. Journal of Physiology **17**, 129–134 (1894)

Hökfelt T. Neuronal communication through multiple coexisting messengers. In: Synaptic Function (G.M. Edelman, W.E. Gall, W.M. Cowan Eds.) pp. 179–211. John Wiley & Sons, New York 1987

Hösli E., Hösli L. Autoradiographic localization of the uptake of $[^3H]$-GABA and $[^3H]$L-glutamic acid in neurones and glial cells of cultured dorsal root ganglia. Neuroscience Letters **7**, 173–176 (1978)

Hösli L., Andrès P.F., Hösli E. Neuron-glia interactions: indirect effect of GABA on cultured glial cells. Experimental Brain Research **33**, 425–434 (1978)

Hof P.R., Duan H., Page T.L., Einstein M., Wicinski B., He Y., Erwin J.M., Morrison J.H. Age-related changes in GluR2 and NMDAR1 glutamate receptor subunit protein immunoreactivity in cortico-cortically projecting neurons in macaque and patas monkeys. Brain Research **928**, 175–186 (2002)

Hoffman P.N., Koo E.H., Muma N.A., Griffin J.W., Price D.L. Role of neurofilaments in the control of axonal caliber in myelinated nerve fibers. In: Intrinsic Determinants of Neuronal Form and Function (R.J. Lasek, M.M. Black Eds.) pp. 389–402. Alan R. Liss, New York 1988

Hoffman P.N., Lasek R.J. The slow component of axonal transport. Identification of major structural polypeptides of the axon and their generality among mammalian neurons. Journal of Cell Biology **66**, 351–366 (1975)

Hollenbeck P.J., Saxton W.M. The axonal transport of mitochondria. Journal of Cell Science **118**, 5411–5419 (2005)

Holloway R.L.Jr. Dendritic branching: some preliminary results of training and complexity in rat visual cortex. Brain Research **2**, 393–396 (1966)

Holmes W.R. Is the function of dendritic spines to concentrate calcium? Brain Research **519**, 338–342 (1990)

Holmgren E. Weitere Mitteilungen über den Bau der Nervenzellen. Anatomischer Anzeiger **16**, 388–397 (1899)

Holmgren E. Beiträge zur Morphologie der Zelle. I. Nervenzellen. Anatomische Hefte **18**, 267–325 (1901)

Holmgren E. Einige Worte über das "Trophospongium" verschiedener Zellarten. Anatomischer Anzeiger **20**, 433–440 (1902)

Holtmaat A., Svoboda K. Experience-dependent structural synaptic plasticity in the mammalian brain. Nature Reviews Neuroscience **10**, 647–658 (2009)

Holtzman E. Peroxisomes in nervous tissue. Annals of the New York Academy of Sciences **386**, 523–525 (1982)

Holtzman E., Novikoff A.B., Villaverde H. Lysosomes and GERL in normal and chromatolytic neurons of the rat ganglion nodosum. Journal of Cell Biology **33**, 419–435 (1967)

Holtzman E., Peterson E.R. Uptake of protein by mammalian neurons. Journal of Cell Biology **40**, 863–869 (1969)

Holz A., Weber W. Periodisch auftretende Querstrukturen in Nervenfasern des Bulbus olfactorius der Elritze *Phoxinus laevis*. Experientia **26**, 1349–1350 (1970)

Honig M.G., Hume R.I. Fluorescent carbocyanine dyes allow living neurons of identified origin to be studied in long-term cultures. Journal of Cell Biology **103**, 171–187 (1986)

Horridge G.A. The origins of the nervous system. In: The Structure and Function of Nervous Tissue (G.H. Bourne Ed.) **1**, 1–31. Academic Press, New York 1968

Horridge G.A., Chapman R.A. Sheaths of the motor axons of the crab *Carcinus*. Quarterly Journal of Microscopical Science **105**, 175–181 (1964)

Horstmann E. Die Faserglia des Selachiergehirns. Zeitschrift für Zellforschung und mikroskopische Anatomie **39**, 588–617 (1954)

Hosoya Y., Fujita T. Scanning electron microscope observation of intraventricular macrophages (Kolmer cells) in the rat brain. Archivum Histologicum Japonicum **35**, 133–140 (1973)

Hsu K., Terakawa S. Fenestration in the myelin sheath of nerve fibers of the shrimp: A novel node of excitation for saltatory conduction. Journal of Neurobiology **30**, 397–409 (1996)

Hu X., Viesselmann C., Nam S., Merriam E., Dent E.W. Activity-dependent dynamic microtubule invasion of dendritic spines. Journal of Neuroscience **28**, 13094–13105 (2008)

Huang T.Y., Cherkas P.S., Rosenthal D.W., Hanani M. Dye coupling among satellite glial cells in mammalian dorsal root ganglia. Brain Research **1036**, 42–49 (2005)

Huang T.Y., Hanani M., Ledda M., De Palo S., Pannese E. Aging is associated with an increase in dye coupling and in gap junction number in satellite glial cells of murine dorsal root ganglia. Neuroscience **137**, 1185–1192 (2006)

Hubbard J.I., Kwanbunbumpen S. Evidence for the vesicle hypothesis. Journal of Physiology **194**, 407–420 (1968)

Huber G.C. Studies on the neuroglia. American Journal of Anatomy **1**, 45–61 (1901)

Humbertson A.Jr., Zimmermann E., Leedy M. A chronological study of mitotic activity in satellite cell hyperplasia associated with chromatolytic neurons. Zeitschrift für Zellforschung und mikroskopische Anatomie **100**, 507–515 (1969)

Huntington H.W., Terry R.D. The origin of the reactive cells in cerebral stab wounds. Journal of Neuropathology and Experimental Neurology **25**, 646–653 (1966)

Huxley A.F., Stämpfli R. Evidence for saltatory conduction in peripheral myelinated nerve fibres. Journal of Physiology **108**, 315–339 (1949)

Hydén H., Pigon A. A cytophysiological study of the functional relationship between oligodendroglial cells and nerve cells of Deiters' nucleus. Journal of Neurochemistry **6**, 57–72 (1960)

Iadecola C., Nedergaard M. Glial regulation of the cerebral micro-vasculature. Nature Neuroscience **10**, 1369–1376 (2007)

Ichimura T., Hashimoto P.H. Structural components in the synaptic cleft captured by freeze-substitution and deep etching of directly frozen cerebellar cortex. Journal of Neurocytology **17**, 3–12 (1988)

Imamoto K., Leblond C.P. Presence of labeled monocytes, macrophages and microglia in a stab wound of the brain following an injection of bone marrow cells labeled with ^{3}H-uridine into rats. Journal of Comparative Neurology **174**, 255–280 (1977)

Imamoto K., Paterson J.A., Leblond C.P. Radioautographic investigation of gliogenesis in the corpus callosum of young rats. I. Sequential changes in oligodendrocytes. Journal of Comparative Neurology **180**, 115–138 (1978)

Inestrosa N.C., Alvarez J. Axons grow in the aging rat but fast tranport and acetylcholinesterase content remain unchanged. Brain Research **441**, 331–338 (1988)

Inomata K., Okada T., Seguchi H., Ogawa K. The confronting cisternae in neurons in the substantia gelatinosa of the spinal cord of rats and Japanese monkeys. Cell and Tissue Research **232**, 487–492 (1983)

Irwin S.A., Patel B., Idupulapati M., Harris J.B., Crisostomo R.A., Larsen B.P., Kooy F., Willems P.J., Cras P., Kozlowski P.B., Swain R.A., Weiler I.J., Greenough W.T. Abnormal dendritic spine characteristics in the temporal and visual cortices of patients with fragile-X syndrome: A quantitative examination. American Journal of Medical Genetics **98**, 161–167 (2001)

Ishii H., Suzuki K., Aikawa H., Nonaka I. Internodal microvillus-like Schwann cell fingers in myelinated fibres in mouse spinal roots. Journal of Neurocytology **14**, 261–267 (1985)

Israel M., Gautron J., Lesbats B. Isolement des vésicules synaptiques de l'organe électrique de la Torpille et localisation de l'acétylcholine à leur niveau. Comptes Rendus des Séances de l'Académie des Sciences **266**, 273–275 (1968)

Iversen L.L., Bloom F.E. Studies of the uptake of ^{3}H-GABA and [^{3}H]glycine in slices and homogenates of rat brain and spinal cord by electron microscopic autoradiography. Brain Research **41**, 131–143 (1972)

Jacob J.M. Fast axonal transport rates are unchanged in 6- and 24-month F344 rats. Brain Research **699**, 154–156 (1995)

Jacobs A.J., Kamholz J., Selzer M.E. The single lamprey neurofilament subunit (NF-180) lacks multiphosphorylation repeats and is expressed selectively in projection neurons. Molecular Brain Research **29**, 43–52 (1995)

Jacobs B., Driscoll L., Schall M. Life-span dendritic and spine changes in areas 10 and 18 of human cortex: A quantitative Golgi study. Journal of Comparative Neurology **386**, 661–680 (1997)

Jacobs J.M., Macfarlane R.M., Cavanagh J.B. Vascular leakage in the dorsal root ganglia of the rat, studied with horseradish peroxidase. Journal of the Neurological Sciences **29**, 95–107 (1976)

Jacobson M. Developmental Neurobiology (2nd ed.). Plenum Press, New York 1978

Jacobson M. Foundations of Neuroscience. Plenum Press, New York and London 1993

Jalenques I., Albuisson E., Despres G., Romand R. Distribution of glial fibrillary acidic protein (GFAP) in the cochlear nucleus of adult and aged rats. Brain Research **686**, 223–232 (1995)

James D.W., Tresman R.L. Synaptic profiles in the outgrowth from chick spinal cord *in vitro*. Zeitschrift für Zellforschung und mikroskopische Anatomie **101**, 598–606 (1969)

James K.A.C., Bray J.J., Morgan I.G., Austin L. The effect of colchicine on the transport of axonal protein in the chicken. Biochemical Journal **117**, 767–771 (1970)

Jan Y.-N., Jan L.Y. Branching out: mechanisms of dendritic arborization. Nature Reviews Neuroscience **11**, 316–328 (2010)

Jensen H., Holtet L., Hoen R. Nerve-Purkinje fiber relationship in the moderator band of bovine and caprine heart. Cell and Tissue Research **188**, 11–18 (1978)

Jessen K.R., Mirsky R. Nonmyelin-forming Schwann cells coexpress surface proteins and intermediate filaments not found in myelin-forming cells: a study of Ran-2, A5E3 antigen and glial fibrillary acidic protein. Journal of Neurocytology **13**, 923–934 (1984)

Jessen K.R., Thorpe R., Mirsky R. Molecular identity, distribution and heterogeneity of glial fibrillary acidic protein: an immunoblotting and immunohistochemical study of Schwann cells, satellite cells, enteric glia and astrocytes. Journal of Neurocytology **13**, 187–200 (1984)

Jiang M., Lee C.L., Smith K.L., Swann J.W. Spine loss and other persistent alterations of hippocampal pyramidal cell dendrites in a model of early-onset epilepsy. Journal of Neuroscience **18**, 8356–8368 (1998)

Jimenez-Andrade J.M., Peters C.M., Mejia N.A., Ghilardi J.R., Kuskowski M.A., Mantyh P.W. Sensory neurons and their supporting cells located in the trigeminal, thoracic and lumbar ganglia differentially express markers of injury following intravenous administration of paclitaxel in the rat. Neuroscience Letters **405**, 62–67 (2006)

Johnson B.M., Miao M., Sadun A.A. Age-related decline of human optic nerve axon populations. Age **10**, 5–9 (1987)

Johnson F., Bottjer S.W. Hormone-induced changes in identified cell populations of the higher vocal center in male canaries. Journal of Neurobiology **24**, 400–418 (1993)

Johnson M.I., Higgins D., Ard M.D. Astrocytes induce dendritic development in cultured sympathetic neurons. Developmental Brain Research **47**, 289–292 (1989)

Jones E.G., Hartman B.K. Recent advances in neuroanatomical methodology. Annual Review of Neuroscience **1**, 215–296 (1978)

Jones E.G., Powell T.P.S. Synapses on the axon hillocks and initial segments of pyramidal cell axons in the cerebral cortex. Journal of Cell Science **5**, 495–507 (1969)

Julien J.-P., Mushynski W.E. The distribution of phosphorylation sites among identified proteolytic fragments of mammalian neurofilaments. Journal of Biological Chemistry **258**, 4019–4025 (1983)

Jung H.J., Raine C.S., Suzuki K. Schwann cells and peripheral nervous system myelin in the rat retina. Acta Neuropathologica **44**, 245–247 (1978)

Kabaso D., Coskren P.J., Henry B.I., Hof P.R., Wearne S.I. The electrotonic structure of pyramidal neurons contributing to prefrontal cortical circuits in macaque monkeys is significantly altered in aging. Cerebral Cortex **19**, 2248–2268 (2009)

Kagayama M. The fine structure of the monkey submandibular gland with a special reference to intra-acinar nerve endings. American Journal of Anatomy **131**, 185–196 (1971)

Kaiserman-Abramof I.R., Palay S.L. Fine structural studies of the cerebellar cortex in a mormyrid fish. In:

Neurobiology of Cerebellar Evolution and Development (R. Llinás Ed.) pp. 171–205. AMA-ERF Institute for Biomedical Research, Chicago 1969

Kaiserman-Abramof I.R., Peters A. Some aspects of the morphology of Betz cells in the cerebral cortex of the cat. Brain Research **43**, 527–546 (1972)

Kamiguchi H., Lemmon V. A neuronal form of the cell adhesion molecule L1 contains a tyrosine-based signal required for sorting to the axonal growth cone. Journal of Neuroscience **18**, 3749–3756 (1998)

Kanaka C., Ohno K., Okabe A., Kuriyama K., Itoh T., Fukuda A., Sato K. The differential expression patterns of messenger RNAs encoding K-Cl cotransporters (KCC1,2) and Na-K-2Cl cotransporter (NKCC1) in the rat nervous system. Neuroscience **104**, 933–946 (2001)

Kane C.J.M., Sims T.J., Gilmore S.A. Astrocytes in the aged rat spinal cord fail to increase GFAP mRNA following sciatic nerve axotomy. Brain Research **759**, 163–165 (1997)

Kaneko A. Electrical connexions between horizontal cells in the dogfish retina. Journal of Physiology **213**, 95–105 (1971)

Kaplan M.R., Meyer-Franke A., Lambert S., Bennett V., Duncan I.D., Levinson S.R., Barres B.A. Induction of sodium channel clustering by oligodendrocytes. Nature **386**, 724–728 (1997)

Kaplan M.S. Proliferation of epithelial cells in the adult primate choroid plexus. Anatomical Record **197**, 495–502 (1980)

Kappers J.A. Beitrag zur experimentellen Untersuchung von Funktion und Herkunft der Kolmerschen Zellen des Plexus chorioideus beim Axolotl und Meerschweinchen. Zeitschrift für Anatomie und Entwicklungsgeschichte **117**, 1–19 (1953)

Kappers J.A. The pineal organ: an introduction. In: The Pineal Gland (G.E.W. Wolstenholme, J. Knight Eds.) pp. 3–34. Churchill Livingstone, Edinburgh 1971

Karabelas A., Purpura D.P. Evidence for autapses in the substantia nigra. Brain Research **200**, 467–473 (1980)

Káradóttir R., Attwell D. Neurotransmitter receptors in the life and death of oligodendrocytes. Neuroscience **145**, 1426–1438 (2007)

Karlsson M., Hildebrand C. Postnatal development and aging of the rat ventral root L5: electron microscopic and immunohistochemical studies. Journal of Comparative Neurology **364**, 211–218 (1996)

Karlsson U. Comparison of the myelin period of peripheral and central origin by electron microscopy. Journal of Ultrastructure Research **15**, 451–468 (1966a)

Karlsson U. Three-dimensional studies of neurons in the lateral geniculate nucleus of the rat. I. Organelle organization in the perikaryon and its proximal branches. Journal of Ultrastructure Research **16**, 429–481 (1966b)

Katz B., Miledi R. Transmitter leakage from motor nerve endings. Proceedings of the Royal Society of London. Series B **196**, 59–72 (1977)

Kawaguchi Y. Physiological, morphological, and histochemical characterization of three classes of interneurons in rat neo-striatum. Journal of Neuroscience **13**, 4908–4923 (1993)

Kawai Y., Senba E. Correlation between dendrodendritic synapses of adrenergic type and synaptically evoked hyperpolarization in the sympathetic ganglion of adult rats. Neuroscience **68**, 925–935 (1995)

Kawamata T., Ninomiya T., Toriyabe M., Yamamoto J., Niiyama Y., Omote K., Namiki A. Immunohistochemical analysis of acid-sensing ion channel 2 expression in rat dorsal root ganglion and effects of axotomy. Neuroscience **143**, 175–187 (2006)

Kawamura Y., Okazaki H., O'Brien P.C., Dyck P.J. Lumbar motoneurons of man: I. Number and diameter histogram of alpha and gamma axons of ventral root. Journal of Neuropathology and Experimental Neurology **36**, 853–860 (1977)

Keast J.R., Anderson T.M. Glutamate and aspartate immunoreactivity in dorsal root ganglion cells supplying visceral and somatic targets and evidence for peripheral axonal transport. Journal of Comparative Neurology **424**, 577–587 (2000)

Keast J.R., Saunders R.J. Testosterone has potent, selective effects on the morphology of pelvic autonomic neurons which control the bladder, lower bowel and internal reproductive organs of the male rat. Neuroscience **85**, 543–556 (1998)

Keirstead H.S., Blakemore W.F. Identification of post-mitotic oligodendrocytes incapable of remyelination within the demyelinated adult spinal cord. Journal of Neuropathology and Experimental Neurology **56**, 1191–1201 (1997)

Kemali M., Sada E. Myelinated cell bodies in the habenular nuclei of the frog. Brain Research **54**, 355–359 (1973)

Kerns J.M., Peters A. Ultrastructure of a large ventrolateral dendritic bundle in the rat ventral horn. Journal of Neurocytology **3**, 533–555 (1974)

Kershman J. Genesis of microglia in the human brain. Archives of Neurology and Psychiatry **41**, 24–50 (1939)

Keuker J.I.H., de Biurrun G., Luiten P.G.M., Fuchs E. Preservation of hippocampal neuron numbers and hippocampal subfield volumes in behaviorally characterized aged tree shrews. Journal of Comparative Neurology **468**, 509–517 (2004)

Key A.H., Retzius G. Studien in der Anatomie des Nervensystems und des Bindegewebes. 2 Hälfte. Samson & Wallin, Stockholm 1876

Kidd M. Electron microscopy of the inner plexiform layer of the retina in the cat and the pigeon. Journal of Anatomy **96**, 179–187 (1962)

Kidd M. Paired helical filaments in electron microscopy of Alzheimer's disease. Nature **197**, 192–193 (1963)

Kim S.U., Masurovsky E.B., Benitez H.H., Murray M.R. Histochemical studies of the intranuclear rodlet in neurons of chicken sympathetic and sensory ganglia. Histochemie **24**, 33–40 (1970)

King J.S. A light and electron microscopic study of perineuronal glial cells and processes in the rabbit neocortex. Anatomical Record **161**, 111–124 (1968)

King J.S., Martin G.F., Bowman M.H. The direct spinal area of the inferior olivary nucleus: an electron microscopic study. Experimental Brain Research **22**, 13–24 (1975)

Kitamura T. The origin of brain macrophages. Some considerations on the microglia theory of del Rio-Hortega. Acta Pathologica Japonica **23**, 11–26 (1973)

Klatzo I., Miquel J., Tobias C., Haymaker W. Effects of alpha particle radiation on the rat brain, including vascular permeability and glycogen studies. Journal of Neuropathology and Experimental Neurology **20**, 459–483 (1961)

Kleim J.A., Vij K., Ballard D.H., Greenough W.T. Learning-dependent synaptic modifications in the cerebellar cortex of the adult rat persist for at least four weeks. Journal of Neuroscience **17**, 717–721 (1997)

Klein C. Innervation des cellules du pancréas endocrine du Poisson Téléostéen *Xiphophorus helleri* H. Zeitschrift für Zellforschung und mikroskopische Anatomie **113**, 564–580 (1971)

Klemm M.F., Van Helden D.F., Luff S.E. Ultrastructural analysis of sympathetic neuromuscular junctions on mesenteric veins of the guinea pig. Journal of Comparative Neurology **334**, 159–167 (1993)

Klinkerfuss G.H. An electron microscopic study of the ependyma and subependymal glia of the lateral ventricle of the cat. American Journal of Anatomy **115**, 71–100 (1964)

Knafo S., Grossman Y., Barkai E., Benshalom G. Olfactory learning is associated with increased spine density along apical dendrites of pyramidal neurons in the rat piriform cortex. European Journal of Neuroscience **13**, 633–638 (2001)

Knight D.S., Ellison J.P., Hibbs R.G., Hyman A.L., Kadowitz P.J. A light and electron microscopic study of the innervation of pulmonary arteries in the cat. Anatomical Record **201**, 513–521 (1981)

Knott G.W., Quairiaux C., Genoud C., Welker E. Formation of dendritic spines with GABAergic synapses induced by whisker stimulation in adult mice. Neuron **34**, 265–273 (2002)

Knowles F., Vollrath L. Synaptic contacts between neurosecretory fibres and pituicytes in the pituitary of the eel. Nature **206**, 1168–1169 (1965)

Knowles R.B., Sabry J.H., Martone M.E., Deerinck T.J., Ellisman M.H., Bassell G.J., Kosik K.S. Translocation of RNA granules in living neurons. Journal of Neuroscience **16**, 7812–7820 (1996)

Knox C.A., Kokmen E., Dyck P.J. Morphometric alteration of rat myelinated fibers with aging. Journal of Neuropathology and Experimental Neurology **48**, 119–139 (1989)

Kobayashi K., Fukuoka T., Yamanaka H., Dai Y., Obata K., Tokunaga A., Noguchi K. Neurons and glial cells differentially express P2Y receptor mRNAs in the rat dorsal root ganglion and spinal cord. Journal of Comparative Neurology **498**, 443–454 (2006)

Koch C., Zador A. The function of dendritic spines: Devices subserving biochemical rather than electrical compartmentalization. Journal of Neuroscience **13**, 413–422 (1993)

Koehler R.C., Roman R.J., Harder D.R. Astrocytes and the regulation of cerebral blood flow. Trends in Neurosciences **32**, 160–169 (2009)

Koelliker A. Histologische Studien an Batrachierlarven. Zeitschrift für wissenschaftliche Zoologie **43**, 1–40 (1886)

Koelliker A. Handbuch der Gewebelehre des Menschen. VI Aufl. Band 2. W. Engelmann, Leipzig 1896

Koelliker A. Die Entwicklung der Elemente des Nervensystems. Zeitschrift für wissenschaftliche Zoologie **82**, 1–38 (1905)

Koenig E. Synthetic mechanisms in the axon. II. RNA in myelin-free axons of the cat. Journal of Neurochemistry **12**, 357–361 (1965)

Koenig E., Martin R., Titmus M., Sotelo-Silveira J.R. Cryptic peripheral ribosomal domains distributed intermittently along mammalian myelinated axons. Journal of Neuroscience **20**, 8390–8400 (2000)

Koenig H. The synthesis and peripheral flow of axoplasm. Transactions of the American Neurological Association **83**, 162–164 (1958)

Koenig J.H., Kosaka T., Ikeda K. The relationship between the number of synaptic vesicles and the amount of transmitter released. Journal of Neuroscience **9**, 1937–1942 (1989)

Kohama S.G., Goss J.R., Finch C.E., McNeill T.H. Increases of glial fibrillary acidic protein in the aging female mouse brain. Neurobiology of Aging **16**, 59–67 (1995)

Kohno K. Neurotubules contained within the dendrite and axon of Purkinje cell of frog. Bulletin of Tokyo Medical and Dental University **11**, 411–426 (1964)

Kohno K. Symmetrical axo-axonic synapses in the axon cap of the goldfish Mauthner cell. Brain Research **23**, 255–258 (1970)

Kolb H. The organization of the outer plexiform layer in the retina of the cat: electron microscopic observations. Journal of Neurocytology **6**, 131–153 (1977)

Kolb H., Famiglietti E.V. Rod and cone pathways in the inner plexiform layer of cat retina. Science **186**, 47–49 (1974)

Kollros J.J. Hormonal control of the size of mesencephalic fifth nucleus in amphibians. In: Thyroid Hormones and Brain Development (G.D. Grave Ed.) pp. 119–136. Raven Press, New York 1977

Kolmer W. Über eine eigenartige Beziehung von Wanderzellen zu den Chorioidealplexus des Gehirns der Wirbeltiere. Anatomischer Anzeiger **54**, 15–19 (1921)

Kolmer W. Über einen supraependymalen Nervenplexus in den Hirnventrikeln des Affen. Zeitschrift für Anatomie und Entwicklungsgeschichte **93**, 182–187 (1930)

Kolmer W. Über das Sagittalorgan, ein zentrales Sinnesorgan der Wirbeltiere, insbesondere beim Affen. Zeitschrift für Zellforschung und mikroskopische Anatomie **13**, 236–248 (1931)

Komiya Y. Slowing with age of the rate of slow axonal flow in bifurcating axons of rat dorsal root ganglion cells. Brain Research **183**, 477–480 (1980)

Komuro T., Baluk P., Burnstock G. An ultrastructural study of neurons and non-neuronal cells in the myenteric plexus of the rabbit colon. Neuroscience **7**, 1797–1806 (1982)

Koneff H. Beiträge zur Kenntnis in den Nervenzellen der peripheren Ganglien. Mittheilungen der Naturforschenden Gesellschaft in Bern, Nr. 1143–1168, S. 13–44 (1887)

Konigsmark B.W., Sidman R.L. Origin of brain macrophages in the mouse. Journal of Neuropathology and Experimental Neurology **22**, 643–676 (1963)

Konur S., Rabinowitz D., Fenstermaker V.L., Yuste R. Systematic regulation of spine sizes and densities in pyramidal neurons. Journal of Neurobiology **56**, 95–112 (2003)

Kordeli E., Lambert S., Bennett V. Ankyrin$_G$. Journal of Biological Chemistry **270**, 2352–2359 (1995)

Korn H., Sotelo C., Bennett M.V.L. The lateral vestibular nucleus of the toadfish *Opsanus tau*: ultrastructural and electrophysiological observations with special reference to electrotonic transmission. Neuroscience **2**, 851–884 (1977)

Korn H., Sotelo C., Crepel F. Electrotonic coupling between neurons in the rat lateral vestibular nucleus. Experimental Brain Research **16**, 255–275 (1973)

Korr H. Proliferation of different cell types in the brain. Advances in Anatomy, Embryology and Cell Biology **61**, 1–72 (1980)

Korr H., Schultze B., Maurer W. Autoradiographic investigations of glial proliferation in the brain of adult mice. II. Cycle time and mode of proliferation of neuroglia and endothelial cells. Journal of Comparative Neurology **160**, 477–490 (1975)

Korr I.M., Appeltauer G.S.L. The time-course of axonal transport of neuronal proteins to muscle. Experimental Neurology **43**, 452–463 (1974)

Korr I.M., Wilkinson P.N., Chornock F.W. Axonal delivery of neuroplasmic components to muscle cells. Science **155**, 342–345 (1967)

Korte G.E., Rosenbluth J. Ependymal astrocytes in the frog cerebellum. Anatomical Record **199**, 267–279 (1981)

Kosaka T. Ruffed cell: a new type of neuron with a distinctive initial unmyelinated portion of the axon in the olfactory bulb of the goldfish (*Carassius auratus*): II. Fine structure of the ruffed cell. Journal of Comparative Neurology **193**, 119–145 (1980)

Kosaka T. Gap junctions between non-pyramidal cell dendrites in the rat hippocampus (CA1 and CA3 regions). Brain Research **271**, 157–161 (1983a)

Kosaka T. Neuronal gap junctions in the polymorph layer of the rat dentate gyrus. Brain Research **277**, 347–351 (1983b)

Kosaka T., Kosaka K. Intraglomerular dendritic link connected by gap junctions and chemical synapses in the mouse main olfactory bulb: electron microscopic serial section analyses. Neuroscience **131**, 611–625 (2005)

Kossel A.H., Williams C.V., Schweizer M., Kater S.B. Afferent innervation influences the development of dendritic branches and spines via both activity-dependent and non-activity-dependent mechanisms. Journal of Neuroscience **17**, 6314–6324 (1997)

Kostyuk P., Verkhratsky A. Calcium stores in neurons and glia. Neuroscience **63**, 381–404 (1994)

Kozlowski G.P., Scott D.E., Krobisch Dudley G. Scanning electron microscopy of the third ventricle of sheep. Zeitschrift für Zellforschung und mikroskopische Anatomie **136**, 169–176 (1973)

Krahn V. The pia mater at the site of the entry of blood vessels into the central nervous system. Anatomy and Embryology **164**, 257–263 (1982)

Kramer-Zucker A.G., Olale F., Haycraft C.J., Yoder B.K., Schier A.F., Drummond I.A. Cilia-driven fluid flow in the zebrafish pronephros, brain and Kupffer's vesicle is required for normal organogenesis. Development **132**, 1907–1921 (2005)

Krause W. Über die Endigung der Muskelnerven. Zeitschrift für rationelle Medicin **18**, 136–160 (1863)

Kraus-Ruppert R., Laissue J., Bürki H., Odartchenko N. Kinetic studies on glial, Schwann and capsular cells labelled with [^{3}H] thymidine in cerebrospinal tissue of young mice. Journal of the Neurological Sciences **26**, 555–563 (1975)

Kreutzberg G.W. Autoradiographische Untersuchung über die Beteiligung von Gliazellen an der axonalen Reaktion im Facialiskern der Ratte. Acta Neuropathologica **7**, 149–161 (1966)

Kreutzberg G.W. Neuronal dynamics and axonal flow. IV. Blockage of intra-axonal enzyme transport by colchicine. Proceeding of the National Academy of Sciences of the U.S.A. **62**, 722–728 (1969)

Kreutzberg G.W. 100 years of Nissl staining. Trends in Neurosciences **7**, 236–237 (1984)

Kriebel M.E., Bennett M.V.L., Waxman S.G., Pappas G.D. Oculomotor neurons in fish: electrotonic coupling and multiple sites of impulse initiation. Science **166**, 520–524 (1969)

Krisch B., Leonhardt H., Oksche A. Compartments and perivascular arrangement of the meninges covering the cerebral cortex of the rat. Cell and Tissue Research **238**, 459–474 (1984)

Kristensson K., Nennesmo I., Persson L., Lycke E. Neuron to neuron transmission of herpes simplex virus. Transport of virus from skin to brainstem nuclei. Journal of the Neurological Sciences **54**, 149–156 (1982)

Kristensson K., Olsson Y. Uptake and retrograde axonal transport of peroxidase in hypoglossal neurones. Electron microscopical localization in the neuronal perikaryon. Acta Neuropathologica **19**, 1–9 (1971)

Kristol C., Akert K., Sandri C., Wyss U.R., Bennett M.V.L., Moor H. The Ranvier nodes in the neurogenic electric organ of the knifefish *Sternarchus*: a freeze-etching study on the distribution of membrane-associated particles. Brain Research **125**, 197–212 (1977)

Krnjević K., Miledi R. Motor units in the rat diaphragm. Journal of Physiology **140**, 427–439 (1958)

Kruger L., Maxwell D.S. Electron microscopy of oligodendrocytes in normal rat cerebrum. American Journal of Anatomy **118**, 411–436 (1966)

Kruger L., Maxwell D.S. Comparative fine structure of vertebrate neuroglia: teleosts and reptiles. Journal of Comparative Neurology **129**, 115–142 (1967)

Kruger L., Maxwell D.S. Cytoplasmic laminar bodies in the striate cortex. Journal of Ultrastructure Research **26**, 387–390 (1969)

Kubota K., Hioki K. Zytologische Untersuchungen der Mantelzellen im menschlichen Spinalganglion. Okajimas Folia Anatomica Japonica **22**, 111–126 (1943)

Kuczmarski E.R., Rosenbaum J.L. Studies on the organization and localization of actin and myosin in neurons. Journal of Cell Biology **80**, 356–371 (1979)

Kühne W. Über die peripherischen Endorgane der motorischen Nerven. Engelmann, Leipzig 1862

Kühne W. Neue Untersuchungen über motorische Nervenendigung. Zeitschrift für Biologie **23**, 1–148 (1886)

Kuffler S.W. Electric potential changes at an isolated nerve-muscle junction. Journal of Neurophysiology **5**, 18–26 (1942a)

Kuffler S.W. Further study on transmission in an isolated nerve-muscle fibre preparation. Journal of Neurophysiology **5**, 309–322 (1942b)

Kusano K. Electrical activity and structural correlates of giant nerve fibers in *Kuruma* shrimp (*Penaeus japonicus*). Journal of Cellular Physiology **68**, 361–383 (1966)

Kyösola K., Rechardt L. Electron microscopic observations on the innervation of the sphincter of Oddi in the dog. Cell and Tissue Research **161**, 167–176 (1975)

Lacas-Gervais S., Guo J., Strenzke N., Scarfone E., Kolpe M., Jahkel M., De Camilli P., Moser T., Rasband M.N., Solimena M. βIVΣ1 spectrin stabilizes the nodes of Ranvier and axon initial segments. Journal of Cell Biology **166**, 983–990 (2004)

Lafarga M., Casafont I., Bengoechea R., Tapia O., Berciano M.T. Cajal's contribution to the knowledge of the neuronal cell nucleus. Chromosoma **118**, 437–443 (2009)

Lafarga M., Palacios G. Intranuclear rodlets in retrochiasmatic area neurons of the hypothalamus of the rat. Experientia **33**, 1368–1369 (1977)

Lampert P.W., Schochet S.S.Jr. Demyelination and remyelination in lead neuropathy. Electron microscopic studies. Journal of Neuropathology and Experimental Neurology **27**, 527–545 (1968)

Landfield P.W., Rose G., Sandles L., Wohlstadter T.C., Lynch G. Patterns of astroglial hypertrophy and neuronal degeneration in the hippocampus of aged, memory-deficient rats. Journal of Gerontology **32**, 3–12 (1977)

Landis D.M.D., Reese T.S. Differences in membrane structure between excitatory and inhibitory synapses in the cerebellar cortex. Journal of Comparative Neurology **155**, 93–126 (1974a)

Landis D.M.D., Reese T.S. Arrays of particles in freeze-fractured astrocytic membranes. Journal of Cell Biology **60**, 316–320 (1974b)

Landis D.M.D., Reese T.S. Regional organization of astrocytic membranes in cerebellar cortex. Neuroscience **7**, 937–950 (1982)

Landis D.M.D., Reese T.S., Raviola E. Differences in membrane structure between excitatory and inhibitory components of the reciprocal synapse in the olfactory bulb. Journal of Comparative Neurology **155**, 67–92 (1974)

Landon D.N., Hall S. The myelinated nerve fibre. In: The Peripheral Nerve (D.N. Landon Ed.) pp. 1–105. Chapman and Hall, London 1976

Landon D.N., Langley O.K. The local chemical environment of nodes of Ranvier: a study of cation binding. Journal of Anatomy **108**, 419–432 (1971)

Lane N.J. Distribution of phosphatases in the Golgi region and associated structures of the thoracic ganglionic neurons in the grasshopper, *Melanoplus differentialis*. Journal of Cell Biology **37**, 89–104 (1968)

Lanterman A.J. Ueber den feineren Bau der markhaltigen Nervenfasern. Archiv für mikroskopische Anatomie **13**, 1–8 (1877)

Lapham L.W. The tetraploid DNA content of normal human Purkinje cells and its development during the perinatal period: a quantitative cytochemical study. Excerpta Medica International Congress Series No. 100. Proceedings of the 5th International Congress of Neuropathology, pp. 445–449 (1965)

Lapham L.W. Tetraploid DNA content of Purkinje neurons of human cerebellar cortex. Science **159**, 310–312 (1968)

Larramendi L.M.H., Fickenscher L., Lemkey-Johnston N. Synaptic vesicles of inhibitory and excitatory terminals in the cerebellum. Science **156**, 967–969 (1967)

Larramendi L.M.H., Victor T. Synapses on the Purkinje cell spines in the mouse. An electronmicroscopic study. Brain Research **5**, 15–30 (1967)

Lasek R.J. Axoplasmic streaming in the cat dorsal root ganglion cell and the rat ventral motoneuron. Anatomical Record **154**, 373–374 (1966)

Lasek R.J. The distribution of nucleic acids in the giant axon of the squid (*Loligo pealii*). Journal of Neurochemistry **17**, 103–109 (1970)

Lasek R.J., Brady S.T. Attachment of transported vesicles to microtubules in axoplasm is facilitated by AMP-PNP. Nature **316**, 645–647 (1985)

Lasek R.J., Dabrowski C., Nordlander R. Analysis of axoplasmic RNA from invertebrate giant axons. Nature New Biology **244**, 162–165 (1973)

Lasek R.J., Dower W.J. Aplysia californica: analysis of nuclear DNA in individual nuclei of giant neurons. Science **172**, 278–280 (1971)

Lasek R.J., Krishnan N., Kaiserman-Abramof I.R. Identification of the subunit proteins of 10-nm neurofilaments isolated from axoplasm of squid and *Myxicola* giant axons. Journal of Cell Biology **82**, 336–346 (1979)

Lasek R.J., Phillips L., Katz M.J., Autilio-Gambetti L. Function and evolution of neurofilament proteins. Annals of the New York Academy of Sciences **455**, 462–478 (1985)

LaVail J.H., LaVail M.M. The retrograde intraaxonal transport of horseradish peroxidase in the chick visual system: a light and electron microscopic study. Journal of Comparative Neurology **157**, 303–358 (1974)

LaVail J.H., Rapisardi S., Sugino I.K. Evidence against the smooth endoplasmic reticulum as a continuous channel for the retrograde axonal transport of horseradish peroxidase. Brain Research **191**, 3–20 (1980)

Lawson L.J., Perry V.H., Gordon S. Turnover of resident microglia in the normal adult mouse brain. Neuroscience **48**, 405–415 (1992)

Leal S., Andrade J.P., Paula-Barbosa M.M., Madeira M.D. Arcuate nucleus of the hypothalamus: effects of age and sex. Journal of Comparative Neurology **401**, 65–88 (1998)

LeBeux Y.J. An ultrastructural study of a cytoplasmic filamentous body, termed nematosome, in the neurons of the rat and cat substantia nigra. Zeitschrift für Zellforschung und mikroskopische Anatomie **133**, 289–325 (1972a)

LeBeux Y.J. Subsurface cisterns and lamellar bodies: particular forms of the endoplasmic reticulum in the neurons. Zeitschrift für Zellforschung und mikroskopische Anatomie **133**, 327–352 (1972b)

LeBeux Y.J., Willemot J. An ultrastructural study of the microfilaments in rat brain by means of heavy meromyosin labeling. I. The perikaryon, the dendrites and the axon. Cell and Tissue Research **160**, 1–36 (1975)

Ledbetter M.C., Porter. K.R. The morphology of microtubules of plant cells. Science **144**, 872–874 (1964)

Ledda M., Altieri L., De Palo S., Pannese E. The Golgi apparatus of spinal ganglion neurons: quantitative changes with aging. Journal of Submicroscopic Cytology and Pathology **38**, 109–116 (2006)

Ledda M., Barni L., Altieri L., Pannese E. Amount and distribution of lipofuscin in nerve and satellite cells from spinal ganglia of young adult and aged rabbits. Journal of Submicroscopic Cytology and Pathology **31**, 237–246 (1999)

Ledda M., Barni L., Altieri L., Pannese E. Decrease in the nucleo-cytoplasmic volume ratio of rabbit spinal ganglion neurons with age. Neuroscience Letters **286**, 171–174 (2000)

Ledda M., Barni L., Altieri L., Pannese E. Changes with age in the Golgi apparatus of rabbit spinal ganglion neurons. Tissue & Cell **33**, 342–348 (2001)

Ledda M., Barni L., Altieri L., Pannese E. The Golgi apparatus of satellite cells associated with spinal ganglion neurons: changes with age in the rabbit. Journal of Submicroscopic Cytology and Pathology **35**, 267–270 (2003)

Lee M.K., Cleveland D.W. Neuronal intermediate filaments. Annual Review of Neuroscience **19**, 187–217 (1996)

Lee V.M.-Y., Balin B.J., Otvos L.Jr., Trojanowski J.Q. A68: A major subunit of paired helical filaments and derivatized forms of normal tau. Science **251**, 675–678 (1991)

Lein P., Johnson M., Guo X., Rueger D., Higgins D. Osteogenic protein-1 induces dendritic growth in rat sympathetic neurons. Neuron **15**, 597–605 (1995)

Lemke G.E., Brockes J.P. Identification and purification of glial growth factor. Journal of Neuroscience **4**, 75–83 (1984)

Lendahl U., Zimmerman L.B., McKay R.D.G. CNS stem cells express a new class of intermediate filament protein. Cell **60**, 585–595 (1990)

Lendvai B., Stern E.A., Chen B., Svoboda K. Experience-dependent plasticity of dendritic spines in the developing rat barrel cortex *in vivo*. Nature **404**, 876–881 (2000)

Lenhossék M. von Der feinere Bau des Nervensystems im Lichte neuester Forschungen. Fischer's Medicinische Buchhandlung H. Kornfeld, Berlin 1893

Lenhossék M. von Centrosom und Sphäre in den Spinalganglienzellen des Frosches. Archiv für mikroskopische Anatomie und Entwicklungs- geschichte **46**, 345–369 (1895)

Lenhossék M. von Beiträge zur Kenntniss der Zwischenzellen des Hodens. Archiv für Anatomie und Physiologie, S. 65–85 (1897)

Lenn N.J., Reese T.S. The fine structure of nerve endings in the nucleus of the trapezoid body and the ventral cochlear nucleus. American Journal of Anatomy **118**, 375–390 (1966)

Leonhardt H. Über ependymale Tanycyten des III. Ventrikels beim Kaninchen in elektronenmikroskopischer Betrachtung. Zeitschrift für Zellforschung und mikroskopische Anatomie **74**, 1–11 (1966)

Leonhardt H. Intraventrikuläre markhaltige Nervenfasern nahe der Apertura lateralis ventriculi quarti des Kaninchengehirns. Zeitschrift für Zellforschung und mikroskopische Anatomie **84**, 1–8 (1968)

Leonhardt H. Subependymale Basalmembranlabyrinthe im Hinterhorn des Seitenventrikels des Kaninchengehirns. Zur Frage des Liquor- abflusses. Zeitschrift für Zellforschung und mikroskopische Anatomie **105**, 595–604 (1970)

Leonhardt H., Lindemann B. Über ein supraependymales Nervenzellen-, Axon- und Gliazellsystem. Eine raster- und transmissions- elektronenmikroskopische Untersuchung am IV. Ventrikel (Apertura lateralis) des Kaninchengehirns. Zeitschrift für Zellforschung und mikroskopische Anatomie **139**, 285–302 (1973)

Leonhardt H., Lindner E. Marklose Nervenfasern im III. und IV. Ventrikel des Kaninchen- und Katzengehirns. Zeitschrift für Zellforschung und mikroskopische Anatomie **78**, 1–18 (1967)

Leonhardt H., Prien H. Eine weitere Art intraventrikulärer kolbenförmiger Axonendigungen aus dem IV. Ventrikel des Kaninchengehirns. Zeitschrift für Zellforschung und mikroskopische Anatomie **92**, 394–399 (1968)

Leranth C., Hajszan T., MacLusky N.J. Androgens increase spine synapse density in the CA1 hippocampal subfield of ovariectomized female rats. Journal of Neuroscience **24**, 495–499 (2004)

Letourneau P.C. Actin in axons: stable scaffolds and dynamic filaments. Results and Problems in Cell Differentiation **48**, 65–90 (2009)

Leuba G. Aging of dendrites in the cerebral cortex of the mouse. Neuropathology and Applied Neurobiology **9**, 467–475 (1983)

Leuba G., Kraftsik R. Changes in volume, surface estimate, three-dimensional shape and total number of neurons of the human primary visual cortex from midgestation until old age. Anatomy and Embryology **190**, 351–366 (1994)

Le Vay S. On the neurons and synapses of the lateral geniculate nucleus of the monkey, and the effects of eye enucleation. Zeitschrift für Zellforschung und mikroskopische Anatomie **113**, 396–419 (1971)

Lever J.D., Graham J.D.P., Irvine G., Chick W.J. The vesiculated axons in relation to arteriolar smooth muscle in the pancreas. A fine structural and quantitative study. Journal of Anatomy **99**, 299–313 (1965)

Lever J.D., Spriggs T.L.B., Graham J.D.P. A formol-fluorescence, fine-structural and autoradiographic study of the adrenergic innervation of the vascular tree in the intact and sympathecto- mized pancreas of the cat. Journal of Anatomy **103**, 15–34 (1968)

Lever J.D., Spriggs T.L.B., Ivens C., Joó F., Mottram D.R., Graham J.D.P. Vasomotor innervation in the hind limb muscles of the normal and 6-hydroxydopamine treated cat. A formol fluorescence, histochemical and fine structural study. Zeitschrift für Zellforschung und mikroskopische Anatomie **122**, 389–396 (1971)

Levi G. Studi sulla grandezza delle cellule. 1. Ricerche comparative sulla grandezza delle cellule del Mammiferi. Archivio Italiano di Anatomia e di Embriologia **5**, 291–358 (1906)

Levi G. I gangli cerebrospinali. Studi di Istologia comparata e di Istogenesi. Archivio Italiano di Anatomia e di Embriologia **7** Suppl., 1–392 (1908)

Levi G. Trattato di Istologia (3ª ed.). U.T.E.T., Torino 1946

Levi-Montalcini R. The nerve growth factor. Annals of the New York Academy of Sciences **118**, 149–168 (1964)

Levine M.S., Adinolfi A.M., Fisher R.S., Hull C.D., Guthrie D., Buchwald N.A. Ultrastructural alterations in caudate nucleus in aged cats. Brain Research **440**, 267–279 (1988)

Lewis P.D. The fate of the subependymal cell in the adult rat brain, with a note on the origin of microglia. Brain **91**, 721–736 (1968)

Lewis S.A., Cowan N.J. Genetics, evolution, and expression of the 68,000-mol-wt neurofilament protein: isolation of a cloned cDNA probe. Journal of Cell Biology **100**, 843–850 (1985)

Li J., Vause C.V., Durham P.L. Calcitonin gene-related peptide stimulation of nitric oxide synthesis and release from trigeminal ganglion glial cells. Brain Research **1196**, 22–32 (2008)

Lieberman A.R. Microtubule-associated smooth endoplasmic reticulum in the frog's brain. Zeitschrift für Zellforschung und mikroskopische Anatomic **116**, 564–577 (1971)

Lieberman A.R. Neurons with presynaptic perikarya and presynaptic dendrites in the rat lateral geniculate nucleus. Brain Research **59**, 35–59 (1973)

Liem R.K.H., Yen S.-H., Salomon G.D., Shelanski M.L. Intermediate filaments in nervous tissues. Journal of Cell Biology **79**, 637–645 (1978)

Liesi P., Dahl D., Vaheri A. Laminin is produced by early rat astrocytes in primary culture. Journal of Cell Biology **96**, 920–924 (1983)

Lillie R.S. Factors affecting transmission and recovery in the passive iron nerve model. Journal of General Physiology **7**, 473–507 (1925)

Lindsey J.D., Landfield P.W., Lynch G. Early onset and topographical distribution of hypertrophied astrocytes in hippocampus of aging rats: A quantitative study. Journal of Gerontology **34**, 661–671 (1979)

Ling E.A. Study in the changes of the proportions and numbers of the various glial cell types in the spinal cord of neonatal and young adult rats. Acta Anatomica **96**, 188–195 (1976)

Ling E.A. Electron microscopic study of macrophages appearing in a stab wound of the brain of rats following intravenous injection of carbon particles. Archivum Histologicum Japonicum **42**, 41–50 (1979)

Ling E.A. The origin and nature of microglia. Advances in Cellular Neurobiology **2**, 33–82 (1981)

Ling E.A., Kaur C., Lu J. The origin, nature and some functional considerations of intraventricular macrophages with special reference to the epiplexus cells. Microscopy Research and Technique **41**, 43–56 (1998)

Ling E.A., Leblond C.P. Investigation of glial cells in semithin sections. II. Variation with age in the numbers of the various glial cell types in rat cortex and corpus callosum. Journal of Comparative Neurology **149**, 73–82 (1973)

Ling E.A., Paterson J.A., Privat A., Mori S., Leblond C.P. Investigation of glial cells in semithin sections. I. Identification of glial cells in the brain of young rats. Journal of Comparative Neurology **149**, 43–72 (1973)

Liu R.P.C., Hamilton B.L. Intranuclear bodies in neurons of the periaqueductal gray matter in the cat. American Journal of Anatomy **147**, 139–145 (1976)

Liu X.-S., Tilwalli S., Ye G.-l., Lio P.A., Pasternak J.F., Trommer B.L. Morphologic and electrophysiologic maturation in developing dentate gyrus granule cells. Brain Research **856**, 202–212 (2000)

Liu Y.-Y., Morris J.F., Ju G. Synaptic relationship of substance P-like-immunoreactive nerve fibers with gland cells of the anterior pituitary in the rat. Cell and Tissue Research **285**, 227–234 (1996)

Livingston R.B., Pfenninger K., Moor H., Akert K. Specialized paranodal and interparanodal glial-axonal junctions in the peripheral and central nervous system: a freeze-etching study. Brain Research **58**, 1–24 (1973)

Llewellyn-Smith I.J., Pilowsky P., Minson J.B., Chalmers J. Synapses on axons of sympathetic preganglionic neurons in rat and rabbit thoracic spinal cord. Journal of Comparative Neurology **354**, 193–208 (1995)

Llinás R., Hillman D.E. Physiological and morphological organization of the cerebellar circuits in various vertebrates. In: Neurobiology of Cerebellar Evolution and Development (R. Llinás Ed.) pp. 43–73. AMA-ERF Institute for Biomedical Research, Chicago 1969

Lodin Z., Booher J., Kasten F.H. Phase-contrast cinematographic study of dissociated neurons from embryonic chick dorsal root ganglia cultured in the Rose chamber. Experimental Cell Research **60**, 27–39 (1970)

Lodin Z., Faltin J., Booher J., Hartman J., Sensenbrenner M. Formation of intercellular contacts in cultures of dissociated neurons from embryonic chicken dorsal root ganglia. Neurobiology **3**, 376–390 (1973)

Lomax A.E., Bertrand P.P., Furness J.B. Electrophysiological characteristics distinguish three classes of neuron in submucosal ganglia of the guinea-pig distal colon. Neuroscience **103**, 245–255 (2001)

Long J.M., Kalehua A.N., Muth N.J., Calhoun M.E., Jucker M., Hengemihle J.M., Ingram D.K., Mouton P.R. Stereological analysis of astrocyte and microglia in aging mouse hippocampus. Neurobiology of Aging **19**, 497–503 (1998)

Lopez-Verrilli M.A., Picou F., Court F.A. Schwann cell-derived exosomes enhance axonal regeneration in the peripheral nervous system. GLIA **61**, 1795–1806 (2013)

Lorez H.P., Richards J.G. Distribution of indolealkylamine nerve terminals in the ventricles of the rat brain. Zeitschrift für Zellforschung und mikroskopische Anatomie **144**, 511–522 (1973)

Lu Y., Inokuchi H., McLachlan E.M., Li J.-S., Higashi H. Correlation between electrophysiology and morphology of three groups of neuron in the dorsal commissural nucleus of lumbosacral spinal cord of mature rats studied in vitro. Journal of Comparative Neurology **437**, 156–169 (2001)

Lucas K. The "all or none" contraction of the amphibian skeletal muscle fibre. Journal of Physiology **38**, 113–133 (1909)

Luduena R.F., Woodward D.O. α- and β-tubulin: separation and partial sequence analysis. Annals of the New York Academy of Sciences **253**, 272–283 (1975)

Ludwig M., Pittman Q.J. Talking back: dendritic neurotransmitter release. Trends in Neurosciences **26**, 255–261 (2003)

Ludwin S.K. Reaction of oligodendrocytes and astrocytes to trauma and implantation. A combined autoradiographic and immunohistochemical study. Laboratory Investigation **52**, 20–30 (1985)

Ludwin S.K., Kosek J.C., Eng L.F. The topographical distribution of S-100 and GFA proteins in the adult rat brain: an immunohisto- chemical study using horseradish peroxidase-labelled antibodies. Journal of Comparative Neurology **165**, 197–208 (1976)

Lübke J., Markram H., Frotscher M., Sakmann B. Frequency and dendritic distribution of autapses established by layer 5 pyramidal neurons in the developing rat neocortex: comparison with synaptic innervation of adjacent neurons of the same class. Journal of Neuroscience **16**, 3209–3218 (1996)

Luff S.E., McLachlan E.M., Hirst G.D.S. An ultrastructural analysis of the sympathetic neuromuscular junctions on arterioles of the submucosa of the guinea pig ileum. Journal of Comparative Neurology **257**, 578–594 (1987)

Lugaro E. Sulle funzioni della nevroglia. Rivista di Patologia Nervosa e Mentale **12**, 225–233 (1907)

Lugaro E. Su alcune questioni generali d'anatomia e fisiologia del sistema nervoso. Rivista di Patologia Nervosa e Mentale **22**, 185–239 (1917)

Lund R.D. Synaptic patterns of the superficial layers of the superior colliculus of the rat. Journal of Comparative Neurology **135**, 179–208 (1969)

Lund R.D., Westrum L.E. Synaptic vesicle differences after primary formalin fixation. Journal of Physiology **185**, 7P-9P (1966)

Luo J., Shook B.A., Daniels S.B., Conover J.C. Subventricular zone-mediated ependyma repair in the adult mammalian brain. Journal of Neuroscience **28**, 3804–3813 (2008)

Lüscher C., Streit J., Quadroni R., Lüscher H.-R. Action potential propagation through embryonic dorsal root ganglion cells in culture. I. Influence of the cell morphology on propagation properties. Journal of Neurophysiology **72**, 622–633 (1994)

Luse S.A. Formation of myelin in the central nervous system of mice and rats, as studied with the electron microscope. Journal of Biophysical and Biochemical Cytology **2**, 777–784 (1956)

Lushnikova I., Skibo G., Muller D., Nikonenko I. Synaptic potentiation induces increased glial coverage of excitatory synapses in CA1 hippocampus. Hippocampus **19**, 753–762 (2009)

Lyon M.J., King J.M. Aging rat vestibular ganglion: II. Quantitative electron microscopic evaluation. Annals of Otology, Rhinology & Laryngology **106**, 753–758 (1997)

Mackenzie M.L., Ghabriel M.N., Allt G. Nodes of Ranvier and Schmidt- Lanterman incisures: an *in vivo* lanthanum tracer study. Journal of Neurocytology **13**, 1043–1055 (1984)

MacVicar B.A., Dudek F.E. Electrotonic coupling between granule cells of rat dentate gyrus:

Physiological and anatomical evidence. Journal of Neurophysiology **47**, 579–592 (1982)

Madeira M.D., Sousa N., Santer R.M., Paula-Barbosa M.M., Gundersen H.J.G. Age and sex do not affect the volume, cell numbers, or cell size of the suprachiasmatic nucleus of the rat: an unbiased stereological study. Journal of Comparative Neurology **361**, 585–601 (1995)

Magalhães M.M. Intranuclear bodies in cells of rabbit and rat retina. Experimental Cell Research **47**, 628–632 (1967)

Maglione M., Tress O., Haas B., Karram K., Trotter J., Willecke K., Kettenmann H. Oligodendrocytes in mouse corpus callosum are coupled via gap junction channels formed by connexin47 and connexin32. Glia **58**, 1104–1117 (2010)

Malatesta P., Hartfuss E., Götz M. Isolation of radial glial cells by fluorescent-activated cell sorting reveals a neuronal lineage. Development **127**, 5253–5263 (2000)

Mann D.M.A., Yates P.O. Polyploidy in the human nervous system. I. The DNA content of neurones and glia of the cerebellum. Journal of the Neurological Sciences **18**, 183–196 (1973)

Mann D.M.A., Yates P.O., Stamp J.E. The relationship between lipofuscin pigment and ageing in the human nervous system. Journal of Neurological Sciences **37**, 83–93 (1978)

Mann G. Histological changes induced in sympathetic, motor, and sensory nerve cells by functional activity. Journal of Anatomy and Physiology **29**, 100–108 (1894)

Mannen H. Contribution to the quantitative study of the nervous tissue. A new method for measurement of the volume and surface area of neurons. Journal of Comparative Neurology **126**, 75–90 (1966)

Manuelidis L., Manuelidis E.E. On the DNA content of cerebellar Purkinje cells *in vivo* and *in vitro*. Experimental Neurology **43**, 192–206 (1974)

Marchbanks R.M. The vesicular hypothesis questioned. Trends in Neurosciences **1**, 83–84 (1978)

Mareš V., Schultze B., Maurer W. Stability of DNA in Purkinje cell nuclei of the mouse. Journal of Cell Biology **63**, 665–674 (1974)

Marinesco G. Sur la présence des corpuscules acidophiles paranucléolaires dans les cellules du *locus niger* et du *locus coeruleus*. Comptes Rendus des Séances de l'Académie des Sciences (Paris) **135**, 1000–1002 (1902)

Marinesco G. La cellule nerveuse. Tome Ier. O. Doin et Fils, Paris 1909

Marin-Padilla M. Structural abnormalities of the cerebral cortex in human chromosomal aberrations: a Golgi study. Brain Research **44**, 625–629 (1972)

Marin-Padilla M. Structural organization of the cerebral cortex (motor area) in human chromosomal aberrations. A Golgi study. I. D_1 (13–15) trisomy, Patau syndrome. Brain Research **66**, 375–391 (1974)

Marin-Padilla M. Pyramidal cell abnormalities in the motor cortex of a child with Down's syndrome. A Golgi study. Journal of Comparative Neurology **167**, 63–82 (1976)

Markus E.J., Petit T.L. Neocortical synaptogenesis, aging, and behavior: lifespan development in the motor-sensory system of the rat. Experimental Neurology **96**, 262–278 (1987)

Markus E.J., Petit T.L., LeBoutillier J.C. Synaptic structural changes during development and aging. Developmental Brain Research **35**, 239–248 (1987)

Marner L., Nyengaard J.R., Tang Y., Pakkenberg B. Marked loss of myelinated nerve fibers in the human brain with age. Journal of Comparative Neurology **462**, 144–152 (2003)

Marszalek J.R., Williamson T.L., Lee M.K., Xu Z., Hoffman P.N., Becher M.W., Crawford T.O., Cleveland D.W. Neurofilament subunit NF-H modulates axonal diameter by selectively slowing neurofilament transport. Journal of Cell Biology **135**, 711–724 (1996)

Martin A.R., Pilar G. Dual mode of synaptic transmission in the avian ciliary ganglion. Journal of Physiology **168**, 443–463 (1963)

Martin A.R., Ringham G.L. Synaptic transfer at a vertebrate central nervous system synapse. Journal of Physiology **251**, 409–426 (1975)

Martin J.R., Webster H. deF. Mitotic Schwann cells in developing nerve: their changes in shape, fine structure, and axon relationships. Developmental Biology **32**, 417–431 (1973)

Martin X., Dolivo M. Neuronal and transneuronal tracing in the trigeminal system of the rat using the herpes virus suis. Brain Research **273**, 253–276 (1983)

Martinelli C., Sartori P., De Palo S., Ledda M., Pannese E. Increase in number of the gap junctions between satellite neuroglial cells during lifetime: An ultrastructural study in rabbit spinal ganglia from youth to extremely advanced age. Brain Research Bulletin **67**, 19–23 (2005)

Martinelli C., Sartori P., De Palo S., Ledda M., Pannese E. The perineuronal glial tissue of spinal ganglia. Quantitative changes in the rabbit from youth to extremely advanced age. Anatomy and Embryology **211**, 455–463 (2006a)

Martinelli C., Sartori P., Ledda M., Pannese E. A study of mitochondria in spinal ganglion neurons during life: Quantitative changes from youth to extremely advanced age. Tissue and Cell **38**, 93–98 (2006b)

Martinelli C., Sartori P., Ledda M., Pannese E. Mitochondria in perineuronal satellite cell sheaths of rabbit spinal ganglia: Quantitative changes during life. Cells Tissues Organs **186**, 141–146 (2007)

Masaki T., Matsumura K., Hirata A., Yamada H., Hase A., Shimizu T., Yorifuji H., Motoyoshi K., Kamakura K. Expression of dystroglycan complex in satellite cells of dorsal root ganglia. Acta Neuropathologica **101**, 174–178 (2001)

Mason A., Larkman A. Correlations between morphology and electrophysiology of pyramidal neurons in slices of rat visual cortex. II. Electrophysiology. Journal of Neuroscience **10**, 1415–1428 (1990)

Massa P.T., Mugnaini E. Cell junctions and intramembrane particles of astrocytes and oligodendrocytes: a freeze-fracture study. Neuroscience 7, 523–538 (1982)

Masurovsky E.B., Benitez H.H., Kim S.U., Murray M.R. Origin, development, and nature of intranuclear rodlets and associated bodies in chicken sympathetic neurons. Journal of Cell Biology **44**, 172–191 (1970)

Matsuda S., Kobayashi N., Terashita T., Shimokawa T., Shigemoto K., Mominoki K., Wakisaka H., Saito S., Miyawaki K., Saito K., Kushihata F., Chen J., Gao S.-Y., Li C.-Y., Wang M., Fujiwara T. Phylogenetic investigation of Dogiel's pericellular nests and Cajal's initial glomeruli in the dorsal root ganglion. Journal of Comparative Neurology **491**, 234–245 (2005)

Matsumoto A. Synaptic changes in the perineal motoneurons of aged male rats. Journal of Comparative Neurology **400**, 103–109 (1998)

Matsumoto A., Arnold A.P., Micevych P.E. Gap junctions between lateral spinal motoneurons in the rat. Brain Research **495**, 362–366 (1989)

Matsumoto A., Okada R., Arai Y. Synaptic changes in the hypothalamic arcuate nucleus of old male rats. Experimental Neurology **78**, 583–590 (1982)

Matsumoto E., Rosenbluth J. Plasma membrane structure at the axon hillock, initial segment and cell body of frog dorsal root ganglion cells. Journal of Neurocytology **14**, 731–747 (1985)

Matsumoto E., Rosenbluth J. Structure of the satellite cell sheath around the cell body, axon hillock, and initial segment of frog dorsal root ganglion cells. Anatomical Record **215**, 182–191 (1986)

Matsumura K., Yamada H., Shimizu T., Campbell K.P. Differential expression of dystrophin, utrophin and dystrophin-associated proteins in peripheral nerve. FEBS Letters **334**, 281–285 (1993)

Matthews M.A., Duncan D. A quantitative study of morphological changes accompanying the initiation and progress of myelin production in the dorsal funiculus of the rat spinal cord. Journal of Comparative Neurology **142**, 1–22 (1971)

Matthews M.R., Raisman G. A light and electron microscopic study of the cellular response to axonal injury in the superior cervical ganglion of the rat. Proceedings of the Royal Society of London, Series B **181**, 43–79 (1972)

Maturana H.R. The fine anatomy of the optic nerve of Anurans. An electron microscope study. Journal of Biophysical and Biochemical Cytology **7**, 107–120 (1960)

Matus A. Actin-based plasticity in dendritic spines. Science **290**, 754–758 (2000)

Matus A., Bernhardt R., Hawkes R. Evidence for a specialized cyto- skeleton in neuronal dendrites. Biology of the Cell **45**, 432 (1982)

Matus A., Huber G., Bernhardt R. Neuronal microdifferentiation. Cold Spring Harbor Symposia on Quantitative Biology **48**, 775–782 (1983)

Matyash V., Kettenmann H. Heterogeneity in astrocyte morphology and physiology. Brain Research Reviews **63**, 2–10 (2010)

Maxwell D.J., Réthelyi M. Ultrastructure and synaptic connections of cutaneous afferent fibres in the spinal cord. Trends in Neurosciences **10**, 117–123 (1987)

Maxwell D.S., Kruger L. The fine structure of astrocytes in the cerebral cortex and their response to focal injury produced by heavy ionizing particles. Journal of Cell Biology **25**, 141–157 (1965)

Maynard E.A., Schultz R.L., Pease D.C. Electron microscopy of the vascular bed of rat cerebral cortex. American Journal of Anatomy **100**, 409–434 (1957)

McCarthy K.D., Partlow L.M. Neuronal stimulation of [^{3}H]thymidine incorporation by primary cultures of highly purified non-neuronal cells. Brain Research **114**, 415–426 (1976)

McCormick D.A., Connors B.W., Lighthall J.W., Prince D.A. Comparative electrophysiology of pyramidal and sparsely spiny stellate neurons of the neocortex. Journal of Neurophysiology **54**, 782–806 (1985)

McKenna O.C., Rosenbluth J. Cytological evidence for catecholamine-containing sensory cells bordering the ventricle of the toad hypothalamus. Journal of Comparative Neurology **154**, 133–148 (1974)

McKeon R.J., Schreiber R.C., Rudge J.S., Silver J. Reduction of neurite outgrowth in a model of glial scarring following CNS injury is correlated with the expression of inhibitory molecules on reactive astrocytes. Journal of Neuroscience **11**, 3398–3411 (1991)

McLean J.H., Hopkins D.A. A light and electron microscopic study of the dorsal motor nucleus of the vagus nerve in the cat. Journal of Comparative Neurology **195**, 157–175 (1981)

McMahan U.J. Fine structure of synapses in the dorsal nucleus of the lateral geniculate body of normal and blinded rats. Zeitschrift für Zellforschung und mikroskopische Anatomie **76**, 116–146 (1967)

McMahan U.J., Purves D. Visual identification of two kinds of nerve cells and their synaptic contacts in a living autonomic ganglion of the mudpuppy (*Necturus maculosus*). Journal of Physiology **254**, 405–425 (1976)

McQuarrie I.G., Brady S.T., Lasek R.J. Retardation in the slow axonal transport of cytoskeletal elements during maturation and aging. Neurobiology of Aging **10**, 359–365 (1989)

McTigue D.M., Tripathi R.B. The life, death, and replacement of oligodendrocytes in the adult CNS. Journal of Neurochemistry **107**, 1–19 (2008)

Meek J., Hafmans T.G.M., Han V., Bell C.C., Grant K. Myelinated dendrites in the mormyrid electrosensory lobe. Journal of Comparative Neurology **431**, 255–275 (2001)

Melendez-Vasquez C.V., Rios J.C., Zanazzi G., Lambert S., Bretscher A., Salzer J.L. Nodes of Ranvier form in association with ezrin-radixin-moesin (ERM)-positive Schwann cell processes. Proceedings of the National Academy of Sciences of the U.S.A. **98**, 1235–1240 (2001)

Mencarelli C., Magi B.,, Marzocchi B., Contorni M., Pallini V. Evolutionary trends of neurofilament

proteins in fish. Comparative Biochemistry and Physiology **100B**, 733–740 (1991)
Merrick S.E., Pleasure S.J., Lurie D.I., Pijak D.S., Selzer M.E., Lee V.M.-Y. Glial cells of the lamprey nervous system contain keratin-like proteins. Journal of Comparative Neurology **355**, 199–210 (1995)
Merrill D.A., Chiba A.A., Tuszynski M.H. Conservation of neuronal number and size in the entorhinal cortex of behaviorally characterized aged rats. Journal of Comparative Neurology **438**, 445–456 (2001)
Merrillees N.C.R. The nervous environment of individual smooth muscle cells of the guinea pig vas deferens. Journal of Cell Biology **37**, 794–817 (1968)
Merrillees N.C.R., Burnstock G., Holman M.E. Correlation of fine structure and physiology of the innervation of smooth muscle in the guinea pig vas deferens. Journal of Cell Biology **19**, 529–550 (1963)
Mertz K., Koscheck T., Schilling K. Brain-derived neurotrophic factor modulates dendritic morphology of cerebellar basket and stellate cells: an *in vitro* study. Neuroscience **97**, 303–310 (2000)
Mesulam M.-M. (Ed.) Tracing Neural Connections with Horseradish Peroxidase. John Wiley & Sons, Chichester 1982
Meszler R.M., Pappas G.D., Bennett M.V.L. Morphology of the electromotor system in the spinal cord of the electric eel, *Electrophorus electricus*. Journal of Neurocytology **3**, 251–261 (1974)
Metuzals J., Tasaki I. Subaxolemmal filamentous network in the giant nerve fiber of the squid (*Loligo pealei* L.) and its possible role in excitability. Journal of Cell Biology **78**, 597–621 (1978)
Meyer D., Birchmeier C. Multiple essential functions of neuregulin in development. Nature **378**, 386–390 (1995)
Meyer S. Die subcutane Methylenblauinjection, ein Mittel zur Darstellung der Elemente des Centralnervensystems von Säugethieren. Archiv für mikroskopische Anatomie und Entwicklungsgeschichte **46**, 282–290 (1895)
Miani N. Proximo-distal movement along the axon of protein synthesized in the perikaryon of regenerating neurons. Nature **185**, 541 (1960)
Miani N. Evidence of a proximo-distal movement along the axon of phospholipid synthesized in the nerve-cell body. Nature **193**, 887–888 (1962)
Miani N. Transport of S-100 protein in mammalian nerve fibers and transneuronal signals. Acta Neuropathologica Suppl. V, 104–108 (1971)
Michailov G.V., Sereda M.W., Brinkmann B.G., Fischer T.M., Haug B., Birchmeier C., Role L., Lai C., Schwab M.H., Nave K.-A. Axonal neuregulin-1 regulates myelin sheath thickness. Science **304**, 700–703 (2004)
Miller K.E., Richards B.A., Kriebel R.M. Glutamine-, glutamine synthetase-, glutamate dehydrogenase- and pyruvate carboxylase-immunoreactivities in the rat dorsal root ganglion and peripheral nerve. Brain Research **945**, 202–211 (2002)
Miller K.R., Streit W.J. The effects of aging, injury and disease on microglial function: a case for cellular senescence. Neuron Glia Biology **3**, 245–253 (2007)
Miller R., Varon S., Kruger L., Coates P.W., Orkand P.M. Formation of synaptic contacts on dissociated chick embryo sensory ganglion cells *in vitro*. Brain Research **24**, 356–358 (1970)
Miller R.H., Abney E.R., David S., ffrench-Constant C., Lindsay R., Patel R., Stone J., Raff M.C. Is reactive gliosis a property of a distinct subpopulation of astrocytes? Journal of Neuroscience **6**, 22–29 (1986)
Miller T.M., Heuser J.E. Endocytosis of synaptic vesicle membrane at the frog neuromuscular junction. Journal of Cell Biology **98**, 685–698 (1984)
Millhouse O.E. Light and electron microscopic studies of the ventricular wall. Zeitschrift für Zellforschung und mikroskopische Anatomie **127**, 149–174 (1972)
Mirsky R., Jessen K.R., Schachner M., Goridis C. Distribution of the adhesion molecules N-CAM and L1 on peripheral neurons and glia in adult rats. Journal of Neurocytology **15**, 799–815 (1986)
Miyamoto T., Morita K., Takemoto D., Takeuchi K., Kitano Y., Miyakawa T., Nakayama K., Okamura Y., Sasaki H., Miyachi Y., Furuse M., Tsukita S. Tight junctions in Schwann cells of peripheral myelinated axons: a lesson from claudin-19-deficient mice. Journal of Cell Biology **169**, 527–538 (2005)
Mizuhira V., Ozawa H. On the fine structure of nerve myelin by means of glutaraldehyde fixation. Journal of Electron Microscopy **16**, 169–172 (1967)
Mohammed H.A., Santer R.M. Total neuronal numbers of rat lumbosacral primary afferent neurons do not change with age. Neuroscience Letters **304**, 149–152 (2001)
Mohr E., Richter D. Axonal mRNAs: functional significance in vertebrates and invertebrates. Journal of Neurocytology **29**, 783–791 (2000)
Monneron A., Bernhard W. Fine structural organization of the interphase nucleus in some mammalian cells. Journal of Ultrastructure Research **27**, 266–288 (1969)
Moore K.L., Barr M.L. Morphology of the nerve cell nucleus in mammals, with special reference to the sex chromatin. Journal of Comparative Neurology **98**, 213–231 (1953)
Moore S.A., Peterson R.G., Felten D.L., O'Connor B.L. Ultrastructural axonal pathology in experimentally diabetic and aging control rats. Brain Research Bulletin **8**, 317–323 (1982)
Morales R., Duncan D. Multilaminated bodies and other unusual configurations of endoplasmic reticulum in the cerebellum of the cat. An electron microscopic study. Journal of Ultrastructure Research **15**, 480–489 (1966)
Morales R., Duncan D. Prismatic and other unusual arrays of mitochondrial cristae in astrocytes of cats and hamsters. Anatomical Record **171**, 545–558 (1971)
Morales R., Duncan D., Rehmet R. A distinctive laminated cytoplasmic body in the lateral geniculate body neurons of the cat. Journal of Ultrastructure Research **10**, 116–123 (1964)

Moran D.T., Chapman K.M., Ellis R.A. The fine structure of cockroach campaniform sensilla. Journal of Cell Biology **48**, 155–173 (1971)

Moran D.T., Rowley J.C.III, Varela F.G. Ultrastructure of the grasshopper proximal femoral chordotonal organ. Cell and Tissue Research **161**, 445–457 (1975)

Morell P. (Ed.) Myelin (2nd ed.). Plenum Press, New York 1984

Morest K.D. Dendrodendritic synapses of cells that have axons: the fine structure of the Golgi type II cell in the medial geniculate body of the cat. Zeitschrift für Anatomie und Entwicklungsgeschichte **133**, 216–246 (1971)

Mori S., Leblond C.P. Identification of microglia in light and electron microscopy. Journal of Comparative Neurology **135**, 57–80 (1969a)

Mori S., Leblond C.P. Electron microscopic features and proliferation of astrocytes in the corpus callosum of the rat. Journal of Comparative Neurology **137**, 197–226 (1969b)

Mori S., Leblond C.P. Electron microscopic identification of three classes of oligodendrocytes and a preliminary study of their proliferative activity in the corpus callosum of young rats. Journal of Comparative Neurology **139**, 1–30 (1970)

Morris J.H., Hudson A.R., Weddell G. A study of degeneration and regeneration in the divided rat sciatic nerve based on electron microscopy. IV. Changes in fascicular microtopography, perineurium and endoneurial fibroblasts. Zeitschrift für Zellforschung und mikroskopische Anatomie **124**, 165–203 (1972)

Morrison J.H., Baxter M.G. The ageing cortical synapse: hallmarks and implications for cognitive decline. Nature Reviews Neuroscience **13**, 240–250 (2012)

Morrison J.H., Hof P.R. Life and death of neurons in the aging brain. Science **278**, 412–419 (1997)

Mosch B., Morawski M., Mittag A., Lenz D., Tarnok A., Arendt T. Aneuploidy and DNA replication in the normal human brain and Alzheimer's disease. Journal of Neuroscience **27**, 6859–6867 (2007)

Moser M.-B., Trommald M., Andersen P. An increase in dendritic spine density on hippocampal CA1 pyramidal cells following spatial learning in adult rats suggests the formation of new synapses. Proceedings of the National Academy of Sciences of the U.S.A. **91**, 12673–12675 (1994)

Moses H.L., Ganote C.E., Beaver D.L., Schuffman S.S. Light and electron microscopic studies of pigment in human and rhesus monkey substantia nigra and locus coeruleus. Anatomical Record **155**, 167–184 (1966)

Mouren-Mathieu A.-M., Colonnier M. The molecular layer of the adult cat cerebellar cortex after lesion of the parallel fibers: an optic and electron microscope study. Brain Research **16**, 307–323 (1969)

Mouton P.R., Long J.M., Lei D.-L., Howard V., Jucker M., Calhoun M.E., Ingram D.K. Age and gender effects on microglia and astrocyte numbers in brains of mice. Brain Research **956**, 30–35 (2002)

Mudge A.W. Schwann cells induce morphological transformation of sensory neurones *in vitro*. Nature **309**, 367–369 (1984)

Müller E., Studien über Neuroglia. Archiv für mikroskopische Anatomie und Entwicklungsgeschichte **55**, 11–62 (1900)

Mugnaini E. On the occurrence of filamentous rodlets in neurons and glia cells of *Myxine glutinosa* (L.). Sarsia **29**, 221–232 (1967)

Mugnaini E. The histology and cytology of the cerebellar cortex. In: The Comparative Anatomy and Histology of the Cerebellum: The Human Cerebellum, Cerebellar Connections and Cerebellar Cortex (O. Larsell, J. Jansen Eds.) pp. 201–262. University of Minnesota Press, Minneapolis 1972

Mugnaini E. Cell junctions of astrocytes, ependyma, and related cells in the mammalian central nervous system, with emphasis on the hypothesis of a generalized functional syncytium of supporting cells. In: Astrocytes. Development, Morphology, and Regional Specialization of Astrocytes (S. Fedoroff, A. Vernadakis Eds.) vol. 1, pp. 329–371. Academic Press, Orlando 1986

Mugnaini E., Osen K.K., Schnapp B., Friedrich V.L.Jr. Distribution of Schwann cell cytoplasm and plasmalemmal vesicles (caveolae) in peripheral myelin sheaths. An electron microscopic study with thin sections and freeze-fracturing. Journal of Neurocytology **6**, 647–668 (1977)

Mugnaini E., Walberg F. Ultrastructure of neuroglia. Ergebnisse der Anatomie und Entwicklungsgeschichte **37**, 193–236 (1964)

Mugnaini E., Walberg F., Brodal A. Mode of termination of primary vestibular fibres in the lateral vestibular nucleus. An experimental electron microscopical study in the cat. Experimental Brain Research **4**, 187–211 (1967a)

Mugnaini E., Walberg F., Hauglie-Hanssen E. Observations on the fine structure of the lateral vestibular nucleus (Deiters' nucleus) in the cat. Experimental Brain Research **4**, 146–186 (1967b)

Muir A.R., Peters A. Quintuple-layered membrane junctions at terminal bars between endothelial cells. Journal of Cell Biology **12**, 443–448 (1962)

Multani P., Myers R.H., Blume H.W., Schomer D.L., Sotrel A. Neocortical dendrite pathology in human partial epilepsy: A quantitative Golgi study. Epilepsia **35**, 728–736 (1994)

Mungai J.M. Dendritic patterns in the somatic sensory cortex of the cat. Journal of Anatomy **101**, 403–418 (1967)

Murphy D.D., Segal M. Regulation of dendritic spine density in cultured rat hippocampal neurons by steroid hormones. Journal of Neuroscience **16**, 4059–4068 (1996)

Nabeshima S., Reese T.S., Landis D.M.D., Brightman M.W. Junctions in the meninges and marginal glia. Journal of Comparative Neurology **164**, 127–170 (1975)

Nadelhaft I. Microtubule densities and total numbers in selected axons of the crayfish abdominal nerve cord. Journal of Neurocytology **3**, 73–86 (1974)

Nadol J.B.Jr. Comparative anatomy of the cochlea and auditory nerve in mammals. Hearing Research **34**, 253–266 (1988)

Nageotte J. Etude sur la greffe des ganglions rachidiens; variations et tropismes du neurone sensitif. Anatomischer Anzeiger **31**, 225–245 (1907)

Nageotte J. Incisures de Schmidt-Lanterman et protoplasma des cellules de Schwann. Comptes Rendus des Séances et Mémoires de la Société de Biologie **68**, 39–42 (1910a)

Nageotte J. Sur une nouvelle formation de la gaine de myéline: le double bracelet épineux de l'étranglement annulaire. Comptes Rendus des Séances de l'Académie des Sciences **150**, 123–126 (1910b)

Nageotte J. L'organisation de la matière dans ses rapports avec la vie. Librairie F. Alcan, Paris 1922

Nagy J.I., Dudek F.E., Rash J.E. Update on connexins and gap junctions in neurons and glia in the mammalian nervous system. Brain Research Reviews **47**, 191–215 (2004)

Nagy J.I., Rash J.E. Connexins and gap junctions of astrocytes and oligodendrocytes in the CNS. Brain Research Reviews **32**, 29–44 (2000)

Naka K.I., Christensen B.N. Direct electrical connections between transient amacrine cells in the catfish retina. Science **214**, 462–464 (1981)

Nakai Y. Electron microscopic observations on synapse-like contacts between pituicytes and different types of nerve fibers in the anuran pars nervosa. Zeitschrift für Zellforschung und mikroskopische Anatomie **110**, 27–39 (1970)

Nakajima Y. Electron microscope observations on the nerve fibers of *Cristaria plicata*. Zeitschrift für Zellforschung und mikroskopische Anatomie **54**, 262–274 (1961)

Nakajima Y. Fine structure of the synaptic endings on the Mauthner cell of the goldfish. Journal of Comparative Neurology **156**, 375–402 (1974)

Nakajima Y., Reese T.S. Inhibitory and excitatory synapses in crayfish stretch receptor organs studied with direct rapid-freezing and freeze-substitution. Journal of Comparative Neurology **213**, 66–73 (1983)

Nakajima Y., Tuttle R., Masuko S., Betz H., Pfeiffer F. Synapses on the Mauthner cell of the goldfish: thin section, freeze-fracture. and immunocytochemical studies. Journal of Electron Microscopy Technique **6**, 143–153 (1987)

Nakamura S., Akiguchi I., Kameyama M., Mizuno N. Age-related changes of pyramidal cell basal dendrites in layers III and V of human motor cortex: a quantitative Golgi study. Acta Neuropathologica **65**, 281–284 (1985)

Nakamura Y., Takeda M., Angelides K.J., Tada K., Hariguchi S., Nishimura T. Assembly, disassembly, and exchange of glial fibrillary acidic protein. Glia **4**, 101–110 (1991)

Nakayama Y., Kohno K. Number and polarity of the ependymal cilia in the central canal of some vertebrates. Journal of Neurocytology **3**, 449–458 (1974)

Nambu A., Llinás R. Morphology of globus pallidus neurons: its correlation with electrophysiology in guinea pig brain slices. Journal of Comparative Neurology **377**, 85–94 (1997)

Napolitano L.M., Scallen T.J. Observations on the fine structure of peripheral nerve myelin. Anatomical Record **163**, 1–6 (1969)

Napper R.M.A., Harvey R.J. Quantitative study of the Purkinje cell dendritic spines in the rat cerebellum. Journal of Comparative Neurology **274**, 158–167 (1988)

Nathaniel E.J.H., Nathaniel D.R. Fine structure of the neurons of the posterior horn in the rat spinal cord. Anatomical Record **155**, 629–642 (1966)

Nathaniel E.J.H., Nathaniel D.R. The reactive astrocyte. Advances in Cellular Neurobiology **2**, 249–301 (1981)

Nathaniel E.J.H., Pease D.C. Degenerative changes in rat dorsal roots during Wallerian degeneration. Journal of Ultrastructure Research **9**, 511–532 (1963)

Němeček St., Wolff J. Light and electron microscopic evidence of complex synapses (glomeruli) in oliva inferior (cat). Experientia **25**, 634–635 (1969)

Nemiloff A. Über die Beziehung der sog. "Zellen der Schwannschen Scheide" zum Myelin in den Nervenfasern von Säugetieren. Archiv für mikroskopische Anatomie und Entwicklungsgeschichte **76**, 329–348 (1910)

Neuhaus J., Wolburg H. Heterogeneity of astrocytic membranes in the optic nerve and spinal cord of the lizard, Anolis carolinensis. Cell and Tissue Research **242**, 185–190 (1985)

Neutra M., Leblond C.P. Synthesis of the carbohydrate of mucus in the Golgi complex as shown by electron microscope radioautography of goblet cells from rats injected with glucose-H^3. Journal of Cell Biology **30**, 119–136 (1966a)

Neutra M., Leblond C.P. Radioautographic comparison of the uptake of galactose-H^3 and glucose-H^3 in the Golgi region of various cells secreting glycoproteins or mucopolysaccharides. Journal of Cell Biology **30**, 137–150 (1966b)

Niblock M.M., Brunso-Bechtold J.K., Riddle D.R. Insulin-like growth factor I stimulates dendritic growth in primary somatosensory cortex. Journal of Neuroscience **20**, 4165–4176 (2000)

Nicaise G. The gliointerstitial system of molluscs. International Review of Cytology **34**, 251–332 (1973)

Nichols N.R., Day J.R., Laping N.J., Johnson S.A., Finch C.E. GFAP mRNA increases with age in rat and human brain. Neurobiology of Aging **14**, 421–429 (1993)

Nielsen S., Smith B.L., Christensen E.I., Agre P. Distribution of the aquaporin CHIP in secretory and resorptive epithelia and capillary endothelia. Proceedings of the National Academy of Sciences of the U.S.A. **90**, 7275–7279 (1993)

Niessing K. Über systemartige Zusammenhänge der Neuroglia im Grosshirn und über ihre funktionelle Bedeutung. Gegenbaurs Morphologisches Jahrbuch **78**, 537–584 (1936)

Nilsson O. The relationship between nerves and smooth muscle cells in the rat iris. I. The dilatator muscle.

Zeitschrift für Zellforschung und mikroskopische Anatomie **64**, 166–171 (1964)

Nimmerjahn A., Kirchhoff F., Helmchen F. Resting microglial cells are highly dynamic surveillants of brain parenchyma in vivo. Science **308**, 1314–1318 (2005)

Nishiyama A., Watanabe M., Yang Z., Bu J. Identity, distribution, and development of polydendrocytes: NG2-expressing glial cells. Journal of Neurocytology **31**, 437–455 (2002)

Nishiyama A., Yu M., Drazba J.A., Tuohy V.K. Normal and reactive NG2+ glial cells are distinct from resting and activated microglia. Journal of Neuroscience Research **48**, 299–312 (1997)

Nissl F. Über den Zusammenhang von Zellstruktur und Zellfunktion in der centralen Nervenzelle mit Demonstration. Tageblatt der 61. Versammlung deutscher Naturforscher und Ärzte in Köln **61**, 194–197 (1889)

Nissl F. Ueber die sogenannten Granula der Nervenzellen. Neurologisches Centralblatt **13**, 676–685, 781–789, 810–814 (1894)

Nissl F. Ueber einige Beziehungen zwischen Nervenzellenerkrankungen und gliösen Erscheinungen bei verschiedenen Psychosen. Archiv für Psychiatrie und Nervenkrankheiten **32**, 656–676 (1899)

Nixon R.A. The slow axonal transport of cytoskeletal proteins. Current Opinion in Cell Biology **10**, 87–92 (1998)

Nixon R.A., Paskevich P.A., Sihag R.K., Thayer C.Y. Phosphorylation on carboxyl terminus domains of neurofilament proteins in retinal ganglion cell neurons in vivo: influences on regional neurofilament accumulation, interneurofilament spacing, and axon caliber. Journal of Cell Biology **126**, 1031–1046 (1994)

Nixon R.A., Shea T.B. Dynamics of neuronal intermediate filaments: A developmental perspective. Cell Motility and the Cytoskeleton **22**, 81–91 (1992)

Noack W., Dumitrescu L., Schweichel J.U. Scanning and electron microscopical investigations of the surface structures of the lateral ventricles in the cat. Brain Research **46**, 121–129 (1972)

Noetzel H., Rox J. Autoradiographische Untersuchungen über Zellteilung und Zellentwicklung im Gehirn der erwachsenen Maus und des erwachsenen Rhesus-Affen nach Injektion von radioaktivem Thymidin. Acta Neuropathologica **3**, 326–342 (1964)

Nordlander R.H., Singer M. The role of ependyma in regeneration of the spinal cord in the urodele amphibian tail. Journal of Comparative Neurology **180**, 349–374 (1978)

Norenberg M.D. Astrocyte responses to CNS injury. Journal of Neuropathology and Experimental Neurology **53**, 213–220 (1994)

Novikoff A.B. Biochemical and staining reactions of cytoplasmic constituents. In: Developing Cell Systems and Their Control (D. Rudnick Ed.) pp. 167–203. Ronald Press, New York 1960

Novikoff A.B. Enzyme localization and ultrastructure of neurons. In: The Neuron (H. Hydén Ed.) pp. 255–318. Elsevier, Amsterdam 1967

Novikoff A.B., Quintana N., Villaverde H., Forschirm R. Nucleoside phosphatase and cholinesterase activities in dorsal root ganglia and peripheral nerve. Journal of Cell Biology **29**, 525–545 (1966)

Novikoff P.M., Novikoff A. B., Quintana N., Hauw J.-J. Golgi apparatus, GERL, and lysosomes of neurons in rat dorsal root ganglia, studied by thick section and thin section cytochemistry. Journal of Cell Biology **50**, 859–886 (1971)

Nurgali K., Stebbing M.J., Furness J.B. Correlation of electrophysiological and morphological characteristics of enteric neurons in the mouse colon. Journal of Comparative Neurology **468**, 112–124 (2004)

Oberheim N.A., Wang X., Goldman S., Nedergaard M. Astrocytic complexity distinguishes the human brain. Trends in Neurosciences **29**, 547–553 (2006)

O'Callaghan J.P., Miller D.B. The concentration of glial fibrillary acidic protein increases with age in the mouse and rat brain. Neurobiology of Aging **12**, 171–174 (1991)

Ochoa J., Mair W.G. The normal sural nerve in man. II. Changes in the axons and Schwann cells due to aging. Acta Neuropathologica **13**, 217–239 (1969)

Ochoa J., Vial J.D. Behaviour of pheripheral nerve structures in chronic neuropathies, with special reference to the Schwann cell. Journal of Anatomy **102**, 95–111 (1967)

Ochs S., Burger E. Movement of substance proximo-distally in nerve axons as studied with spinal cord injections of radioactive phosphorus. American Journal of Physiology **194**, 499–506 (1958)

Oehmichen M., Huber H. Reactive microglia with membrane features of mononuclear phagocytes. Journal of Neuropathology and Experimental Neurology **35**, 30–39 (1976)

Ogata J., Budzilovich G.N., Cravioto H. A study of rod-like structures (Hirano bodies) in 240 normal and pathological brains. Acta Neuropathologica **21**, 61–67 (1972)

Ogata T. Structure of motor endplates in the different fiber types of vertebrate skeletal muscles. Archives of Histology and Cytology **51**, 385–424 (1988)

Ohara O., Gahara Y., Miyake T., Teraoka H., Kitamura T. Neurofilament deficiency in quail caused by nonsense mutation in neurofilament-L gene. Journal of Cell Biology **121**, 387–395 (1993)

Ohgoh M., Kimura M., Ogura H., Katayama K., Nishizawa Y. Apoptotic cell death of cultured cerebral cortical neurons induced by withdrawal of astroglial trophic support. Experimental Neurology **149**, 51–63 (1998)

Ohnishi A., O'Brien P.C., Dyck P.J. Studies to improve fixation of human nerves. V. Effect of temperature, fixative and $CaCl_2$ on density of microtubules and neurofilaments. Journal of Neuropathology and Experimental Neurology **35**, 167–179 (1976)

Okabe S., Miyasaka H., Hirokawa N. Dynamics of the neuronal intermediate filaments. Journal of Cell Biology **121**, 375–386 (1993)

Okado N., Yokota N. Axoglial synaptoid contacts in the neural lobe of the human fetus. Anatomical Record **202**, 117–124 (1982)

Okamoto K., Hirai S. Fine structures of Marinesco body and nuclear body in the substantia nigra. Clinical Neurology (Tokyo) **21**, 781–789 (1981)

Okamura Y., Tsukita S. Morphology of freeze-substituted myelinated axon in mouse peripheral nerves. Brain Research **383**, 146–158 (1986)

Oksche A. Sensory and glandular elements of the pineal organ. In: The Pineal Gland (G.E.W. Wolstenholme, J. Knight Eds.) pp. 127–146. Churchill Livingstone, Edinburgh 1971

Oksche A., Kirschstein H. Entstehung und Ultrastruktur der Biondi-Körper in den Plexus chorioidei des Menschen (Biopsiematerial). Zeitschrift für Zellforschung und mikroskopische Anatomie **124**, 320–341 (1972)

Olmos G., Naftolin F., Perez J., Tranque P.A., Garcia-Segura L.M. Synaptic remodeling in the rat arquate nucleus during the estrous cycle. Neuroscience **32**, 663–667 (1989)

Olsson Y. Studies on vascular permeability in peripheral nerves. I. Distribution of circulating fluorescent serum albumin in normal, crushed and sectioned rat sciatic nerve. Acta Neuropathologica **7**, 1–15 (1966)

Olsson Y. Vascular permeability in the peripheral nervous system. In: Peripheral Neuropathy (P.J. Dyck, P.K. Thomas, E.H. Lambert, R. Bunge Eds.) 2nd ed., vol. I, pp. 579–597. W.B. Saunders, Philadelphia 1984

Olsson Y., Reese T.S. Permeability of vasa nervorum and perineurium in mouse sciatic nerve studied by fluorescence and electron microscopy. Journal of Neuropathology and Experimental Neurology **30**, 105–119 (1971)

O'Malley A., O'Connell C., Regan C.M. Ultrastructural analysis reveals avoidance conditioning to induce a transient increase in hippocampal dentate spine density in the 6 hour post-training period of consolidation. Neuroscience **87**, 607–613 (1998)

Ong W.Y., Levine J.M. A light and electron microscopic study of NG2 chondroitin sulfate proteoglycan-positive oligodendrocyte precursor cells in the normal and kainate-lesioned rat hippocampus. Neuroscience **92**, 83–95 (1999)

Ono S., Inoue K., Mannen T., Mitake S., Shirai T., Kanda F., Jinnai K., Takahashi K. Intracytoplasmic inclusion bodies of the thalamus and the substantia nigra, and Marinesco bodies in myotonic dystrophy: a quantitative morphological study. Acta Neuropathologica **77**, 350–356 (1989)

Oosaki T., Ishii S. Junctional structure of smooth muscle cells. The ultrastructure of the regions of junction between smooth muscle cells in the rat small intestine. Journal of Ultrastructure Research **10**, 567–577 (1964)

Oppenheim R.W., Chu-Wang I.-W., Maderdrut J.L. Cell death of motoneurons in the chick embryo spinal cord. III. The differentiation of motoneurons prior to their induced degeneration following limb-bud removal. Journal of Comparative Neurology **177**, 87–112 (1978)

Orkand R.K. Glial cells. In: Handbook of Physiology, sect. 1, vol. I (E.R. Kandel Ed.) pp. 855–875. American Physiological Society, Bethesda 1977

Orkand R.K., Nicholls J.G., Kuffler S.W. Effect of nerve impulses on the membrane potential of glial cells in the central nervous system of amphibia. Journal of Neurophysiology **29**, 788–806 (1966)

Osinchak J. Electron microscopic localization of acid phosphatase and thiamine pyrophosphatase activity in hypothalamic neurosecretory cells of the rat. Journal of Cell Biology **21**, 35–47 (1964)

Overton E. Ueber die allgemeinen osmotischen Eigenschaften der Zelle, ihre vermutlichen Ursachen und ihre Bedeutung für die Physiologie. Vierteljahrsschrift der Naturforschenden Gesellschaft in Zürich **44**, 88–135 (1899)

Pachter J.S., Liem R.K.H. α-Internexin, a 66-kD intermediate filament-binding protein from mammalian central nervous system. Journal of Cell Biology **101**, 1316–1322 (1985)

Pakkenberg B., Gundersen H.J.G. Neocortical neuron number in humans: effect of sex and age. Journal of Comparative Neurology **384**, 312–320 (1997)

Palacios-Prü E.L., Mendoza R.V., Palacios L. *In vitro* and *in situ* formation of neuron-glia junctions. Experimental Neurology **82**, 541–549 (1983)

Palade G.E., Palay S.L. Electron microscope observations of interneuronal and neuromuscular synapses. Anatomical Record **118**, 335–336 (1954)

Palay S.L. Synapses in the central nervous system. Journal of Biophysical and Biochemical Cytology **2** Suppl., 193–202 (1956)

Palay S.L. Contributions of electron microscopy to neuroanatomy. In: New Research Techniques of Neuroanatomy (W.F. Windle Ed.) pp. 5–16. C.C. Thomas, Springfield 1957

Palay S.L. The morphology of synapses in the central nervous system. Experimental Cell Research Suppl. 5, 275–293 (1958a)

Palay S.L. An electron microscopical study of neuroglia. In: Biology of Neuroglia (W.F. Windle Ed.) pp. 24–38. C.C. Thomas, Springfield 1958b

Palay S.L. The structural basis for neural action. In: Brain Function (M.A.B. Brazier Ed.) vol. 2, pp. 69–108. University of California Press, Berkeley 1964

Palay S.L., Chan-Palay V. Cerebellar Cortex. Cytology and Organization. Springer-Verlag, Berlin 1974

Palay S.L., Chan-Palay V. General morphology of neurons and neuroglia. In: Handbook of Physiology, sect. 1, vol. I (E.R. Kandel Ed.) pp. 5–37. American Physiological Society, Bethesda 1977

Palay S.L., Palade G.E. The fine structure of neurons. Journal of Biophysical and Biochemical Cytology **1**, 69–88 (1955)

Palay S.L., Sotelo C., Peters A., Orkand P.M. The axon hillock and the initial segment. Journal of Cell Biology **38**, 193–201 (1968)

Palay S.L., Wissig S.L. Secretory granules and Nissl substance in fresh supraoptic neurones of the rabbit. Anatomical Record **116**, 301–313 (1953)

Palmer L.M., Stuart G.J. Site of action potential initiation in layer 5 pyramidal neurons. Journal of Neuroscience **26**, 1854–1863 (2006)

Pannese E. L'organizzazione strutturale delle lamine gliali periventricolari dell'uomo in condizioni normali e patologiche. Zeitschrift für Zellforschung und mikroskopische Anatomie **45**, 137–151 (1956)

Pannese E. L'architettura gliale della sostanza bianca del midollo spinale di vertebrati. 1°. Indagini anatomo-comparative. Zeitschrift für Anatomie und Entwicklungsgeschichte **120**, 173–197 (1957)

Pannese E. L'architettura gliale della sostanza bianca del midollo spinale di vertebrati. 2°. Indagini morfogenetiche. Conclusioni generali. Zeitschrift für Anatomie und Entwicklungsgeschichte **120**, 279–301 (1958a)

Pannese E. Sulla picnosi fisiologica e sui caratteri del nucleo dei glyociti dell'uomo. Sistema Nervoso **10**, 124–133 (1958b)

Pannese E. Observations on the morphology, submicroscopic structure and biological properties of satellite cells (s.c.) in sensory ganglia of mammals. Zeitschrift für Zellforschung und mikroskopische Anatomie **52**, 567–597 (1960)

Pannese E. Investigations on the ultrastructural changes of the spinal ganglion neurons in the course of axon regeneration and cell hypertrophy. I. Changes during axon regeneration. Zeitschrift für Zellforschung und mikroskopische Anatomie **60**, 711–740 (1963a)

Pannese E. Investigations on the ultrastructural changes of the spinal ganglion neurons in the course of axon regeneration and cell hypertrophy. II. Changes during cell hypertrophy and comparison between the ultrastructure of nerve cells of the same type under different functional conditions. Zeitschrift für Zellforschung und mikroskopische Anatomie **61**, 561–586 (1963b)

Pannese E. Number and structure of perisomatic satellite cells of spinal ganglia under normal conditions or during axon regeneration and neuronal hypertrophy. Zeitschrift für Zellforschung und mikroskopische Anatomie **63**, 568–592 (1964)

Pannese E. Developmental changes of the endoplasmic reticulum and ribosomes in nerve cells of the spinal ganglia of the domestic fowl. Journal of Comparative Neurology **132**, 331–364 (1968)

Pannese E. Unusual membrane-particle complexes within nerve cells of the spinal ganglia. Journal of Ultrastructure Research **29**, 334–342 (1969a)

Pannese E. Electron microscopical study on the development of the satellite cell sheath in spinal ganglia. Journal of Comparative Neurology **135**, 381–422 (1969b)

Pannese E. The histogenesis of the spinal ganglia. Advances in Anatomy Embryology and Cell Biology **47** (5) 1–97 (1974)

Pannese E. The response of the satellite and other non-neuronal cells to the degeneration of neuroblasts in chick embryo spinal ganglia. Cell and Tissue Research **190**, 1–14 (1978)

Pannese E. The satellite cells of the sensory ganglia. Advances in Anatomy Embryology and Cell Biology **65**, 1–111 (1981)

Pannese E. The black reaction. Brain Research Bulletin **41**, 343–349 (1996)

Pannese E. The Golgi stain: invention, diffusion and impact on Neurosciences. Journal of the History of the Neurosciences **8**, 132–140 (1999)

Pannese E. Perikaryal surface specializations of neurons in sensory ganglia. International Review of Cytology **220**, 1–34 (2002)

Pannese E. The contribution of Camillo Golgi to our understanding of the structure of the nervous system. Archives Italiennes de Biologie **145**, 111–115 (2007)

Pannese E. The structure of the perineuronal sheath of satellite glial cells (SGCs) in sensory ganglia. Neuron Glia Biology **6**, 3–10 (2010)

Pannese E. Morphological changes in nerve cells during normal aging. Brain Structure and Function **216**, 85–89 (2011)

Pannese E., Arcidiacono G., Frattola D., Rigamonti L., Procacci P., Ledda M. Quantitative relationships between axoplasm and Schwann cell sheath in unmyelinated nerve fibres. An electron microscope study. Journal of Anatomy **159**, 49–56 (1988b)

Pannese E., Arcidiacono G., Rigamonti L., Procacci P., Ledda M. Stability at low temperatures of neuronal microtubules in spinal ganglia and dorsal roots of the lizard (*Lacerta muralis*). Journal of Ultrastructure Research **79**, 18–30 (1982)

Pannese E., Arcidiacono G., Rigamonti L., Procacci P., Ledda M. Microtubules in unmyelinated and myelinated axons: a quantitative analysis. Journal of Ultrastructure Research **89**, 240–248 (1984a)

Pannese E., Bianchi R., Calligaris B., Ventura R., Weibel E.R. Quantitative relationships between nerve and satellite cells in spinal ganglia. An electron microscopical study. I. Mammals. Brain Research **46**, 215–234 (1972)

Pannese E., Bianchi R., Gioia M., Ventura R. A quantitative electron microscope study of the perikaryal projections of sensory ganglion neurons. II. Gecko and lizard. Journal of Comparative Neurology **240**, 212–218 (1985)

Pannese E., Ferrannini E. Nuclear pyknosis in neuroglia cells of normal mammals. Acta Neuropathologica **8**, 309–319 (1967)

Pannese E., Gioia M., Carandente O., Ventura R. A quantitative electron microscope study of the perikaryal projections of sensory ganglion neurons. I. Cat and rabbit. Journal of Comparative Neurology **214**, 239–250 (1983)

Pannese E., Ledda M. Ribosomes in myelinated axons of the rabbit spinal ganglion neurons. Journal of Submicroscopic Cytology and Pathology **23**, 33–38 (1991)

Pannese E., Ledda M., Arcidiacono G., Frattola D., Rigamonti L., Procacci P. Mitotic Schwann cells in normal mature spinal roots. La Cellule **74**, 171–178 (1987a)

Pannese E., Ledda M., Arcidiacono G., Frattola D., Rigamonti L., Procacci P. Qualitative and quantitative observations on the structure of the Schwann cells in myelinated fibres. Acta Anatomica **131**, 314–326 (1988a)

Pannese E., Ledda M., Arcidiacono G., Rigamonti L. Clusters of nerve cell bodies enclosed within a common connective tissue envelope in the spinal ganglia of the lizard and rat. Cell and Tissue Research **264**, 209–214 (1991)

Pannese E., Ledda M., Arcidiacono G., Rigamonti L., Procacci P. Density and distribution of microtubules in the axons of the lizard dorsal roots. Journal of Submicroscopic Cytology **13**, 169–181 (1981)

Pannese E., Ledda M., Arcidiacono G., Rigamonti L., Procacci P. A comparison of the density of microtubules in the central and peripheral axonal branches of the pseudounipolar neurons of lizard spinal ganglia. Anatomical Record **208**, 595–605 (1984c)

Pannese E, Ledda M., Cherkas P.S., Huang T.Y., Hanani M. Satellite cell reactions to axon injury of sensory ganglion neurons: Increase in number of gap junctions and formation of bridges connecting previously separate perineuronal sheaths. Anatomy and Embryology **206**, 337–347 (2003)

Pannese E., Ledda M., Conte V., Procacci P. The perikaryal projections of rabbit spinal ganglion neurons. A comparison of thin section reconstructions and scanning microscopy views. Anatomy and Embryology **181**, 427–432 (1990a)

Pannese E., Ledda M., Conte V., Procacci P., Matsuda S. Scanning electron-microscope observations of the perikaryal projections of rabbit spinal ganglion neurons after enzymatic removal of connective tissue and satellite cells. Cell and Tissue Research **260**, 167–173 (1990b)

Pannese E., Ledda M., Matsuda S. Nerve fibres with myelinated and unmyelinated portions in dorsal spinal roots. Journal of Neurocytology **17**, 693–700 (1988d)

Pannese E., Luciano L., Iurato S., Reale E. Cholinesterase activity in spinal ganglia neuroblasts: A histochemical study at the electron microscope. Journal of Ultrastructural Research **36**, 46–67 (1971)

Pannese E., Luciano L., Iurato S., Reale E. Lysosomes in normal and degenerating neuroblasts of the chick embryo spinal ganglia. A cytochemical and quantitative study by electron microscopy. Acta Neuropathologica **36**, 209–220 (1976)

Pannese E., Luciano L., Iurato S., Reale E. Intercellular junctions and other membrane specializations in developing spinal ganglia: a freeze-fracture study. Journal of Ultrastructure Research **60**, 169–180 (1977)

Pannese E., Luciano L., Reale E. Intercellular junctions in developing spinal ganglia. Zoon **6**, 129–138 (1978)

Pannese E., Procacci P. Ultrastructural localization of NGF receptors in satellite cells of the rat spinal ganglia. Journal of Neurocytology **31**, 755–763 (2002)

Pannese E., Procacci P., Ledda M. Ultrastructural localization of actin in the cell body of rat spinal ganglion neurons. Anatomy and Embryology **194**, 527–531 (1996a)

Pannese E., Procacci P., Ledda M., Arcidiacono G., Frattola D., Rigamonti L. Association between microtubules and mitochondria in myelinated axons of *Lacerta muralis*. A quantitative analysis. Cell and Tissue Research **245**, 1–8 (1986)

Pannese E., Procacci P., Ledda M., Arcidiacono G., Rigamonti L. A quantitative study of microtubules in motor and sensory axons. Acta Anatomica **118**, 193–200 (1984c)

Pannese E., Procacci P., Ledda M., Arcidiacono G., Rigamonti L. Internodal microvilli of Schwann cells of myelinated fibres in lizard spinal roots project onto unmyelinated axons. Journal of Neurocytology **18**, 295–302 (1989)

Pannese E., Procacci P., Ledda M., Conte V. Age-related reduction of the satellite cell sheath around spinal ganglion neurons in the rabbit. Journal of Neurocytology **25**, 137–146 (1996b)

Pannese E., Procacci P., Ledda M., Conte V. The percentage of nerve cell bodies arranged in clusters decreases with age in the spinal ganglia of adult rabbits. Anatomy and Embryology **187**, 331–334 (1993)

Pannese E., Rigamonti L., Procacci P., Ledda M., Arcidiacono G., Frattola D. An electron microscope study of quantitative relationships between axon and Schwann cell sheath in myelinated fibres of peripheral nerves. Anatomy and Embryology **175**, 423–430 (1987b)

Pannese E., Rigamonti L., Procacci P., Ledda M., Arcidiacono G., Frattola D. The structure of Schwann cells in unmyelinated fibres. A qualitative and quantitative electron microscope study. Journal of Submicroscopic Cytology and Pathology **20**, 325–333 (1988c)

Pannese E., Ventura R., Bianchi R. Quantitative relationships between nerve and satellite cells in spinal ganglia: an electron microscopical study. II. Reptiles. Journal of Comparative Neurology **160**, 463–476 (1975)

Papasozomenos S.Ch., Yoon M., Crane R., Autilio-Gambetti L., Gambetti P. Redistribution of proteins of fast axonal transport following administration of β,β^1-iminodipropionitrile: a quantitative autoradiographic study. Journal of Cell Biology **95**, 672–675 (1982)

Park M.K., Ashby M.C., Erdemli G., Petersen O.H., Tepikin A.V. Perinuclear, perigranular and sub-plasmalemmal mitochondria have distinct functions in the regulation of cellular calcium transport. EMBO Journal **20**, 1863–1874 (2001)

Parnavelas J.G., Lynch G., Brecha N., Cotman C.W., Globus A. Spine loss and regrowth in hippocampus following deafferentation. Nature **248**, 71–73 (1974)

Parpura V., Basarsky T.A., Liu F., Jeftinija K., Jeftinija S., Haydon P.G. Glutamate-mediated astrocyte-neuron signalling. Nature **369**, 744–747 (1994)

Parysek L.M., Goldman R.D. Distribution of a novel 57 kDa intermediate filament (IF) protein in the nervous system. Journal of Neuroscience **8**, 555–563 (1988)

Pasic T.R., Rubel E.W. Rapid changes in cochlear nucleus cell size following blockade of auditory nerve electrical activity in gerbils. Journal of Comparative Neurology **283**, 474–480 (1989)

Paterson J.A., Privat A., Ling E.A., Leblond C.P. Investigation of glial cells in semithin sections. III. Transformation of subependymal cells into glial cells, as shown by radioautography after ^{3}H-thymidine injection into the lateral ventricle of the brain of young rats. Journal of Comparative Neurology **149**, 83–102 (1973)

Patrizi G., Poger M. The ultrastructure of the nuclear periphery. The zonula nucleum limitans. Journal of Ultrastructure Research **17**, 127–136 (1967)

Paul E. Histochemische Studien an den Plexus chorioidei, an der Paraphyse und am Ependym von *Rana temporaria* L. Zeitschrift für Zellforschung und mikroskopische Anatomie **91**, 519–546 (1968)

Paula-Barbosa M. Specialized membrane junctions in the avian cerebellum. Experientia **32**, 633–636 (1976)

Pazour G.J., Witman G.B. The vertebrate primary cilium is a sensory organelle. Current Opinion in Cell Biology **15**, 105–110 (2003)

Pearce J., Govind C.K. Reciprocal axo-axonal synapses between the common inhibitor and excitor motoneurons in crustacean limb muscles. Journal of Neurocytology **22**, 259–265 (1993)

Pearse B.M.F. Clathrin: A unique protein associated with intracellular transfer of membrane by coated vesicles. Proceedings of the National Academy of Sciences of the U.S.A. **73**, 255–1259 (1976)

Pebusque M.-J., Robaglia A., Seïte R. Diurnal rhythm of nucleolar volume in sympathetic neurons of the rat superior cervical ganglion. European Journal of Cell Biology **24**, 128–130 (1981)

Peinado M.A., Martinez M., Pedrosa J.A., Quesada A., Peinado J.M. Quantitative morphological changes in neurons and glia in the frontal lobe of the aging rat. Anatomical Record **237**, 104–108 (1993)

Pellerin L., Bouzier-Sore A.-K., Aubert A., Serres S., Merle M., Costalat R., Magistretti P.J. Activity-dependent regulation of energy metabolism by astrocytes: An update. Glia **55**, 1251–1262 (2007)

Pelvig D.P., Pakkenberg H., Stark A.K., Pakkenberg B. Neocortical glial cell numbers in human brains. Neurobiology of Aging **29**, 1754–1762 (2008)

Penfield W. Neuroglia and microglia. The interstitial tissue of the central nervous system. In: Special Cytology (E.V. Cowdry Ed.) pp. 1031–1068. P.B. Hoeber, New York 1928

Peracchia C. A system of parallel septa in crayfish nerve fibers. Journal of Cell Biology **44**, 125–133 (1970)

Perea G., Araque A. Glia modulates synaptic transmission. Brain Research Reviews **63**, 93–102 (2010)

Pereda A.E., Rash J.E., Nagy J.I., Bennett M.V.L. Dynamics of electrical transmission at club endings on the Mauthner cells. Brain Research Reviews **47**, 227–244 (2004)

Perišić M., Cuénod M. Synaptic transmission depressed by colchicine blockade of axoplasmic flow. Science **175**, 1140–1142 (1972)

Perri V., Sacchi O., Raviola E., Raviola G. Evaluation of the number and distribution of synaptic vesicles at cholinergic nerve-endings after sustained stimulation. Brain Research **39**, 526–529 (1972)

Perry V.H., Hume D.A., Gordon S. Immunohistochemical localization of macrophages and microglia in the adult and developing mouse brain. Neuroscience **15**, 313–326 (1985)

Perry V.H., Matyszak M.K., Fearn S. Altered antigen expression of microglia in the aged rodent CNS. Glia **7**, 60–67 (1993)

Peruzzo B., Pastor F.E., Blázquez J.L., Amat P., Rodríguez E.M. Polarized endocytosis and transcytosis in the hypothalamic tanycytes of the rat. Cell and Tissue Research **317**, 147–164 (2004)

Pesonen N. Über die intraependymalen Nervenelemente. Anatomischer Anzeiger **90**, 193–223 (1940)

Pestronk A., Drachman D.B., Griffin J.W. Effects of aging on nerve sprouting and regeneration. Experimental Neurology **70**, 65–82 (1980)

Peters A. The structure of myelin sheaths in the central nervous system of *Xenopus laevis* (Daudin). Journal of Biophysical and Biochemical Cytology **7**, 121–126 (1960)

Peters A. A radial component of central myelin sheaths. Journal of Biophysical and Biochemical Cytology **11**, 733–735 (1961)

Peters A. The structure of the dorsal root nerves of amphioxus. An electron microscope study. Journal of Comparative Neurology **121**, 287–304 (1963)

Peters A. Observations on the connexions between myelin sheaths and glial cells in the optic nerves of young rats. Journal of Anatomy **98**, 125–134 (1964)

Peters A. The node of Ranvier in the central nervous system. Quarterly Journal of Experimental Physiology **51**, 229–236 (1966)

Peters A. The surface fine structure of the choroid plexus and ependymal lining of the rat lateral ventricle. Journal of Neurocytology **3**, 99–108 (1974)

Peters A. Age-related changes in oligodendrocytes in monkey cerebral cortex. Journal of Comparative Neurology **371**, 153–163 (1996)

Peters A. The effects of normal aging on myelin and nerve fibers: a review. Journal of Neurocytology **31**, 581–593 (2002)

Peters A. A fourth type of neuroglial cell in the adult central nervous system. Journal of Neurocytology **33**, 345–357 (2004)

Peters A., Kaiserman-Abramof I.R. The small pyramidal neuron of the rat cerebral cortex. The perikaryon, dendrites and spines. American Journal of Anatomy **127**, 321–356 (1970)

Peters A., Morrison J.H., Rosene D.L., Hyman B.T. Are neurons lost from the primate cerebral cortex during normal aging? Cerebral Cortex **8**, 295–300 (1998a)

Peters A., Moss M.B., Sethares C. Effects of aging on myelinated nerve fibers in monkey primary visual cortex. Journal of Comparative Neurology **419**, 364–376 (2000)

Peters A., Muir A.R. The relationship between axons and Schwann cells during development of peripheral nerves in the rat. Quarterly Journal of Experimental Physiology **44**, 117–130 (1959)

Peters A., Palay S.L. The morphology of laminae A and A_1 of the dorsal nucleus of the lateral geniculate body of the cat. Journal of Anatomy **100**, 451–486 (1966)

Peters A., Palay S.L., Webster H. deF. The Fine Structure of the Nervous System: Neurons and Their Supporting Cells (3rd ed.) Oxford University Press, New York and Oxford 1991

Peters A., Proskauer C.C. The ratio between myelin segments and oligodendrocytes in the optic nerve of the adult rat. Anatomical Record **163**, 243 (1969)

Peters A., Proskauer C.C. Synaptic relationships between a multipolar stellate cell and a pyramidal neuron in the rat visual cortex. A combined Golgi-electron microscope study. Journal of Neurocytology **9**, 163–183 (1980)

Peters A., Proskauer C.C., Kaiserman-Abramof I.R. The small pyramidal neuron of the rat cerebral cortex. The axon hillock and initial segment. Journal of Cell Biology **39**, 604–619 (1968)

Peters A., Sethares C. Aging and the Meynert cells in rhesus monkey primary visual cortex. Anatomical Record **236**, 721–729 (1993)

Peters A., Sethares C. The effects of age on the cells in layer 1 of primate cerebral cortex. Cerebral Cortex **12**, 27–36 (2002)

Peters A., Sethares C. Is there remyelination during aging of the primate central nervous system? Journal of Comparative Neurology **460**, 238–254 (2003)

Peters A., Sethares C. Oligodendrocytes, their progenitors and other neuroglial cells in the aging primate cerebral cortex. Cerebral Cortex **14**, 995–1007 (2004)

Peters A., Sethares C., Luebke J.I. Synapses are lost during aging in the primate prefrontal cortex. Neuroscience **152**, 970–981 (2008a)

Peters A., Sethares C., Moss M.B. The effects of aging on layer 1 in area 46 of prefrontal cortex in the rhesus monkey. Cerebral Cortex **8**, 671–684 (1998b)

Peters A., Sethares C., Moss M.B. How the primate fornix is affected by age. Journal of Comparative Neurology **518**, 3962–3980 (2010)

Peters A., Vaughan D.W. Central nervous system. In: Aging and Cell Structure (J.E. Johnson Jr. Ed.) vol. I, pp. 1–34. Plenum Press, New York and London 1981

Peters A., Vaughn J.E. Microtubules and filaments in the axons and astrocytes of early postnatal rat optic nerves. Journal of Cell Biology **32**, 113–119 (1967)

Peters A., Verderosa A., Sethares C. The neuroglial population in the primary visual cortex of the aging rhesus monkey. Glia **56**, 1151–1161 (2008b)

Peyronnard J.-M., Aguayo A.J., Bray G.M. Schwann cell internuclear distances in normal and regenerating unmyelinated nerve fibers. Archives of Neurology **29**, 56–59 (1973)

Pfenninger K., Akert K., Moor H., Sandri C. The fine structure of freeze-fractured presynaptic membranes. Journal of Neurocytology **1**, 129–149 (1972)

Phillips L.L., Autilio-Gambetti L., Lasek R.J. Bodian's silver method reveals molecular variation in the evolution of neurofilament proteins. Brain Research **278**, 219–223 (1983)

Pilar G., Landmesser L. Ultrastructural differences during embryonic cell death in normal and peripherally deprived ciliary ganglia. Journal of Cell Biology **68**, 339–356 (1976)

Pilcz A. Beitrag zur Lehre von der Pigmententwickelung in den Nervenzellen. Arbeiten aus dem Institut für Anatomie und Physiologie des Centralnervensystems an der Wiener Universität (H. Obersteiner Hrsg.) III Heft, S. 123–139 (1895)

Pinault D., Smith Y., Deschênes M. Dendrodendritic and axoaxonic synapses in the thalamic reticular nucleus of the adult rat. Journal of Neuroscience **17**, 3215–3233 (1997)

Pinching A.J. Synaptic connexions in the glomerular layer of the olfactory bulb. Journal of Physiology **210**, 14P-15P (1970)

Pinching A.J. Myelinated dendritic segments in the monkey olfactory bulb. Brain Research **29**, 133–138 (1971)

Pinching A.J., Powell T.P.S. The neuropil of the glomeruli of the olfactory bulb. Journal of Cell Science **9**, 347–377 (1971a)

Pinching A.J., Powell T.P.S. The neuropil of the periglomerular region of the olfactory bulb. Journal of Cell Science **9**, 379–409 (1971b)

Pineda A., Maxwell D.S., Kruger L. The fine structure of neurons and satellite cells in the trigeminal ganglion of cat and monkey. American Journal of Anatomy **121**, 461–488 (1967)

Piper M., Holt C. RNA translation in axons. Annual Review of Cell and Developmental Biology **20**, 505–523 (2004)

Platani M., Goldberg I., Swedlow J.R., Lamond A.I. In vivo analysis of Cajal body movement, separation, and joining in live human cells. Journal of Cell Biology **151**, 1561–1574 (2000)

Pleasure S.J., Selzer M.E., Lee V.M.-Y. Lamprey neurofilaments combine in one subunit the features of each mammalian NF triplet protein but are highly phosphorylated only in large axons. Journal of Neuroscience **9**, 698–709 (1989)

Pocock J.M., Kettenmann H. Neurotransmitter receptors on microglia. Trends in Neurosciences **30**, 527–535 (2007)

Poe B.H., Linville C., Brunso-Bechtold J. Age-related decline of presumptive inhibitory synapses in the sensorimotor cortex as revealed by the physical disector. Journal of Comparative Neurology **439**, 65–72 (2001)

Pomerat C.M., Hendelman W.J., Raiborn C.W.Jr., Massey J.F. Dynamic activities of nervous tissue *in vitro*. In: The Neuron (H. Hydén Ed.) pp. 119–178. Elsevier, Amsterdam 1967

Pomeroy S.L., Zurakowski D., Khoxayo S., Endara M., Dikkes P. Postnatal addition of satellite cells to parasympathetic neurons. Journal of Comparative Neurology **375**, 518–525 (1996)

Popken G.J., Farel P.B. Sensory neuron number in neonatal and adult rats estimated by means of stereologic and profile-based methods. Journal of Comparative Neurology **386**, 8–15 (1997)

Porter E.L., Hart V.W. Reflex contractions of an all-or-none character in the spinal cat. American Journal of Physiology **66**, 391–403 (1923)

Porter J.T., McCarthy K.D. Hippocampal astrocytes *in situ* respond to glutamate released from synaptic terminals. Journal of Neuroscience **16**, 5073–5081 (1996)

Porter W.T. The presence of Ranvier's constrictions in the spinal cord of vertebrates. Quarterly Journal of Microscopical Science **31**, 91–98 (1890)

Portier M.-M., de Néchaud B., Gros F. Peripherin, a new member of the intermediate filament protein family. Developmental Neuroscience **6**, 335–344 (1984)

Potter H.D. The distribution of neurofibrils coextensive with microtubules and neurofilaments in dendrites and axons of the tectum, cerebellum, and pallium of the frog. Journal of Comparative Neurology **143**, 385–410 (1971)

Pozzo-Miller L.D., Pivovarova N.B., Leapman R.D., Buchanan R.A., Reese T.S., Andrews S.B. Activity-dependent calcium sequestration in dendrites of hippocampal neurons in brain slices. Journal of Neuroscience **17**, 8729–8738 (1997)

Prenant A. Notes cytologiques. III. Cristalloïdes intranucléaires des cellules nerveuses sympathiques chez les mammifères. Archives d'Anatomie microscopique **1**, 366–373 (1897)

Prentice F.D., Wood J.G. Adrenergic innervation of cat adrenal medulla. Anatomical Record **181**, 689–704 (1975)

Price D.L. The response of amphibian glial cells to axonal transection. Journal of Neuropathology and Experimental Neurology **31**, 267–277 (1972)

Price J.L., Powell T.P.S. The synaptology of the granule cells of the olfactory bulb. Journal of Cell Science **7**, 125–155 (1970)

Price T.J., Hargreaves K.M., Cervero F. Protein expression and mRNA cellular distribution of the NKCC1 cotransporter in the dorsal root and trigeminal ganglia of the rat. Brain Research **1112**, 146–158 (2006)

Prineas J.W., McLeod J.G. Chronic relapsing polyneuritis. Journal of the Neurological Sciences **27**, 427–458 (1976)

Privat A., Leblond C.P. The subependymal layer and neighboring region in the brain of the young rat. Journal of Comparative Neurology **146**, 277–302 (1972)

Privat A., Rataboul P. Fibrous and protoplasmic astrocytes. In: Astrocytes (S. Fedoroff, A. Vernadakis Eds.) **1**, 105–129. Academic Press, Orlando 1986

Procacci P., Magnaghi V., Pannese E. Perineuronal satellite cells in mouse spinal ganglia express the gap junction protein connexin43 throughout life with decline in old age. Brain Research Bulletin **75**, 562–569 (2008)

Protopapas A.D., Bower J.M. Physiological characterization of layer III non-pyramidal neurons in piriform (olfactory) cortex of rat. Brain Research **865**, 1–11 (2000)

Pumplin D.W., Reese T.S., Llinás R. Are the presynaptic membrane particles the calcium channels? Proceedings of the National Academy of Sciences of the U.S.A. **78**, 7210–7213 (1981)

Purkinje J.E. Ueber Flimmerbewegungen im Gehirn. Müller's Archiv für Anatomie, Physiologie und wissenschaftliche Medicin, S. 289–290 (1836)

Purkinje J.E. Neueste Untersuchungen aus der Nerven- und Hirn- anatomie. In: Bericht über die Versammlung deutscher Naturforscher und Aerzte in Prag im September 1837 (K. Sternberg, J.V. von Krombholz Hrsg.) S. 177–180. G. Haase, Prag 1838

Purpura D.P. Dendritic spine "dysgenesis" and mental retardation. Science **186**, 1126–1128 (1974)

Purves D. Body and Brain. A Trophic Theory of Neural Connections. Harvard University Press, Cambridge 1988

Purves D., Lichtman J.W. Geometrical differences among homologous neurons in mammals. Science **228**, 298–302 (1985)

Purves D., Voyvodic J.T. Imaging mammalian nerve cells and their connections over time in living animals. Trends in Neurosciences **10**, 398–404 (1987)

Pysh J.J., Wiley R.G. Synaptic vesicle depletion and recovery in cat sympathetic ganglia electrically stimulated in vivo. Journal of Cell Biology **60**, 365–374 (1974)

Qian J., Bull M.S., Levitt P. Target-derived astroglia regulate axonal outgrowth in a region-specific manner. Developmental Biology **149**, 278–294 (1992)

Raff M.C., Miller R.H. Glial cell development in the rat optic nerve. Trends in Neurosciences **7**, 469–472 (1984)

Raff M.C., Mirsky R., Fields K.L., Lisak R.P., Dorfman S.H., Silberberg D.H., Gregson N.A., Leibowitz S., Kennedy M.C. Galactocerebroside is a specific cell-surface antigenic marker for oligodendrocytes in culture. Nature **274**, 813–816 (1978)

Rafols J.A., Cheng H.W., McNeill T.H. Golgi study of the mouse striatum: age-related dendritic changes in different neuronal populations. Journal of Comparative Neurology **279**, 212–227 (1989)

Rafols J.A., Goshgarian H.G. Spinal tanycytes in the adult rat: a correlative Golgi gold-toning study. Anatomical Record **211**, 75–86 (1985)

Rafols J.A., Valverde F. The structure of the dorsal lateral geniculate nucleus in the mouse. A Golgi and electron microscopic study. Journal of Comparative Neurology **150**, 303–332 (1973)

Raine C.S. On the occurrence of Schwann cells within the normal central nervous system. Journal of Neurocytology **5**, 371–380 (1976)

Raine C.S. On the association between perinodal astrocytic processes and the node of Ranvier in the C.N.S. Journal of Neurocytology **13**, 21–27 (1984)

Raine C.S., Ghetti B., Shelanski M.L. On the association between microtubules and mitochondria within axons. Brain Research **34**, 389–393 (1971)

Rainnie D.G., Asprodini E.K., Shinnick-Gallagher P. Intracellular recordings from morphologically identified neurons of the basolateral amygdala. Journal of Neurophysiology **69**, 1350–1362 (1993)

Rakic P. Intrinsic and extrinsic factors influencing the shape of neurons and their assembly into neuronal circuits. In: Frontiers in Neurology and Neuroscience Research (P. Seeman, G. Brown Eds.) pp. 112–132. University of Toronto Press, Toronto 1974

Rakic P. Elusive radial glial cells: Historical and evolutionary perspective. Glia **43**, 19–32 (2003a)

Rakic P. Developmental and evolutionary adaptations of cortical radial glia. Cerebral Cortex **13**, 541–549 (2003b)

Rakic P., Sidman R.L. Sequence of developmental abnormalities leading to granule cell deficit in cerebellar cortex of weaver mutant mice. Journal of Comparative Neurology **152**, 103–132 (1973a)

Rakic P., Sidman R.L. Organization of cerebellar cortex secondary to deficit of granule cells in weaver mutant mice. Journal of Comparative Neurology **152**, 133–162 (1973b)

Rall W. Dendritic spines and synaptic potency. In: Studies in Neurophysiology (R. Porter Ed.) pp. 203–209. Cambridge University Press, Cambridge 1978

Rall W., Segev I. Functional possibilities for synapses on dendrites and dendritic spines. In: Synaptic Function (G.M. Edelman, W.E. Gall, W.M. Cowan Eds.) John Wiley & Sons, New York 1987

Rall W., Shepherd G.M., Reese T.S., Brightman M.W. Dendrodendritic synaptic pathway for inhibition in the olfactory bulb. Experimental Neurology **14**, 44–56 (1966)

Ralston H.J.III The organization of the substantia gelatinosa Rolandi in the cat lumbosacral spinal cord. Zeitschrift für Zellforschung und mikroskopische Anatomie **67**, 1–23 (1965)

Ralston H.J.III, Herman M.M. The fine structure of neurons and synapses in the ventrobasal thalamus of the cat. Brain Research **14**, 77–97 (1969)

Rambourg A., Marraud A., Chretien M. Tri-dimensional structure of the forming face of the Golgi apparatus as seen in the high voltage electron microscope after osmium impregnation of the small nerve cells in the semilunar ganglion of the trigeminal nerve. Journal of Microscopy **97**, 49–57 (1973)

Rami A., Patel A.J., Rabié A. Thyroid hormone and development of the rat hippocampus: morphological alterations in granule and pyramidal cells. Neuroscience **19**, 1217–1226 (1986)

Ramón-Cueto A., Valverde F. Olfactory bulb ensheathing glia: a unique cell type with axonal growth-promoting properties. Glia **14**, 163–173 (1995)

Ramón-Moliner E. The morphology of dendrites. In: The Structure and Function of Nervous Tissue (G.H. Bourne Ed.) **1**, 205–267. Academic Press, New York 1968

Ramón y Cajal S. Estructura de los centros nerviosos de las aves. Revista Trimestral de Histología Normal y Patológica **1**, 1–10 (1888)

Ramón y Cajal S. Sobre la existencia de terminaciones nerviosas pericelulares en los ganglios nerviosos raquidianos. Pequeñas Comunicaciones Anatómicas, Barcelona, pp. 1–5 (1890)

Ramón y Cajal S. Sur la structure de l'écorce cérébrale de quelques mammifères. La Cellule **7**, 123–176 (1891)

Ramón y Cajal S. Les épines collatérales des cellules du cerveau colorées au bleu de méthylène. Revista trimestral micrográfica **1**, 5–19 (1896a)

Ramón y Cajal S. Le bleu de méthylène dans les centres nerveux. Revista trimestral micrográfica **1**, 21–82 (1896b)

Ramón y Cajal S. Estructura del protoplasma nervioso. Revista trimestral micrográfica **1**, 83–113 (1896c)

Ramón y Cajal S. Un sencillo método de coloración selectiva del retículo protoplásmico y sus efectos en los diversos organos nerviosos. Trabajos del Laboratorio de Investigaciones biológicas de la Universidad de Madrid **2**, 129–221 (1903)

Ramón y Cajal S. Histologie du système nerveux de l'homme et des vertébrés. Vol. 1. Maloine, Paris 1909

Ramón y Cajal S. El núcleo de las células piramidales del cerebro humano y de algunos mamíferos. Trabajos del Laboratorio de Investigaciones biológicas de la Universidad de Madrid **8**, 27–62 (1910)

Ramón y Cajal S. El aparato endocelular de Golgi de la célula de Schwann y algunas observaciones sobre la estructura de los tubos nerviosos. Trabajos del Laboratorio de Investigaciones biológicas de la Universidad de Madrid **10**, 221–246 (1912)

Ramón y Cajal S. Contribución al conocimiento de la neuroglia del cerebro humano. Trabajos del Laboratorio de Investigaciones biológicas de la Universidad de Madrid **11**, 255–315 (1913)

Ramón y Cajal S. Degeneration and Regeneration of the Nervous System. Vol. 1 and 2. Oxford University Press, London 1928

Ramón y Cajal S. Les preuves objectives de l'unité anatomique des cellules nerveuses. Travaux du Laboratoire de Recherches biologiques de l' Université de Madrid **29**, 1–137 (1934)

Ranvier L. Contributions à l'histologie et à la physiologie des nerfs périphériques. Comptes Rendus des Séances de l'Academie des Sciences **73**, 1168–1171 (1871a)

Ranvier L. Sur la distribution des étranglements annulaires des tubes nerveux. Comptes Rendus des Séances de la Société de Biologie **23**, 185–186 (1871b)

Ranvier L. De la dégénérescence des nerfs après leur section. Comptes Rendus des Séances de l'Academie des Sciences **75**, 1831–1835 (1872)

Ranvier L. Sur les éléments conjonctifs de la moelle épinière. Comptes Rendus des Séances de l'Academie des Sciences **77**, 1299–1302 (1873)

Ranvier L. Leçons sur l'histologie du Système nerveux. 2 Tomes. F. Savy, Paris 1878

Ranvier L. De la névroglie. Archives de Physiologie normale et pathologique 3e S., **1**, 177–185 (1883)

Rao K.S., Krinke G. Changes with age in the number and size of myelinated axons in the rat L4 dorsal spinal root. Acta Anatomica **117**, 187–192 (1983)

Rao M.V., Engle L.J., Mohan P.S., Yuan A., Qiu D., Cataldo A., Hassinger L., Jacobsen S., Lee V.M.-Y., Andreadis A., Julien J.-P., Bridgman P.C., Nixon R.A. Myosin Va binding to neurofilaments is essential for correct myosin Va distribution and transport and neurofilament density. Journal of Cell Biology **159**, 279–289 (2002)

Rapisardi S.C., Miles T.P. Synaptology of retinal terminals in the dorsal lateral geniculate nucleus of the cat. Journal of Comparative Neurology **223**, 515–534 (1984)

Rasband M.N., Trimmer J.S., Peles E., Levinson S.R., Shrager P. K^+ channel distribution and clustering in developing and hypomyelinated axons of the optic nerve. Journal of Neurocytology **28**, 319–331 (1999)

Rascher G., Wolburg H. The blood–brain barrier in the aging brain. In: Neuroglia in the Aging Brain (J. de Vellis Ed.) pp. 305–320. Humana Press, Totowa, N.J. 2001

Rash J.E., Duffy H.S., Dudek F.E., Bilhartz B.L., Whalen L.R., Yasumura T. Grid-mapped freeze-fracture analysis of gap junctions in gray and white matter of adult rat central nervous system, with evidence for a "panglial syncytium" that is not coupled to neurons. Journal of Comparative Neurology **388**, 265–292 (1997)

Rash J.E., Yasumura T., Dudek F.E., Nagy J.I. Cell-specific expression of connexins and evidence of restricted gap junctional coupling between glial cells and between neurons. Journal of Neuroscience **21**, 1983–2000 (2001)

Rash J.E., Yasumura T., Hudson C.S., Agre P., Nielsen S. Direct immunogold labeling of aquaporin-4 in square arrays of astrocyte and ependymocyte plasma membranes in rat brain and spinal cord. Proceedings of the National Academy of Sciences of the U.S.A. **95**, 11981–11986 (1998)

Raviola E., Gilula N.B. Gap junctions between photoreceptor cells in the vertebrate retina. Proceedings of the National Academy of Sciences of the U.S.A. **70**, 1677–1681 (1973)

Raviola E., Gilula N.B. Intramembrane organization of specialized contacts in the outer plexiform layer of the retina. A freeze-fracture study in monkeys and rabbits. Journal of Cell Biology **65**, 192–222 (1975)

Raviola G., Raviola E. Light and electron microscopic observations on the inner plexiform layer of the rabbit retina. American Journal of Anatomy **120**, 403–426 (1967)

Reale E., Luciano L., Spitznas M. Zonulae occludentes of the myelin lamellae in the nerve fibre layer of the retina and in the optic nerve of the rabbit: a demonstration by the freeze-fracture method. Journal of Neurocytology **4**, 131–140 (1975)

Reese T.S., Brightman M.W. Olfactory surface and central olfactory connexions in some vertebrates. In: Taste and Smell in Vertebrates (G.E.W. Wolstenholme, J. Knight Eds.) pp. 115–149. J. & A. Churchill, London 1970

Reese T.S., Karnovsky M.J. Fine structural localization of a blood–brain barrier to exogenous peroxidase. Journal of Cell Biology **34**, 207–217 (1967)

Reger J.F. Electron microscopy of the motor end-plate in rat intercostal muscle. Anatomical Record **122**, 1–10 (1955)

Rehen S.K., McConnell M.J., Kaushal D., Kingsbury M.A., Yang A.H., Chun J. Chromosomal variation in neurons of the developing and adult mammalian nervous system. Proceedings of the National Academy of Sciences of the U.S.A. **98**, 13361–13366 (2001)

Reich F. Ueber eine neue Granulation in den Nervenzellen. Archiv für Anatomie und Physiologie. Physiologische Abtheilung, S. 208–214 (1903)

Reich F. Über den zelligen Aufbau der Nervenfaser auf Grund mikro- histiochemischer Untersuchungen. I. Die chemischen Bestandteile des Nervenmarks, ihr mikrochemisches und färberisches Verhalten. Journal für Psychologie und Neurologie **8**, 244–273 (1907)

Reichardt L.F., Prokop A. The role of extracellular matrix in nervous system development and maintenance. Developmental Neurobiology **71**, 883–888 (2011)

Reichardt L.F., Tomaselli K.J. Extracellular matrix molecules and their receptors: functions in neural development. Annual Review of Neuroscience **14**, 531–570 (1991)

Reichel W., Hollander J., Clark J.H., Strehler B.L. Lipofuscin pigment accumulation as a function of age and distribution in rodent brain. Journal of Gerontology **23**, 71–78 (1968)

Reichenbach A., Bringmann A. New functions of Müller cells. Glia **61**, 651–678 (2013)

Reichert F., Saada A., Rotshenker S. Peripheral nerve injury induces Schwann cells to express two macrophage phenotypes: phagocytosis and the galactose-specific lectin MAC-2. Journal of Neuroscience **14**, 3231–3245 (1994)

Reichert K.E. Ueber das Verhalten der Nervenfasern bei dem Verlauf, der Vertheilung und Endigung in einem Hautmuskel des Frosches (Rana temporaria). Müller's Archiv für Anatomie, Physiologie und wissenschaftliche Medicin, S. 29–73 (1851)

Remahl S., Hildebrand C. Myelinated non-axonal neuronal elements in the feline olfactory bulb lack sites with a nodal structural differentiation. Brain Research **325**, 1–11 (1985)

Remak R. Vorläufige Mittheilung microscopischer Beobachtungen über den innern Bau der Cerebrospinalnerven und über die Entwickelung ihrer Formelemente. Müller's Archiv für Anatomie, Physiologie und wissenschaftliche Medicin, S. 145–161 (1836)

Remak R. Weitere microscopische Beobachtungen über die Primitivfasern des Nervensystems der Wirbelthiere. Froriep's neue Notizen aus dem Gebiete der Natur- und Heilkunde **3**, 36–41 (1837)

Remak R. Observationes anatomicae et microscopicae de systematis nervosi structura. Dissertatio. Berolini 1838

Remak R. Ueber den Inhalt der Nervenprimitivröhren. Müller's Archiv für Anatomie, Physiologie und wissenschaftliche Medicin, S. 197–201 (1843)

Remak R. Neurologische Erläuterungen. Müller's Archiv für Anatomie, Physiologie und wissenschaftliche Medicin, S. 463–472 (1844)

Remak R. Ueber den Bau der grauen Säulen im Rückenmarke der Säugethiere. Göschen's deutsche Klinik **7**, 295 (1855)

Rende M., Muir D., Ruoslahti E., Hagg T., Varon S., Manthorpe M. Immunolocalization of ciliary neurotrophic factor in adult rat sciatic nerve. Glia **5**, 25–32 (1992)

Réthelyi M., Lozsádi D. Synapses upon the axon origin of dorsal horn neurons in the rat spinal cord. Neuroscience Letters **117**, 20–25 (1990)

Réthelyi M., Szentágothai J. The large synaptic complexes of the substantia gelatinosa. Experimental Brain Research **7**, 258–274 (1969)

Retzius G. Ueber den Bau der Oberflächenschicht der Grosshirnrinde beim Menschen und bei den Säugethieren. Biologiska Föreningens Förhandlingar **3**, 90–102 (1891)

Retzius G. Zur Frage von den freien Nervenendigungen in den Spinalganglien. Biologische Untersuchungen, N.F. **6**, 59–61 (1894)

Revel J.-P., Hamilton D.W. The double nature of the intermediate dense line in peripheral nerve myelin. Anatomical Record **163**, 7–16 (1969)

Reyher C.K.H., Lübke J., Larsen W.J., Hendrix G.M., Shipley M.T., Baumgarten H.G. Olfactory bulb granule cell aggregates: morphological evidence for interperikaryal electrotonic coupling via gap junctions. Journal of Neuroscience **11**, 1485–1495 (1991)

Reyners H., Gianfelici de Reyners E., Maisin J.-R. The beta astrocyte: a newly recognized radiosensitive glial cell type in the cerebral cortex. Journal of Neurocytology **11**, 967–983 (1982)

Riccio R.V., Matthews M.A. Effects of intraocular tetrodotoxin on dendritic spines in the developing rat visual cortex: a Golgi analysis. Developmental Brain Research **19**, 173–182 (1985)

Richardson K.C. Electronmicroscopic observations on Auerbach's plexus in the rabbit, with special reference to the problem of smooth muscle innervation. American Journal of Anatomy **103**, 99–135 (1958)

Richardson K.C. Studies on the structure of autonomic nerves in the small intestine, correlating the silver-impregnated image in light microscopy with the permanganate-fixed ultrastructure in electronmicroscopy. Journal of Anatomy **94**, 457–472 (1960)

Richardson K.C. The fine structure of autonomic nerve endings in smooth muscle of the rat vas deferens. Journal of Anatomy **96**, 427–442 (1962)

Richardson K.C. The fine structure of the albino rabbit iris with special reference to the identification of adrenergic and cholinergic nerves and nerve endings in its intrinsic muscles. American Journal of Anatomy **114**, 173–184 (1964)

Rieske-Shows E., Tetzlaff W., Czonkowska A., Gräber M., Kreutzberg G.W. Microglia in culture. In: Glial-Neuronal Communication in Development and Regeneration (H.H. Althaus, W. Seifert Eds.) pp. 41–51. Springer-Verlag, Berlin 1987

Ringo D.L. The arrangement of subunits in flagellar fibers. Journal of Ultrastructure Research **17**, 266–277 (1967)

Rinne U.K. Ultrastructure of the median eminence of the rat. Zeitschrift für Zellforschung und mikroskopische Anatomie **74**, 98–122 (1966)

Río Hortega P. del El "tercer elemento" de los centros nerviosos. I. La microglía en estado normal. Boletín de la Sociedad Española de Biología **9**, 68–82 (1919)

Río Hortega P. del La glía de escasas radiaciones (oligodendroglía). Boletín de la Real Sociedad Española de Historia natural **21**, 63–92 (1921)

Río Hortega P. del Tercera aportación al conocimiento morfológico e interpretación funcional de la oligodendroglía. Memorias de la Real Sociedad Española de Historia natural **14**, 5–122 (1928)

Río Hortega P. del Concepts histogénique, morphologique, physiologique et physio-pathologique de la microglie. Revue Neurologique **37**, 956–986 (1930)

Río Hortega P. del, Penfield W. Cerebral cicatrix. The reaction of neuroglia and microglia to brain wounds. Bulletin of the Johns Hopkins Hospital **41**, 278–303 (1927)

Ritchie J.M., Rogart R.B. Density of sodium channels in mammalian myelinated nerve fibers and nature of the axonal membrane under the myelin sheath. Proceedings of the National Academy of Sciences of the U.S.A. **74**, 211–215 (1977)

Rivera-Pomar J.M. Die Ultrastruktur der Kapillaren in der Area postrema der Katze. Zeitschrift für Zellforschung und mikroskopische Anatomie **75**, 542–554 (1966)

Roberts D.E., Killiany R.J., Rosene D.L. Neuron numbers in the hypothalamus of the normal aging rhesus monkey: stability across the adult lifespan and between the sexes. Journal of Comparative Neurology **520**, 1181–1197 (2012)

Robertson A., Day B., Pollock M., Collier P. The neuropathy of elderly mice. Acta Neuropathologica **86**, 163–171 (1993)

Robertson J.D. Ultrastructure of two invertebrate synapses. Proceedings of the Society for Experimental Biology and Medicine **82**, 219–223 (1953)

Robertson J.D. Recent electron microscope observations on the ultrastructure of the crayfish median-to-motor giant synapse. Experimental Cell Research **8**, 226–229 (1955)

Robertson J.D. The ultrastructure of a reptilian myoneural junction. Journal of Biophysical and Biochemical Cytology **2**, 381–394 (1956)

Robertson J.D. The ultrastructure of nodes of Ranvier in frog nerve fibres. Journal of Physiology **137**, 8P-9P (1957)

Robertson J.D. The ultrastructure of Schmidt-Lanterman clefts and related shearing defects of the myelin sheath. Journal of Biophysical and Biochemical Cytology **4**, 39–46 (1958)

Robertson J.D. The ultrastructure of cell membranes and their derivatives. Biochemical Society Symposia No. 16, pp. 3–43. Cambridge University Press, London 1959a

Robertson J.D. Preliminary observations on the ultrastructure of nodes of Ranvier. Zeitschrift für Zellforschung und mikroskopische Anatomie **50**, 553–560 (1959b)

Robertson J.D., Bodenheimer T.S., Stage D.E. The ultrastructure of Mauthner cell synapses and nodes in goldfish brains. Journal of Cell Biology **19**, 159–199 (1963)

Robertson W.F. The normal histology and pathology of the neuroglia (in relation specially to mental diseases). Journal of Mental Science **43**, 733–752 (1897)

Robertson W.F. On a new method of obtaining a black reaction in certain tissue elements of the central nervous system (platinum method). Scottish Medical and Surgical Journal **4**, 23–30 (1899)

Robertson W.F. A microscopic demonstration of the normal and pathological histology of mesoglia cells. Journal of Mental Science **46**, 724 (1900)

Robitaille R. Modulation of synaptic efficacy and synaptic depression by glial cells at the frog neuromuscular junction. Neuron **21**, 847–855 (1998)

Robitaille R., Adler E.M., Charlton M.P. Strategic location of calcium channels at transmitter release sites of frog neuromuscular synapses. Neuron **5**, 773–779 (1990)

Robitaille R., Bourque M.-J., Vandaele S. Localization of L-type Ca^{2+} channels at perisynaptic glial cells of the frog neuromuscular junction. Journal of Neuroscience **16**, 148–158 (1996)

Roessmann U., Friede R.L. Entry of labeled monocytic cells into the central nervous system. Acta Neuropathologica **10**, 359–362 (1968)

Roessmann U., Velasco M.E., Sindely S.D., Gambetti P. Glial fibrillary acidic protein (GFAP) in ependymal cells during development. An immunocytochemical study. Brain Research **200**, 13–21 (1980)

Rogers J., Zornetzer S.F., Bloom F.E. Senescent pathology of cerebellum: Purkinje neurons and their parallel fiber afferents. Neurobiology of Aging **2**, 15–25 (1981)

Rogers J., Zornetzer S.F., Bloom F.E., Mervis R.E. Senescent microstructural changes in rat cerebellum. Brain Research **292**, 23–32 (1984)

Roncoroni L. Su un nuovo reperto nel nucleo delle cellule nervose. Archivio di Psichiatria, Scienze penali ed Antropologia criminale **16**, 447–450 (1895)

Roots B.I. Phylogenetic development of astrocytes. In: Astrocytes (S. Fedoroff, A. Vernadakis Eds.) **1**, 1–34. Academic Press, Orlando 1986

Rosenbluth J. The fine structure of acoustic ganglia in the rat. Journal of Cell Biology **12**, 329–359 (1962a)

Rosenbluth J. Subsurface cisterns and their relationship to the neuronal plasma membrane. Journal of Cell Biology **13**, 405–421 (1962b)

Rosenbluth J. The visceral ganglion of *Aplysia californica*. Zeitschrift für Zellforschung und mikroskopische Anatomie **60**, 213–236 (1963)

Rosenbluth J. Redundant myelin sheaths and other ultrastructural features of the toad cerebellum. Journal of Cell Biology **28**, 73–93 (1966)

Rosenbluth J. Intramembranous particle distribution at the node of Ranvier and adjacent axolemma in myelinated axons of the frog brain. Journal of Neurocytology **5**, 731–745 (1976)

Rosenbluth J. Role of glial cells in the differentiation and function of myelinated axons. International Journal of Developmental Neuroscience **6**, 3–24 (1988)

Rosenbluth J. Multiple functions of the paranodal junction of myelinated nerve fibers. Journal of Neuroscience Research **87**, 3250–3258 (2009)

Rosenbluth J., Palay S.L. The fine structure of nerve cell bodies and their myelin sheaths in the eighth nerve ganglion of the goldfish. Journal of Biophysical and Biochemical Cytology **9**, 853–877 (1961)

Rosenbluth J., Wissig S.L. The distribution of exogenous ferritin in toad spinal ganglia and the mechanism of its uptake by neurons. Journal of Cell Biology **23**, 307–325 (1964)

Rosenthal J.F. De formatione granulosa in nervis aliisque partibus organismi animalis. Dissertatio inauguralis anatomico-physiologica. Friedlaender, Vratislaviae 1839

Roslansky P.F., Cornell-Bell A., Rice R.V., Adelman W.J.Jr. Polypeptide composition of squid neurofilaments. Proceedings of the National Academy of Sciences of the U.S.A. **77**, 404–408 (1980)

Rossi A., Palombi F. A neurofibrillar body in the perikaryon of nervus terminalis ganglion cells in teleosts. Zeitschrift für Zellforschung und mikroskopische Anatomie **93**, 395–403 (1969)

Roubos E.W., Moorer-Van Delft C.M. Synaptology of the central nervous system of the freshwater snail *Lymnaea stagnalis* (L.), with particular reference to neurosecretion. Cell and Tissue Research **198**, 217–235 (1979)

Rouget C. Mémoire sur le développement, la structure et les proprietés physiologiques des capillaires sanguins et lymphatiques. Archives de Physiologie Normale et Pathologique **5**, 603–663 (1873)

Rouget M. Note sur la terminaison des nerfs moteurs dans les muscles chez les reptiles, les oiseaux et le mammifères. Comptes Rendus des Séances de l'Academie des Sciences **55**, 548–551 (1862)

Rovainen C.M. Synaptic interactions of identified nerve cells in the spinal cord of the sea lamprey. Journal of Comparative Neurology **154**, 189–206 (1974)

Roy S., Coffee P., Smith G., Liem R.K.H., Brady S.T., Black M.M. Neurofilaments are transported rapidly but intermittently in axons: implications for slow axonal transport. Journal of Neuroscience **20,** 6849–6861 (2000)

Rozanski G.M., Kim H., Li Q., Wong F.K., Stanley E.F. Slow chemical transmission between dorsal root ganglion neuron somata. European Journal of Neuroscience **36**, 3314–3321 (2012)

Rudelli R.D., Brown W.T., Wisniewski K., Jenkins E.C., Laure-Kamionowska M., Connell F., Wisniewski H.M. Adult fragile X syndrome. Clinico-neuropathologic findings. Acta Neuropathologica **67**, 289–295 (1985)

Rungger-Brändle E., Achtstätter T., Franke W.W. An epithelium-type cytoskeleton in a glial cell: astrocytes of amphibian optic nerves contain cytokeratin filaments and are connected by desmosomes. Journal of Cell Biology **109**, 705–716 (1989)

Ruskell G.L. The fine structure of nerve terminations in the lacrimal glands of monkeys. Journal of Anatomy **103**, 65–76 (1968)

Rustioni A., Sotelo C. Synaptic organization of the nucleus gracilis of the cat. Experimental identification of dorsal root fibers and cortical afferents. Journal of Comparative Neurology **155**, 441–468 (1974)

Rutherford J.G., Gwyn D.G. Gap junctions in the inferior olivary nucleus of the squirrel monkey, *Saimiri sciureus*. Brain Research **128**, 374–378 (1977)

Rutishauser U., Jessell T.M. Cell adhesion molecules in vertebrate neural development. Physiological Reviews **68**, 819–857 (1988)

Sabatini D.D., Bensch K., Barrnett R.J. Cytochemistry and electron microscopy. The preservation of cellular ultrastructure and enzymatic activity by aldehyde fixation. Journal of Cell Biology **17**, 19–58 (1963)

Sabel B.A., Stein D.G. Extensive loss of subcortical neurons in the aging rat brain. Experimental Neurology **73**, 507–516 (1981)

Safaei R., Fischer I. Turnover of cytoskeletal proteins in vivo. Brain Research **533**, 83–90 (1990)

Saito K. The initial segment of DSCT (dorsal spinocerebellar tract) neurons in the cat. Journal of Electron Microscopy **21**, 325–326 (1972)

Sakuma E., Wang H.J., Asai Y., Tamaki D., Amano K., Mabuchi Y., Herbert D.C., Soji T. Gap junctional communication between the satellite cells of rat dorsal root ganglia. Acta Anatomica Nipponica **76**, 297–302 (2001)

Salzer J.L., Bunge R.P., Glaser L. Studies of Schwann cell proliferation. III. Evidence for the surface localization of the neurite mitogen. Journal of Cell Biology **84**, 767–778 (1980)

Samorajski T., Friede R.L. A quantitative electron microscopic study of myelination in the pyramidal tract of rat. Journal of Comparative Neurology **134**, 323–338 (1968)

Samorajski T., Friede R.L., Reimer P.R. Hypomyelination in the quaking mouse. A model for the analysis of disturbed myelin formation. Journal of Neuropathology and Experimental Neurology **29**, 507–523 (1970)

Samorajski T., Keefe J.R., Ordy J.M. Intracellular localization of lipofuscin age pigments in the nervous system. Journal of Gerontology **19**, 262–276 (1964)

Samorajski T., Ordy J.M., Keefe J.R. The fine structure of lipofuscin age pigment in the nervous system of aged mice. Journal of Cell Biology **26**, 779–795 (1965)

Samorajski T., Ordy J.M., Rady-Reimer P. Lipofuscin pigment accumulation in the nervous system of aging mice. Anatomical Record **160**, 555–574 (1968)

Samuels A.J., Boyarsky L.L., Gerard R.W., Libet B., Brust M. Distribution, exchange and migration of phosphate compounds in the nervous system. American Journal of Physiology **164**, 1–5 (1951)

Sánchez I., Hassinger L., Paskevich P.A., Shine H.D., Nixon R.A. Oligodendroglia regulate the regional expansion of axon caliber and local accumulation of neurofilaments during development independently of myelin formation. Journal of Neuroscience **16**, 5095–5105 (1996)

Sandelin M., Zabihi S., Liu L., Wicher G., Kozlova E.N. Metastasis-associated S100A4 (Mts1) protein is expressed in subpopulations of sensory and autonomic neurons and in Schwann cells of the adult rat. Journal of Comparative Neurology **473**, 233–243 (2004)

Sandell J.H., Peters A. Effects of age on nerve fibers in the rhesus monkey optic nerve. Journal of Comparative Neurology **429**, 541–553 (2001)

Sandell J.H., Peters A. Effects of age on the glial cells in the rhesus monkey optic nerve. Journal of Comparative Neurology **445**, 13–28 (2002)

Sandell J.H., Peters A. Disrupted myelin and axon loss in the anterior commissure of the aged rhesus monkey. Journal of Comparative Neurology **466**, 14–30 (2003)

Sandow S.L., Whitehouse D., Hill C.E. Specialised sympathetic neuroeffector associations in rat iris arterioles. Journal of Anatomy **192**, 45–57 (1998)

Sandri C., Van Buren J.M., Akert K. Membrane morphology of the vertebrate nervous system. A study with freeze-etch technique. Progress in Brain Research **46**, 1–384 (1977)

Sandritter W., Nováková V., Pilny J., Kiefer G. Cytophotometrische Messungen des Nukleinsäure- und Proteingehaltes von Ganglienzellen der Ratte während der postnatalen Entwicklung und im Alter. Zeitschrift für Zellforschung und mikroskopische Anatomie **80**, 145–152 (1967)

Santolaya R.C. Nucleolus-like bodies in the neuronal cytoplasm of the mouse arcuate nucleus. Zeitschrift für Zellforschung und mikroskopische Anatomie **146**, 319–328 (1973)

Santolaya R.C., Rodriguez Echandia E.L. The surface of the choroid plexus cell under normal and experimental conditions. Zeitschrift für Zellforschung und mikroskopische Anatomie **92**, 43–51 (1968)

Sarnat H.B., Campa J.F., Lloyd J.M. Inverse prominence of ependyma and capillaries in the spinal cord of vertebrates: a comparative histochemical study. American Journal of Anatomy **143**, 439–450 (1975)

Schaffer K. Beitrag zur Histologie der Ammonshornformation. Archiv für mikroskopische Anatomie **39**, 611–632 (1892)

Scharf J.H. Sensible Ganglien. In: Handbuch der mikroskopischen Anatomie des Menschen (M. von Möllendorff, W. Bargmann Hrsg.) Bd. 4/3, S. 14–15, 290–300. Springer-Verlag, Berlin 1958

Scharrer B. Neurosecretion. Comparative and evolutionary aspects. Progress in Brain Research **45**, 125–137 (1976)

Scharrer B., Kater S.B. Neurosecretion. XV. An electron microscopic study of the corpora cardiaca of *Periplaneta americana* after experimentally induced hormone release. Zeitschrift für Zellforschung und mikroskopische Anatomie **95**, 177–186 (1969)

Scharrer E., Brown S. Neurosecretion. XII. The formation of neurosecretory granules in the earthworm, *Lumbricus terrestris* L. Zeitschrift für Zellforschung und mikroskopische Anatomie **54**, 530–540 (1961)

Scheff S.W., Price D.A., Sparks D.L. Quantitative assessment of possible age-related change in synaptic numbers in the human frontal cortex. Neurobiology of Aging **22**, 355–365 (2001)

Scheibel A.B. Structural aspects of the aging brain: spine systems and the dendritic arbor. In: Alzheimer's Disease: Senile Dementia and Related Disorders (R. Katzman, R.D. Terry, K.L. Bick Eds.), Aging **7**, 353–373. Raven Press, New York 1978

Scheibel M.E., Crandall P.H., Scheibel A.B. The hippocampal-dentate complex in temporal lobe epilepsy. Epilepsia **15**, 55–80 (1974)

Scheibel M.E., Lindsay R.D., Tomiyasu U., Scheibel A.B. Progressive dendritic changes in aging human cortex. Experimental Neurology **47**, 392–403 (1975)

Scheibel M.E., Lindsay R.D., Tomiyasu U., Scheibel A.B. Progressive dendritic changes in the aging human limbic system. Experimental Neurology **53**, 420–430 (1976)

Scheibel M.E., Tomiyasu U., Scheibel A.B. The aging human Betz cell. Experimental Neurology **56**, 598–609 (1977)

Scherer S.S., Arroyo E.J. Recent progress on the molecular organization of myelinated axons. Journal of the Peripheral Nervous System **7**, 1–12 (2002)

Scherer S.S., Xu T., Crino P., Arroyo E.J., Gutmann D.H. Ezrin, radixin, and moesin are components of Schwann cell microvilli. Journal of Neuroscience Research **65**, 150–164 (2001)

Schiefferdecker P. Beiträge zur Kenntniss des Baus der Nervenfasern. Archiv für mikroskopische Anatomie **30**, 435–494 (1887)

Schiffer D., Giordana M.T., Migheli A., Giaccone G., Pezzotta S., Mauro A. Glial fibrillary acidic protein and vimentin in the experimental glial reaction of the rat brain. Brain Research **374**, 110–118 (1986)

Schikorski T., Stevens C.F. Quantitative ultrastructural analysis of hippocampal excitatory synapses. Journal of Neuroscience **17**, 5858–5867 (1997)

Schlaepfer W.W., Freeman L.A. Neurofilament proteins of rat peripheral nerve and spinal cord. Journal of Cell Biology **78**, 653–662 (1978)

Schlaepfer W.W., Myers F.K. Relationship of myelin internode elongation and growth in the rat sural nerve. Journal of Comparative Neurology **147**, 255–266 (1973)

Schleiden M.J. Beiträge zur Phytogenesis. Müller's Archiv für Anatomie, Physiologie und wissenschaftliche Medicin, S. 137–176 (1838)

Schmidt H.D. On the construction of the dark or double-bordered nerve fibre. Monthly Microscopical Journal **11**, 200–221 (1874)

Schmidt J.W., Catterall W.A. Biosynthesis and processing of the αsubunit of the voltage-sensitive sodium channel in rat brain. Cell **46**, 437–445 (1986)

Schmidt R.S. Sexual dimorphism in succinic dehydrogenase staining of toad pretrigeminal nucleus. Experimental Brain Research **45**, 447–450 (1982a)

Schmidt R.S. Masculinization of toad pretrigeminal nucleus by androgens. Brain Research **244**, 190–192 (1982b)

Schmidt W.J. Die Bausteine des Tierkörpers in polarisiertem Lichte. F. Cohen, Bonn 1924

Schmidt W.J. Doppelbrechung und Feinbau der Markscheide der Nervenfasern. Zeitschrift für Zellforschung und mikroskopische Anatomie **23**, 657–676 (1936)

Schmidt W.J. Die Doppelbrechung von Karyoplasma, Zytoplasma und Metaplasma. Gebrüder Borntraeger, Berlin 1937

Schmitt F.O. The structure of the axon filaments of the giant nerve fibers of Loligo and Myxicola. Journal of Experimental Zoology **113**, 499–515 (1950)

Schmitt F.O., Bear R.S. The optical properties of vertebrate nerve axons as related to fiber size. Journal of Cellular and Comparative Physiology **9**, 261–273 (1937)

Schmitt F.O., Bear R.S., Clark G.L. X-ray diffraction studies on nerve. Radiology **25**, 131–151 (1935)

Schmitt F.O., Bear R.S., Ponder E. Optical properties of the red cell membrane. Journal of Cellular and Comparative Physiology **9**, 89–92 (1936)

Schmitt F.O., Geren B.B. The fibrous structure of the nerve axon in relation to the localization of "neurotubules". Journal of Experimental Medicine **91**, 499–504 (1950)

Schnapp B., Mugnaini E. The myelin sheath: electron microscopic studies with thin sections and freeze-fracture. In: Golgi Centennial Symposium: Perspectives in Neurobiology (M. Santini Ed.) pp. 209–233. Raven Press, New York 1975

Schnapp B., Mugnaini E. Membrane architecture of myelinated fibers as seen by freeze-fracture. In: Physiology and Pathobiology of Axons (S.G. Waxman Ed.) pp. 83–123. Raven Press, New York 1978

Schnapp B.J., Reese T.S. Dynein is the motor for retrograde axonal transport of organelles. Proceedings of the National Academy of Sciences of the U.S.A. **86**, 1548 1552 (1989)

Schnitzer J., Franke W.W., Schachner M. Immunocytochemical demonstration of vimentin in astrocytes and ependymal cells of developing and adult mouse nervous system. Journal of Cell Biology **90**, 435–447 (1981)

Schnitzlein H.N., Brown H.K. Spinal motoneurons of the goldfish (*Carassius auratus*). Brain, Behavior and Evolution **12**, 207–228 (1975)

Schoch S., Gundelfinger E.D. Molecular organization of the presynaptic active zone. Cell and Tissue Research **326**, 379–391 (2006)

Schon F., Kelly J.S. Autoradiographic localisation of [^{3}H] GABA and [^{3}H]glutamate over satellite glial cells. Brain Research **66**, 275–288 (1974a)

Schon F., Kelly J.S. The characterisation of [^{3}H]GABA uptake into the satellite glial cells of rat sensory ganglia. Brain Research **66**, 289–300 (1974b)

Schubert P., Kreutzberg G.W. Axonal transport of adenosine and uridine derivatives and transfer to postsynaptic neurons. Brain Research **76**, 526–530 (1974)

Schubert P., Lux H.D., Kreutzberg G.W. Single cell isotope injection technique, a tool for studying axonal and dendritic transport. Acta Neuropathologica Suppl. V, 179–186 (1971)

Schüz A. Comparison between the dimensions of dendritic spines in the cerebral cortex of newborn and adult guinea pigs. Journal of Comparative Neurology **244**, 277–285 (1986)

Schultz R.L. Macroglial identification in electron micrographs. Journal of Comparative Neurology **122**, 281–295 (1964)

Schultze M. Allgemeines über die Structurelemente des Nervensystems. In: Handbuch der Lehre von den Geweben des Menschen und der Thiere (S. Stricker Hrsg.) S. 108–136. W. Engelmann, Leipzig 1871

Schulz R. Ueber arteficielle, cadaveröse und pathologische Veränderungen des Rückenmarks. Neurologisches Centralblatt **2**, 529–536, 553–559 (1883)

Schuster T. Elektronenmikroskopische Untersuchungen am Nc. n. oculomotorii von *Salmo irideus* (Gibbons 1855). Zeitschrift für mikroskopisch-anatomische Forschung **87**, 730–764 (1973)

Schwab M.E., Caroni P. Oligodendrocytes and CNS myelin are nonpermissive substrates for neurite growth and fibroblast spreading *in vitro*. Journal of Neuroscience **8**, 2381–2393 (1988)

Schwab M.E., Thoenen H. Selective trans-synaptic migration of tetanus toxin after retrograde axonal transport in peripheral sympathetic nerves: a comparison with nerve growth factor. Brain Research **122**, 459–474 (1977)

Schwab M.E., Thoenen H. Retrograde axonal transport. In: Handbook of Neurochemistry (A. Lajtha Ed.) 2nd ed. **5**, 381–404. Plenum Press, New York 1983

Schwalbe G. Ueber den Bau der Spinalganglien nebst Bemerkungen über die sympathischen Ganglienzellen. Archiv für mikroskopische Anatomie **4**, 45–72 (1868)

Schwann Th. Mikroskopische Untersuchungen über die Uebereinstimmung in der Struktur und dem Wachsthum der Thiere und Pflanzen. Verlag der Sander'schen Buchhandlung (G. Reimer), Berlin 1839

Schwob J.E., Farber N.B., Gottlieb D.I. Neurons of the olfactory epithelium in adult rats contain vimentin. Journal of Neuroscience **6**, 208–217 (1986)

Scott D.E., Kozlowski G.P., Krobisch Dudley G. A comparative ultrastructural analysis of the third cerebral ventricle of the north American mink (*Mustela vison*). Anatomical Record **175**, 155–168 (1973)

Scott D.E., Van Dyke D.H., Paull W.K., Kozlowski G.P. Ultrastructural analysis of the human cerebral ventricular system. III. The choroid plexus. Cell and Tissue Research **150**, 389–397 (1974)

Sechrist J.W. Neurocytogenesis. I. Neurofibrils, neurofilaments, and the terminal mitotic cycle. American Journal of Anatomy **124**, 117–133 (1969)

Sedlacek M., Horak M., Vyklický L. Jr. Morphology and physiology of lamina I neurons of the caudal part of the trigeminal nucleus. Neuroscience **147**, 325–333 (2007)

Segal M., Greenberger V., Korkotian E. Formation of dendritic spines in cultured striatal neurons depends on excitatory afferent activity. European Journal of Neuroscience **17**, 2573–2585 (2003)

Seïte R. Etude ultrastructurale de divers types d'inclusions nucléaires dans les neurones sympathiques du Chat. Journal of Ultrastructure Research **30**, 152–165 (1970)

Seïte R., Escaig J., Couineau S. Microfilaments et microtubules nucléaires et organisation ultrastructurale des batonnets intranucléaires des neurones sympathiques. Journal of Ultrastructure Research **37**, 449–478 (1971)

Seïte R., Vuillet-Luciani J., Vio M., Cataldo C. Sur la présence d'inclusions nucléaires dans certains neurones du noyau caudé du rat: répartition, fréquence et organisation ultrastructurale. Biologie Cellulaire **30**, 73–76 (1977a)

Seïte R., Zerbib R., Vuillet-Luciani J., Vio M. Nuclear inclusions in sympathetic neurons: a quantitative and ultrastructural study in the superior cervical and celiac ganglia of the cat. Journal of Ultrastructure Research **61**, 254–259 (1977b)

Sétáló G., Székely G. The presence of membrane specializations indicative of somato-dendritic synaptic junctions in the optic tectum of the frog. Experimental Brain Research **4**, 237–242 (1967)

Sha L., Miller S.M., Szurszewski J.H. Morphology and electrophysiology of neurons in dog paraventricular nucleus: in vitro study. Brain Research **1010**, 95–107 (2004)

Shapovalov A.I. Interneuronal synapses with electrical, dual and chemical mode of transmission in vertebrates. Neuroscience **5**, 1113–1124 (1980)

Sharma A.K., Bajada S., Thomas P.K. Age changes in the tibial and plantar nerves of the rat. Journal of Anatomy **130**, 417–428 (1980)

Sheffield L.G., Berman N.E.J. Microglial expression of MHC class II increases in normal aging of nonhuman primates. Neurobiology of Aging **19**, 47–55 (1998)

Shelton M.K., McCarthy K.D. Astrocytes *in situ* exhibit functional neurotransmitter receptors. In: Neuroglia in the Aging Brain (J. de Vellis Ed.) pp. 59–76. Humana Press, Totowa, N.J. 2001

Sheng M.H.-T. The postsynaptic specialization. In: Synapses (W.M. Cowan, T.C. Südhof, C.F. Stevens Eds.) pp. 315–355. Johns Hopkins University Press, Baltimore and London 2001

Shepherd G.M. The dendritic spine: A multifunctional integrative unit. Journal of Neurophysiology **75**, 2197–2210 (1996)

Sherrington C.S. Remarks on some aspects of reflex inhibition. Proceedings of the Royal Society of London. Series B **97**, 519–545 (1925)

Shimizu N., Hamuro Y. Deposition of glycogen and changes in some enzymes in brain wounds. Nature **181**, 781–782 (1958)

Shimizu Y. The satellite cells in cultures of dissociated spinal ganglia. Zeitschrift für Zellforschung und mikroskopische Anatomie **67**, 185–195 (1965)

Shinder V., Devor M. Structural basis of neuron-to-neuron cross-excitation in dorsal root ganglia. Journal of Neurocytology **23**, 515–531 (1994)

Shinoda K., Nagano M., Osawa Y. An aromatase-associated cytoplasmic inclusion, the "stigmoid body", in the rat brain: II. Ultrastructure (with a review of its history and nomenclature). Journal of Comparative Neurology **329**, 1–19 (1993)

Shinowara N.L., Beutel W.B., Revel J.-P. Comparative analysis of junctions in the myelin sheath of central and peripheral axons of fish, amphibians and mammals: a freeze-fracture study using complementary replicas. Journal of Neurocytology **9**, 15–38 (1980)

Shipley M.T., Ennis M. Functional organization of olfactory system. Journal of Neurobiology **30**, 123–176 (1996)

Siegesmund K.A., Dutta C.R., Fox C.A. The ultrastructure of the intranuclear rodlet in certain nerve cells. Journal of Anatomy **98**, 93–97 (1964)

Silver R., Silverman A.-J., Vitković L., Lederhendler I.I. Mast cells in the brain: evidence and functional significance. Trends in Neurosciences **19**, 25–31 (1996)

Silverman W.F., Sladek J.R.Jr. Ultrastructural changes in magnocellular neurons from the supraoptic nucleus of aged rats. Developmental Brain Research **58**, 25–34 (1991)

Simard M., Nedergaard M. The neurobiology of glia in the context of water and ion homeostasis. Neuroscience **129**, 877–896 (2004)

Simon A., Oláh S., Molnár G., Szabadics J., Tamás G. Gap-junctional coupling between neurogliaform cells and various interneuron types in the neocortex. Journal of Neuroscience **25**, 6278–6285 (2005)

Simpson F.O., Devine C.E. The fine structure of autonomic neuromuscular contacts in arterioles of sheep renal cortex. Journal of Anatomy **100**, 127–137 (1966)

Singer M., Nordlander R.H., Egar M. Axonal guidance during embryogenesis and regeneration in the spinal cord of the newt: the blueprint hypothesis of neuronal pathway patterning. Journal of Comparative Neurology **185**, 1–22 (1979)

Singer S.J., Nicolson G.L. The fluid mosaic model of the structure of cell membranes. Science **175**, 720–731 (1972)

Sjöstrand F.S. Electron-microscopic demonstration of a membrane structure isolated from nerve tissue. Nature **165**, 482–483 (1950)

Sjöstrand F.S. The ultrastructure of the inner segments of the retinal rods of the guinea pig eye as revealed by electron microscopy. Journal of Cellular and Comparative Physiology **42**, 45–70 (1953a)

Sjöstrand F.S. The ultrastructure of the retinal rod synapses of the guinea pig eye. Journal of Applied Physics **24**, 1422 (1953b)

Sjöstrand F.S. The lamellated structure of the nerve myelin sheath as revealed by high resolution electron microscopy. Experientia **9**, 68–69 (1953c)

Sjöstrand F.S. Ultrastructure of retinal rod synapses of the guinea pig eye as revealed by three-dimensional reconstructions from serial sections. Journal of Ultrastructure Research **2**, 122–170 (1958)

Sjöstrand F.S., Rhodin J. The ultrastructure of the proximal convoluted tubules of the mouse kidney as revealed by high resolution electron microscopy. Experimental Cell Research **4**, 426–456 (1953)

Sjöstrand J. Proliferative changes in glial cells during nerve regeneration. Zeitschrift für Zellforschung und mikroskopische Anatomie **68**, 481–493 (1965)

Skoff R.P. The fine structure of pulse labeled (^{3}H-thymidine cells) in degenerating rat optic nerve. Journal of Comparative Neurology **161**, 595–612 (1975)

Skoff R.P., Vaughn J.E. An autoradiographic study of cellular proliferation in degenerating rat optic nerve. Journal of Comparative Neurology **141**, 133–156 (1971)

Slifer E.H., Sekhon S.S. The femoral chordotonal organs of a grasshopper, Orthoptera, Acrididae. Journal of Neurocytology **4**, 419–438 (1975)

Sloane J.A., Hollander W., Rosene D.L., Moss M.B., Kemper T., Abraham C.R. Astrocytic hypertrophy and altered GFAP degradation with age in subcortical white matter of the rhesus monkey. Brain Research **862**, 1–10 (2000)

Sloper J.J. Dendro-dendritic synapses in the primate motor cortex. Brain Research **34**, 186–192 (1971)

Sloper J.J. Gap junctions between dendrites in the primate neocortex. Brain Research **44**, 641–646 (1972)

Sloper J.J., Powell T.P.S. Observations on the axon initial segment and other structures in the neocortex using conventional staining and ethanolic phosphotungstic acid. Brain Research **50**, 163–169 (1973)

Smart I. The subependymal layer of the mouse brain and its cell production as shown by radioautography after thymidine-H^3 injection. Journal of Comparative Neurology **116**, 325–347 (1961)

Smart I., Leblond C.P. Evidence for division and transformations of neuroglia cells in the mouse brain, as derived from radioautography after injection of thymidine H3. Journal of Comparative Neurology **116**, 349–368 (1961)

Smith D.E. The distribution of neurofilaments and microtubules in Clarke's nucleus in the kitten. Anatomical Record **172**, 408 (1972)

Smith D.S., Järlfors U., Beránek R. The organization of synaptic axoplasm in the lamprey (*Petromyzon marinus*) central nervous system. Journal of Cell Biology **46**, 199–219 (1970)

Smith D.S., Järlfors U., Cameron B.F. Morphological evidence for the participation of microtubules in axonal transport. Annals of the New York Academy of Sciences **253**, 472–506 (1975)

Smith I.W., Mikesh M., Lee Y.i., Thompson W.J. Terminal Schwann cells participate in the competition underlying neuromuscular synapse elimination. Journal of Neuroscience **33**, 17724–17736 (2013)

Smith J.M., O'Leary J.L., Harris A.B., Gay A.J. Ultrastructural features of the lateral geniculate nucleus of the cat. Journal of Comparative Neurology **123**, 357–378 (1964)

Smith R.S. Microtubule and neurofilament densities in amphibian spinal root nerve fibers: relationship to axoplasmic transport. Canadian Journal of Physiology and Pharmacology **51**, 798–806 (1973)

Snider W.D. The dendritic complexity and innervation of submandibular neurons in five species of mammals. Journal of Neuroscience **7**, 1760–1768 (1987)

Snider W.D. Nerve growth factor enhances dendritic arborization of sympathetic ganglion cells in developing mammals. Journal of Neuroscience **8**, 2628–2634 (1988)

Snipes G.J., Suter U. Molecular anatomy and genetics of myelin proteins in the peripheral nervous system. Journal of Anatomy **186**, 483–494 (1995)

Soffer D., Raine C.S. Morphologic analysis of axo-glial membrane specializations in the demyelinated central nervous system. Brain Research **186**, 301–313 (1980)

Sofroniew M.V., Vinters H.V. Astrocytes: biology and pathology. Acta Neuropathologica **119**, 7–35 (2010)

Sogn C.J.L., Puchades M., Gundersen V. Rare contacts between synapses and microglial processes containing high levels of Iba1 and actin – a postembedding immunogold study in the healthy rat brain. European Journal of Neuroscience **38**, 2030–2040 (2013)

Somjen G.G. Electrophysiology of neuroglia. Annual Review of Physiology **37**, 163–190 (1975)

Somogyi P., Hámori J. A quantitative electron microscopic study of the Purkinje cell axon initial segment. Neuroscience **1**, 361–365 (1976)

Sonnhof U., Richter D.W., Taugner R. Electrotonic coupling between frog spinal motoneurons. An electrophysiological and morphological study. Brain Research **138**, 197–215 (1977)

Sosa J.M., De Zorrilla N.B. Morphological variations of the Golgi apparatus in spinal ganglion nerve cells, related to ageing. Acta Anatomica **64**, 475–497 (1966)

Sotelo C. Permanence of postsynaptic specializations in the frog sympathetic ganglion cells after denervation. Experimental Brain Research **6**, 294–305 (1968)

Sotelo C. Ultrastructural aspects of the cerebellar cortex of the frog. In: Neurobiology of Cerebellar Evolution and Development (R. Llinás Ed.) pp. 327–371. AMA-ERF Institute for Biomedical Research, Chicago 1969

Sotelo C. Morphological correlates of electrotonic coupling between neurons in mammalian nervous system. In: Golgi Centennial Symposium: Perspectives in Neurobiology (M. Santini Ed.) pp. 355–365. Raven Press, New York 1975

Sotelo C., Angaut P. The fine structure of the cerebellar central nuclei in the cat. I. Neurons and neuroglial cells. Experimental Brain Research **16**, 410–430 (1973)

Sotelo C., Gentschev T., Zamora A.J. Gap junctions in ventral cochlear nucleus of the rat. A possible new example of electrotonic junctions in the mammalian C.N.S. Neuroscience **1**, 5–7 (1976)

Sotelo C., Gotow T., Wassef M. Localization of glutamic-acid-decarboxylase-immunoreactive axon terminals in the inferior olive of the rat, with special emphasis on anatomical relations between GABAergic synapses and dendrodendritic gap junctions. Journal of Comparative Neurology **252**, 32–50 (1986)

Sotelo C., Hillman D.E., Zamora A.J., Llinás R. Climbing fiber deafferentation: its action on Purkinje cell dendritic spines. Brain Research **98**, 574–581 (1975a)

Sotelo C., Llinás R. Specialized membrane junctions between neurons in the vertebrate cerebellar cortex. Journal of Cell Biology **53**, 271–289 (1972)

Sotelo C., Llinás R., Baker R. Structural study of inferior olivary nucleus of the cat: morphological correlates of electrotonic coupling. Journal of Neurophysiology **37**, 541–559 (1974)

Sotelo C., Palay S.L. The fine structure of the lateral vestibular nucleus in the rat. I. Neurons and neuroglial cells. Journal of Cell Biology **36**, 151–179 (1968)

Sotelo C., Palay S.L. The fine structure of the lateral vestibular nucleus in the rat. II. Synaptic organization. Brain Research **18**, 93–115 (1970)

Sotelo C., Palay S.L. Altered axons and axon terminals in the lateral vestibular nucleus of the rat. Possible example of axonal remodeling. Laboratory Investigation **25**, 653–671 (1971)

Sotelo C., Réthelyi M., Szabo T. Morphological correlates of electrotonic coupling in the magnocellular mesencephalic nucleus of the weakly electric fish *Gymnotus carapo*. Journal of Neurocytology **4**, 587–607 (1975b)

Sotelo C., Taxi J. Ultrastructural aspects of electrotonic junctions in the spinal cord of the frog. Brain Research **17**, 137–141 (1970)

Sotelo-Silveira J.R., Calliari A., Kun A., Koenig E., Sotelo J.R. RNA trafficking in axons. Traffic **7**, 508–515 (2006)

Sottile V., Li M., Scotting P.J. Stem cell marker expression in the Bergmann glia population of the adult mouse brain. Brain Research **1099**, 8–17 (2006)

Špaček J. Three-dimensional reconstructions of astroglia and oligodendroglia cells. Zeitschrift für Zellforschung und mikroskopische Anatomie **112**, 430–442 (1971)

Špaček J. Three-dimensional analysis of dendritic spines. II. Spine apparatus and other cytoplasmic components. Anatomy and Embryology **171**, 235–243 (1985a)

Špaček J. Three-dimensional analysis of dendritic spines. III. Glial sheath. Anatomy and Embryology **171**, 245–252 (1985b)

Spassky N., Merkle F.T., Flames N., Tramontin A.D., García-Verdugo J.M., Alvarez-Buylla A. Adult ependymal cells are postmitotic and are derived from radial glial cells during embryogenesis. Journal of Neuroscience **25**, 10–18 (2005)

Speidel C.C. Studies of living nerves. II. Activities of ameboid growth cones, sheath cells, and myelin segments, as revealed by prolonged observation of individual nerve fibers in frog tadpoles. American Journal of Anatomy **52**, 1–79 (1933)

Sporrong B., Kannisto P., Owman Ch., Sjöberg N.-O., Walles B. Histochemistry and ultrastructure of adrenergic and acetylcholinesterase-containing nerves supplying follicles and endocrine cells in the guinea-pig ovary. Cell and Tissue Research **240**, 505–511 (1985)

Srebro Z. The ultrastructure of gliosomes in the brains of amphibia. Journal of Cell Biology **26**, 313–322 (1965)

Steer J.M. Some observations on the fine structure of rat dorsal spinal nerve roots. Journal of Anatomy **109**, 467–485 (1971)

Steers W.D., Kolbeck S., Creedon D., Tuttle J.B. Nerve growth factor in the urinary bladder of the adult regulates neuronal form and function. Journal of Clinical Investigation **88**, 1709–1715 (1991)

Stefanelli A., Caravita S. Ultrastructural features of the synaptic complex of the vestibular nuclei of *Lampetra planeri* (Bloch). Zeitschrift für Zellforschung und mikroskopische Anatomie **108**, 282–296 (1970)

Steinert P.M., Roop D.R. Molecular and cellular biology of intermediate filaments. Annual Review of Biochemistry **57**, 593–625 (1988)

Stensaas L.J. The ultrastructure of astrocytes, oligodendrocytes, and microglia in the optic nerve of urodele amphibians (*A. punctatum, T. pyrrhogaster, T. viridescens*). Journal of Neurocytology **6**, 269–286 (1977)

Stensaas L.J., Gilson B.C. Ependymal and subependymal cells of the caudato-pallial junction in the lateral ventricle of the neonatal rabbit. Zeitschrift für Zellforschung und mikroskopische Anatomie **132**, 297–322 (1972)

Stensaas L.J., Stensaas S.S. Astrocytic neuroglial cells, oligodendrocytes and microgliacytes in the spinal cord of the toad. I. Light microscopy. II. Electron microscopy. Zeitschrift für Zellforschung und mikroskopische Anatomie **84**, 473–489; **86**, 184–213 (1968)

Stenwig A.E. The origin of brain macrophages in traumatic lesions, Wallerian degeneration, and retrograde degeneration. Journal of Neuropathology and Experimental Neurology **31**, 696–704 (1972)

Stephenson J.L., Byers M.R. GFAP immunoreactivity in trigeminal ganglion satellite cells after tooth injury in rats. Experimental Neurology **131**, 11–22 (1995)

Sterling P. Anatomy and physiology of goldfish oculomotor system. I. Structure of abducens nucleus. Journal of Neurophysiology **40**, 557–572 (1977)

Sterling P., Matthews G. Structure and function of ribbon synapses. Trends in Neurosciences **28**, 20–29 (2005)

Stern J.E., Armstrong W.E. Reorganization of the dendritic trees of oxytocin and vasopressin neurons of the rat supraoptic nucleus during lactation. Journal of Neuroscience **18**, 841–853 (1998)

Sternberger L.A., Sternberger N.H. Monoclonal antibodies distinguish phosphorylated and nonphosphorylated forms of neurofilaments *in situ*. Proceedings of the National Academy of Sciences of the U.S.A. **80**, 6126–6130 (1983)

Sternberger N.H., Itoyama Y., Kies M.W., Webster H.deF. Immunocytochemical method to identify basic protein in myelin-forming oligodendrocytes of newborn rat C.N.S. Journal of Neurocytology **7**, 251–263 (1978)

Stevenson J.A., Yoon M.G. Morphology of radial glia, ependymal cells, and periventricular neurons in the optic tectum of goldfish (*Carassius auratus*). Journal of Comparative Neurology **205**, 128–138 (1982)

Steward O. Polyribosomes at the base of dendritic spines of central nervous system neurons. Their possible role in synapse construction and modification. Cold Spring Harbor Symposia on Quantitative Biology **48**, 745–759 (1983)

Steward O., Levy W.B. Preferential localization of polyribosomes under the base of dendritic spines in granule cells of the dentate gyrus. Journal of Neuroscience **2**, 284–291 (1982)

Steward O., Schuman E.M. Protein synthesis at synaptic sites on dendrites. Annual Review of Neuroscience **24**, 299–325 (2001)

Stewart P.A., Hayakawa E.M. Interendothelial junctional changes underlie the developmental 'tightening' of the blood–brain barrier. Developmental Brain Research **32**, 271–281 (1987)

Stewart P.A., Wiley M.J. Developing nervous tissue induces formation of blood–brain barrier characteristics in invading endothelial cells: a study using quail-chick transplantation chimeras. Developmental Biology **84**, 183–192 (1981)

Streit W.J., Graeber M.B., Kreutzberg G.W. Functional plasticity of microglia: a review. Glia **1**, 301–307 (1988)

Strocchi P., Dahl D., Gilbert J.M. Studies on the biosynthesis of intermediate filament proteins in the rat CNS. Journal of Neurochemistry **39**, 1132–1141 (1982)

Stromska D.P., Ochs S. Axoplasmic transport in aged rats. Experimental Neurology **77**, 215–224 (1982)

Sturrock R.R. Histogenesis of the anterior limb of the anterior commissure of the mouse brain. III. An electron microscopic study of gliogenesis. Journal of Anatomy **117**, 37–53 (1974)

Sturrock R.R. Changes in neuroglia and myelination in the white matter of aging mice. Journal of Gerontology **31**, 513–522 (1976)

Sturrock R.R. Microglia in the human embryonic optic nerve. Journal of Anatomy **139**, 81–91 (1984)

Suetsugu M., Mehraein P. Spine distribution along the apical dendrites of the pyramidal neurons in Down's syndrome. A quantitative Golgi study. Acta Neuropathologica **50**, 207–210 (1980)

Suyeoka O. Small myelinated perikarya in the cerebellar granular layer of mammals including man. Experientia **24**, 472–473 (1968)

Swartz F.J., Bhatnagar K.P. Are CNS neurons polyploid? A critical analysis based upon cytophotometric study of the DNA content of cerebellar and olfactory bulbar neurons of the bat. Brain Research **208**, 267–281 (1981)

Swift H. Nucleic acids and cell morphology in Dipteran salivary glands. In: The Molecular Control of Cellular Activity (J.M. Allen Ed.) pp. 73–125. McGraw-Hill Book Company, New York 1962

Syková E. The extracellular space in the CNS: its regulation, volume and geometry in normal and pathological neuronal function. Neuroscientist **3**, 28–41 (1997)

Sylvia A.L., Rosenthal M. Effects of age on brain oxidative metabolism in vivo. Brain Research **165**, 235–248 (1979)

Szabadics J., Lorincz A., Tamás G. β and γ frequency synchronization by dendritic GABAergic synapses and gap junctions in a network of cortical interneurons. Journal of Neuroscience **21**, 5824–5831 (2001)

Szabo T., Sakata H., Ravaille M. An electrotonically coupled pathway in the central nervous system of some Teleost fish, Gymnotidae and Mormyridae. Brain Research **95**, 459–474 (1975)

Szentágothai J. Glomerular synapses, complex synaptic arrangements, and their operational significance. In: The Neurosciences. Second Study Program (F.O. Schmitt Ed.) pp. 427–443. Rockefeller University Press, New York 1970

Tabira T., Cullen M.J., Reier P.J., Webster H.deF. An experimental analysis of interlamellar tight junctions in amphibian and mammalian C.N.S. myelin. Journal of Neurocytology **7**, 489–503 (1978)

Takahashi K. Special tubular structures in nerve fibers in the granular layer of cerebellum of Syrian hamster. Journal of Electron Microscopy **18**, 312–314 (1969)

Takahashi K., Hama K. Some observations on the fine structure of the synaptic area in the ciliary ganglion of the chick. Zeitschrift für Zellforschung und mikroskopische Anatomie **67**, 174–184 (1965)

Takeda M., Tanimoto T., Kadoj Y., Nasu M., Takahashi M., Kitagawa J., Matsumoto S. Enhanced excitability of nociceptive trigeminal ganglion neurons by satellite glial cytokine following peripheral inflammation. Pain **129**, 155–166 (2007)

Takeda S., Okabe S., Funakoshi T., Hirokawa N. Differential dynamics of neurofilament-H protein and neurofilament-L protein in neurons. Journal of Cell Biology **127**, 173–185 (1994)

Takei K., Mignery G.A., Mugnaini E., Südhof T.C., De Camilli P. Inositol 1,4,5-trisphosphate receptor causes formation of ER cisternal stacks in transfected fibroblasts and in cerebellar Purkinje cells. Neuron **12**, 327–342 (1994)

Takei K., Stukenbrok H., Metcalf A., Mignery G.A., Südhof T.C., Volpe P., De Camilli P. Ca^{2+} stores in Purkinje neurons: endoplasmic reticulum subcompartments demonstrated by the heterogeneous distribution of the $InsP_3$ receptor, Ca^{2+}-ATPase, and calsequestrin. Journal of Neuroscience **12**, 489–505 (1992)

Takeichi M. The fine structure of ependymal cells. 1. The fine structure of ependymal cells in the kitten. Archivum Histologicum Japonicum **26**, 483–505 (1966)

Tamás G., Buhl E.H., Somogyi P. Massive autaptic self-innervation of GABAergic neurons in cat visual cortex. Journal of Neuroscience **17**, 6352–6364 (1997)

Tan A.M., Zhang W., Levine J.M. NG2: a component of the glial scar that inhibits axon growth. Journal of Anatomy **207**, 717–725 (2005)

Tanaka K., Webster H.deF. Myelinated fiber regeneration after crush injury is retarded in sciatic nerves of aging mice. Journal of Comparative Neurology **308**, 180–187 (1991)

Tandler B., Ross L.L. Observations of nerve terminals in human labial salivary glands. Journal of Cell Biology **42**, 339–343 (1969)

Tani E., Itagaki T., Nakano M. Tight junctions of oligodendrocytes. Cell and Tissue Research **184**, 139–142 (1977)

Tani E., Nishiura M., Higashi N. Freeze-fracture studies of gap junctions of normal and neoplastic astrocytes. Acta Neuropathologica **26**, 127–138 (1973)

Tao-Cheng J.-H., Nagy Z., Brightman M.W. Tight junctions of brain endothelium *in vitro* are enhanced by astroglia. Journal of Neuroscience **7**, 3293–3299 (1987)

Tapscott S.J., Bennett G.S., Holtzer H. Neuronal precursor cells in the chick neural tube express neurofilament proteins. Nature **292**, 836–838 (1981)

Tasaki I., Takeuchi T. Der am Ranvierschen Knoten entstehende Aktionsstrom und seine Bedeutung für die Erregungsleitung. Pflügers Archiv für die gesamte Physiologie **244**, 696–711 (1941)

Tasaki I., Takeuchi T. Weitere Studien über den Aktionsstrom der markhaltigen Nervenfaser und über die elektrosaltatorische Übertragung des Nervenimpulses. Pflügers Archiv für die gesamte Physiologie **245**, 764–782 (1942)

Tashiro A., Yuste R. Structure and molecular organization of dendritic spines. Histology and Histopathology **18**, 617–634 (2003)

Tatsuoka H., Reese T.S. New structural features of synapses in the anteroventral cochlear nucleus prepared by direct freezing and freeze-substitution. Journal of Comparative Neurology **290**, 343–357 (1989)

Taugner R. Myelinated dendrites in the spinal cord of frogs (*Rana esculenta*). Anatomy and Embryology **159**, 81–84 (1980)

Taveggia C., Zanazzi G., Petrylak A., Yano Y., Rosenbluth J., Einheber S., Xu X., Esper R.M., Loeb J.A., Shrager P., Chao M.V., Falls D.L., Role L., Salzer J.L. Neuregulin-1 type III determines the ensheathment fate of axons. Neuron **47**, 681–694 (2005)

Taxi J. Etude, au microscope électronique, de l'innervation du muscle lisse intestinal, comparée à celle de quelques autres muscles lisses de mammifères. Archives de Biologie **75**, 301–328 (1964)

Taxi J. Contribution a l'étude des connexions des neurones moteurs du système nerveux autonome. Annales des Sciences Naturelles, Zoologie **7**, 413–674 (1965)

Tello F. Las neurofibrillas en los vertebrados inferiores. Trabajos del Laboratorio de Investigaciones biológicas de la Universidad de Madrid **3**, 113–151 (1904)

Tennyson V.M. The fine structure of the developing nervous system. In: Developmental Neurobiology (W.A. Himwich Ed.) pp. 47–116. C.C. Thomas, Springfield 1970

Tennyson V.M., Pappas G.D. Electronmicroscope studies of the developing telencephalic choroid plexus in normal and hydrocephalic rabbits. In: Disorders of the Developing Nervous System (W.S. Fields, M.M. Desmond Eds.) pp. 267–318. C.C. Thomas, Springfield 1961

Tennyson V.M., Pappas G.D. An electron microscope study of ependymal cells of the fetal, early postnatal and adult rabbit. Zeitschrift für Zellforschung und mikroskopische Anatomie **56**, 595–618 (1962)

Tergast P. Ueber das Verhältniss von Nerve und Muskel. Archly für mikroskopische Anatomie **9**, 36–46 (1873)

Terni T. Sulla correlazione fra ampiezza del territorio di innervazione e volume delle cellule gangliari. 1°. Ricerche sui ganglii spinali della coda nei Chelonii. Anatomischer Anzeiger **47**, 369–386 (1914)

Terni T. Sulla correlazione fra ampiezza del territorio di innervazione e grandezza delle cellule gangliari. 2°. Ricerche sui gangli̇̈ spinali che innervano la coda rigenerata, nei Sauri̇̈ (*Gongylus ocellatus*). Archivio Italiano di Anatomia e di Embriologia **17**, 507–543 (1920)

Terry L.C., Bray G.M., Aguayo A.J. Schwann cell multiplication in developing rat unmyelinated nerves. A radioautographic study. Brain Research **69**, 144–148 (1974)

Terry R.D., Deteresa R., Hansen L.A. Neocortical cell counts in normal human adult aging. Annals of Neurology **21**, 530–539 (1987)

Tetzlaff W. The development of a zonula occludens in peripheral myelin of the chick embryo. A freeze-fracture study. Cell and Tissue Research **189**, 187–201 (1978)

Thaemert J.C. Fine structure of neuromuscular relationships in mouse heart. Anatomical Record **163**, 575–586 (1969)

Thaemert J.C. Atrioventricular node innervation in ultrastructural three dimensions. American Journal of Anatomy **128**, 239–264 (1970)

Thippeswamy T., Mckay J.S., Morris R., Quinn J., Wong L.-F., Murphy D. Glial-mediated neuroprotection: evidence for the protective role of the NO-cGMP pathway via neuron-glial communication in the peripheral nervous system. Glia **49**, 197–210 (2005)

Thoenen H., Barde Y.-A. Physiology of nerve growth factor. Physiological Reviews **60**, 1284–1335 (1980)

Thoenen H., Otten U., Oesch F. Axoplasmic transport of enzymes involved in the synthesis of noradrenaline: relationship between the rate of transport and subcellular distribution. Brain Research **62**, 471–475 (1973)

Thomas P.K., King R.H.M., Sharma A.K. Changes with age in the peripheral nerves of the rat. An ultrastructural study. Acta Neuropathologica **52**, 1–6 (1980)

Thomas P.K., Young J.Z. Internode lengths in the nerves of fishes. Journal of Anatomy **83**, 336–350 (1949)

Thomas W.E. Brain macrophages: on the role of pericytes and perivascular cells. Brain Research Reviews **31**, 42–57 (1999)

Thomson A.M., Armstrong W.E. Biocytin-labelling and its impact on late 20th century studies of cortical circuitry. Brain Research Reviews **66**, 43–53 (2011)

Tiedge H., Brosius J. Translational machinery in dendrites of hippocampal neurons in culture. Journal of Neuroscience **16**, 7171–7181 (1996)

Tigges J., Herndon J.G., Peters A. Axon terminals on Betz cell somata of area 4 in rhesus monkey throughout adulthood. Anatomical Record **232**, 305–315 (1992)

Tigges M., Tigges J. Distribution and morphology of myelinated perikarya and dendrites in the olfactory bulb of primates. Journal of Neurocytology **9**, 825–834 (1980)

Tilney L.G., Bryan J., Bush D.J., Fujiwara K., Mooseker M.S., Murphy D.B., Snyder D.H. Microtubules: evidence for 13 protofilaments. Journal of Cell Biology **59**, 267–275 (1973)

Tilney L.G., Hiramoto Y., Marsland D. Studies on the microtubules in Heliozoa. III. A pressure analysis of the role of these structures in the formation and maintenance of the axopodia of *Actinosphaerium nucleofilum* (Barrett). Journal of Cell Biology **29**, 77–95 (1966)

Tilney L.G., Porter K.R. Studies on the microtubules in Heliozoa. II. The effect of low temperature on these structures in the formation and maintenance of the axopodia. Journal of Cell Biology **34**, 327–343 (1967)

Tisdale A.D., Nakajima Y. Fine structure of synaptic vesicles in two types of nerve terminals in crayfish stretch receptor organs: influence of fixation methods. Journal of Comparative Neurology **165**, 369–386 (1976)

Tokuyasu K., Yamada E. The fine structure of the retina studied with the electron microscope. IV. Morphogenesis of outer segments of retinal rods. Journal of Biophysical and Biochemical Cytology **6**, 225–230 (1959)

tom Dieck S., Altrock W.D., Kessels M.M., Qualmann B., Regus H., Brauner D., Fejtová A., Bracko O., Gundelfinger E.D., Brandstätter J.H. Molecular dissection of the photoreceptor ribbon synapse: physical interaction of Bassoon and RIBEYE is essential for the assembly of the ribbon complex. Journal of Cell Biology **168**, 825–836 (2005)

Toran-Allerand C.D., Hashimoto K., Greenough W.T., Saltarelli M. Sex steroids and the development of the newborn mouse hypothalamus and preoptic area in vitro: III. Effects of estrogen on dendritic differentiation. Developmental Brain Research **7**, 97–101 (1983)

Torri-Tarelli F., Grohovaz F., Fesce R., Ceccarelli B. Temporal coincidence between synaptic vesicle fusion and quantal secretion of acetylcholine. Journal of Cell Biology **101**, 1386–1399 (1985)

Torvik A., Skjörten F. Electron microscopic observations on nerve cell regeneration and degeneration after axon lesions. II. Changes in the glial cells. Acta Neuropathologica **17**, 265–282 (1971)

Tourneux F., Le Goff R. Note sur les étranglements des tubes nerveux de la moelle épinière. Journal de l'Anatomie et de la Physiologie normales et pathologiques de l'homme et des animaux **11**, 403–404 (1875)

Trachtenberg J.T., Chen B.E., Knott G.W., Feng G., Sanes J.R., Welker E., Svoboda K. Long-term *in vivo* imaging of experience-dependent synaptic plasticity in adult cortex. Nature **420**, 788–794 (2002)

Tramontin A.D., García-Verdugo J.M., Lim D.A., Alvarez-Buylla A. Postnatal development of radial glia and the ventricular zone (VZ): a continuum of the neural stem cell compartment. Cerebral Cortex **13**, 580–587 (2003)

Trapp B.D., Andrews S.B., Wong A., O'Connell M., Griffin J.W. Co-localization of the myelin-associated glycoprotein and the microfilament components, F-actin and spectrin, in Schwann cells of myelinated nerve fibres. Journal of Neurocytology **18**, 47–60 (1989)

Trendelenburg G., Dirnagl U. Neuroprotective role of astrocytes in cerebral ischemia: focus on ischemic preconditioning. Glia **51**, 307–320 (2005)

Tretjakoff D. Die zentralen Sinnesorgane bei Petromyzon. Archiv für mikroskopische Anatomie (I) **83**, 68–117 (1913)

Trommald M., Jensen V., Andersen P. Analysis of dendritic spines in rat CA1 pyramidal cells intracellularly filled with a fluorescent dye. Journal of Comparative Neurology **353**, 260–274 (1995)

Tropea M., Johnson M.I., Higgins D. Glial cells promote dendritic development in rat sympathetic neurons in vitro. Glia **1**, 380–392 (1988)

Troy C.M., Brown K., Greene L.A., Shelanski M.L. Ontogeny of the neuronal intermediate filament protein, peripherin, in the mouse embryo. Neuroscience **36**, 217–237 (1990)

Trujillo-Cenóz O. Some aspects of the structural organization of the arthropod ganglia. Zeitschrift für Zellforschung und mikroskopische Anatomie **56**, 649–682 (1962)

Tsukita S., Ishikawa H. Three-dimensional distribution of smooth endoplasmic reticulum in myelinated axons. Journal of Electron Microscopy **25**, 141–149 (1976)

Tsukita S., Ishikawa H. The cytoskeleton in myelinated axons: serial section study. Biomedical Research **2**, 424–437 (1981)

Tuczinski H.-J., Friede R.L. Internodal length in ventral roots of bovine spinal nerves varies independently of fibre calibre. Journal of Anatomy **138**, 423–433 (1984)

Uchiyama H., Ito H. Fiber connections and synaptic organization of the preoptic retinopetal nucleus in the filefish (Balistidae, Teleostei). Brain Research **298**, 11–24 (1984)

Uchizono K. Characteristics of excitatory and inhibitory synapses in the central nervous system of the cat. Nature **207**, 642–643 (1965)

Uemura E. Age-related changes in prefrontal cortex of *Macaca mulatta*: synaptic density. Experimental Neurology **69**, 164–172 (1980)

Uemura E. Age-related changes in the subiculum of *Macaca mulatta*: synaptic density. Experimental Neurology **87**, 403–411 (1985a)

Uemura E. Age-related changes in the subiculum of *Macaca mulatta*: dendritic branching pattern. Experimental Neurology **87**, 412–427 (1985b)

Uga S., Ikui H. Membrane modification occurring between neurons and Schwann cells. Journal of Electron Microscopy **17**, 155 (1968)

Ulfhake B., Kellerth J.-O. A quantitative light microscopic study of the dendrites of cat spinal α-motoneurons after intracellular staining with horseradish peroxidase. Journal of Comparative Neurology **202**, 571–583 (1981)

Ulibarri C., Popper P., Micevych P.E. Motoneurons dorsolateral to the central canal innervate perineal muscles in the Mongolian gerbil. Journal of Comparative Neurology **356**, 225–237 (1995)

Unsicker K. Zur Innervation der Nebennierenrinde vom Goldhamster. Eine fluoreszenz- und elektronenmikroskopische Studie. Zeitschrift für Zellforschung und mikroskopische Anatomie **95**, 608–619 (1969)

Unsicker K. On the innervation of the rat and pig adrenal cortex. Zeitschrift für Zellforschung und mikroskopische Anatomie **116**, 151–156 (1971)

Uylings H.B.M., Kuypers K., Diamond M.C., Veltman W.A.M. Effects of differential environments on plasticity of dendrites of cortical pyramidal neurons in adult rats. Experimental Neurology **62**, 658–677 (1978)

Valdivia O. Methods of fixation and the morphology of synaptic vesicles. Journal of Comparative Neurology **142**, 257–274 (1971)

Vale R.D., Reese T.S., Sheetz M.P. Identification of a novel force-generating protein, kinesin, involved in microtubule-based motility. Cell **42**, 39–50 (1985)

Valentin G. Über den Verlauf und die letzten Enden der Nerven. Verhandlungen der Kaiserlichen Leopoldinisch-Carolinischen Akademie der Naturforscher **18**, 51–240 (1836)

Valentin G. Ueber die Scheiden der Ganglienkugeln und deren Fortsetzungen. Müller's Archiv für Anatomie, Physiologie und wissenschaftliche Medicin, S. 139–164 (1839)

Valverde F. Apical dendritic spines of the visual cortex and light deprivation in the mouse. Experimental Brain Research **3**, 337–352 (1967)

Valverde F. Rate and extent of recovery from dark rearing in the visual cortex of the mouse. Brain Research **33**, 1–11 (1971)

Valverde F., Lopez-Mascaraque L. Neuroglial arrangements in the olfactory glomeruli of the hedgehog. Journal of Comparative Neurology **307**, 658–674 (1991)

van den Bosch de Aguilar P., Goemaere-Vanneste J., Klosen P., Terao E. Ageing changes of spinal ganglion neurons. In: Development and Involution of Neurones (K. Fujisawa, Y. Morimatsu Eds.) pp. 109–150. Japan Scientific Societies Press, Tokyo 1992

van den Bosch de Aguilar P., Vanneste J. Ageing of the spinal ganglion neurons in the rat: a radioautographic study following injection of [^{3}H]lysine. Neuroscience Letters **18**, 225–230 (1980)

van den Bosch de Aguilar P., Vanneste J. Etude ultrastructurale des neurones ganglionnaires spinaux au cours du vieillissement chez le rat. Acta Anatomica **110**, 59–70 (1981)

Van der Loos H. Fine structure of synapses in the cerebral cortex. Zeitschrift für Zellforschung und mikroskopische Anatomie **60**, 815–825 (1963)

Van der Loos H. The 'improperly' oriented pyramidal cell in the cerebral cortex and its possible bearing on problems of neuronal growth and cell orientation. Bulletin of the Johns Hopkins Hospital **117**, 228–250 (1965)

Van der Loos H. The history of the neuron. In: The Neuron (H. Hydén Ed.) pp. 1–47. Elsevier, Amsterdam 1967

Van der Loos H., Glaser E.M. Autapses in neocortex cerebri: synapses between a pyramidal cell's axon and its own dendrites. Brain Research **48**, 355–360 (1972)

van der Want J.J.L., Gramsbergen A., Ijkema-Paassen J., de Weerd H., Liem R.S.B. Dendro-dendritic connections between motoneurons in the rat spinal cord: an electron microscopic investigation. Brain Research **779**, 342–345 (1998)

Van Deurs B. Structural aspects of brain barriers, with special reference to the permeability of the cerebral endothelium and choroidal epithelium. International Review of Cytology **65**, 117–191 (1980)

Van Gehuchten A. La structure des centres nerveux. La moëlle épinière et le cervelet. La Cellule **7**, 79–122 (1891)

Van Gehuchten A. Nouvelles recherches sur les ganglions cérébro-spinaux. La Cellule **8**, 233–253 (1892)

Van Harreveld A. On the force and size of motor units in the rabbit's sartorius muscle. American Journal of Physiology **151**, 96–106 (1947)

Van Harreveld A. The extracellular space in the vertebrate central nervous system. In: The Structure and Function of Nervous Tissue (G.H. Bourne Ed.) **4**, 447–511. Academic Press, New York 1972

Van Leeuwen F.W., Pool C.W., Sluiter A.A. Enkephalin immunoreactivity in synaptoid elements on glial cells in the rat neural lobe. Neuroscience **8**, 229–241 (1983)

Van Leeuwenhoek A. (1718) Cited by Van der Loos H. (1967)

Vanneste J., van den Bosch de Aguilar P. Mitochondrial alterations in the spinal ganglion neurons in ageing rats. Acta Neuropathologica **54**, 83–87 (1981)

Varon S. Glia, nerve growth factor and ganglionic metabolism. In: Neurotransmission (L. Ahtee, Ed.) pp. 275–284. Pergamon Press, Oxford 1976

Varon S., Raiborn C. Dissociation of chick embryo spinal ganglia and the effects on cell yields by the mouse 7s nerve growth factor protein. Neurobiology **2**, 106–122 (1972)

Varon S., Raiborn C., Tyszka E. *In vitro* studies of dissociated cells from newborn mouse dorsal root ganglia. Brain Research **54**, 51–63 (1973)

Vaughan D.W. Age-related deterioration of pyramidal cell basal dendrites in rat auditory cortex. Journal of Comparative Neurology **171**, 501–516 (1977)

Vaughan D.W. Effects of advancing age on peripheral nerve regeneration. Journal of Comparative Neurology **323**, 219–237 (1992)

Vaughan D.W., Peters A. Neuroglial cells in the cerebral cortex of rats from young adulthood to old age: an electron microscope study. Journal of Neurocytology **3**, 405–429 (1974)

Vaughan D.W., Vincent J.M. Ultrastructure of neurons in the auditory cortex of ageing rats: a morphometric study. Journal of Neurocytology **8**, 215–228 (1979)

Vaughn J.E., Pease D.C. Electron microscopic studies of Wallerian degeneration in rat optic nerves. II. Astrocytes, oligodendrocytes and adventitial cells. Journal of Comparative Neurology **140**, 207–226 (1970)

Vaughn J.E., Peters A. Electron microscopy of the early postnatal development of fibrous astrocytes. American Journal of Anatomy **121**, 131–152 (1967)

Vaughn J.E., Peters A. The morphology and development of neuroglial cells. In: Cellular Aspects of Neural Growth and Differentiation (D.C. Pease Ed.) pp. 103–140. University of California Press, Berkeley 1971.

Vega J.A., Del Valle-Soto M.E., Calzada B., Alvarez-Mendez J.C. Immunohistochemical localization of S-100 protein subunits (α and β) in dorsal root ganglia of the rat. Cellular and Molecular Biology **37**, 173–181 (1991)

Ventura R., Harris K.M. Three-dimensional relationships between hippocampal synapses and astrocytes. Journal of Neuroscience **19**, 6897–6906 (1999)

Verbavatz J.-M., Ma T., Gobin R., Verkman A.S. Absence of orthogonal arrays in kidney, brain and muscle from transgenic knockout mice lacking water channel aquaporin-4. Journal of Cell Science **110**, 2855–2860 (1997)

Verdú E., Ceballos D., Vilches J.J., Navarro X. Influence of aging on peripheral nerve function and regeneration. Journal of the Peripheral Nervous System **5**, 191–208 (2000)

Verkhratsky A., Steinhäuser C. Ion channels in glial cells. Brain Research Reviews **32**, 380–412 (2000)

Verma P., Chierzi S., Codd A.M., Campbell D.S., Meyer R.L., Holt C.E., Fawcett J.W. Axonal protein synthesis and degradation are necessary for efficient growth cone regeneration. Journal of Neuroscience **25**, 331–342 (2005)

Vesin M.-F., Urade Y., Hayaishi O., Droz B. Neuronal and glial prostaglandin D synthase isozymes in chick dorsal root ganglia: A light and electron microscopic immunocytochemical study. Journal of Neuroscience **15**, 470–476 (1995)

Viancour T.A., Forman D.S. Polarity orientations of microtubules in squid and lobster axons. Journal of Neurocytology **16**, 69–75 (1987)

Viancour T.A., Kreiter N.A. Vesicular fast axonal transport rates in young and old rat axons. Brain Research **628**, 209–217 (1993)

Vigh B., Vigh-Teichmann I. Actual problems of the cerebrospinal fluid-contacting neurons. Microscopy Research and Technique **41**, 57–83 (1998)

Vigh B., Vigh-Teichmann I., Aros B. Ultrastruktur der Liquorkontaktneurone des Zentralkanals des Rückenmarkes vom Karpfen (*Cyprinus carpio*). Zeitschrift für Zellforschung und mikroskopische Anatomie **122**, 301–309 (1971)

Vigh-Teichmann I., Vigh B., Koritsánszky S. Liquorkontaktneurone im Nucleus lateralis tuberis von Fischen. Zeitschrift für Zellforschung und mikroskopische Anatomie **105**, 325–338 (1970)

Villegas G.M. Ultrastructure of the human retina. Journal of Anatomy **98**, 501–513 (1964)

Villegas G.M., Villegas R. Morphogenesis of the Schwann channels in the squid nerve. Journal of Ultrastructure Research **8**, 197–205 (1963)

Villegas G.M., Villegas R. Extracellular pathways in the peripheral nerve fibres: Schwann-cell-layer permeability to thorium dioxide. Biochimica et Biophysica Acta **88**, 231–233 (1964)

Virchow R. Ueber das granulirte Ansehen der Wandungen der Gehirnventrikel. Allgemeine Zeitschrift für Psychiatrie **3**, 242–250 (1846)

Virchow R. Ueber eine im Gehirn und Rückenmark des Menschen aufgefundene Substanz mit der chemischen Reaction der Cellulose. Archiv für pathologische Anatomie und Physiologie und für klinische Medicin **6**, 135–138 (1854a)

Virchow R. Ueber das ausgebreitete Vorkommen einer dem Nervenmark analogen Substanz in den thierischen Geweben. Archiv für pathologische Anatomie und Physiologie und für klinische Medicin. **6**, 562–572 (1854b)

Vit J.-P., Jasmin L., Bhargava A., Ohara P.T. Satellite glial cells in the trigeminal ganglion as a determinant of orofacial neuropathic pain. Neuron Glia Biology **2**, 247–257 (2006)

Vizoso A.D., Young J.Z. Internode length and fibre diameter in developing and regenerating nerves. Journal of Anatomy **82**, 110–134 (1948)

Vollrath L., Huss H. The synaptic ribbons of the guinea-pig pineal gland under normal and experimental conditions. Zeitschrift für Zellforschung und mikroskopische Anatomie **139**, 417–429 (1973)

Voyvodic J.T. Peripheral target regulation of dendritic geometry in the rat superior cervical ganglion. Journal of Neuroscience **9**, 1997–2010 (1989)

Wagner R. Sympathischer Nerv, Ganglienstructur und Nervenendigungen. In: Handwörterbuch der Physiologie (R. Wagner Hrsg.) **3**, I, 360–406. Vieweg, Braunschweig 1846

Wake H., Moorhouse A.J., Jinno S., Kohsaka S., Nabekura J. Resting microglia directly monitor the functional state of synapses *in vivo* and determine the fate of ischemic terminals. Journal of Neuroscience **29**, 3974–3980 (2009)

Waksman B.H. Experimental study of diphtheritic polyneuritis in the rabbit and guinea pig. III. The blood-nerve barrier in the rabbit. Journal of Neuropathology and Experimental Neurology **20**, 35–77 (1961)

Walberg F. A special type of synaptic vesicles in boutons in the inferior olive. Journal of Ultrastructure Research **12**, 237 (1965a)

Walberg F. Axoaxonic contacts in the cuneate nucleus, probable basis for presynaptic depolarization. Experimental Neurology **13**, 218–231 (1965b)

Walberg F. Elongated vesicles in terminal boutons of the central nervous system, a result of aldehyde fixation. Acta Anatomica **65**, 224–235 (1966)

Waldeyer H.W.G. Über einige neuere Forschungen im Gebiete der Anatomie des Centralnervensystems. Deutsche medicinische Wochenschrift **17**, 1213–1218, 1244–1246, 1267–1269, 1287–1289, 1331–1332, 1352–1356 (1891)

Waller A.V. Experiments on the section of the glossopharyngeal and hypoglossal nerves of the frog, and observations of the alterations produced thereby in the structure of their primitive fibres. Philosophical Transactions of the Royal Society of London **140**, 423–429 (1850)

Waller A.V. Examen des altérations qui ont lieu dans les filets d'origine du nerf pneumogastrique et des nerfs rachidiens, par suite de la section de ces nerfs au-dessus de leur ganglions. Comptes Rendus des Séances de l'Académie des Sciences **34**, 842–847 (1852a)

Waller A.V. Huitième mémoire sur le système nerveux. Comptes Rendus des Séances de l'Académie des Sciences **35**, 561–564 (1852b)

Walz W. Controversy surrounding the existence of discrete functional classes of astrocytes in adult gray matter. Glia **31**, 95–103 (2000)

Walz W., Hertz L. Functional interactions between neurons and astrocytes. II. Potassium homeostasis at the cellular level. Progress in Neurobiology **20**, 133–183 (1983)

Walz W., Lang M.K. Immunocytochemical evidence for a distinct GFAP-negative subpopulation of astrocytes in the adult rat hippocampus. Neuroscience Letters **257**, 127–130 (1998)

Wang H., Kunkel D.D., Martin T.M., Schwartzkroin P.A., Tempel B.L. Heteromultimeric K^+ channels in terminal and juxtaparanodal regions of neurons. Nature **365**, 75–79 (1993)

Wang L., Brown A. Rapid movement of microtubules in axons. Current Biology **12**, 1496–1501 (2002)

Wang L., Ho C.-l., Sun D., Liem R.K.H., Brown A. Rapid movement of axonal neurofilaments interrupted by prolonged pauses. Nature Cell Biology **2**, 137–141 (2000)

Wang T.-J., Chen J.-R., Wang Y.-J., Tseng G.-F. The cytoarchitecture and soma-dendritic arbors of the pyramidal neurons of aged rat sensorimotor cortex: an intracellular dye injection study. Neuroscience **158**, 776–785 (2009)

Warchol J.B. Intranuclear microfilament bundles in the ependymal cells of the third ventricle of the rat. Cell and Tissue Research **194**, 353–359 (1978)

Warner F.D., Satir P. The substructure of ciliary microtubules. Journal of Cell Science **12**, 313–326 (1973)

Washburn M.S., Moises H.C. Electrophysiological and morphological properties of rat basolateral amygdaloid neurons *in vitro*. Journal of Neuroscience **12**, 4066–4079 (1992)

Watson M.L. The nuclear envelope. Its structure and relation to cytoplasmic membranes. Journal of Biophysical and Biochemical Cytology **1**, 257–270 (1955)

Waxman S.G., Black J.A. Freeze-fracture ultrastructure of the perinodal astrocyte and associated glial junctions. Brain Research **308**, 77–87 (1984)

Waxman S.G., Dichter M.A., Hartwieg E.A., Matheson J.K. Recapitulation of normal neuro-glial relations in dissociated cell cultures of dorsal root ganglia. Brain Research **122**, 344–350 (1977)

Waxman S.G., Pappas G.D. An electron microscopic study of synaptic morphology in the oculomotor nuclei of three inframammalian species. Journal of Comparative Neurology **143**, 41–72 (1971)

Waxman S.G., Pappas G.D., Bennett M.V.L. Morphological correlates of functional differentiation of nodes of Ranvier along single fibers in the neurogenic electric organ of the knife fish *Sternarchus*. Journal of Cell Biology **53**, 210–224 (1972)

Weick M., Cherkas P.S., Härtig W., Pannicke T., Uckermann O., Bringmann A., Tal M., Reichenbach A., Hanani M. P2 receptors in satellite glial cells in trigeminal ganglia of mice. Neuroscience **120**, 969–977 (2003)

Weigert C. Bemerkungen über des Neurogliagerüst des menschlichen Centralnervensystems. Anatomischer Anzeiger **5**, 543–551 (1890)

Weigert C. Beiträge zur Kenntnis der normalen menschlichen Neuroglia. Festschrift zum fünfzigjährigen Jubiläum des ärztlichen Vereins. A. Weisbrod, Frankfurt a.M. 1895

Weiler M., Roed I.S., Whittaker V.P. The kinetics of acetylcholine turnover in a resting cholinergic nerve terminal and the magnitude of the cytoplasmic compartment. Journal of Neurochemistry **38**, 1187–1191 (1982)

Weinberg H.J., Spencer P.S. Studies on the control of myelinogenesis. I. Myelination of regenerating axons after entry into a foreign unmyelinated nerve. Journal of Neurocytology **4**, 395–418 (1975)

Weinberg H.J., Spencer P.S. Studies on the control of myelinogenesis. II. Evidence for neuronal regulation of myelin production. Brain Research **113**, 363–378 (1976)

Weiss P. Endoneurial edema in constricted nerve. Anatomical Record **86**, 491–522 (1943)

Weiss P. In vitro transformation of spindle cells of neural origin into macrophages. Anatomical Record **88**, 205–221 (1944)

Weiss P. Neuronal dynamics. Neurosciences Research Program Bulletin **5**, 371–415 (1967)

Weiss P., Hiscoe H.B. Experiments on the mechanism of nerve growth. Journal of Experimental Zoology **107**, 315–395 (1948)

Weiss P., Wang H. Transformation of adult Schwann cells into macrophages. Proceedings of the Society for Experimental Biology and Medicine **58**, 273–275 (1945)

Weiss P.A., Mayr R. Neuronal organelles in neuroplasmic ("axonal") flow. II. Neurotubules. Acta Neuropathologica Suppl. V, 198–206 (1971)

Wen G.Y., Wiśniewski H.M., Kascsak R.J. Biondi ring tangles in the choroid plexus of Alzheimer disease and normal aging brains: a quantitative study. Brain Research **832**, 40–46 (1999)

Wendell-Smith C.P., Blunt M.J., Baldwin F. The ultrastructural characterization of macroglial cell types. Journal of Comparative Neurology **127**, 219–240 (1966)

Wenzel J., Lammert G., Meyer U., Krug M. The influence of long-term potentiation on the spatial relationship between astrocyte processes and potentiated synapses in the dentate gyrus neuropil of rat brain. Brain Research **560**, 122–131 (1991)

Wernig A., Herrera A.A. Sprouting and remodelling at the nerve-muscle junction. Progress in Neurobiology **27**, 251–291 (1986)

Westergaard E. The Lateral Cerebral Ventricles and the Ventricular Walls. Doctoral dissertation, University of Aarhus 1970

Westergaard E. The fine structure of nerve fibers and endings in the lateral cerebral ventricles of the rat. Journal of Comparative Neurology **144**, 345–354 (1972)

Westra J.W., Rivera R.R., Bushman D.M., Yung Y.C., Peterson S.E., Barral S., Chun J. Neuronal DNA content variation (DCV) with regional and individual differences in the human brain. Journal of Comparative Neurology **518**, 3981–4000 (2010)

Westrum L.E. Synaptic contacts on axons in the cerebral cortex. Nature **210**, 1289–1290 (1966)

Westrum L.E. Observations on initial segments of axons in the prepyriform cortex of the rat. Journal of Comparative Neurology **139**, 337–356 (1970)

Westrum L.E., Blackstad T.W. An electron microscopic study of the stratum radiatum of the rat hippocampus (regio superior, CA 1) with particular emphasis on synaptology. Journal of Comparative Neurology **119**, 281–310 (1962)

Wetmore C., Olson L. Neuronal and nonneuronal expression of neurotrophins and their receptors in sensory and sympathetic ganglia suggest new intercellular trophic interactions. Journal of Comparative Neurology **353**, 143–159 (1995)

White E.L. Synaptic organization in the olfactory glomerulus of the mouse. Brain Research **37**, 69–80 (1972)

Whittaker V.P., Michaelson I.A., Kirkland R.J.A. The separation of synaptic vesicles from nerve-ending particles ('synaptosomes'). Biochemical Journal **90**, 293–303 (1964)

Wickelgren W.O., Leonard J.P., Grimes M.J., Clark R.D. Ultrastructural correlates of transmitter release in presynaptic areas of lamprey reticulospinal axons. Journal of Neuroscience **5**, 1188–1201 (1985)

Wiedenmann B., Franke W.W. Identification and localization of synaptophysin, an integral membrane glycoprotein of Mr 38,000 characteristic of presynaptic vesicles. Cell **41**, 1017–1028 (1985)

Wigley R., Hamilton N., Nishiyama A., Kirchhoff F., Butt A.M. Morphological and physiological interactions of NG2-glia with astrocytes and neurons. Journal of Anatomy **210**, 661–670 (2007)

Wilkinson R., Leaver C., Simmons A., Pereira R.A. Restricted replication of herpes simplex virus in satellite glial cell cultures clonally derived from adult mice. Journal of NeuroVirology **5**, 384–391 (1999)

Willey T.J. The ultrastructure of the cat olfactory bulb. Journal of Comparative Neurology **152**, 211–232 (1973)

Willey T.J., Schultz R.L. Intranuclear inclusions in neurons of the cat primary olfactory system. Brain Research **29**, 31–45 (1971)

Williams D.W., Truman J.W. Remodeling dendrites during insect metamorphosis. Journal of Neurobiology **64**, 24–33 (2005)
Williams P.L., Wendell-Smith C.P. Some additional parametric variations between peripheral nerve fibre populations. Journal of Anatomy **109**, 505–526 (1971)
Wilson A.J., Furness J.B., Costa M. The fine structure of the submucous plexus of the guinea-pig ileum. I. The ganglia, neurons, Schwann cells and neuropil. Journal of Neurocytology **10**, 759–784 (1981)
Windebank A.J., Wood P., Bunge R.P., Dyck P.J. Myelination determines the caliber of dorsal root ganglion neurons in culture. Journal of Neuroscience **5**, 1563–1569 (1985)
Wiśniewski H.M., Terry R.D. Morphology of the aging brain, human and animal. Progress in Brain Research **40**, 167–186 (1973)
Witkovsky P., Roberts B.L. Electron microscopic observations of the mesencephalic nucleus of the fifth nerve in the Selachian brain. Journal of Neurocytology **5**, 643–660 (1976)
Witkovsky P., Stell W.K. Retinal structure in the smooth dogfish *Mustelus canis*: electron microscopy of serially sectioned bipolar cell synaptic terminals. Journal of Comparative Neurology **150**, 147–168 (1973)
Wittkowski W. Ependymokrinie und Rezeptoren in der Wand des Recessus infundibularis der Maus und ihre Beziehung zum kleinzelligen Hypothalamus. Zeitschrift für Zellforschung und mikroskopische Anatomie **93**, 530–546 (1969)
Wolburg H., Kästner R., Kurz-Isler G. Lack of orthogonal particle assemblies and presence of tight junctions in astrocytes of the goldfish (*Carassius auratus*). A freeze-fracture study. Cell and Tissue Research **234**, 389–402 (1983)
Wolff J. Beiträge zur Ultrastruktur der Kapillaren in der normalen Grosshirnrinde. Zeitschrift für Zellforschung und mikroskopische Anatomie **60**, 409–431 (1963)
Wolff J.R., Rickmann M., Chronwall B.M. Axo-glial synapses and GABA- accumulating glial cells in the embryonic neocortex of the rat. Cell and Tissue Research **201**, 239–248 (1979)
Wong M.T.T. Somato-dendritic and dendro-dendritic synapses in the squirrel monkey lateral geniculate nucleus. Brain Research **20**, 135–139 (1970)
Wong P.C., Marszalek J., Crawford T.O., Xu Z., Hsieh S.-T., Griffin J.W., Cleveland D.W. Increasing neurofilament subunit NF-M expression reduces axonal NF-H, inhibits radial growth, and results in neurofilamentous accumulation in motor neurons. Journal of Cell Biology **130**, 1413–1422 (1995)
Wong T.P., Campbell P.M., Ribeiro-da-Silva A., Cuello A.C. Synaptic numbers across cortical laminae and cognitive performance of the rat during ageing. Neuroscience **84**, 403–412 (1998)
Wong T.P., Marchese G., Casu M.A., Ribeiro-da-Silva A., Cuello A.C., De Koninck Y. Loss of presynaptic and postsynaptic structures is accompanied by compensatory increase in action potential-dependent synaptic input to layer V neocortical pyramidal neurons in aged rats. Journal of Neuroscience **20**, 8596–8606 (2000)
Wong-Riley M.T.T. Terminal degeneration and glial reactions in the lateral geniculate nucleus of the squirrel monkey after eye removal. Journal of Comparative Neurology **144**, 61–92 (1972)
Wood P.M., Bunge R.P. Evidence that sensory axons are mitogenic for Schwann cells. Nature **256**, 662–664 (1975)
Woodham P., Anderson P.N., Nadim W., Turmaine M. Satellite cells surrounding axotomised rat dorsal root ganglion cells increase expression of a GFAP-like protein. Neuroscience Letters **98**, 8–12 (1989)
Woolley C.S., McEwen B.S. Roles of estradiol and progesterone in regulation of hippocampal dendritic spine density during the estrous cycle in the rat. Journal of Comparative Neurology **336**, 293–306 (1993)
Wouterlood F.G., Barbas-Henry H.A., Lohman A.H.M. Mixed-junction axon terminals on identified principal abducens motoneurons in the monitor lizard. Brain Research **463**, 198–203 (1988)
Wray S.H. Innervation ratios for large and small limb muscles in the baboon. Journal of Comparative Neurology **137**, 227–250 (1969)
Wu C., Ivanova E., Cui J., Lu Q., Pan Z.-H. Action potential generation at an axon initial segment-like process in the axonless retinal AII amacrine cell. Journal of Neuroscience **31**, 14654–14659 (2011)
Wu H.-H., Bellmunt E., Scheib J.L., Venegas V., Burkert C., Reichardt L.F., Zhou Z., Fariñas I., Carter B.D. Glial precursors clear sensory neuron corpses during development via Jedi-1, an engulfment receptor. Nature Neuroscience **12**, 1534–1541 (2009)
Wuerker R.B., Kirkpatrick J.B. Neuronal microtubules, neurofilaments, and microfilaments. International Review of Cytology **33**, 45–75 (1972)
Wuerker R.B., Palay S.L. Neurofilaments and microtubules in anterior horn cells of the rat. Tissue and Cell **1**, 387–402 (1969)
Wyburn G.M. The capsule of spinal ganglion cells. Journal of Anatomy **92**, 528–533 (1958)
Wyckoff R.W.G., Young J.Z. The motorneuron surface. Proceedings of the Royal Society of London. Series B **144**, 440–450 (1956)
Xi M.-C., Liu R.-H., Engelhardt J.K., Morales F.R., Chase M.H. Changes in the axonal conduction velocity of pyramidal tract neurons in the aged cat. Neuroscience **92**, 219–225 (1999)
Xie F., Zheng B. White matter inhibitors in CNS axon regeneration failure. Experimental Neurology **209**, 302–312 (2008)
Xu Z., Marszalek J.R., Lee M.K., Wong P.C., Folmer J., Crawford T.O., Hsieh S.-T., Griffin J.W., Cleveland D.W. Subunit composition of neurofilaments specifies axonal diameter. Journal of Cell Biology **133**, 1061–1069 (1996)
Yamada H., Miyake S. Elektronenmikroskopische Untersuchungen an Nervenfasern in menschlichen Schweissdrüsen. Zeitschrift für Zellforschung und mikroskopische Anatomie **52**, 129–139 (1960)
Yamada K.M., Spooner B.S., Wessells N.K. Axon growth: roles of microfilaments and microtubules.

Proceedings of the National Academy of Sciences of the U.S.A. **66**, 1206–1212 (1970)

Yamagata M., Sanes J.R., Weiner S.A. Synaptic adhesion molecules. Current Opinion in Cell Biology **15**, 621–632 (2003)

Yamasaki H., Itakura C., Mizutani M. Hereditary hypotrophic axonopathy with neurofilament deficiency in a mutant strain of the Japanese quail. Acta Neuropathologica **82**, 427–434 (1991)

Yamauchi A., Burnstock G. Nerve-myoepithelium and nerve-glandular epithelium contacts in the lacrimal gland of the sheep. Journal of Cell Biology **34**, 917–919 (1967)

Yamauchi A., Burnstock G. An electron microscopic study on the innervation of the trout heart. Journal of Comparative Neurology **132**, 567–588 (1968)

Yamauchi A., Chiba T. Adrenergic and cholinergic innervation of the turtle heart ventricle. Zeitschrift für Zellforschung und mikroskopische Anatomie **143**, 485–493 (1973)

Yamauchi A., Fujimaki Y., Yokota R. Reciprocal synapses between cholinergic postganglionic axon and adrenergic interneuron in the cardiac ganglion of the turtle. Journal of Ultrastructure Research **50**, 47–57 (1975)

Yen S.-H., Fields K.L. Antibodies to neurofilament, glial filament, and fibroblast intermediate filament proteins bind to different cell types of the nervous system. Journal of Cell Biology **88**, 115–126 (1981)

Young D. Fine structure of the sensory cilium of an insect auditory receptor. Journal of Neurocytology **2**, 47–58 (1973)

Young H.M. The ultrastructure of the intestinal nerve of Remak in the domestic fowl. Cell and Tissue Research **260**, 601–616 (1990)

Young J.A.C., Brown D.A., Kelly J.S., Schon F. Autoradiographic localization of sites of [^{3}H] γ aminobutyric acid accumulation in peripheral autonomic ganglia. Brain Research **63**, 479–486 (1973)

Yuan A., Rao M.V., Sasaki T., Chen Y., Kumar A., Veeranna, Liem R.K.H., Eyer J., Peterson A.C., Julien J.-P., Nixon R.A. α-Internexin is structurally and functionally associated with the neurofilament triplet proteins in the mature CNS. Journal of Neuroscience **26**, 10006–10019 (2006)

Yuan A., Sasaki T., Kumar A., Peterhoff C.M., Rao M.V., Liem R.K., Julien J.-P., Nixon R.A. Peripherin is a subunit of peripheral nerve neurofilaments: implications for differential vulnerability of CNS and peripheral nervous system axons. Journal of Neuroscience **32**, 8501–8508 (2012)

Yuan A., Sasaki T., Rao M.V., Kumar A., Kanumuri V., Dunlop D.S., Liem R.K., Nixon R.A. Neurofilaments form a highly stable stationary cytoskeleton after reaching a critical level in axons. Journal of Neuroscience **29**, 11316–11329 (2009)

Zeglio P. Ricerche sulla distribuzione del pigmento giallo nel sistema nervoso dell'uomo nelle varie eta. Archivio Italiano di Anatomia e di Embriologia **35**, 371–396 (1936)

Zelander T., Ekholm R., Edlund Y. The ultrastructural organization of the rat exocrine pancreas. III. Intralobular vessels and nerves. Journal of Ultrastructure Research **7**, 84–101 (1962)

Zelená J. Ribosome-like particles in myelinated axons of the rat. Brain Research **24**, 359–363 (1970)

Zenker W., Högl E. The prebifurcation section of the axon of the rat spinal ganglion cell. Cell and Tissue Research **165**, 345–363 (1976)

Zenker W., Mayr R., Gruber H. Axoplasmic organelles: quantitative differences between ventral and dorsal root fibres of the rat. Experientia **29**, 77–78 (1973)

Zenker W., Mayr R., Gruber H. Neurotubules: different densities in peripheral motor and sensory nerve fibres. Experientia **31**, 318–320 (1975)

Zhang H., Mei X., Zhang P., Ma C., White F.A., Donnelly D.F., Lamotte R.H. Altered functional properties of satellite glial cells in compressed spinal ganglia. Glia **57**, 1588–1599 (2009)

Zhang J.-H., Sampogna S., Morales F.R., Chase M.H. Age-related intra-axonal accumulation of neurofilaments in the dorsal column nuclei of the cat brainstem: a light and electron microscopic immunohistochemical study. Brain Research **797**, 333–338 (1998)

Zhang X., Chen Y., Wang C., Huang L.-Y.M. Neuronal somatic ATP release triggers neuron-satellite glial cell communication in dorsal root ganglia. Proceedings of the National Academy of Sciences of the U.S.A. **104**, 9864–9869 (2007)

Zhang X.-F., Han P., Faltynek C.R., Jarvis M.F., Shieh C.-C. Functional expression of $P2X_7$ receptors in non-neuronal cells of rat dorsal root ganglia. Brain Research **1052**, 63–70 (2005)

Zheng J.-Q., Kelly T.K., Chang B., Ryazantsev S., Rajasekaran A.K., Martin K.C., Twiss J.L. A functional role for intra-axonal protein synthesis during axonal regeneration from adult sensory neurons. Journal of Neuroscience **21**, 9291–9303 (2001)

Zhou X.F., Rush R.A., McLachlan E.M. Differential expression of the p75 nerve growth factor receptor in glia and neurons of the rat dorsal root ganglia after peripheral nerve transection. Journal of Neuroscience **16**, 2901–2911 (1996)

Zhuang X., Silverman A.-J., Silver R. Brain mast cell degranulation regulates blood-brain barrier. Journal of Neurobiology **31**, 393–403 (1996)

Zimmermann H. Neuro-neuronal and neuro-glial contacts in the epithelium of the saccus vasculosus of *Perca fluviatilis* (Teleostei): a finestructural-cytochemical study. Progress in Brain Research **38**, 271–277 (1972)

Zimmermann K. Der feinere Bau der Blutkapillaren. Zeitschrift für Anatomie und Entwicklungsgeschichte **68**, 29–109 (1923)

Ziskin J.L., Nishiyama A., Rubio M., Fukaya M., Bergles D.E. Vesicular release of glutamate from unmyelinated axons in white matter. Nature Neuroscience **10**, 321–330 (2007)

Zuo Y., Lin A., Chang P., Gan W.-B. Development of long-term dendritic spine stability in diverse regions of cerebral cortex. Neuron 46, 181–189 (2005)

Index[1]

[1] Items prefixed by Greek letters are indexed according to the first letters of the English segment of their names F=figures, T=tables.

E. Pannese, *Neurocytology: Fine Structure of Neurons, Nerve Processes, and Neuroglial Cells*, 2nd Edition, DOI 10.1007/978-3-319-06856-5, © Springer International Publishing Switzerland 2015

B

C

D

Q

R

S

MIX
Papier aus verantwortungsvollen Quellen
Paper from responsible sources
FSC® C105338
FSC
www.fsc.org

If you have any concerns about our products,
you can contact us on
ProductSafety@springernature.com

In case Publisher is established outside the EU,
the EU authorized representative is:
Springer Nature Customer Service Center GmbH
Europaplatz 3, 69115 Heidelberg, Germany

Printed by Libri Plureos GmbH
in Hamburg, Germany